Engine Performance Diagnosis and Tune-Up

By Chek-Chart,
a Division of
The H. M. Goushā Company

Ken Layne, CGAM, *Editor*
Sydnie W. Changelon, *Contributing Editor*
Gordon Clark, *Managing Editor*

Canfield Press San Francisco
A Harper & Row publication
New York Hagerstown London

Acknowledgments

In producing this series of manuals for automobile mechanics, Chek-Chart has made extensive use of the help and knowledge of the car manufacturing and aftermarket industry. It is a highly complex, fast-changing field, and we are grateful for the help provided by the following companies and organizations:

- Allen Testproducts
- American Motors Corporation
- Atlas Supply Company
- Borg-Warner Corporation
- Champion Spark Plug Company
- Chevron Research Company
- Chrysler Corporation
- Clayton Manufacturing Company
- Cummins Engine Company, Inc.
- Ford Motor Company
- Fram Corporation, a Bendix Company
- General Motors Corporation
  - Buick Motor Division
  - Cadillac Motor Car Division
  - Chevrolet Motor Division
  - Oldsmobile Division
  - Pontiac Division
  - AC-Delco Division
  - Delco-Remy Division
  - Harrison Radiator Division
  - Saginaw Steering Gear Division
- International Harvester Company
- Marquette Mfg. Co., a Division of Applied Power Inc.
- The Prestolite Company, an Eltra Company
- Robert Bosch Corporation
- South Coast (Calif.) Air Quality Management District
- Stant Manufacturing Company
- Sun Electric Corporation
- Volkswagen of America, Inc.

The authors have made every effort to ensure that the material in this book is as accurate and up-to-date as possible. However, neither Chek-Chart nor Harper & Row nor any related companies can be held responsible for mistakes or omissions, or for changes in procedures or specifications made by the carmakers or suppliers.

This book was reviewed by a selected group of educators from across the nation. Their suggestions and thoughtful comments are an invaluable part of the final book you are reading. These consultants were:

- Larry Bennett, Oakland Community College, Troy, Mich.;
- John Flaherty, Sand Hills Community College, Carthage, N.C.;
- Dallas Hazleton, Fullerton College, Fullerton, Calif.;
- Don Nilson, Chabot College, Hayward, Calif.;
- James E. White, Southern Illinois University, Carbondale, Ill.

For Harper & Row, the sponsoring editor is John A. Woods. The production manager is Laura Argento, the production coordinator is Marion Hartsough, and the production editor is Julie Segedy. The designers are James Stockton and Donna Davis.

At Chek-Chart, Randie Sahim and Ray Lyons participated in the production of this book, under the direction of Elmer M. Thompson. Photographs were produced by Kalton C. Lahue and Gordon Agur, and original art was produced by Jim Geddes, F. J. Zienty, Karen von Felten, and Joanne Fanizzo. Art coordinator is Gerald McEwan. Paul E. Sanderson is the educational advisor for the series, and the entire project is under the general direction of Robert J. Mahaffay and Robert M. Bleiweiss.

Library of Congress Cataloging and Publication Data:

Chek-Chart, 1978
Engine Performance Diagnosis and Tune-Up
(Canfield Press/Chek-Chart Automotive Series)
Includes index.
1. Classroom Manual. 2. Shop Manual.

ISBN: 0-06-454003-0
Library of Congress Catalog Card No.: 78-14618

Contents

On The Cover:
Front — Performance testing on a chassis dynamometer, courtesy of the Clayton Manufacturing Company.
Back — The latest generation of engine analyzers and diagnostic computers, courtesy of FMC Corporation and Sun Electric Corporation.

Introduction to Engine Performance Diagnosis and Tune-Up

Engine Performance Diagnosis and Tune-Up is part of the Harper & Row/Chek-Chart Automotive Series. The package for each course has two volumes, a *Classroom Manual* and a *Shop Manual*.

Other titles in this series include:

- Automotive Electrical Systems
- Fuel Systems and Emission Controls
- Automatic Transmissions
- Heating and Air Conditioning (planned)
- Steering, Suspension, Alignment, Wheels, and Tires (planned)
- Brake Systems (planned)
- Engine Overhaul and Rebuilding (planned)
- Manual Transmissions, Drivelines, and Differentials (planned).

Each book is written to help the instructor teach students to become excellent professional automotive mechanics. The 2-manual texts are the core of a complete learning system that leads a student from basic theories to actual hands-on experience.

The entire series is job-oriented, especially designed for students who intend to work in the car service profession. A student will be able to use the knowledge gained from these books and from the instructor to get and keep a job. Learning the material and techniques in these volumes is a giant leap toward a satisfying, rewarding career.

The books are divided into *Classroom Manuals* and *Shop Manuals* for an improved presentation of the descriptive information and study lessons, along with the practical testing, repair, and overhaul procedures. The manuals are to be used together: the descriptive chapters in the *Classroom Manual* correspond to the application chapters in the *Shop Manual*.

Each book is divided into several parts, and each of these parts is complete by itself. Instructors will find the chapters to be complete, readable, and well thought-out. Students will benefit from the many learning aids included, as well as from the thoroughness of the presentation.

The series was researched and written by the editorial staff of Chek-Chart, and was produced by Harper & Row Publishers. For over 50 years, Chek-Chart has provided car and equipment manufacturers' service specifications to the automotive service field. Chek-Chart's complete, up-to-date automotive data bank was used extensively to prepare this textbook series.

Because of the comprehensive material, the hundreds of high-quality illustrations, and the inclusion of the latest automotive technology, instructors and students alike will find that these books will keep their value over the years. In fact, they will form the core of the master mechanic's professional library.

How To Use This Book

Why Are There Two Manuals?

This 2-volume set — **Engine Performance Diagnosis and Tune-Up** — is not like any other textbook you've ever used before. It is actually two books, the *Classroom Manual* and the *Shop Manual*.

The *Classroom Manual* will teach you what you need to know about diagnosing a car's problems, and what a good tune-up is. The *Shop Manual* will show you how to do the tests and tune-up the car for best performance, mileage, and emission control.

The *Classroom Manual* will be valuable in class and at home, for study and for reference. It has text and pictures that you can use for years to refresh your memory about the basics of engine diagnosis and tune-up.

In the *Shop Manual*, you will learn test procedures, service tips, and tune-up instructions. Use the two manuals together to fully understand why a well-tuned car is so important, and how to get it tuned and keep it tuned.

What's In These Manuals?

There are several aids in the *Classroom Manual* that will help you learn more:

1. The text is broken into short bits for easier understanding and review.

2. Each chapter is fully illustrated with drawings and photographs.

3. Key words in the text are printed in **boldface type** and are defined on the same page and in a glossary at the end of the manual.

4. Review questions are included for each chapter. Use these to test your knowledge.

5. A brief summary of every chapter will help you to review for exams.

6. Every few pages you will find short blocks of "nice to know" information, in addition to the main text.

7. At the back of the *Classroom Manual* there is a sample test, similar to those given for National Institute for Automotive Service Excellence (NIASE) certification. Use it to help you study and to prepare yourself when you are ready to be certified as an expert in one of several areas of automobile mechanics.

The *Shop Manual* has detailed instructions on overhaul, test, and service procedures. These are easy to understand, and many have step-by-step, photo-illustrated explanations that guide you through the procedures. This is what you'll find in the *Shop Manual*:
1. Helpful information tells you how to use and maintain shop tools and test equipment.
2. Safety precautions are detailed.
3. System diagrams help you locate trouble-spots while you learn to read the diagrams.
4. Tips the professionals use are presented clearly and accurately.
5. A full index will help you quickly find what you need.
6. Test procedures and troubleshooting hints will help you work better and faster.

Where Should I Begin?

If you already know something about a car's basic systems and how to keep them operating at peak performance, you may find that parts of this book are a helpful review. If you are just starting in car repair, then the subjects covered in these manuals may be all new to you.

Your instructor will design a course to take advantage of what you already know, and what facilities and equipment are available to work with. You may be asked to take certain chapters of these manuals out of order. That's fine. The important thing is to really understand each subject before you move on to the next.

Study the vocabulary words in boldface type. Use the review questions to help you understand the material. While reading in the *Classroom Manual*, refer to your *Shop Manual* to relate the descriptive text to the service procedures. And when you are working on actual car systems, look back to the *Classroom Manual* to keep the basic information fresh in your mind. Working on such a complicated piece of equipment as a modern car isn't always easy. Use the information in the *Classroom Manual*, the procedures of the *Shop Manual*, and the knowledge of your instructor to help you.

The *Shop Manual* is a good book for work, not just a good workbook. Keep it on hand while you're working on equipment. It folds flat on the workbench and under the car, and can withstand quite a bit of rough handling.

When you do test procedures and overhaul equipment, you will also need a source of accurate manufacturer's specifications. Most auto shops have either the carmaker's annual shop service manuals, which lists these specifications, or an independent guide, such as the **Chek-Chart Car Care Guide**. This unique book, which is updated every year, gives you the complete service instructions, electronic ignition troubleshooting tips, and tune-up information that you need to work on specific cars.

PART ONE

Engine Operation and Tune-Up Requirements

1 Engine Operating Principles

The modern automobile engine has many systems and parts that must work together efficiently for the best operating performance, economy, and emission control. The systems that work to meet these requirements are:

- The engine mechanical parts
- The ignition system
- The fuel system
- The engine lubrication system
- The cooling system
- The emission control systems
- The starting system
- The charging system.

Poor performance or a breakdown in any one of these areas can affect total engine operation. Automotive mechanics must make sure that the systems are operating correctly and, if any problems are found, either correct the problems or warn the car owner of their findings. This process of diagnosis, preventive maintenance, and corrective maintenance is commonly called an engine tune-up.

Before you can test these areas of engine operation, you must understand how all of the engine systems work and how they relate to each other. Only then will you be able to test the systems quickly and accurately and spot faults before they become major problems.

The first five chapters of this *Classroom Manual* examine the various engine systems as they relate to proper engine operation and to the engine tune-up. Some of this material is a preview of subjects that are covered in detail in Parts Two, Three, and Four of this book. You may already have studied some of the subjects in the first five chapters. If so, these chapters will be a quick review to refresh your basic understanding. If this material is new to you, study it thoroughly. In any case, the information in the first five chapters is a necessary foundation for the conclusions reached in Chapter 6.

Chapter 6, "What Is A Tune-Up," introduces the concept of a *complete* tune-up and diagnosis. This chapter examines the trends and requirements that affect today's tune-up mechanic. In a way, Chapter 6 is the heart of this *Classroom Manual*. To get the most out of it, though, you need a thorough understanding of all the systems involved in tune-up, beginning with engine operating principles.

ENGINE OPERATION

All modern automobile engines are called internal combustion engines, because fuel burns inside the engine. The engine converts the thermal energy of the burning fuel to mechanical energy.

Figure 1-1. The engine block contains the cylinders and water jackets for cooling. The crankshaft turns in main bearings at the bottom of the block.

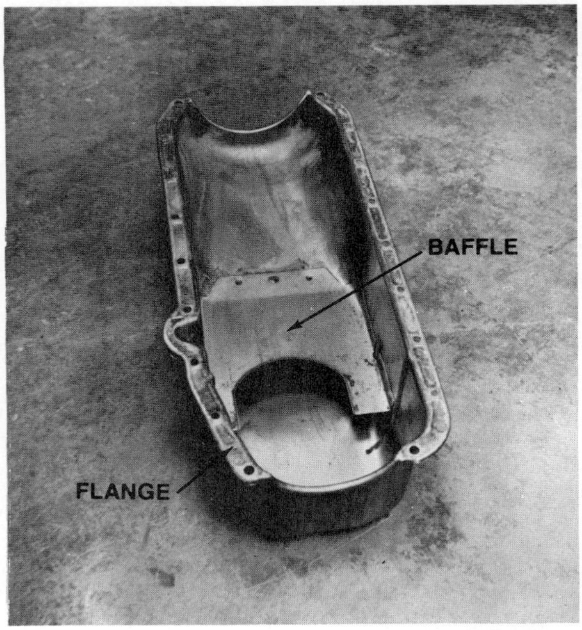

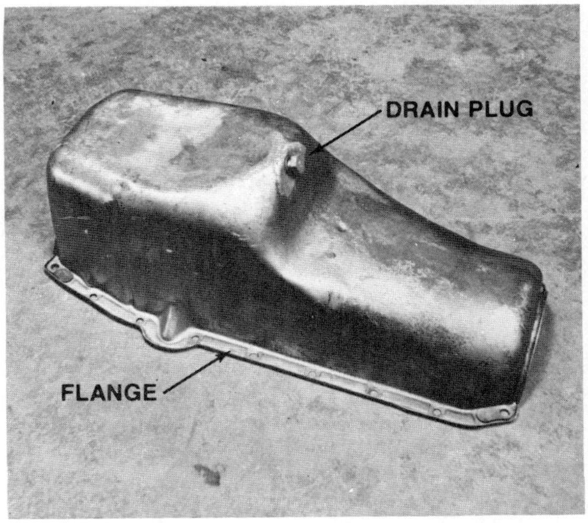

Figure 1-2. The oil pan is fastened to the bottom of the block.

Reciprocating Engine

Except for the Wankel rotary engine, all production automotive engines are the reciprocating, or piston, type. Reciprocating means "up and down" or "back and forth." It is the up and down action of a piston in a cylinder that gives the **reciprocating engine** its name. Almost all engines of this type are built upon a cylinder block, or engine block, figure 1-1. The block is an iron or aluminum casting that has the engine cylinders as well as passages, called water jackets, for coolant circulation. The top of the block is covered with the cylinder head, which contains additional water jackets and which forms the combustion chamber. The bottom of the block is covered with an oil pan, or oil sump, figure 1-2. An example of a major exception to this type of engine construction is the air-cooled Volkswagen engine. It is shown in figure 1-3.

Reciprocating Engine: Also called a piston engine. An engine in which the pistons move up and down or back and forth, as a result of combustion in the top of the cylinder.

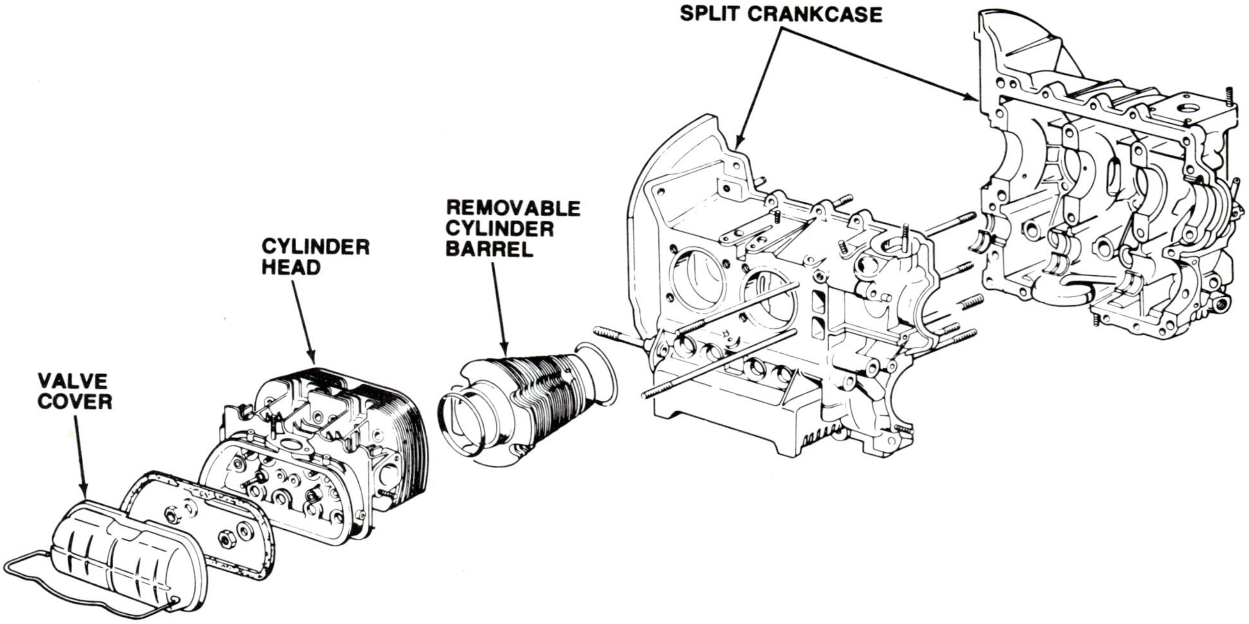

Figure 1-3. The Volkswagen horizontally opposed engine is built with a split crankcase. Individual cylinder castings bolt to the split crankcase, and cylinder heads attach to each pair of cylinders.

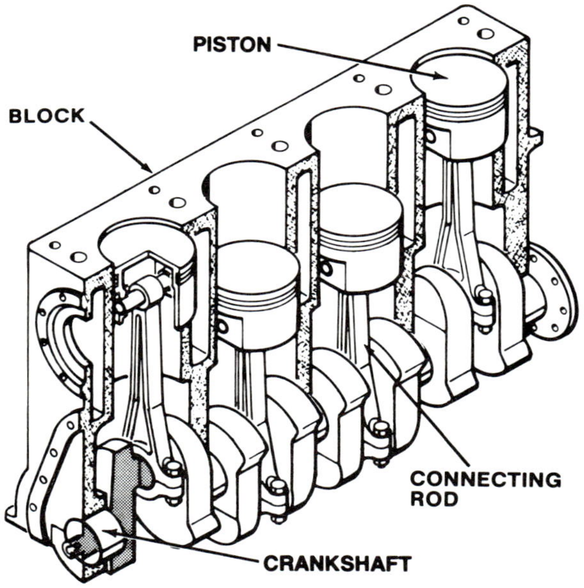

Figure 1-4. The crankshaft changes the reciprocating motion of the pistons to rotating motion.

Power is produced by the reciprocating motion of a piston in a cylinder, but this up and down power must be changed to rotating power to turn the wheels of a car. The piston is attached to the top of a connecting rod by a pin, called a piston pin or a wristpin, figure 1-4. The bottom of the connecting rod is attached to the crankshaft. The connecting rod transmits the up and down motion of the piston to the crankshaft, which changes it to rotating motion. The connecting rod is mounted on the crankshaft with large bearings, called rod bearings. Similar bearings, called main bearings, are used to mount the crankshaft in the block.

Almost all automotive engines are fueled by gasoline. The gasoline is mixed with air and is drawn into the combustion chambers through valves. The mixture is compressed, and ignited by the spark plug. Force from the burning, expanding gases drives the piston down and makes the crankshaft turn.

One complete top-to-bottom or bottom-to-top movement of the piston is called a **stroke**, figure 1-5. One stroke of a piston causes the crankshaft to rotate 180 degrees. Two strokes of the piston cause the crankshaft to make a complete, 360-degree revolution. The terms "engine revolution" and "revolutions per minute (rpm)" refer to the complete, 360-degree revolution of the crankshaft caused by two piston strokes.

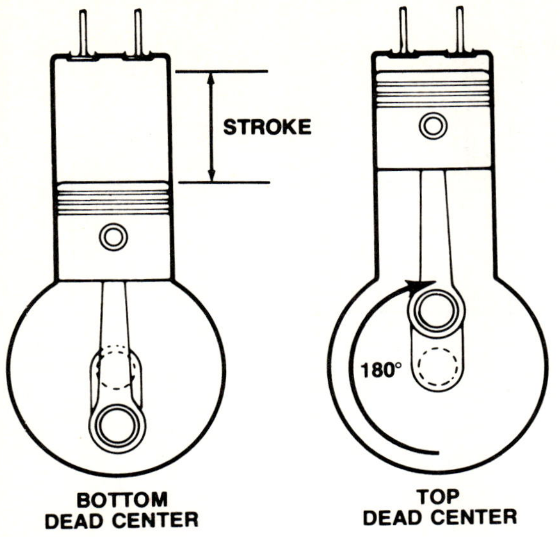

Figure 1-5. One top-to-bottom or bottom-to-top movement of the piston is called a stroke. One piston stroke equals 180 degrees of crankshaft rotation; two strokes equal 360 degrees of crankshaft rotation.

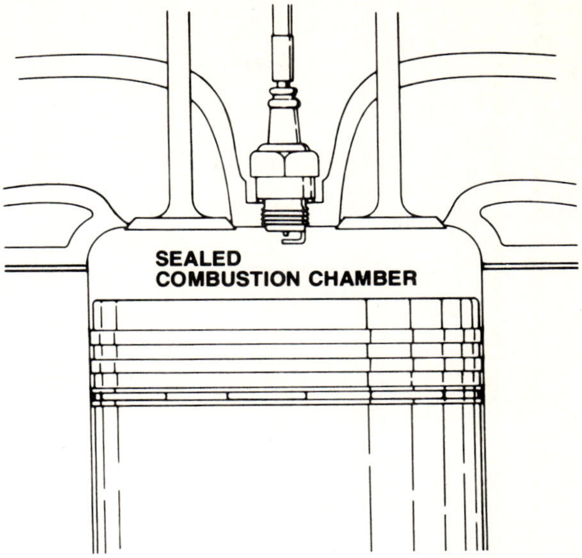

Figure 1-6. For combustion to produce power in an engine, the combustion chamber must be sealed.

Most automobile engines are 4-stroke, spark-ignition, reciprocating engines. Other types of engines include 2-stroke and compression-ignition (diesel) engines. Two-stroke engines are used mainly in motorcycles, lawnmowers, and other types of equipment. They are rarely used in automobiles, and they are not covered in this text. Compression-ignition engines do not need a spark to ignite the air-fuel mixture; their high compression causes extreme heat, which ignites the fuel. We will not discuss them here, because there are relatively few compression-ignition passenger car engines in production today.

THE FOUR-STROKE CYCLE

When gasoline is mixed with oxygen and ignited, it burns. This burning is called combustion and is a way of releasing the energy stored in the air-fuel mixture. To do any useful work in an engine, the combustion energy must be changed to mechanical energy. This is done by confining the combustion to the sealed combustion chamber, figure 1-6, where the combustion energy can work on the movable piston to produce mechanical energy. The combustion chamber must be sealed as tightly as possible for efficient engine operation. Any leakage from the combustion chamber allows part of the combustion energy to leak away without adding to the mechanical energy developed by the piston movement.

Stroke: One complete top-to-bottom or bottom-to-top movement of an engine piston.

■ The Beginnings Of Measurements

The United States now is the last major industrial nation not using the metric system exclusively, and that is changing. Within a decade or so, a complete changeover to metrics will have been accomplished.

Why metrics? Because it is far more precise and universal than the presently used system. And where did this present system come from? Mostly, from old England.

The inch, decreed King Edward II, was equal to the length of three barleycorns placed end to end. Unfortunately, that was not too accurate. King Henry I decided that a yard was the distance from the tip of his nose to the tip of his outstretched thumb. Another measurement, the rod (seldom used today), was the distance of sixteen left feet of men who lined up left-foot-to-left-foot. A furlong was 40 rods, or the length of a furrow in a farmer's field. An acre was said to be the amount of land a team of oxen could plow in one day.

When these measurements were first invented, they seemed to work well enough. But obviously, they were terribly inexact. (What happens if this year's barleycorns are bigger than last year's?) The world needed a new system of measurements, one that would not change from country to country or from year to year. The system finally devised to meet this need is called metrics.

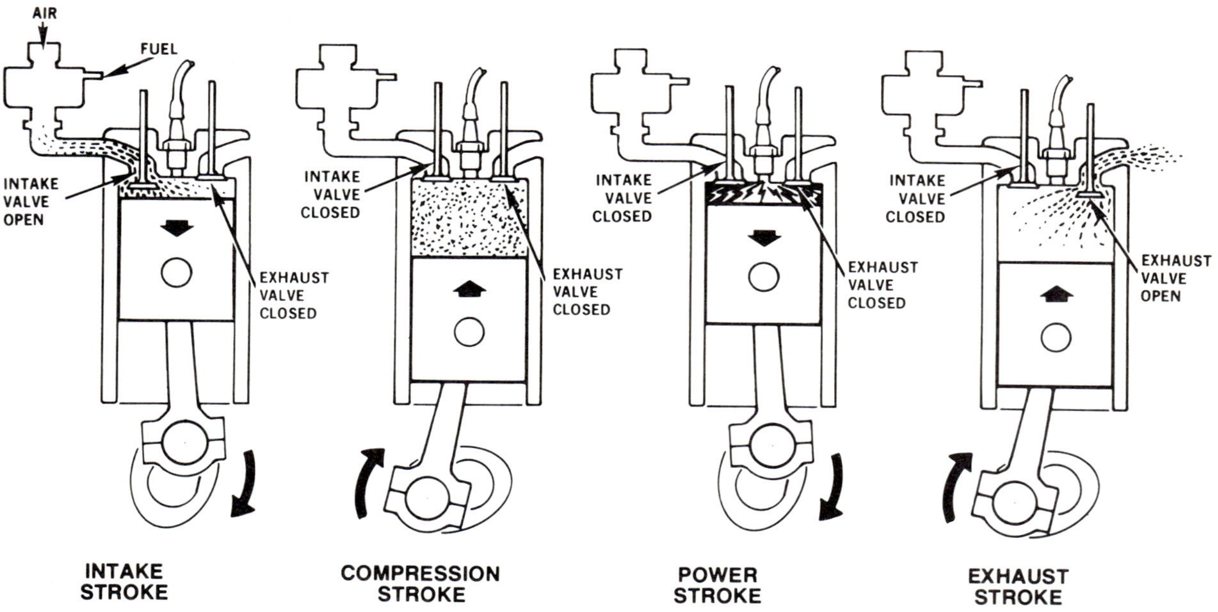

Figure 1-7. The downward movement of the piston draws the air-fuel mixture into the cylinder through the intake valve on the intake stroke. On the compression stroke, the mixture is compressed by the upward movement of the piston with both valves closed. Ignition occurs at the beginning of the power stroke, and combustion drives the piston downward to produce power. On the exhaust stroke, the upward-moving piston forces the burned gases out the open exhaust valve.

Depending on the type of engine, the operating cycle may require either two or four strokes to complete. The **four-stroke engine** is also called the Otto cycle engine, in reference to the German engineer, Dr. Nikolaus Otto, who first applied the principle in 1876.

In the 4-stroke engine, four strokes of the piston in the cylinder are required to complete one operating cycle: two strokes up and two strokes down. Each stroke is named after the action it performs—intake, compression, power, and exhaust—in that order, figure 1-7:

1. Intake stroke: as the piston moves down, a vaporized mixture of fuel and air is drawn into the cylinder through the open intake valve.
2. Compression stroke: the piston returns up, the intake valve closes, the mixture is compressed within the combustion chamber, and ignited by a spark.
3. Power stroke: the expanding gases of combustion force the piston down in the cylinder again. The exhaust valve opens near the bottom of the stroke.
4. Exhaust stroke: the piston returns up with the exhaust valve open, and the burned gases are pushed out to prepare for the next intake stroke. The intake valve usually opens just before the top of the exhaust stroke.

This 4-stroke cycle is continuously repeated in every cylinder as long as the engine remains running.

CYLINDER AND PISTON ARRANGEMENT

Up to this point, we have been talking about a single piston in a single cylinder. While single-cylinder engines are common in motorcycles, outboard motors, and small agricultural implements, automotive engines have more than one cylinder. Most passenger car engines have 4, 6, or 8 cylinders, although engines with 3, 5, 12, and 16 cylinders have been built. In automotive engines, all pistons are attached to a single crankshaft. The more cylinders an engine has, the more power strokes are produced for each revolution. This means that an 8-cylinder engine produces a power stroke twice as often as a 4-cylinder engine. The 8-cylinder engine runs more smoothly because the power strokes are closer together in time and degrees of engine rotation. The cylinders of multicylinder automobile engines are arranged in one of three ways, figure 1-8:

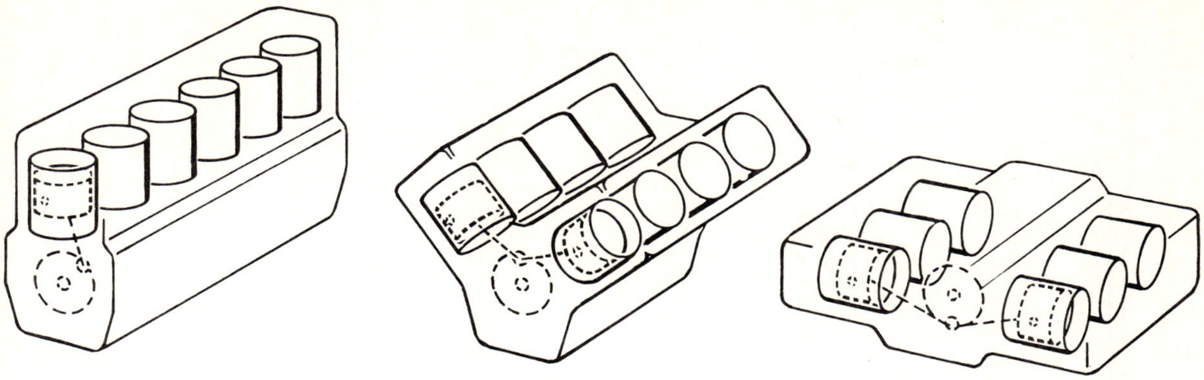

Figure 1-8. Common cylinder arrangements for automotive engines.

1. Inline engines use a single bank of cylinders. Most 4-cylinder and many 6-cylinder engines are of this type.
2. V-type engines use two equal banks of cylinders usually inclined 60 degrees or 90 degrees apart. Most V-type engines are 6- or 8-cylinder, but some V-4 designs have been built.
3. Horizontally opposed, or "pancake," engines use two equal banks of cylinders 180 degrees apart. These space-saving engine designs are usually air-cooled, and are found in Corvairs, Porsches, and Volkswagens. Subaru uses a water-cooled, horizontally opposed, 4-cylinder engine.

As described earlier, each cylinder contains intake and exhaust valves. These are opened by lobes on the camshaft and closed by valve springs, figure 1-9. To coordinate this opening and closing action with piston movement, the camshaft's rotation must be synchronized, or timed, with that of the crankshaft. This synchronization is accomplished in one of three ways: by gears, by a chain and sprockets, or by a timing belt and sprockets. Timing marks on the gears or sprockets are used to synchronize the two shafts with each other, figure 1-10.

A camshaft gear meshes with a driven gear on the ignition distributor to synchronize, or time spark plug firing with piston and valve positions at the beginning of the power stroke. Because each piston is attached to the crankshaft in a predetermined **firing order**, only one power stroke takes place at any instant, regardless of the number of cylinders in the engine. These power strokes follow each other in a rapid sequence, producing a smooth flow of continuous power.

Four-Stroke Engine: The Otto cycle engine. An engine in which a piston must complete four strokes to make up one operating cycle. The strokes are: intake, compression, power, and exhaust.

Firing Order: The order in which combustion occurs in the cylinders of an engine.

■ The Origins Of Metrics

The metric system, although somewhat different from the system we know today, was invented in France in 1793, during the French Revolution. By 1840, France made the use of metrics mandatory throughout the country.

In the United States, patriot and president John Quincy Adams sought to have his new nation adopt the metric system. He introduced a bill in Congress in 1821 to that effect, but it was defeated. At that time, most of the U.S. trade was with England, which did not use metrics.

England has since converted to the international system, and now the United States is the last remaining major industrial nation to still be using its old English inch-pound system. Finally, in 1975, the U.S. Congress declared that this country would indeed switch to metrics, a process that is now underway.

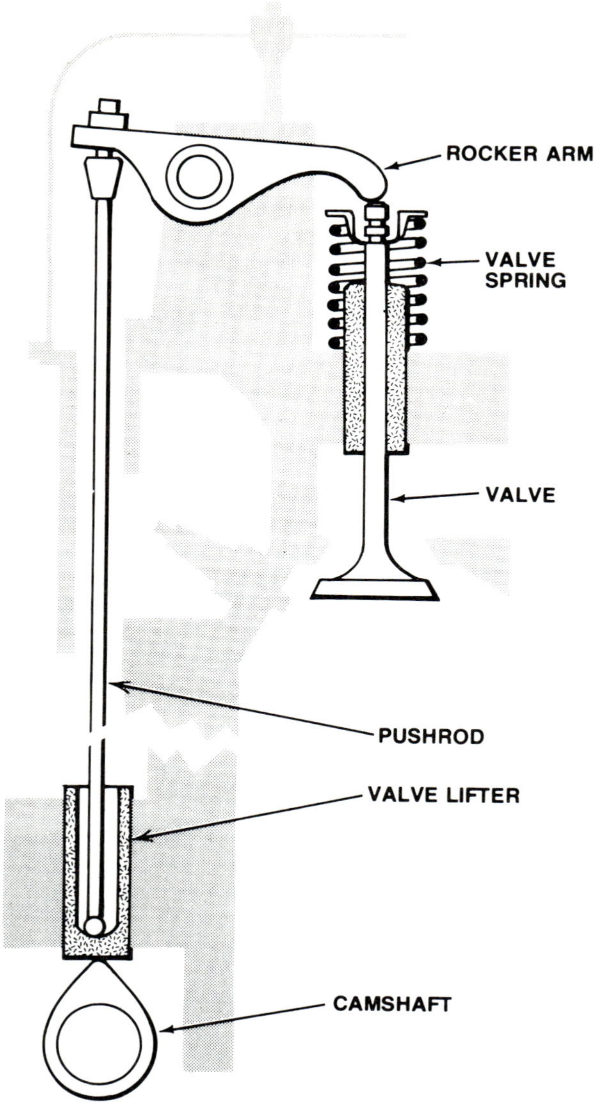

Figure 1-9. Valve mechanism for an overhead-valve engine.

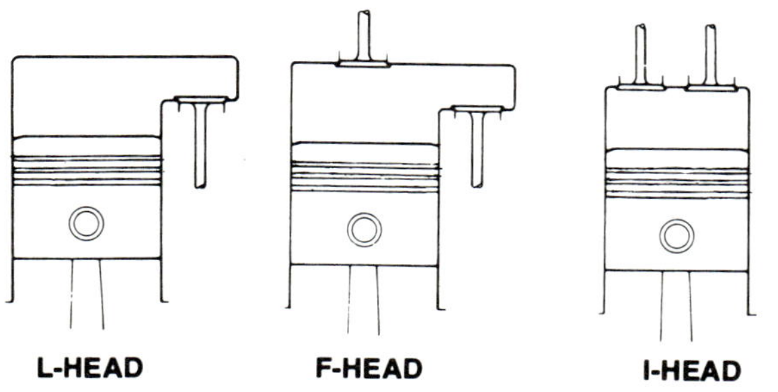

Figure 1-11. Common valve arrangements for four-stroke automotive engines.

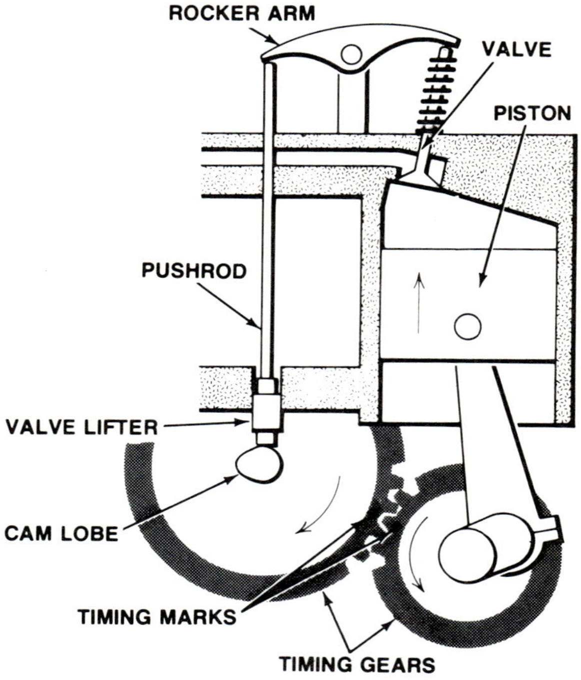

Figure 1-10. Timing marks on the timing gears synchronize valve action with piston movement.

VALVE ARRANGEMENT

Passenger-car engines have the intake and exhaust valves arranged in one of three ways, figure 1-11:

1. The L-head design has both valves located side-by-side in the engine block. Because the cylinder head is rather flat, containing only the combustion chamber, water jacket, and spark plugs, L-head engines are also called "flatheads." Once very common, this valve arrangement has not been used in a domestic engine since the early 1960's.
2. The F-head design locates the intake valve in the cylinder head and exhaust valve in the engine block. A compromise between the L-head and I-head designs, the F-head was last used in the 1971 Jeep.
3. The I-head or overhead-valve design has both the intake and the exhaust valve in the cylinder head. Most modern engines use this design. Such engines can also be overhead-cam, as we will see.

In the overhead-valve design, the camshaft is in the engine block, and the valves are opened by valve lifters, pushrods, and rocker arms, figure 1-12. In the overhead-cam engine, the camshaft is mounted in the head, either above or to one side of the valves. This improves valve action at higher engine speeds. The valves may open directly by means of valve lifters or cam

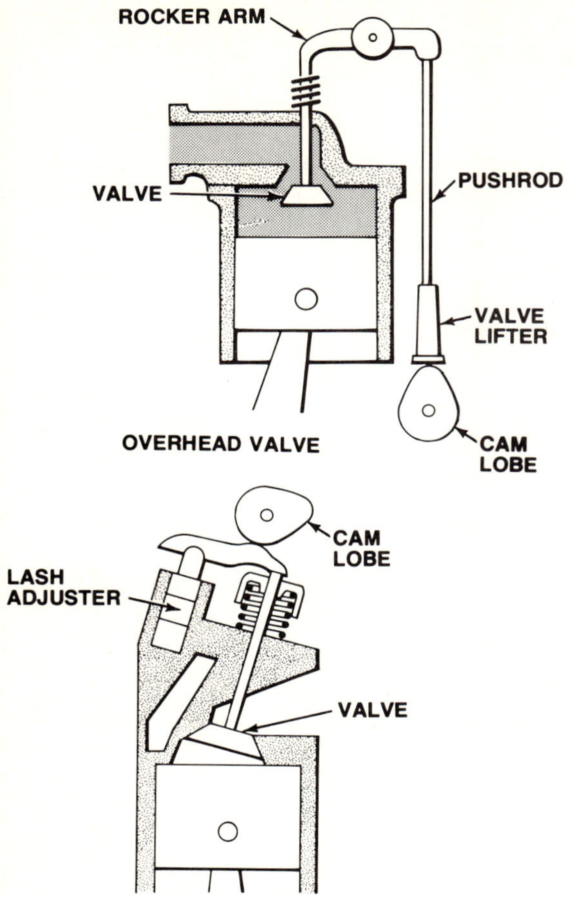

Figure 1-12. In an I-head engine, the valves are operated by a camshaft in the block or in the head.

followers, or through rocker arms. The double-overhead-cam engine has two camshafts, one on each side of the valves. One camshaft operates the intake valves; the other operates the exhaust valves.

Valve lifters can be mechanical or hydraulic. Mechanical lifters are solid metal, as shown in figure 1-12. A hydraulic lifter is a metal cylinder containing a plunger that rides on oil, figure 1-13. A chamber below the plunger fills with engine oil and, as the camshaft lobe lifts the lifter, the chamber is sealed by a check valve. The trapped oil transmits the lifting motion of the camshaft lobe to the valve pushrod. Hydraulic lifters are generally quieter than mechanical lifters and do not need to be adjusted as often, because the amount of oil in the chamber varies to keep the valve adjustment correct.

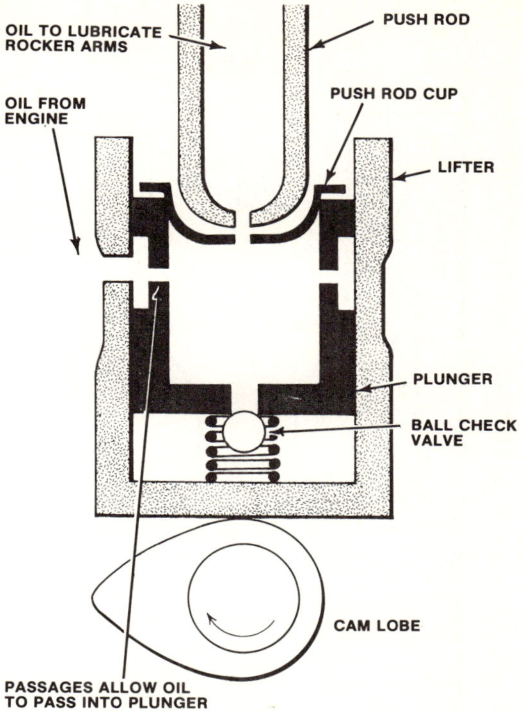

Figure 1-13. A hydraulic valve lifter uses engine oil to transmit motion.

■ What Is The Metric System?

The entire metric system is based on the meter. Originally, the French scientists divided the distance between the North Pole and the equator into ten million equal sections. Each section was called a meter. Now, however, scientists have given the meter a new, more complex (but more accurate) definition: One meter equals the length of 1,650,763.73 wave lengths of the red-orange radiation line produced by an atom called Krypton 86!

The base unit can be divided or multiplied by 10, 100, or 1,000. The prefixes on the names of the units tell which factor (number) to use when multiplying or dividing. These are the most common prefixes, although there are others:

micro=	1/1,000,000	deka=	10
milli=	1/1,000	hecto=	100
centi=	1/100	kilo=	1,000
deci=	1/10	mega=	1,000,000

For example, 1 kilometer equals 1,000 meters, and 1 centimeter equals 1/100 (one one-hundredth) of a meter. Another example: 1,000 millimeters equals 100 centimeters equals 10 decimeters equals 1 meter. These prefixes apply to other units as well, such as the gram (kilogram, centigram, milligram, etc.).

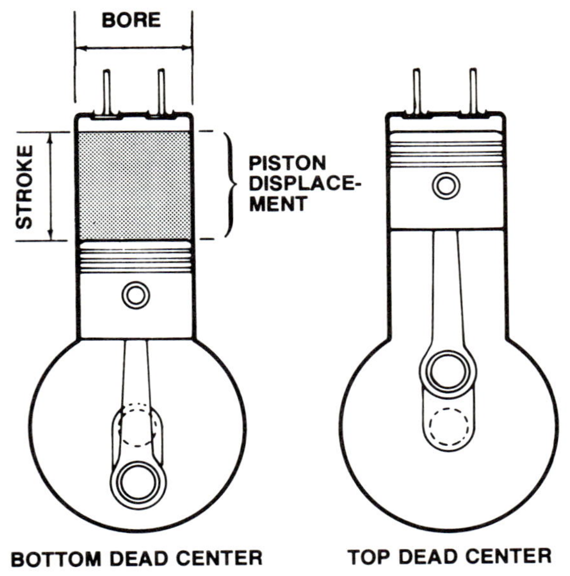

Figure 1-14. Basic engine dimensions.

ENGINE DISPLACEMENT AND COMPRESSION RATIO

Two frequently used engine specifications are engine **displacement** and **compression ratio**. Displacement and compression ratio are related to each other as we will learn in the following paragraphs.

Engine Displacement

Commonly used to indicate engine size, this specification is really a measurement of cylinder volume. The number of cylinders helps determine displacement, but the arrangement of the cylinders or valves does not. Engine displacement is found by multiplying the number of cylinders in the engine by the piston displacement of one cylinder. The total engine displacement is the volume displaced by all the pistons.

To calculate engine displacement, you must begin with the piston displacement of one cylinder. This is the space through which the piston's top surface moves as it travels from the bottom of its stroke (**bottom dead center**) to the top of its stroke (**top dead center**), figure 1-14. It is the volume of air displaced or pushed out from the cylinder by the piston movement. Piston displacement can be figured out as follows:

1. Divide the bore (cylinder diameter) by two. This will give you the radius of the bore.
2. Square the radius by multiplying it by itself (2×2, 3×3, 4×4, etc.).
3. Multiply the square of the radius by 3.1416 (pi or π) to find the area of the cylinder cross section.
4. Multiply the area of the cylinder cross section by the length of the stroke.

You now know the piston displacement for one cylinder. Multiply this by the number of cylinders to determine the total engine displacement. The formula for the complete procedure reads:

$R^2 \times \pi \times$ stroke $\times$ No. of cylinders = displacement

For example, to find the displacement of a V-6 engine with a 3.80-inch bore and a 3.40-inch stroke:

$$\left(\frac{3.80^2}{2}\right) = 3.61 \times 3.1416 \times 3.40 \times 6 = 231 \text{ cu. in.}$$

The displacement is 231 cubic inches. Fractions of an inch are usually not included.

This procedure can be greatly simplified by using a cubic inch displacement chart. You simply locate that point on the chart where the bore and stroke specifications for a given engine intersect, then multiply that figure by the total number of cylinders in the engine to find its displacement.

The greater the engine displacement, the more air-fuel mixture the cylinders can accept, and so the greater the power output (assuming all other factors remain equal).

Metric displacement specifications

When stated in English values, displacement is given in cubic inches; the engine's cubic inch displacement is abbreviated as "cu. in." or "cid." When stated in metric values, displacement is given in cubic centimeters (cc) or in liters (one liter equals 1,000 cc). To convert engine displacement specifications from one value to another, use the following formulas:

- To change cubic centimeters to cubic inches, multiply by 0.061 (cc $\times$ 0.061 = cid).
- To change cubic inches to cubic centimeters, multiply by 16.39 (cid $\times$ 16.39 = cc).
- To change liters to cubic inches, multiply by 61.02 (liters $\times$ 61.02 = cid).

Our 231-cid engine from the previous example is also a 3,786-cc engine ($231 \times 16.39 = 3{,}786$). When expressed in liters, this figure would be rounded off to 3.8 liters.

Compression Ratio

This specification compares the total cylinder volume to the volume of only the combustion chamber, figure 1-15. Total cylinder volume may seem to be the same as piston displacement, but it is not. Total cylinder volume equals piston displacement plus combustion

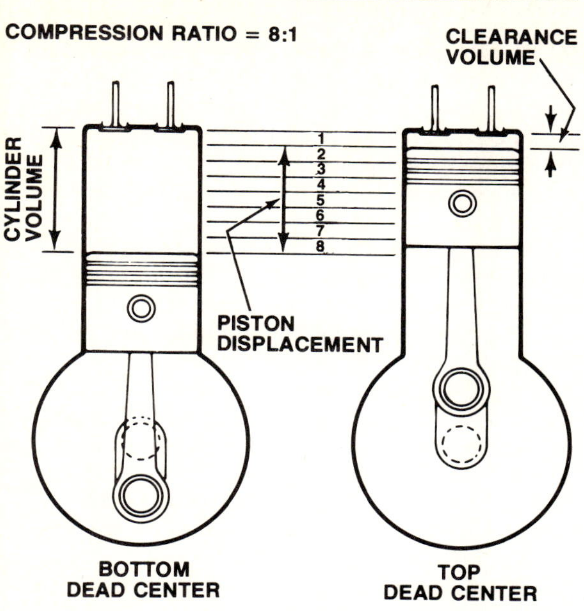

Figure 1-15. Compression ratio is the ratio of the total cylinder volume to the clearance volume.

chamber volume. The combustion chamber volume — with the piston at top dead center — is often called the **clearance volume**.

Compression ratio is the total volume of a cylinder divided by the clearance volume. If the clearance volume is ⅛ of the total cylinder volume, the compression ratio is 8:1. The formula can be written as follows:

$$\frac{\text{Total volume}}{\text{Clearance volume}} = \text{Compression ratio}$$

To determine the compression ratio of an engine in which each piston displaces 31.12 cu. in. and which has a clearance volume of 4.15 cu. in.:

$31.12 + 4.15 = 35.27$ (total cylinder volume)

$$\frac{35.27}{4.15} = 8.498$$

The compression ratio is 8.498:1; this would be rounded off and expressed as a compression ratio of 8.5:1.

In theory, the higher the compression ratio, the more power an engine will develop from a given quantity of fuel. To put it another way, higher compression ratios are supposed to produce better mileage and fuel economy. The reason is that combustion takes place faster because the fuel molecules are more tightly packed and the flame of combustion passes through a smaller area.

But there are practical limits to how high a compression ratio can be. To prevent engine damage, most gasoline-burning engines are restricted to a compression ratio no greater than 11.5:1. Ratios as great as 11:1 were common in the 1960's. Ratios this high, however, make it difficult to control emissions. High-compression engines burn high-octane leaded fuel and create high combustion chamber temperatures. This in turn creates oxides of nitrogen (NO_x), a primary air pollutant. Since 1970, compression ratios have been lowered to an average of 8:1 to permit the use of lower octane, low-lead or unleaded fuel, and to reduce NO_x formation.

IGNITION INTERVAL

Every two strokes of a piston causes the crankshaft to rotate 360 degrees, as we have seen. Therefore, every four strokes of a piston causes the crankshaft to rotate 720 degrees (360 + 360 = 720). Because four strokes of a piston equal one engine operating cycle, one engine operating cycle equals 720 degrees of crankshaft rotation.

During the four strokes of the operating cycle, the spark plug fires only once, at the beginning of the power stroke. In a one-cylinder engine, there would be only one ignition spark every 720 degrees of crankshaft rotation. These 720 degrees are called the **ignition interval**, or **firing interval**, of the engine. It is the number of degrees of crankshaft rotation that occur between ignition sparks.

Displacement: A measurement of the volume of air displaced by a piston as it moves from bottom to top of its stroke. Engine displacement is the piston displacement multiplied by the number of pistons in an engine.

Compression Ratio: The total volume of an engine cylinder divided by its clearance volume.

Bottom Dead Center: The exact bottom of a piston stroke. Abbreviated: bdc.

Top Dead Center: The exact top of a piston's stroke. Also a specification used when tuning an engine. Abbreviated: tdc.

Clearance Volume: The volume of a combustion chamber when the piston is at top dead center.

Ignition Interval (Firing Interval): The number of degrees of crankshaft rotation between ignition sparks.

Figure 1-16. The uneven firing intervals of Buick's early V-6 engine. (Buick)

The more cylinders an engine has, the more power strokes are produced per engine revolution. A 4-cylinder engine produces a power stroke four times as often as a 1-cylinder engine. The 4-cylinder engine has power strokes that are closer together in terms of time and degrees of crankshaft rotation.

A 1-cylinder engine has one power stroke during each engine operation cycle, that is, one power stroke for every two crankshaft revolutions. In a 4-cylinder engine, the operating cycles of each cylinder mean that there are four power strokes for every two crankshaft revolutions.

Common Ignition Intervals

If four power strokes occur during 720 degrees of crankshaft rotation, then one power stroke must occur every 180 degrees ($720 \div 4 = 180$). The ignition system must produce a spark for every power stroke, so it produces a spark every 180 degrees of crankshaft rotation. This means that a 4-cylinder engine has an ignition interval of 180 degrees.

An inline 6-cylinder engine has six power strokes during every 720 degrees of crankshaft rotation, for an ignition interval of 120 degrees ($720 \div 6 = 120$). An 8-cylinder engine has an ignition interval of 90 degrees ($720 \div 8 = 90$).

Unusual Ignition Intervals

Most automotive engines have 4, 6, or 8 cylinders, but some unusual engines are in use today. Foreign manufacturers such as Jaguar and Ferrari produce 12-cylinder engines, with a 60-degree firing interval. Audi of Germany has a 5-cylinder engine, with a 144-degree firing interval.

Other unusual firing intervals result from other engine designs. General Motors has produced two different 6-cylinder engines from 8-cylinder engine blocks. The Buick engine, developed in the 1960's, has alternating 90- and 150-degree firing intervals, figure 1-16. The uneven firing intervals resulted from building a V-6 with a 90-degree crankshaft and block. This engine was modified in mid-1977 by redesigning the crankshaft to provide uniform 120-degree firing intervals, as in an inline six. In 1978, Chevrolet introduced a V-6 engine that fires at alternating 108- and 132-degree intervals.

Spark frequency

In a spark-ignition engine, each power stroke is caused by a spark igniting the air-fuel mixture. Each power stroke needs an individual spark. An 8-cylinder engine, for example, requires four sparks per engine revolution (remember that there are two 360-degree engine revolutions in each 720-degree operating cycle). When the engine is running at about 1,000 rpm, the, ignition system must deliver 4,000 sparks per minute. At high speed, about 4,000 rpm, the ignition system must deliver 16,000 sparks per minute. Precise ignition system performance is needed to meet these demands.

FIRING ORDER

To cause each cylinder in an engine to fire once within 720 degrees of crankshaft rotation and at regular intervals, the pistons and connecting rods are arranged on the crankshaft in a specific order. This is called the firing order, and it varies with different engine designs. Firing orders are designed to reduce the vibration and imbalance created by the power strokes of the pistons.

Engine cylinders are numbered for easy identification. However, the cylinders do not fire in the order in which they are numbered. Straight, or inline, engines are numbered from front to rear, figure 1-17. A typical 4-cylinder engine firing order is 1-3-4-2. That is, the number 1 cylinder power stroke is followed by the number 3 cylinder power stroke, then the number 4 power stroke, and finally the number 2 power stroke. Then the next number 1 power stroke

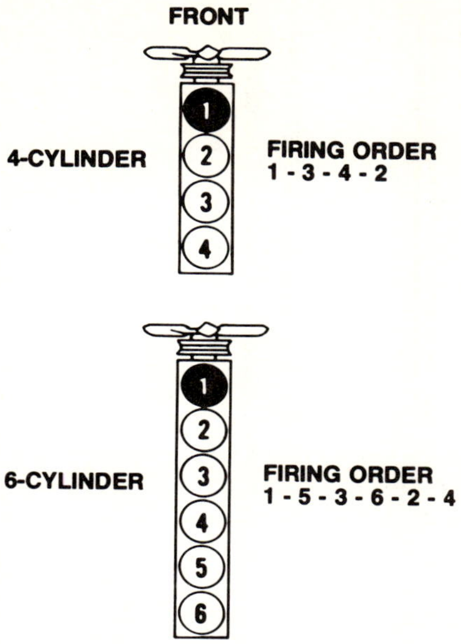

Figure 1-17. Cylinder numbering of an inline engine.

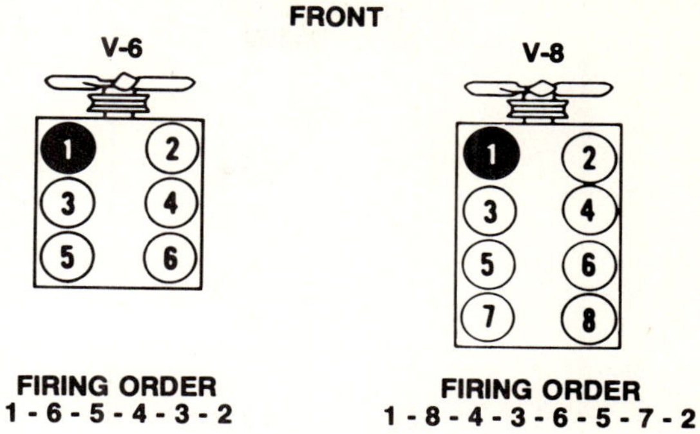

Figure 1-18. American Motors, Chrysler, and most General Motors V-type engines are numbered in this way.

occurs. A few 4-cylinder engines have different firing orders. For example, the English Ford and Pinto 1,600-cc engines fire 1-2-4-3.

The cylinders of an inline 6-cylinder engine also are numbered from front to rear, but do not fire in that order. The firing order for all inline 6-cylinder engines is 1-5-3-6-2-4, figure 1-17.

Except for Ford products and a few GM engines, the cylinders of all domestic V-type engines are numbered the same, figure 1-18. The front cylinder on the left (driver's) side is number 1. The front cylinder on the right (passenger's) side is number 2. Behind number 1 is number 3; behind number 2 is number 4, and so on. The firing order for a V-6 numbered this way is 1-6-5-4-3-2. For a V-8, the firing order is 1-8-4-3-6-5-7-2.

Besides Ford products, the exceptions to this rule for V-type engine cylinder numbering are:

1. Vertical-valve Buick V-8's built before 1967
2. Cadillac V-8's.

The older Buick V-8's have the number 1 cylinder at the front of the right side, figure 1-19, rather than the left side. The firing order is 1-2-7-8-4-5-6-3. Cadillac 425-, 472-, and 500-cid engines are numbered like the older Buicks, figure 1-19, but the firing order is 1-5-6-3-4-2-7-8. Cadillac 390- and 429-cid engines are numbered in the more conventional manner, with the number 1 cylinder at the front left, figure 1-19. However, the firing order for these engines is 1-8-7-2-6-5-4-3.

Ford V-type engines have the number 1 cylinder at the right front, figure 1-20. The numbering continues down the right side, then goes to the left side from front to rear. A Ford V-6 firing order is 1-4-2-5-3-6. Ford V-8 firing orders are 1-3-7-2-6-5-4-8 or 1-5-4-2-6-3-7-8, depending on engine size.

The ignition system must deliver ignition voltage to the correct cylinder at the correct time. To maintain the firing order, the spark plug cables must be attached to the distributor cap in the proper order, figure 1-21.

ENGINE—IGNITION SYNCHRONIZATION

During the engine operating cycle, the intake and exhaust valves open and close at specific times. The ignition system delivers a spark when the piston is near the top of the compression stroke and both valves are closed. These actions must all be coordinated, or engine damage could occur.

■ Sixteen Cylinders

The chances are remote that you'll ever be called on to service a 16-cylinder 1935-37 Cadillac, but in case you do, this is the firing order: 1-8-9-14-3-6-11-2-15-10-7-4-13-12-5-16. The firing order on the 1938-40 V-16 models is: 1-4-9-12-3-16-11-8-15-14-7-6-13-2-5-10. Odd-numbered cylinders are on the left bank, even-numbered cylinders are on the right.

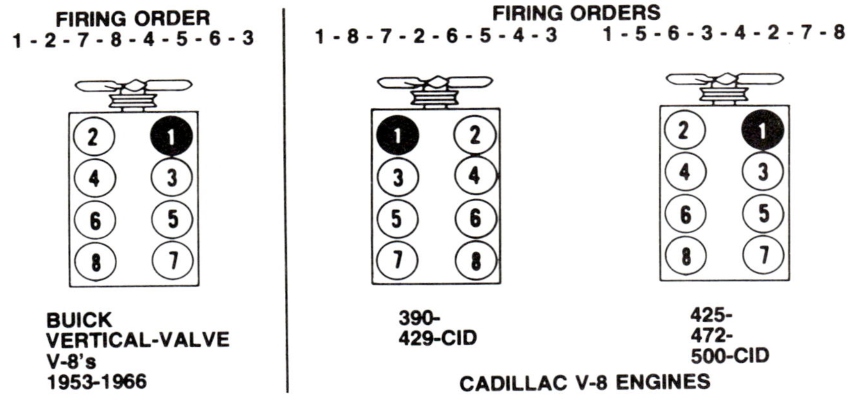

Figure 1-19. Cadillac V-8's and older Buick V-8's have unusual firing orders.

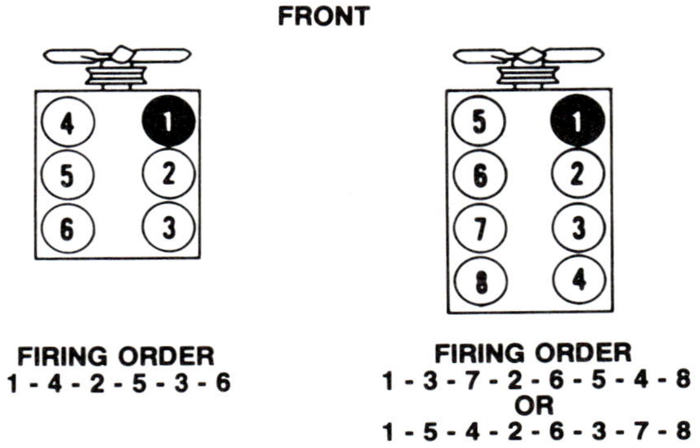

Figure 1-20. Ford numbers its V-type engines in this way.

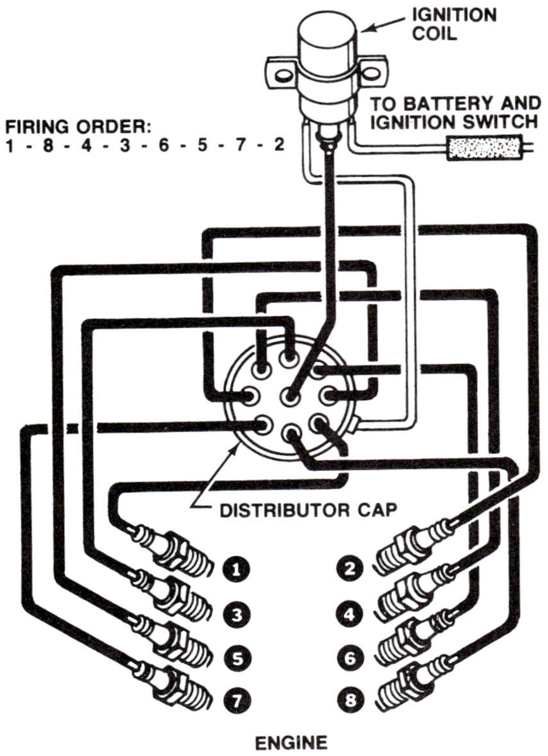

Figure 1-21. The spark plug cables must be connected to the distributor in the proper firing order.

Distributor Drive

The distributor must supply one spark to each cylinder during each cylinder's operating cycle. The distributor cam has as many lobes as the engine has cylinders, or, in a solid-state system, the trigger wheel has as many teeth as the engine has cylinders. One revolution of the distributor shaft will deliver one spark to each cylinder. Each cylinder needs only one spark for each *two* crankshaft revolutions. The distributor shaft must turn at one-half engine crankshaft speed. Therefore, the distributor is driven by the camshaft, which also turns at one-half crankshaft speed.

CRANKSHAFT POSITION

The bottom of the piston stroke is called bottom dead center (bdc), figure 1-22. The top of the piston stroke is called top dead center (tdc). The ignition spark occurs near top dead center, as the compression stroke is ending. As the piston approaches the top of its stroke, it is said to be **before top dead center** (btdc). A spark that occurs before top dead center is called an advanced spark. As the piston passes top dead center and starts down, it is said to be **after top dead center** (atdc). A spark that occurs after top dead center is called a retarded spark.

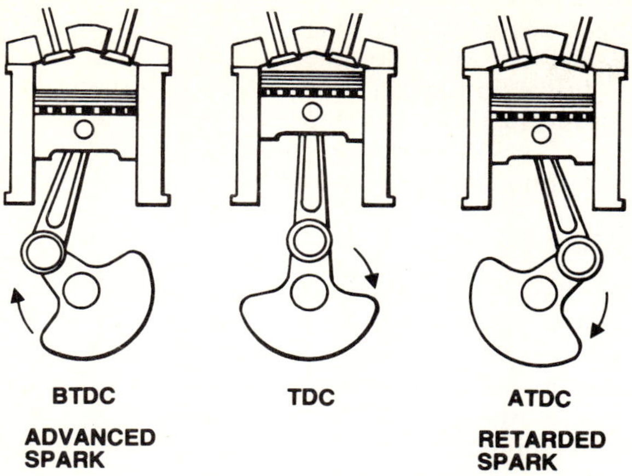

Figure 1-22. Piston position is identified in terms of crankshaft rotation.

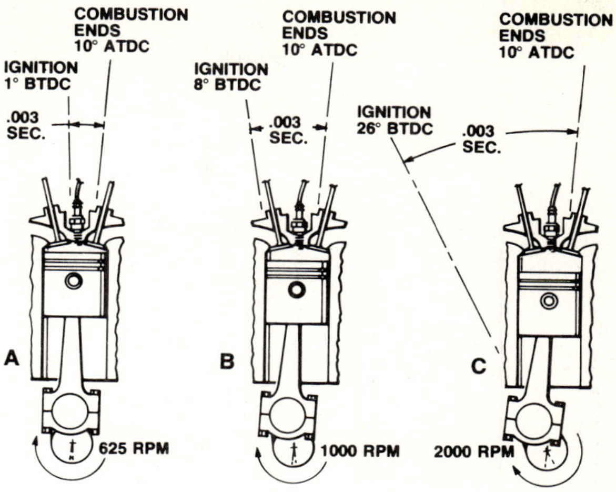

Figure 1-23. As engine speed increases, ignition timing must be advanced.

Burn Time

Approximately 3 milliseconds (0.003 second) elapse from the instant the air-fuel mixture ignites until its combustion is complete. Remember that this burn time is a function of *time* and not of piston travel or crankshaft degrees. The ignition spark must occur early enough so that the combustion pressure reaches its maximum just after top dead center, when the piston is beginning its downward power stroke. Combustion should be completed by about 10° atdc. If the spark occurs too soon before top dead center, the rising piston will be opposed by combustion pressure. If the spark occurs too late, the force on the piston will be reduced. In either case, power will be lost. In extreme cases, the engine could be damaged. Ignition must start at the proper instant for maximum power and efficiency.

Engine Speed

As engine speed increases, piston speed increases. If the air-fuel ratio remains relatively constant, the fuel burning time will remain constant. However, at greater engine speed, the piston will travel farther during this burning time. Ignition timing must be changed to ensure that maximum combustion pressure occurs at the proper piston position.

For example, consider an engine, figure 1-23, that requires 0.003 second for the fuel charge to burn and that achieves maximum power if the burning is completed at 10° atdc.

- At an idle speed of 625 rpm, position A, the crankshaft rotates about 11 degrees in 0.003

Before Top Dead Center: The position of a piston as it nears top dead center. Abbreviated: btdc. Usually expressed in degrees, such as 5° btdc.

After Top Dead Center: The position of a piston after it has passed top dead center. Abbreviated: atdc. Usually expressed in degrees, such as 5° atdc.

■ Touching All The Metric Bases

Metric measurements of weight (mass) and length are becoming familiar to most of us, yet those are only two of seven base units and two supplementary units. Some of these, too, we already know:

Length = Meter
Weight (Mass) = Kilogram
Time = Second
Electric Current = Ampere
Temperature = Kelvin (Celsius)
Amount of Substance = Mole
Luminous Intensity = Candela
Plane Angle = Radian
Solid Angle = Steradian

We already use the second and the ampere. The base temperature unit, the Kelvin, is commonly used on a Celsius scale, where 0° C equals 32° F and 100° C equals 212° F. The term Celsius is now used in place of the term centigrade. The remaining base units are most often found in scientific or mathematical uses.

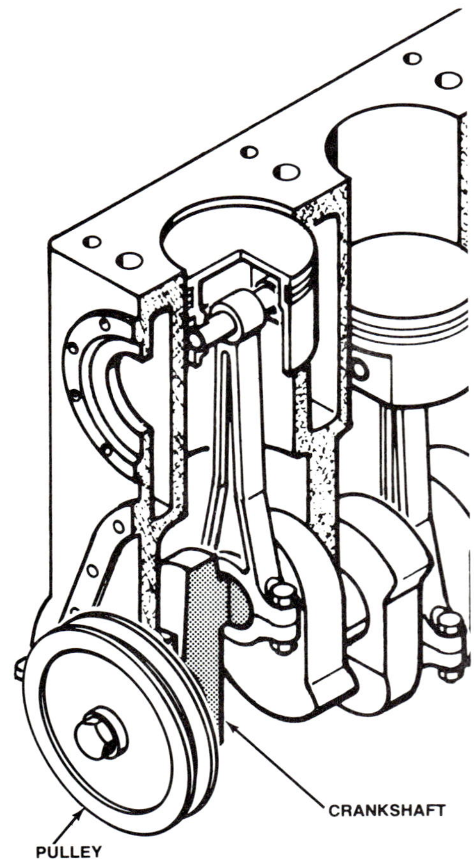

Figure 1-24. Most engines have a pulley bolted to the front end of the crankshaft.

second. Therefore, timing must be set at 1° btdc to allow ample burning time.

- At 1,000 rpm, position B, the crankshaft rotates 18 degrees in 0.003 second. Ignition should begin at 8° btdc.
- At 2,000 rpm, position C, the crankshaft rotates 36 degrees in 0.003 second. Spark timing must be advanced to 26° btdc.

Change in timing is called spark advance, or ignition advance, and is explained in more detail in Chapter 9.

INITIAL TIMING

As we have seen, ignition timing must be set correctly for the engine to run at all. This is called the engine's initial, or basic, timing. Initial timing is the correct setting at a specified engine speed. In figure 1-23, initial timing was 1° btdc. Initial timing is normally within a few degrees of top dead center. For many years, most engines were timed at the specified slow-idle speed for the engine. However, since about 1974 many Ford, Oldsmobile, and Cadillac engines require timing at speeds either above or below the slow-idle speed.

Initial timing can be adjusted to compensate for mechanical wear, slippage, and other factors. We will learn more about setting the ignition timing in the *Shop Manual*.

Timing Marks

We have seen that initial timing is related to crankshaft position. To properly time the engine, we must be able to determine crankshaft position. The crankshaft is completely enclosed in the engine block, but most cars have a pulley and vibration damper bolted to the front of the crankshaft, figure 1-24. This pulley rotates with the crankshaft and can be considered an extension of the shaft.

Marks on the pulley show crankshaft position. For example, when a mark on the pulley is aligned with a mark on the engine block, the number 1 piston is at top dead center.

Timing marks vary widely, even within a manufacturer's product line. There are two common types of timing marks, figure 1-25:

- A mark on the crankshaft pulley and marks, representing degrees of crankshaft position, on the engine block, position A.
- Marks on the pulley, representing degrees of crankshaft position, and a pointer on the engine block, position B.

Some cars have timing marks on the engine flywheel. The flywheel is attached to the end of the crankshaft opposite the pulley. Honda automobiles, some Opels, and water-cooled Volkswagens, Audis, and Porsches have this type of timing mark. Some older domestic engines also have a flywheel timing mark.

Many late-model cars have conventional timing marks, in addition to a special test socket for electromagnetic engine timing. These can be timed with a timing light or, for greater accuracy, with a special test probe that fits into the socket.

SUMMARY

Engine tune-up is a combination of preventive and corrective maintenance. Before you can tune an engine, you must understand how it works.

Most car engines are internal combustion, reciprocating engines. They can also be classed as gasoline-burning, 4-stroke, multiple-cylinder engines. An air-fuel mixture is ignited by a spark inside a sealed combustion chamber. This pushes a piston downward, and a crankshaft converts this up-and-down motion to rotating motion.

Valves at the top of the cylinders open and

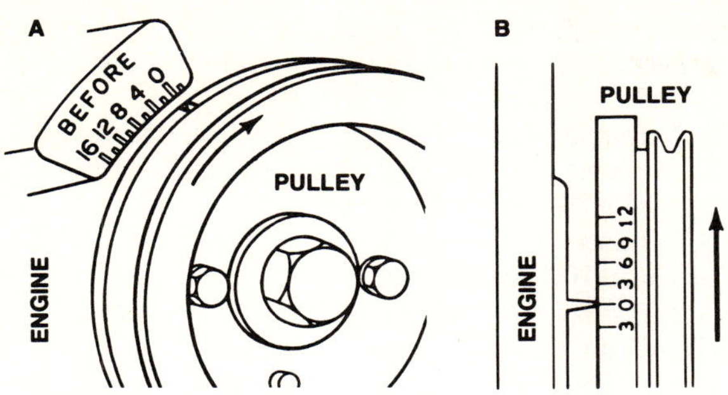

Figure 1-25. Common timing marks.

close to admit the air-fuel mixture and release the exhaust. They are operated by a camshaft and are synchronized with the engine speed. The ignition spark also must be timed to ignite the air-fuel charge at the right instant. The sequence in which the cylinders fire is the firing order. Ignition timing is the synchronization of ignition and engine operation.

■ "I Wouldn't Touch Metric Conversion With A 3.049-meter Pole"

While converting the United States system to metrics will take plenty of time and will cost enormous amounts of money, some measurements are already in metrics.

For instance, cameras and film generally use millimeters (mm) for measuring the size of the lenses or the size of the film. In electronics, we now use the metric system of seconds, volts, watts, amperes and hertz (cycles per second). The drug industry changed over to metrics more than 15 years ago, and the nation's carmakers are not only using some imported engines (with metric designations such as the 2.3-liter engine), but are building some of their own engines and cars with metric measurements.

Mechanics need to know the metric system. This nation's tire manufacturers are now using kilopascals as well as pounds per square inch to indicate tire inflation. And the quart of oil may be a thing of the past, as the liter (which is just a little bit larger) takes over.

Review Questions

Select the single most correct answer.
Compare your answers with the correct answers on page 280.

1. How many strokes of a piston are required to turn the crankshaft through 360°?
   a. One
   b. Two
   c. Four
   d. None of the above

2. The bore is the diameter of the:
   a. Connecting rod
   b. Cylinder
   c. Crankshaft
   d. Combustion chamber

3. The four-stroke cycle operates in which order:
   a. Intake, exhaust, power, compression
   b. Intake, power, exhaust, compression
   c. Compression, power, intake, exhaust
   d. Intake, compression, power, exhaust

4. The distributor is driven by the:
   a. Crankshaft
   b. Rocker arms
   c. Connecting rods
   d. Camshaft

5. A "retarded spark" is one that occurs:
   a. At top dead center
   b. Before top dead center
   c. After top dead center
   d. At bottom dead center

6. The intake and exhaust valves of each cylinder are operated by:
   a. Reduction gears
   b. Engine camshaft
   c. Firing relays
   d. Engine crankshaft

7. Synchronization of camshaft and crankshaft rotations is accomplished by:
   a. Gears
   b. A chain and sprockets
   c. A timing belt and sprockets
   d. Any of the above

8. In an I-head engine:
   a. The intake valves are in the block; the exhaust valves are in the head
   b. All the valves are in the block
   c. All the valves are in the head
   d. None of the above

9. Compression ratio is:
   a. Piston displacement plus clearance volume
   b. Total volume times number of cylinders
   c. Total volume divided by clearance volume
   d. Stroke divided by bore

10. The firing interval of an engine is the number of degrees of crankshaft rotation that:
    a. Take place in a 4-stroke engine
    b. Are required to complete one full stroke
    c. Occur between ignition sparks
    d. All of the above

Chapter

2

Engine Air-Fuel Requirements

Automobile engines run on a mixture of gasoline and air. Gasoline has several advantages as a fuel:

1. Vaporization, or evaporation, occurs easily.
2. It burns quickly, but under control, when mixed with air and ignited.
3. It has a high heat value and produces a large amount of heat energy.
4. It is easy to store, handle, and transport.

Gasoline also has certain disadvantages. The chief disadvantage is that combustion produces air pollutants that are given off into the atmosphere through the engine's exhaust.

As a fuel, however, nothing better than gasoline is now available as a substitute. To better understand how the fuel system works in an engine, we must understand:

- How internal combustion engines use air pressure
- The engine's air-fuel requirements.

AIR PRESSURE — HIGH AND LOW

You can think of an internal combustion engine as a big air pump. As the pistons move up and down in the cylinders, they pump in air and fuel for combustion and pump out the burned exhaust. They do this by creating a difference in air pressure. The air outside an engine has weight and exerts pressure. So does the air inside an engine.

As a piston moves down on an intake stroke with the intake valve open it creates a larger area inside the cylinder for the air to fill. This lowers the air pressure within the engine. Because the pressure inside the engine is lower than the pressure outside, air will flow in through the carburetor to try to fill the low pressure area and equalize the pressure.

We call the low pressure within the engine **vacuum**. You can think of the vacuum as sucking air into the engine, but it is really the higher pressure on the outside that forces air into the low-pressure area inside. The difference in pressure between two areas is called a **pressure differential**. The pressure differential principle has many applications in automotive fuel and emission systems.

An engine pumps exhaust out of its cylinders by creating pressure as a piston moves upward on the exhaust stroke. This creates high pressure in the cylinder, which forces the exhaust toward the lower pressure area outside the engine. Pressure differential can be applied to liquids as well as to air. Fuel pumps work on this principle. The pump creates a low-pressure area in the fuel system that allows the higher pressure of the air and fuel in the tank to force the fuel through the lines to the carburetor.

AIRFLOW REQUIREMENTS

All gasoline automobile engines share certain air-fuel requirements. For example, a 4-stroke engine can take in only a certain amount of air and fuel at any one time. How much it consumes depends upon how much air the engine can take in; this, in turn, depends on four major factors:

1. Engine displacement
2. Maximum engine revolutions per minute (rpm)
3. Carburetor airflow capacity
4. Volumetric efficiency.

The first two factors can be used to figure the engine's airflow requirement. The carburetor airflow capacity must match the engine's requirement. This requirement is measured in cubic feet per minute (cfm). When it is figured, we assume that the engine has 100-percent volumetric efficiency, often called "perfect breathing." The following paragraphs describe volumetric efficiency and how it is related to engine airflow.

Volumetric Efficiency

Volumetric efficiency is a term used to describe the airflow volume actually entering an engine, compared to the engine displacement, which is the maximum volume that it *could* take in. Volumetric efficiency is expressed as a percentage, and it changes with engine speed. For example, an engine might have 75 percent volumetric efficiency at 2,000 rpm. The same engine might be rated at 85 percent at 1,000 rpm and 60 percent at 3,000 rpm.

To find volumetric efficiency, the airflow volume must be measured at a specified temperature and pressure. This is because the airflow volume will increase as pressure increases and as temperature decreases. Standard pressure for measuring volumetric efficiency is **atmospheric pressure** at sea level, which is 14.7 pounds per square inch (psi) or 760 millimeters of mercury (mm Hg). Standard temperature is 0° C or 32° F.

Although we calculate an engine's cubic feet per minute of airflow at a volumetric efficiency of 100 percent, this figure is seldom if ever reached by a stock engine. With a stock engine, you can expect a volumetric efficiency of about 75 percent at maximum speed, or 80 percent at the highest torque, or turning force. A high-performance engine will be about 5 percent more efficient, and a racing engine will add another 10 percent to that.

AIR-FUEL RATIOS

Because liquid fuel will not burn, it must first be changed into a vapor and mixed with air before it is ignited in the cylinders. For most engines, this is done by the carburetor. For fuel-injected engines, the fuel vaporization and mixing with air is done in the intake manifold or in the combustion chamber before ignition.

In either case, there is a direct relationship between an engine's airflow and its fuel requirements. This relationship is called the **air-fuel ratio**.

The air-fuel ratio is the proportion by weight of air and gasoline mixed by the carburetor as required for combustion by the engine. This ratio is important, since there are limits to how rich (with more fuel) or how lean (with less fuel) it can be, and still remain fully combustible for efficient firing. The mixtures with which an engine can operate efficiently range from 8:1 to 18.5:1, figure 2-1. These ratios are usually stated

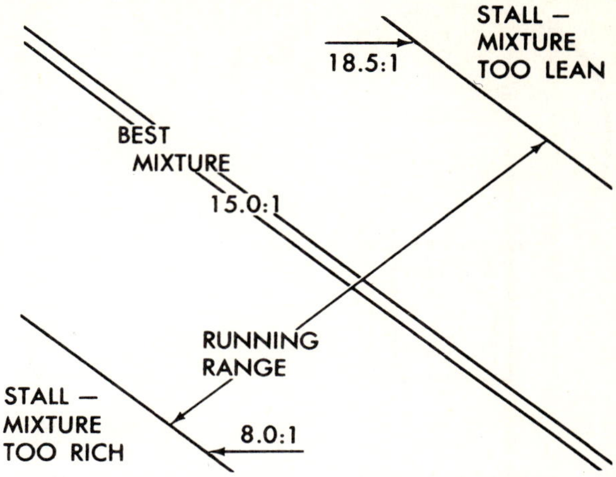

Figure 2-1. Air-fuel ratio limits for a four-stroke gasoline engine. (Chevrolet)

Vacuum: A pressure less than atmospheric pressure.

Pressure Differential: A difference in pressure between two points.

Volumetric Efficiency: The comparison of the *actual* volume of air-fuel mixture drawn into an engine to the *theoretical maximum* volume that could be drawn in. Written as a percentage.

Atmospheric Pressure: The pressure on the earth's surface caused by the weight of air in the atmosphere. At sea level, this pressure is 14.7 psi at 32° F(0° C).

Air-Fuel Ratio: The ratio of air to gasoline in the air-fuel mixture drawn into an engine.

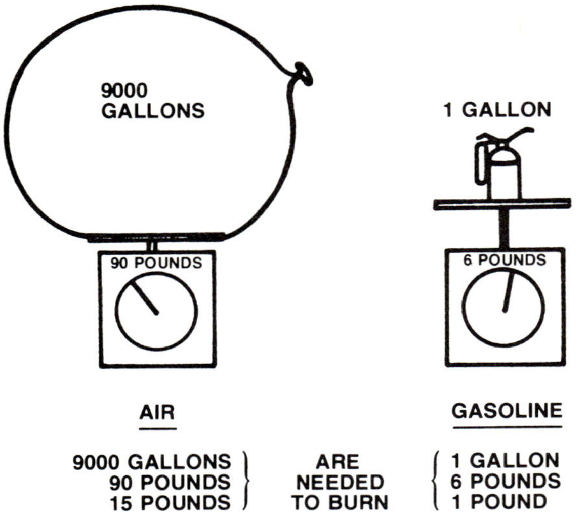

Figure 2-2. The most efficient air-fuel ratio is 15 pounds of air to 1 pound of gasoline (15:1). (Chevrolet)

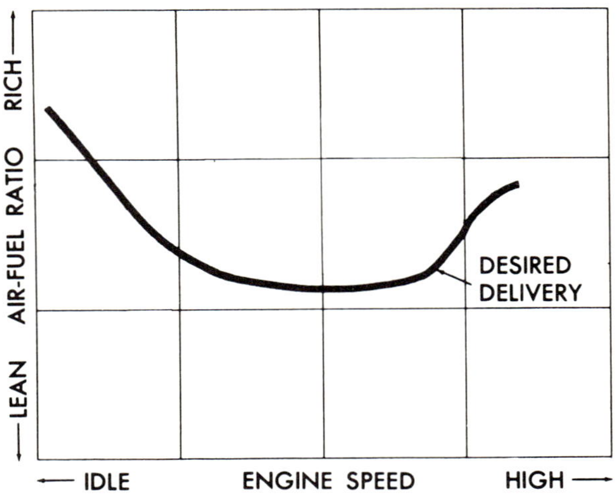

Figure 2-3. The desired air-fuel ratio changes as engine operating conditions change. (Chevrolet)

this way: eight pounds of air combined with one pound of gasoline (8:1) is the richest mixture which an engine can tolerate and still fire regularly; 18.5 pounds of air mixed with one pound of gasoline (18.5:1) is the leanest. Richer or leaner air-fuel ratios will cause the engine to misfire, or simply refuse to run at all.

To get the best engine efficiency and economy, about 9,000 gallons of air are needed to burn one gallon of gasoline. When expressing this in terms of volume, we find the air-fuel ratio to be 9,000:1. Not only are ratios of this size hard to understand, but they are also hard for engineers to use in their designs and experiments. Therefore, *weight* rather than *volume* is used to figure out air-fuel ratios, since it is easier to work with pounds than with gallons.

For example, to convert a volume ratio of 9,000:1 into a more useful weight ratio:

1. 100 gallons of air = 1 pound
2. 9,000 gallons of air = 9000 ÷ 100 = 90 pounds
3. 1 gallon of gasoline = 6 pounds (approximate).

This means that it takes approximately 15 pounds of air to burn one pound of gasoline, so our air-fuel ratio is 15:1, figure 2-2.

Stoichiometric Air-Fuel Ratio

The ideal mixture or ratio at which all the fuel will blend with all of the oxygen in the air and be *completely* burned is called the **stoichiometric ratio** — a chemically perfect combination. In theory, an air-fuel ratio of about 14.7:1 will produce this ratio, but the *exact* ratio at which perfect mixture and complete combustion take place depends on the molecular structure of gasoline, which can vary somewhat.

Interestingly, the stoichiometric ratio is also the point of the highest amounts of oxides of nitrogen (NO_x) emissions in the exhaust. This is important to remember for a tune-up mechanic, since part of the reason for a tune-up is to get the ideal performance and the lowest emissions.

This relationship between the amounts of air and fuel flow in an engine is sometimes called the fuel-air ratio. Because an engine uses far more air than fuel, the fuel-air ratio is always a number less than one, such as 0.0625. The fuel-air ratio is just a different way of expressing the more familiar air-fuel ratio.

Engine Air-Fuel Requirements

An automobile engine will work with the air-fuel mixture ranging from 8:1 to 18.5:1. But the ideal ratio would be one that provides both the most power *and* the most economy, while producing the least emissions. But such a ratio does not exist because the fuel requirements of an engine vary widely depending upon temperature, load, and speed conditions.

Research has proven that the best fuel economy is obtained with a 15:1 to 16:1 ratio, while maximum power output is achieved with a 12.5:1 to 13.5:1 ratio. A rich mixture is required for idle, heavy load, and high speed conditions; a leaner mixture is required for normal cruising and light load conditions. As you can see, no single air-fuel ratio provides the best fuel economy *and* the maximum power output at the same time, figure 2-3.

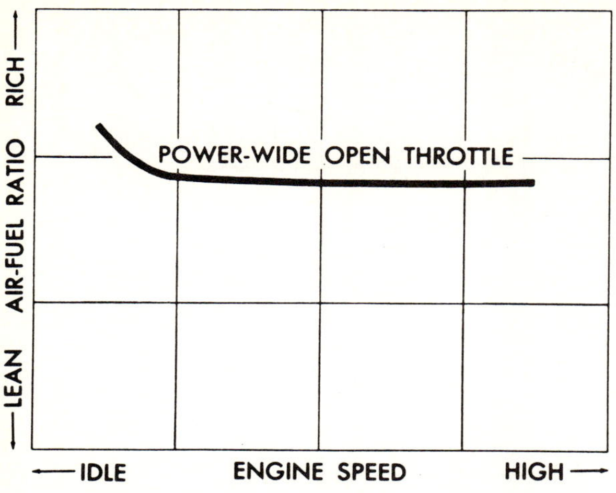

Figure 2-4. The air-fuel ratio needed for maximum power is relatively constant, except at low speed, where it must be slightly richer. (Chevrolet)

Just as outside conditions such as speed, load, temperature, and atmospheric pressure change the engine's fuel requirements, other forces at work inside the engine cause additional variations. Here are three examples:

1. Exhaust gases remain inside the cylinders and dilute the incoming air-fuel mixture, especially during idle.
2. The mixture is imperfect because complete vaporization of the fuel may not take place.
3. Mixture distribution by the intake manifold to each cylinder is not exactly equal; some cylinders may get a richer or leaner mixture than others.

If an engine is to run well under such a wide variety of outside and inside conditions, the carburetor must be able to vary the air-fuel ratio quickly, and to give the best mixture possible for the engine's requirements at a given moment.

The best air-fuel ratio for one engine may not be the best ratio for another, even when the two engines are of the same size and design. Engines are mass produced, and will have slight variations in manifolding, combustion chambers, valve timing, and ignition timing. To accurately determine the best mixture, the engine should be run on a **dynamometer** to measure speed, load, and power requirements for all types of driving conditions.

Power Vs. Economy

If the goal is to get the most power out of an engine, all of the oxygen in the mixture must be burned, because the power output of any engine is limited by the amount of air it can pull in. To be sure that the oxygen combines properly with the available fuel, extra fuel must be provided. This increases the air-fuel ratio (makes it richer in fuel), resulting in some fuel which remains unburned.

This is also true at idle because of the exhaust gases left over in the cylinders. These tend to dilute the incoming mixture since some of the fuel combines with the exhaust. To make certain that the mixture is properly combustible during idle, more fuel must be delivered to make up for the fuel that combines with the exhaust gases. This makes it more difficult to equally distribute the mixture to the cylinders and creates waste material in the form of carbon monoxide, which is emitted into the atmosphere as a pollutant. To get the best fuel economy and the least emissions, the gasoline must be burned as completely as possible in the combustion chamber. This means that the greatest amount of energy (economy) will be produced with the least amount of leftover waste material (emissions). If enough oxygen is to be available to combine with the gasoline, then more air must be provided. This results in a leaner air-fuel mixture (less gasoline) than the ideal ratio.

The air-fuel ratio required to provide maximum power will change very little, except at low speeds, figure 2-4. Reducing speed reduces

Stoichiometric Ratio: An ideal air-fuel mixture for combustion in which all oxygen and all fuel will be completely burned.

Dynamometer: A device used to measure the power of an engine, and to test the engine under simulated road conditions.

■ Poor Driving Means Poor Mileage

Carmakers these days are scrambling to produce cars that use less fuel and are more efficient. Still, the driving patterns of individual drivers can make a big difference in how much fuel the car will use.

Here are some interesting points to remember:

- Stopping and restarting a car engine consumes less fuel than running it at idle for one minute.
- Driving at 55 mph rather than 65 mph reduces fuel consumption by about 12 percent. Cutting the speed to 40-mph is even more economical.
- A minor tune-up will increase mileage by about 10 percent.
- The failure of one spark plug in an 8-cylinder engine can cut 12 percent from the car's mpg. It also can increase HC emissions by as much as 300 percent.
- Turning off the air conditioner can improve gas mileage by 2 or 3 mpg.

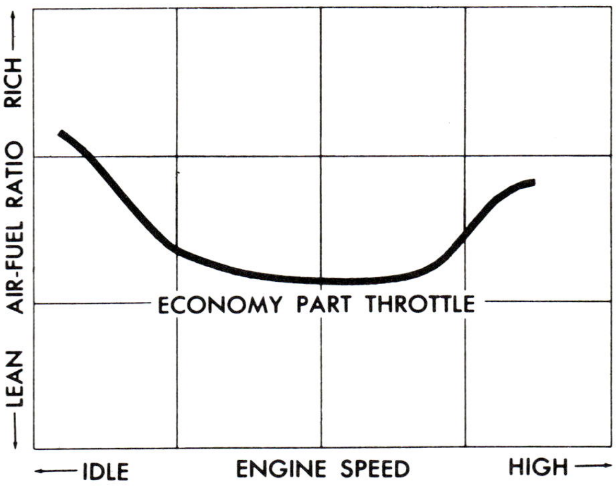

Figure 2-5. The air-fuel ratio for best economy is lean in the middle of the speed range but needs enrichment at high and low speeds. (Chevrolet)

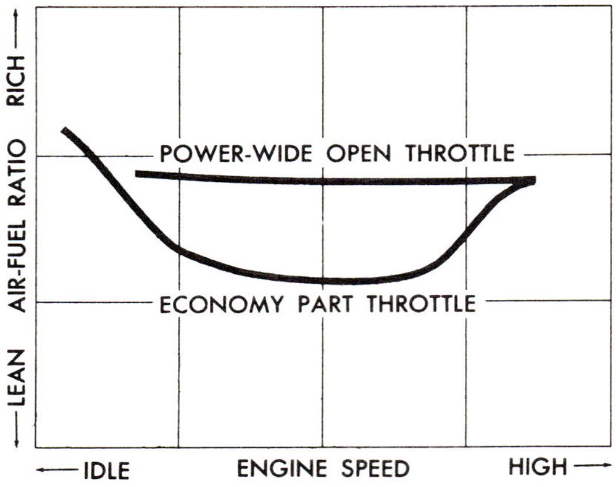

Figure 2-6. The engine must receive lean air-fuel ratios for best economy or rich ratios for maximum power at any given speed. (Chevrolet)

the airflow into the engine. The result is a poorer mixing of the air and fuel, and less efficiency in its distribution to the cylinders. Thus, at low speeds, a slight enrichment of the mixture is needed to make up for this.

The same is true for maximum fuel economy — the leaner air-fuel ratio used will remain virtually the same throughout most of the operating range, figure 2-5. But enrichment will be required during idle and low speeds, as well as during higher speeds and under load — two conditions which require more power.

Enrichment can also occur when it is not required or wanted, as in the case of high-altitude driving. As altitude increases, atmospheric pressure decreases and the air becomes thinner than it is at sea level. The same amount of air actually weighs less and contains less oxygen at higher altitudes. This means that an engine will take in fewer pounds of air and less oxygen. The result is a richer air-fuel ratio, which must be corrected for efficient high-altitude engine operation. Altitude-compensating carburetors correct this problem.

For these reasons, the carburetor must be able to deliver fuel so that the best mileage is provided during normal cruising, with maximum power available whenever the engine is under load, acceleration, or high speed, figure 2-6.

FUEL DISTRIBUTION

Before gasoline can do its job as a fuel, it must be metered, atomized, and distributed to each cylinder in the form of a burnable mixture. To do this, a metering device — a carburetor, figure 2-7 — mixes the gasoline with air in the correct ratio and distributes the mixture as required by engine load, speed, throttle plate position, and operating temperature.

Proper fuel distribution depends on six factors:
1. Correct fuel **volatility**
2. Proper fuel **atomization**
3. Complete fuel **vaporization**
4. Intake manifold passage design
5. Carburetor throttle plate angle
6. Carburetor location on the intake manifold.

We will discuss volatility later in this chapter.

Fuel Atomization and Vaporization

Several things are involved in changing gasoline from a liquid into a combustible form. First, the liquid fuel enters the carburetor, where it is sprayed into the incoming air and atomized (reduced to a mist), figure 2-8. The resulting air-fuel mixture then moves into the intake manifold, where the mist is changed into a vapor.

Vaporization will take place only when the fuel is hot enough to boil. The boiling point is related to pressure: the higher the pressure, the higher the boiling point; the lower the pressure, the lower the boiling point. Because intake manifold pressure is quite a bit less than atmospheric pressure, the boiling point of gasoline drops when it enters the manifold.

Heat from the intake manifold floor combines with heat absorbed from air particles surrounding the fuel particles to begin vaporization. It is helped by raising the temperature of the intake

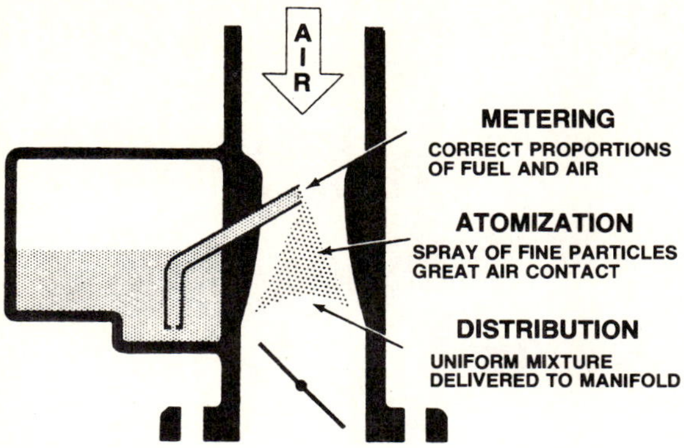

Figure 2-7. The carburetor does the basic job of fuel metering, atomization, and distribution. (Chevrolet)

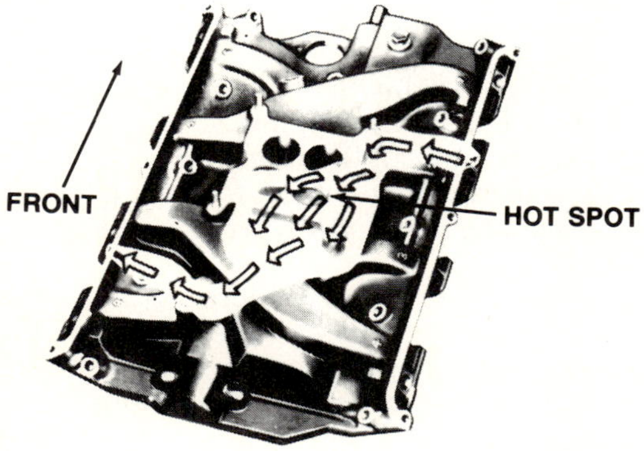

Figure 2-9. Exhaust gases are routed from ports in the cylinder heads through separate passages in the intake manifold to form the manifold hot spot. (Ford)

manifold, since the higher the temperature, the more complete the vaporization will be. This heated area in the intake manifold is called a "hot spot."

Sometimes, there will be poor vaporization. This can be caused by several things:
1. A mixture velocity that is too low
2. A cold manifold or low manifold vacuum
3. Cold incoming air
4. Insufficient fuel volatility
5. Poor manifold design
6. Low carburetor flow capacity.

When there is poor vaporization, too much liquid fuel reaches the cylinders. Some of this extra fuel is given off as unburned hydrocarbons, and some will wash oil from the cylinder walls and cause engine wear. The rest will be carried past the piston rings as blowby gases.

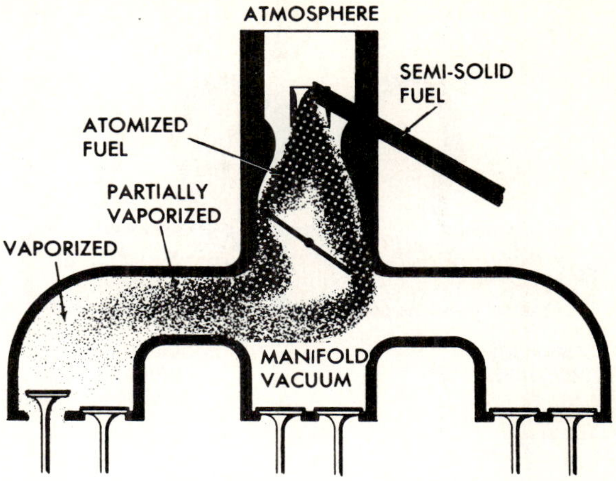

Figure 2-8. Changing liquid fuel to a combustible material is a 2-stage process. First, it is atomized to a mist and mixed with air. The air-fuel mixture then must be vaporized. (Chevrolet)

Intake Manifold Design

The design of an intake manifold has a direct bearing on mixture distribution and volumetric efficiency over the speed range of an engine. The location, size, and surface area of the hot spot on the manifold floor affects vaporization. The hot spot is normally just under the carburetor, figure 2-9. Although the hot spot is usually heated by exhaust, engine coolant is sometimes circulated through passages between the carburetor base and manifold.

Both velocity and heating are also affected by the size of the manifold passages through which the mixture must travel. If the passages are large, the mixture will travel slowly, allowing fuel particles to cling to the manifold walls and avoid vaporization. Small passages create a higher velocity but restrict the travel and distribution of the mixture. The angles at which internal manifold passages turn can be critical, too, figure 2-10. When they are too sharp, fuel tends to separate out of the mixture.

Volatility: The ability of a liquid to change from a liquid to a vapor.

Atomization: Breaking down into small particles or a fine mist.

Vaporization: Changing a liquid, such as gasoline, into a gaseous state.

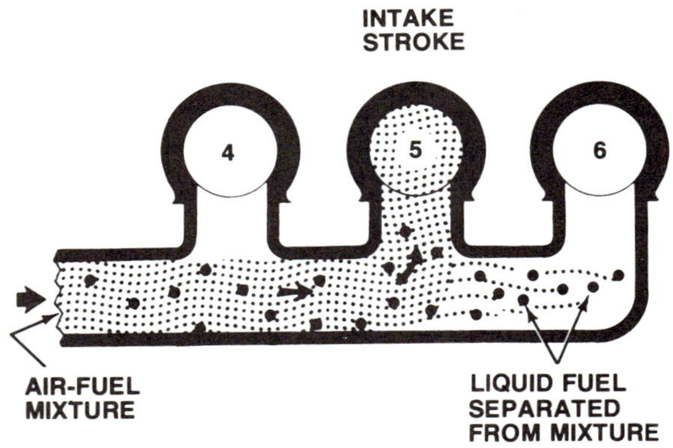

Figure 2-10. An intake manifold with large passages and sharp angles will cause liquid fuel to separate out of the air-fuel mixture. (Chevrolet)

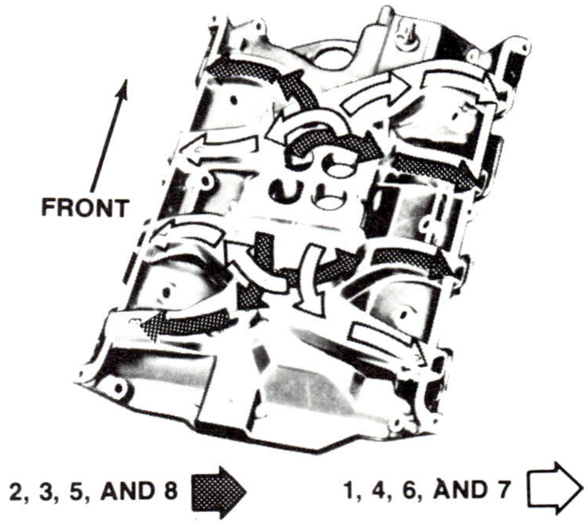

Figure 2-11. Typical V-8 intake manifold fuel passages. (Ford)

The air-fuel mixture should be distributed as evenly as possible among the cylinders. Figure 2-11 shows a typical V-8 intake manifold designed for good distribution to all cylinders. If one or more cylinders receives an overly lean mixture, an increase in the overall mixture will be necessary for that cylinder to fire properly. This will cause the other cylinders to receive a mixture that is too rich. This will raise emissions and lower fuel economy.

Carburetor Size and Placement

Carburetor airflow must be matched to the airflow requirements of the engine. A carburetor that will provide more air and fuel than the engine needs will produce a rich mixture. This can reduce *both* fuel economy *and* power. A carburetor that provides less fuel and air than the engine needs will cause the engine to work harder to provide the power for any speed and load condition. Again, this means that the engine will not be providing the best combination of economy and power.

The location of the carburetor on the intake manifold is important. Incorrect placement in relation to the manifold passages can interfere with proper fuel distribution. If the carburetor is located closer to one cylinder than to the others, improper vaporization and distribution to the cylinders may result.

FUEL COMPOSITION

Gasoline is a clear, colorless liquid — a complex blend of various basic hydrocarbons (hydrogen and carbon). As a fuel, it has good vaporization qualities and is capable of producing tremendous power when combined with oxygen and ignited. Yet it is impossible to accurately predict how a certain blend of gasoline will perform in a particular engine, since no two engines are identical. Remember, mass-produced engines are subject to individual variations in production, which can affect fuel efficiency.

In laboratory tests, oil refiners calculate and measure the characteristics most important in producing gasolines suitable for specific jobs. It is blended to meet particular temperature and altitude conditions. The gasoline you use during summer months is not the same blend available in the winter, nor is the gasoline sold in Denver the same as that sold in Death Valley. In addition to temperature and altitude, refiners must consider several other things during the blending process: volatility, chemical impurities, octane rating, and additives.

Volatility

Volatility is a measure of gasoline's ability to change from a liquid to a vapor and is related to temperature and altitude. The more volatile it is, the more efficiently the gasoline will vaporize. As we've seen, efficient vaporization is needed for even fuel distribution to all of the engine's cylinders and for complete combustion.

Volatility is controlled by blending different hydrocarbons that have different boiling points. In this way, it is possible to produce a fuel with a high boiling point for use in warm weather, and one with a lower boiling point for cold-weather driving. Such blending involves some guesswork about weather conditions, so severe and unexpected temperature changes can cause a number of temperature-related problems ranging from hard starting to vapor lock.

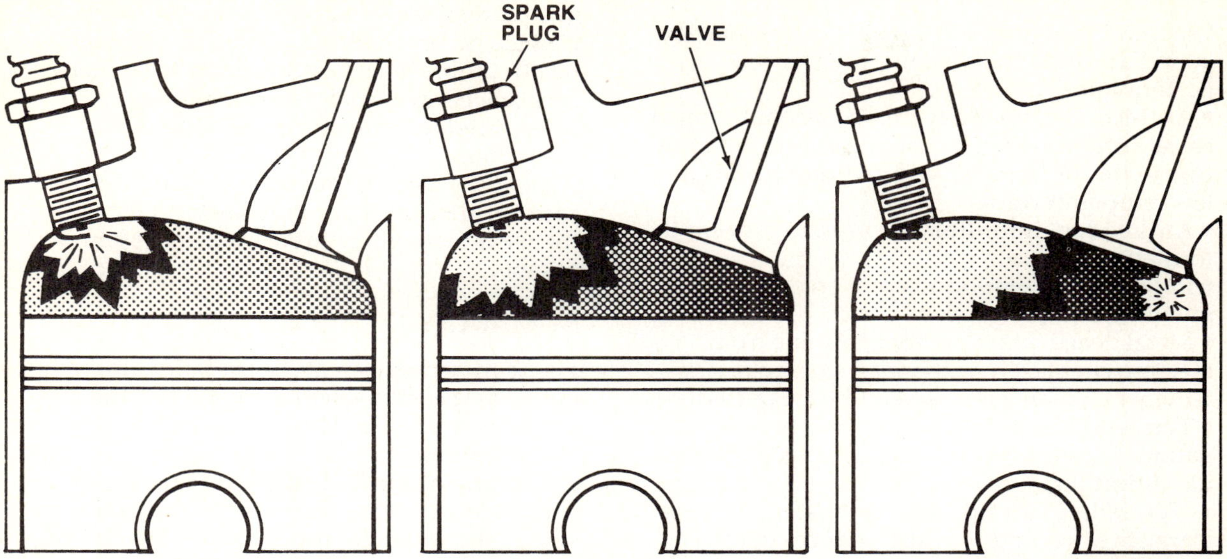

Figure 2-12. Detonation is a secondary ignition of the air-fuel mixture, caused by high cylinder temperatures. It is commonly called "pinging," or "knocking."

Chemical Impurities

Gasoline is refined from crude oil and contains a number of impurities which can harm engines and carburetors. For example, if the sulfur content is too high, some of it may reach the engine crankcase, where it will combine with water to form sulfuric acid. This substance will corrode engine parts, but proper crankcase ventilation helps keep damage to a minimum. Another impurity, gum, tends to form sticky deposits, which will eventually clog carburetor passages and cause piston rings and valves to stick.

To a large extent, the amount of chemical impurities present in gasoline depends upon the type of crude oil used, the refining process, and the oil refiner's desire to keep his production costs low.

Octane Rating

As engine compression pressure reaches a certain level, a great deal of heat is generated as the air-fuel mixture is compressed. Unless gasoline is formulated to hold up under such high pressures and temperatures, there will often be a secondary explosion called **detonation**, figure 2-12. This is popularly called "knocking" or "pinging." Detonation causes a loss of power and overheating of valves, pistons, and spark plugs. The overheating in turn causes more detonation, and may eventually damage the engine.

To prevent detonation, gasoline must have a certain **antiknock value**. This characteristic derives from the type of crude oil and the refining processes used to extract the gasoline. It is measured by an octane rating. Gasolines with a high **octane rating** resist detonation during combustion, while those with a low octane value do not.

There is another cause of knocking called **preignition**. This occurs when deposits which build up on the combustion chamber surfaces get hot enough to ignite the air-fuel mixture before the spark plug can. Knocking or pinging also can be caused by ignition timing that is advanced more than it should be for a particular engine and fuel combination.

Detonation: Also called knocking, pinging. An unwanted explosion of an air-fuel mixture caused by high heat and compression.

Antiknock Value: The characteristic of gasoline that helps prevent detonation or "knocking."

Octane Rating: The measurement of the antiknock value of a gasoline.

Preignition: Ignition of the air-fuel charge in the combustion chamber before the spark plug can ignite the mixture. A cause of "knocking."

Additives

Chemicals not normally present in gasoline are added during refining to improve its performance.

- Anti-icers are specially treated alcohols which act as antifreeze in the gasoline to prevent moisture in the air from causing carburetor icing at low temperatures.
- **Antioxidant inhibitors** are used to prevent the formation of gum.
- Phosphorus compounds prevent spark plug misfiring and preignition.
- Metal deactivators prevent gasoline from reacting chemically with metal storage containers in which it is stored and transported.
- **Tetraethyl lead** (TEL) is used to prevent detonation. Gasoline containing TEL is colored for identification.

Tetraethyl lead aids the combustion process by raising fuel octane which reduces detonation. This allows engines to be built with high compression ratios, which increase power and fuel economy. Lead also prolongs valve life by lubricating valve stems and seats.

Unfortunately, lead has some undesirable effects, as well:

Antioxidant Inhibitors: A gasoline additive used to prevent the formation of gum.

Tetraethyl Lead: A gasoline additive used to help prevent detonation. Abbreviated: TEL.

1. It is poisonous and is given off in the exhaust as a particulate. Lead particulate emissions from a single car are quite small, but they can be a health problem if a large quantity from many cars collects in a small geographic area.
2. Lead reduces the efficiency of catalytic converters.
3. Lead deposits in the engine will dilute crankcase oil, shorten spark plug life, and increase the wear of moving parts in the lower part of the engine.

SUMMARY

Piston movement creates a pressure differential between the air inside the engine and the air outside the engine. This causes airflow into and out of the engine.

The engine air-fuel ratio is commonly measured in terms of the weight of the air taken in versus the weight of the gasoline taken in. Because an engine must operate with the best possible combination of power, economy, and clean exhaust, the desired air-fuel ratio changes as operating conditions change.

The carburetor varies the air-fuel ratio and atomizes the fuel in the airflow to ease vaporization. Manifolds must be designed to provide the best possible combination of vaporization, mixture distribution, and volumetric efficiency. Carburetors and manifolds are carefully matched to each other and to the engine.

Fuel composition also affects engine efficiency. Fuel volatility, purity, octane rating, and additives are considered during fuel blending, as well as the geographical region in which the fuel will be sold.

Review Questions

Select the single most correct answer.
Compare your answers with the correct answers on page 280

1. A disadvantage of gasoline is that it:
   a. Vaporizes easily
   b. Burns quickly
   c. Produces pollutants upon combustion
   d. Has a high heat value

2. Which is *not* a factor in determining airflow requirement:
   a. Engine displacement
   b. Maximum rpm
   c. Carburetor placement
   d. Volumetric efficiency

3. Volumetric efficiency:
   a. Is the ratio of air entering the engine to engine displacement
   b. Decreases as engine speed increases
   c. Is expressed as a percentage
   d. All of the above

4. At maximum speed, the volumetric efficiency of a stock engine is approximately:
   a. 75%
   b. 50%
   c. 10%
   d. None of the above

5. The richest air-fuel ratio that an internal combustion engine can tolerate is about:
   a. 4:1
   b. 2.5:1
   c. 8:1
   d. 18.5:1

6. To burn one pound of gasoline with maximum efficiency requires about:
   a. 8 pounds of air
   b. 15 pounds of air
   c. 18.5 pounds of air
   d. None of the above

7. A rich air-fuel mixture is needed for:
   a. Idle
   b. Heavy load
   c. Acceleration
   d. All of the above

8. Maximum fuel economy requires:
   a. Less air
   b. Leaner air-fuel mixtures
   c. Richer air-fuel mixtures
   d. Higher temperatures

9. Obtaining maximum power results in:
   a. No change in air-fuel ratios
   b. Leaner mixtures
   c. Unburned oxygen
   d. Excess unburned fuel

Chapter

3

Engine Lubrication

The lubrication system circulates motor oil throughout the engine to do a number of jobs. This chapter will tell you:
- The purposes of motor oil
- How it is rated
- How additives help the oil do specific jobs
- How the lubrication system works
- The relationship between lubrication and performance, economy, and emission control.

PURPOSES OF MOTOR OIL

Motor oil in a car engine does five major jobs:
1. It reduces friction between moving parts, which lessens both wear and heat.
2. It acts as a coolant, drawing heat away from the metal of the engine.
3. It carries dirt particles away from moving surfaces, cleaning the engine.
4. It helps seal the combustion chamber by forming a film around the valve guides and between the piston rings and the cylinder wall.
5. It acts as a shock absorber, cushioning engine parts to protect them from the force of combustion.

All of these jobs help keep the engine running smoothly and efficiently. If the motor oil fails to do any one of these, engine performance will be reduced and the engine might be damaged.

MOTOR OIL COMPOSITION AND ADDITIVES

Motor oil is a byproduct of petroleum, as are gasoline, kerosene, and other types of lubricants. To change or improve the performance of a motor oil, manufacturers blend in chemical additives.

Motor Oil Composition

Petroleum-based motor oils contain mostly hydrogen and carbon. They are hydrocarbon compounds, as is gasoline.

Motor Oil Additives

The purpose of a motor oil additive can be to:
1. Replace a property of the oil that was lost during refinement.
2. Strengthen a natural quality already in the oil.
3. Add a certain property that the oil did not naturally have.

A few of the common motor oil additives and their jobs are described in the following paragraphs.

Oxygen tends to combine chemically with hot motor oil. This is called oxidation, and it can leave hard carbon and **varnish** deposits in the

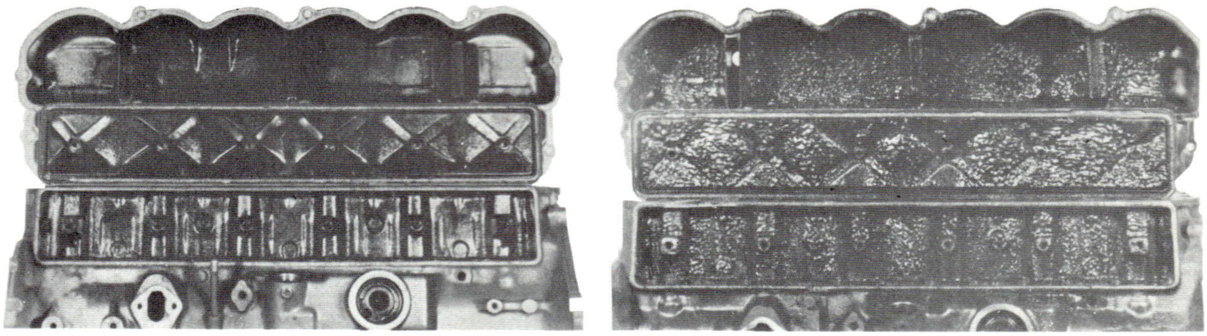

ACCEPTABLE BORDERLINE

Figure 3-1. Sludge deposits in an engine.

engine. **Antioxidants** reduce this high-temperature problem.

In any engine, some of the combustion chamber gases slip past the piston rings and enter the crankcase. This is called **blowby**. Blowby gases contain water vapor and acids that will rust or corrode engine parts. Rust and corrosion preventives are added to motor oil to reduce the effects of blowby gases.

The oil in an engine is constantly being churned by moving parts. This can mix air with the oil and cause it to foam. Foam does not protect moving parts as well as liquid oil. Foam inhibitors in a motor oil will reduce foaming.

Both the water vapor and the fuel in blowby gases tend to mix with cold oil and cause **sludge**, figure 3-1. This thick black deposit clogs oil passages and increases engine friction. **Detergent-dispersants** reduce sludge formation and keep sludge particles suspended in the oil, to be removed when the oil and filter are changed.

Viscosity is the tendency of a liquid to keep from flowing. Some additives help an oil to flow under various temperature conditions. These are called "viscosity index improvers."

Additive precautions
Because different manufacturers use different additives, it is good practice not to mix oil brands within an engine. The chemicals could oppose each other and decrease the cleaning and lubricating abilities of the oil. For the same reason, many oil manufacturers advise against the use of other oil additives that are sold separately.

MOTOR OIL RATINGS

Because engines and operating conditions vary greatly, oil manufacturers blend and sell different types of motor oil. For example, the oil used in a diesel truck is different than that used in a high-performance sports car. Both engines need protection, but most oils cannot serve both extremes.

The two parts of a motor oil's rating are based on its:
- Service classification
- Viscosity.

API Service Classification

The **API service classification** is defined by the American Petroleum Institute (API). An oil is used in standardized laboratory engines and the oil's performance is monitored during specific operating conditions. The areas monitored include resistance to oxidation, corrosion, wear, and carbon formation. On the basis of the oil's performance, a service classification is assigned.

Oils for use in gasoline engines have an "S" followed by a letter from "A" through "E" indicating the service level:
- SA has no performance requirements. It is for use under very mild conditions, and might contain only one or two mild additives.
- SB provides antioxidant and antiscuff protection and resistance to corrosion, and is also for use under mild conditions.
- SC is suitable for use in passenger car engines, and some truck engines, manufactured in 1964 through 1967. It contains additives to protect against high- and low-temperature deposits, wear, rust, and corrosion.
- SD is suitable for use in engines manufactured between 1968 and 1970, and some 1971 engines. It provides more protection than SC, and can be substituted for an SC grade oil.
- SE is currently the highest grade motor oil for gasoline passenger car engines. It should be used in some 1971 engines and all 1972 and later engines, and can be substituted for SC and SD grades. It will lubricate and protect under severe conditions.

Oils for use in diesel engines have a "C" followed by a letter from "A" through "D." The CC and CD grades are used in passenger car

New System		Old System	New System		Old System
SA	=	ML	CA	=	DG
SB	=	MM	CB	=	DM
SC	=	MS (1964)	CC	=	DM
SD	=	MS (1968)	CD	=	DS
SE	=	none			

Figure 3-2. An older rating method had fewer categories and used different letters. It is no longer used. This is how that system corresponds to the "S" and "C" system now used.

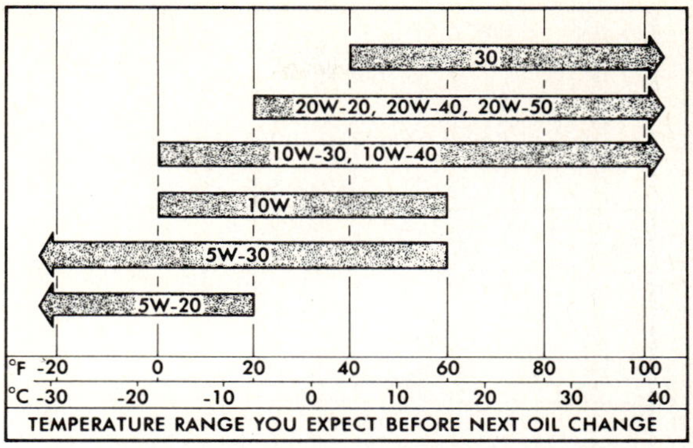

Figure 3-3. Viscosity chart for 1978 Chevrolet passenger car engines. (Chevrolet)

diesel engines.

Figure 3-2 shows an older service classification no longer in use.

One oil may carry more than one API service rating. This means that the oil has passed the tests for two or more categories, such as an SE-CC oil.

SAE Viscosity Rating

All motor oils also receive viscosity ratings from the Society of Automotive Engineers (SAE). A thick oil that does not flow easily has a high viscosity. A thin oil that does flow easily has a low viscosity. The **SAE viscosity ratings** signify the oil's viscosity at a specific temperature. Motor oil viscosity numbers are 5, 10, 20, 30, 40, and 50. The lower numbers represent low-viscosity (thin) oils. The higher numbers represent high-viscosity (thick) oils.

Oils are tested at 0° F (−18° C) to determine their viscosity when cold. The ratings given during the cold test are SAE 5W, 10W, and 20W. The "W" means that the oil can withstand the severe conditions at 0° F. Oils are tested at 210° F (99° C) to determine their viscosity when hot. The ratings given during the hot test are SAE 20, 30, 40, and 50.

When an oil has only one SAE viscosity number it is called a **single-grade** or **straight-grade** oil. These oils can be used safely when the engine operating temperature does not vary greatly. However, if the engine operates under widely varying temperatures, it may benefit from a motor oil that can be effective when hot or cold.

For this reason, oil manufacturers also make **multigrade** oils. These oils are tested at both 0° and 210° F (−18° and 99° C), and have two viscosity numbers. For example, an SAE 20W-40 oil has a viscosity of an SAE 20 when tested cold and a viscosity of an SAE 40 when tested hot. It will protect an engine at either temperature extreme.

Carmakers specify the exact type of oil that should be used under certain operating conditions. Figures 3-3, 3-4, 3-5, and 3-6 show oil charts prepared by specific manufacturers.

Varnish: An undesirable deposit, usually on the engine pistons, formed by the oxidation of fuel and of motor oil.

Antioxidants: Chemicals or compounds added to motor oil to reduce oil oxidation, which leaves carbon and varnish in the engine.

Blowby: Combustion gases that slip past the piston rings into the crankcase; composed of water vapor, acids, and unburned fuel.

Sludge: A thick, black deposit caused by the mixing of blowby and cold oil.

Detergent-dispersant: A chemical added to motor oil to break down and disperse sludge and fuel particles picked up by the oil.

Viscosity: The tendency of a liquid such as oil to resist flowing.

API Service Classification: A system of letters signifying an oil's performance; assigned by the American Petroleum Institute.

SAE Viscosity Rating: A system of numbers signifying an oil's viscosity at a certain temperature; assigned by the Society of Automotive Engineers.

Single Grade (Straight Grade): An oil that has been tested at only one temperature, and so has only one SAE viscosity number.

Multigrade: An oil that has been tested at two different temperatures, and so has two SAE viscosity numbers.

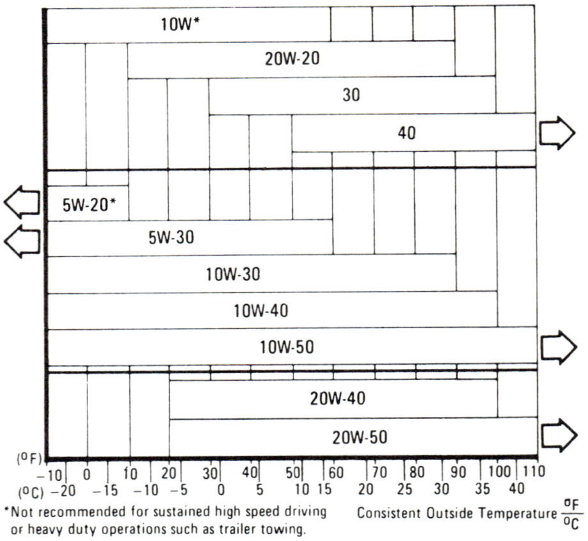

Figure 3-4. Viscosity chart for 1978 Ford passenger car engines. (Ford)

Lowest Temperature Anticipated	Recommended Single Viscosity	Recommended Multi-Viscosity
Above +40° F	SAE 30 or SAE 40	SAE 10W-30, 20W-40, or 10W-40
Above 0° F	SAE 20W-20	SAE 10W-30 or 10W-40
Below 0° F	SAE 10W*	SAE 5W-20 or 5W-30

*Sustained high speeds (above 55 mph) should be avoided when using SAE 10W engine oil since oil consumption may be greater under this condition.

Figure 3-6. Viscosity chart for 1978 AMC passenger car engines. (AMC)

Synthetic Motor Oils

Synthetic-base lubricants have been used in aviation and special-purpose engines for decades. In the 1970's, **synthetic motor oils** have attracted a lot of attention and created controversy.

Before we look at the controversy, we must define "synthetic base." It may be a misleading term, because *all* of the new oils are at least partly derived from petroleum. Of the four major classes of synthetic-base stocks, two are 100-percent petroleum based; one is 60- to 80-percent derived from a petroleum base; and the fourth is 10- to 30-percent derived from a petroleum base. The differences between conventional and synthetic motor oils are in the refining or preparation process and the additives used.

During the refining of conventional oils, no attempt is made to change the molecular struc-

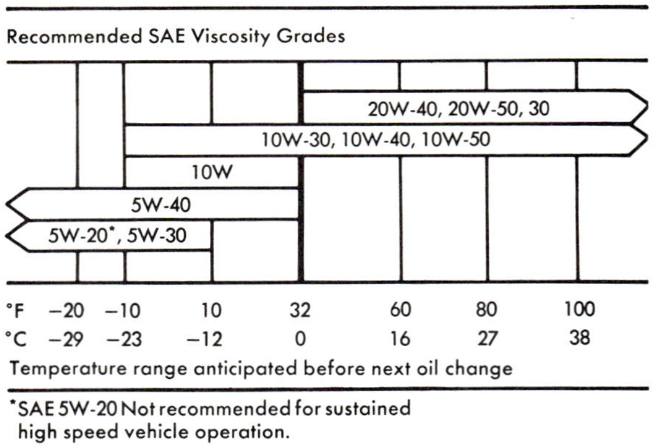

Figure 3-5. Viscosity chart for 1978 Chrysler passenger car engines. (Chrysler)

ture of the crude oil. To make a synthetic oil, chemists carefully combine certain smaller molecules that come from the crude oil and from other sources. The smaller molecules are joined to form the larger molecules needed for motor oil. Chemists can change the synthetic stock by controlling the molecular combinations. After preparation, additives are blended with the synthetic stock just as with conventional stock, although synthetic oils tend to have more additives than do conventional oils.

The SAE and API have not yet set any standards for synthetic oils. However, oil manufacturers have made some performance claims, including statements that synthetics can:

- Last up to 50,000 miles or more (compared to the 6,000- to 12,000-mile replacement intervals for conventional oils)
- Reduce wear on moving parts
- Increase fuel mileage by decreasing engine friction
- Protect an engine throughout a greater-than-normal temperature range.

Much research has been done to investigate these statements. Synthetic oils do tend to have less of a viscosity change through temperature changes. Engine wear tests have had conflicting results, with some engine parts having increased wear with synthetic oil use. The extended drain interval has not been claimed for all synthetic oils.

The energy used to prepare synthetic oils makes them three to eight times as expensive as conventional oils. If an extended service interval is followed, the high original price might be offset. However, conventional oil filters are not made for those service intervals, and the filter might not last the life of the oil. Also, carmakers have not accepted the performance claims of synthetics, and continue to recommend conventional service intervals. Some carmakers

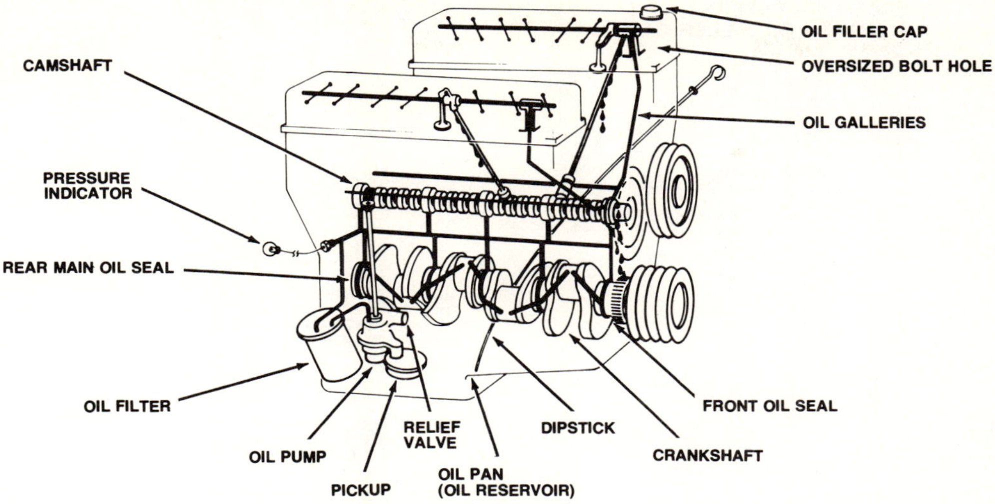

Figure 3-7. The engine oiling system, showing the oil galleries and the oil lines.

have stated that damage to an engine during an extended oil service interval will not be covered by the new-car warranty.

Nevertheless, carmakers are lengthening the oil change interval for conventional motor oils. The 1968 Dodge, for example, required 12 oil changes in 50,000 miles of operation. By contrast, the 1978 Dodge required only six oil changes in the same number of miles.

ENGINE OILING SYSTEM AND PRESSURE REQUIREMENTS

All modern passenger-car engines have a pressurized lubrication system. The parts of this system, figure 3-7, are the:

- Oil reservoir and its ventilation
- Oil pump and pickup
- Pressure relief valve
- Filter
- Galleries and lines
- Indicators.

These parts are described in the following paragraphs.

Reservoir And Ventilation

There must be enough oil in the engine to circulate throughout the system and some reserve so that the oil can cool down before being sent through the system again. All of this oil is kept in the engine oil pan, or sump. Because the oil pan is at the bottom of the engine, the oil drains down into the pan after passing through the engine. The flow of air past the pan when the car is moving helps to keep the oil cool.

Engine oil pan capacities vary according to engine size and design. Small 4-cylinder engines usually carry 3 to 4 quarts of oil. Most V-type engines hold 4 to 5 quarts. Some high-performance and diesel V-type engines require 6 or 7 quarts. Oil is put in the crankcase through

Synthetic Motor Oils: Lubricants formed by artificially combining molecules of petroleum and other materials.

■ Multiple API Classifications

It is possible for an oil manufacturer to blend an oil that can pass both the SE and the CC or CD performance tests. The result would be rated as an SE-CC or SE-CD oil, presumably able to protect a gasoline engine as well as a diesel engine.

General Motors, however, is cautioning the public about the use of these multiple rating oils. For its gasoline engines, it recommends either a straight SE or an SE-CC oil. Use of an SE-CD oil in a gasoline engine, it says, will not provide enough engine wear protection. For its 4-stroke diesel engine, GM recommends a straight CD or SE-CD oil. A CC or SE-CC oil does not protect the engine well enough.

These recommendations were made because, in GM's opinion, the blending of an SE-CD oil must be a compromise between the requirements of a diesel engine and those of a gasoline engine. Under some operating conditions, the compromised oil may not be able to fully protect a gasoline engine.

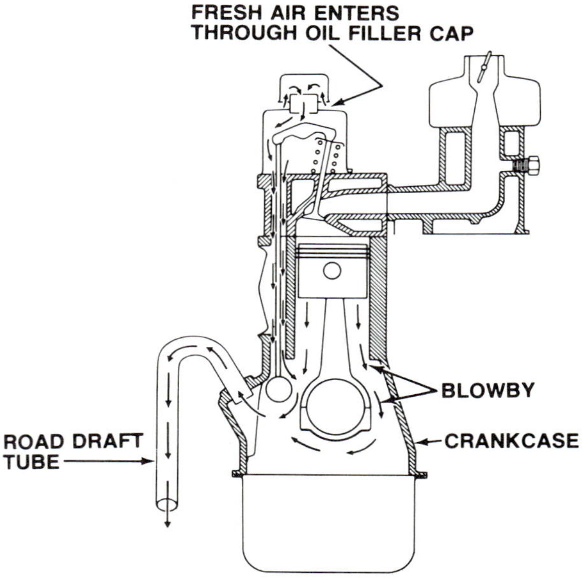

Figure 3-8. An open crankcase ventilation system using a road draft tube.

an oil filler cap at the top of the engine.

We have seen that blowby gases enter the crankcase from the combustion chambers. The gases can mix with motor oil and cause engine damage. Because the combustion chamber gases are under very high pressure, they increase the air pressure within the crankcase. If the crankcase is not vented, the pressure will force oil out of the engine at loosely sealed points such as the oil filler cap, the junction of the oilpan and block, and other points. This is messy, wasteful, and can be dangerous as a fire hazard or if the oil gets on brake or clutch materials. To keep this from happening, the crankcase is ventilated.

On older cars, the oilpan was vented directly to the atmosphere through a **road draft tube**, figure 3-8. Until the 1960's, this was the most common type of crankcase ventilation.

All late-model cars have **positive crankcase ventilation (PCV)** systems, figure 3-9. In a PCV system, clean, filtered air is drawn into the crankcase, and crankcase vapors are recycled to the intake manifold. This keeps blowby from adding to air pollution and provides better crankcase ventilation than the road draft tube does. Crankcase ventilation and PCV systems are explained in detail in Chapter 16.

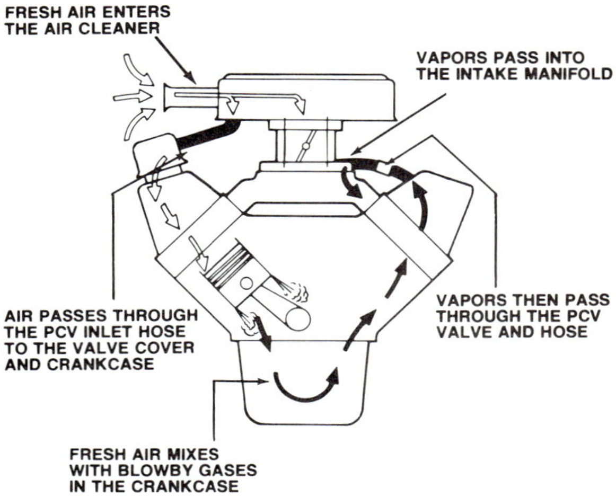

Figure 3-9. The latest PCV systems are totally closed.

Oil Pump And Pickup

The oil pump is a mechanical device that forces motor oil to circulate through the engine. On most engines, the pump is driven by the camshaft through an extension of the distributor shaft, figure 3-10. Some engines may have a crankshaft-driven oil pump, and most overhead-cam engines have a separate engine shaft to drive the oil pump and the distributor.

Two types of oil pumps are in use today:

- The gear type
- The rotor type.

The most common is the gear type, figure 3-11. One gear is driven by gears and a shaft

■ What Makes Smog?

Cars are a major source of air pollution, but they are certainly not the only source. Furthermore, because of the vast strides made in recent years in curtailing exhaust emissions, the modern automobile is becoming less and less the guilty culprit in producing airborne contaminants.

Internal combustion engines produce three major forms of contaminants: hydrocarbons (HC), carbon monoxide (CO), and nitrogen oxides (NO_x). Photochemical smog is caused by sunlight acting on HC and NO_x. (It is not, as some believe, the combination of fog and smoke.) This type of smog was first studied and described in the Los Angeles area in the late 1940's and early 1950's.

The discoverer of the nature of photochemical smog and manmade pollution was Dr. Arie Haagen-Smit, a leader in the early battle for clean air and the first chairman of the California Air Resources Board.

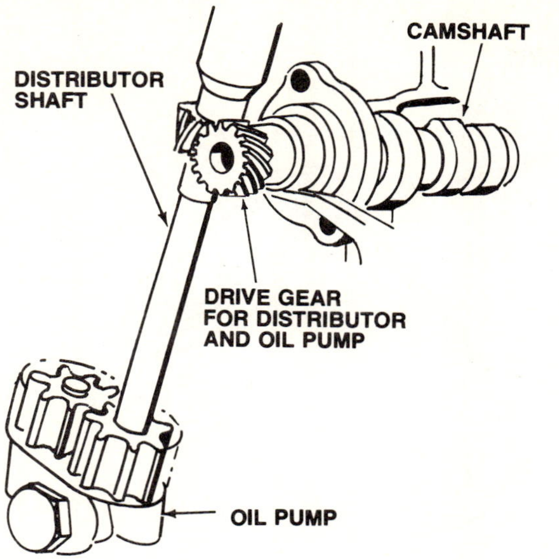

Figure 3-10. The oil pump is driven by the camshaft.

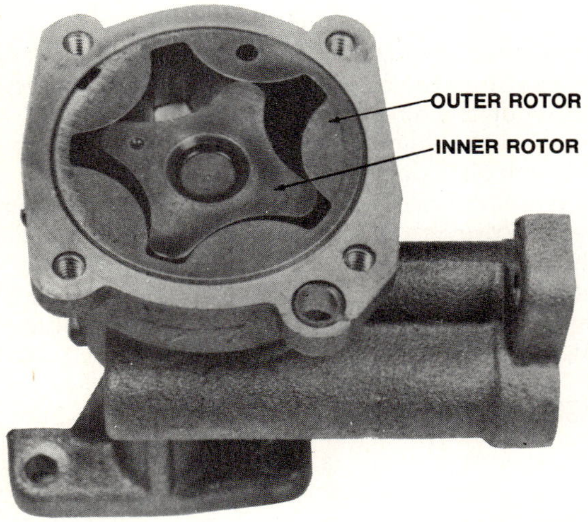

Figure 3-12. A rotor type oil pump.

from the camshaft and is called the drive gear. When the drive gear turns it forces the second gear, called the idler gear, to turn also. As the two gears turn, the oil between the gear teeth is carried along. At the point where the two gears mesh, there is very little room for oil. The oil is forced out of this area and leaves the pump under pressure.

The rotor type works on the same principle of carrying oil from a large area into a smaller area to create pressure. Here, an inner rotor is

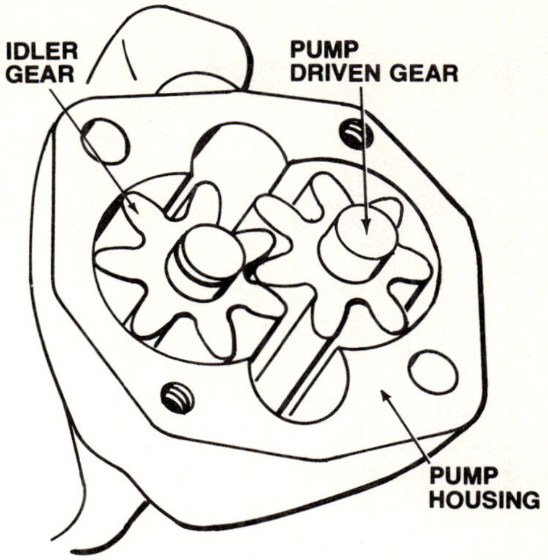

Figure 3-11. A gear type oil pump.

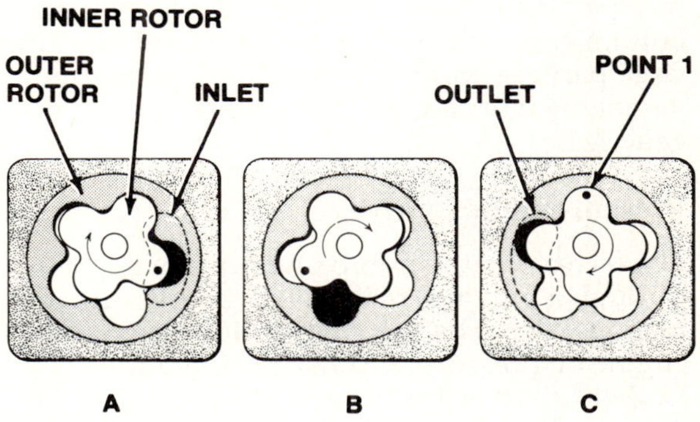

A. Oil is picked up in lobe of outer rotor.
B. Oil is moved in lobe of outer rotor to outlet.
C. Oil is forced out of outlet because the inner and outer rotors mesh too tightly at point 1 and the oil cannot pass through.

Figure 3-13. The principle of operation of the rotor type oil pump.

mounted off-center within an outer rotor, figure 3-12. The inner rotor is driven through gears by the camshaft. The outer rotor is driven by the inner rotor. As the rotors turn, they carry oil from the areas of large clearance to the areas of

Road Draft Tube: The earliest type of crankcase ventilation, which vented blowby gases to the atmosphere.

Positive Crankcase Ventilation (PCV): Late-model crankcase ventilation systems that return blowby gases to the combustion chambers.

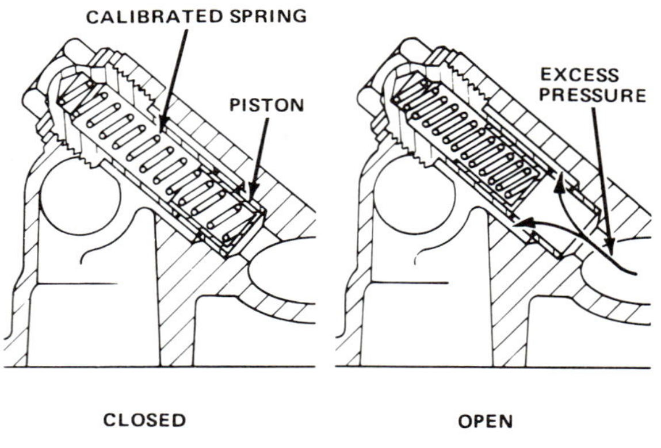

Figure 3-14. The oil pressure relief valve.

small clearance, figure 3-13, forcing the oil to flow from the pump under pressure.

The oil entering the pump comes from the oilpan. An oil pickup tube extends from the pump to the bottom of the pan. A screen at the bottom end of the pickup tube keeps sludge and large particles from entering the oil pump and damaging it or clogging the oil lines and galleries.

Pressure Relief Valve

The oil leaving the pump under pressure passes through a pressure relief valve, or pressure regulating valve. The valve limits the maximum engine oil pressure. It consists of a spring-loaded ball or piston set into an opening in the valve body, figure 3-14. Oil pressure forces the ball or piston to move against spring tension and open the hole in the valve body. Some oil can escape through this hole to decrease the overall system pressure.

The pressure relief valve is usually built into the oil pump housing. The oil that escapes through the relief hole is sent back to the inlet side of the pump to be repressurized. This arrangement of oil flow from the relief valve reduces oil foaming and agitation so that the pump's output is a solid stream.

Filter

Oil leaving the pressure relief valve flows through the oil filter before reaching the rest of the system. The filter is made of paper or cloth fibers that will pass liquid oil but trap dirt.

If the filter becomes plugged with dirt, oil flow is restricted.

To prevent a clogged filter from completely stopping oil flow and damaging the engine, a bypass valve is built into the filter or the filter housing. It allows oil to flow around the filter instead of through it when the oil is too thick to go through the filter due to cold temperatures, or when the filter outlet pressure drops. Oil reaches the engine to protect moving surfaces, but the oil is dirty and will not do a very good job. Oil filters must be replaced at specific intervals to prevent this from happening. Older cars have replaceable oil filter elements inside permanent housings. Most modern engines use spin-on filters that are completely disposable, figure 3-15.

Figure 3-15. A spin-on, completely disposable oil filter.

Galleries And Lines

A modern engine has many surfaces that must always have oil present. To make sure that all of these places receive oil, the lubrication system

■ Splash Oiling

Some older passenger car engines and small gasoline engines do not have an oil pump or a pressurized oiling system. Instead, they depend on the motion of engine parts to splash oil from the oil pan up into the engine.

Special scoops on the connecting rod big ends scoop the oil up and direct it toward critical spots. Some engines may rely on splash oiling for the lower half of the engine and use an oil pump to supply the upper half of the engine.

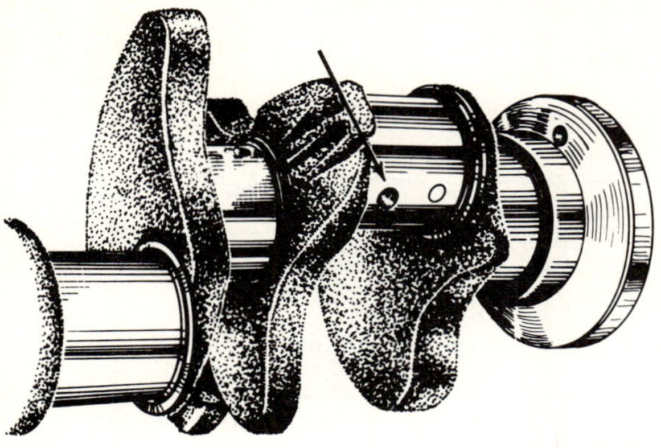
Figure 3-16. The crankshaft has oil galleries drilled in it.

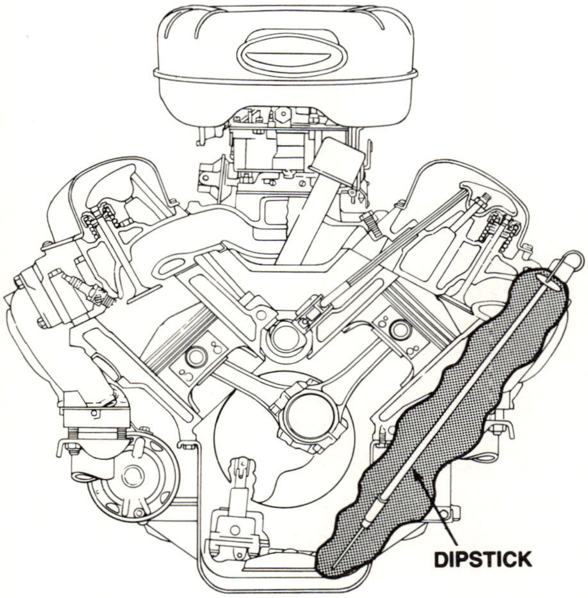

Figure 3-18. The engine dipstick.

includes a network of passages that direct oil to certain parts of the engine, figure 3-7. These passages, called oil galleries, can be drilled through the block, the head, and the crankshaft. They can also be separate tubes, called oil lines, connected to the engine. Not all engines have separate oil lines, but all do have oil galleries.

From the filter, oil flows to a large main gallery. Inline engines and some V-type engines have one main gallery. Other V-type engines have two main galleries, one for each bank of cylinders. From the main gallery, smaller passages direct oil to the camshaft bearings and to the crankshaft main bearings.

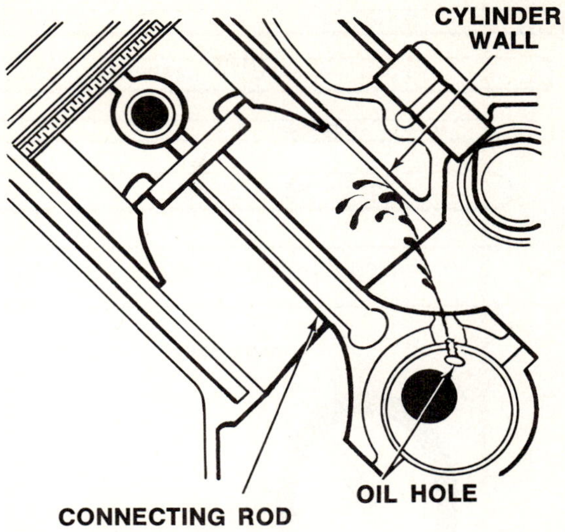

Figure 3-17. Some oil spurts through the connecting rod cap and onto the cylinder walls.

Oil must also reach the crankshaft connecting rod bearings. This is done by drilling holes through the crankshaft, figure 3-16, so that oil at the main bearings can also flow to the rod bearings. The caps that are bolted around the rod bearings have small holes that can line up with similar holes in the crankshaft and the rod bearings. When the holes in the cap and bearing line up with the hole in the crankshaft, figure 3-17, a small stream of oil squirts through and hits the lower cylinder wall. This helps lubricate the cylinder wall and piston.

Different engine designs have different ways of getting oil from the main gallery to the head and valve assemblies. There can be galleries drilled through the block and head, or the oil can travel through hollow valve pushrods. Also, an enlarged head-bolt hole can allow oil flow. In any case, the valve assembly is lubricated. From the valve assembly, the oil drains back down through the engine into the oil pan. The oil drain holes can be placed so that the dripping oil helps lubricate the camshaft.

Indicators

To check the oil level when the engine is at rest, a dipstick is installed in the oilpan. To keep the driver informed of engine oil pressure while the engine is running, carmakers equip cars with a low oil pressure warning lamp or a gauge that indicates the pressure at all times.

Dipstick

When the engine is at rest, almost all of the oil drains down into the oil pan. All engines have a

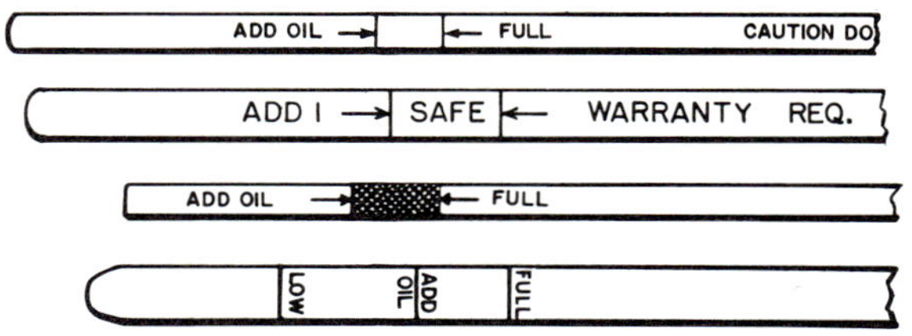

Figure 3-19. Typical dipstick markings.

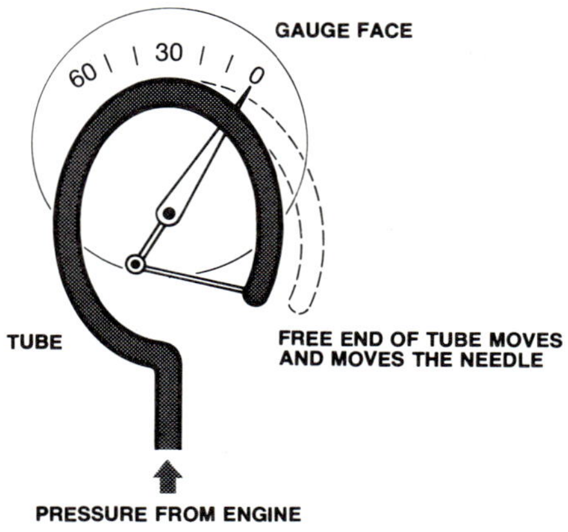

Figure 3-21. A typical mechanical oil pressure gauge.

measuring rod, called a dipstick, that extends from the outside of the engine into the pan, figure 3-18. The dipstick has markings on it that indicate the maximum and minimum oil levels for that engine. When you pull the dipstick out of the pan, a film of oil can be seen on the stick. The level of the film shows the level of the oil in the pan, figure 3-19.

Oil pressure warning lamp

The oil pressure warning lamp lights when the oil pressure is less than a certain amount. This happens during cranking and when there is a problem in the lubrication system. A warning lamp may be able to alert the driver more rapidly than a gauge, because gauge needle movement is less attention-getting than a bright light.

Warning lamps light when a set of electrical contacts close, figure 3-20. The contacts are controlled by a movable diaphragm that is exposed to engine oil pressure. The contacts are closed when oil pressure is low, and the lamp lights. When pressure increases, the diaphragm moves and the contacts open, turning the lamp off.

The contacts and diaphragm are contained within a sending unit. The sending unit is threaded into the engine block or into the oil filter housing.

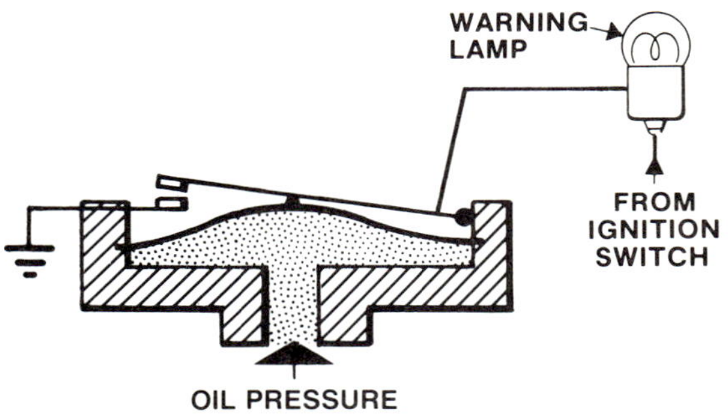

Figure 3-20. A diagram of a low-pressure warning lamp.

Oil pressure gauge

Oil pressure gauges can be operated electrically or mechanically. An electric gauge has a sending unit similar to the warning lamp sending unit. A movable diaphragm varies the amount of electrical current flow through the gauge in proportion to oil pressure. The current flow determines the position of the gauge needle.

Mechanical gauges, figure 3-21, have a tube running from the engine to the gauge. When there is engine oil pressure, oil is forced into the tube and to the gauge. The pressure at the gauge end of the tube is the same as engine oil pressure, and determines the position of the gauge needle.

LUBRICATION EFFECTS ON PERFORMANCE, ECONOMY, AND EMISSION CONTROL

The lubrication system can have direct, measurable effects on performance, economy, and exhaust emissions.

Performance

An engine's performance is related to its mechanical condition. When an engine is new, the moving parts fit together very closely. These close fits are called **tolerances**. As the engine wears, the tolerances increase, figure 3-22. This means, for example, that piston rings do not ride close to the cylinder walls, camshaft lobes lose their original shape, and valves do not fit as accurately into their seats and guides. When an engine wears, or gets "loose," it does not operate as efficiently as it once did, and performance suffers.

The lubrication system helps keep engine wear to a minimum. Using the proper API rating oil will reduce engine wear because the oil's additives are matched to the engine's requirements. Changing the oil and filter at or before recommended intervals will also greatly reduce

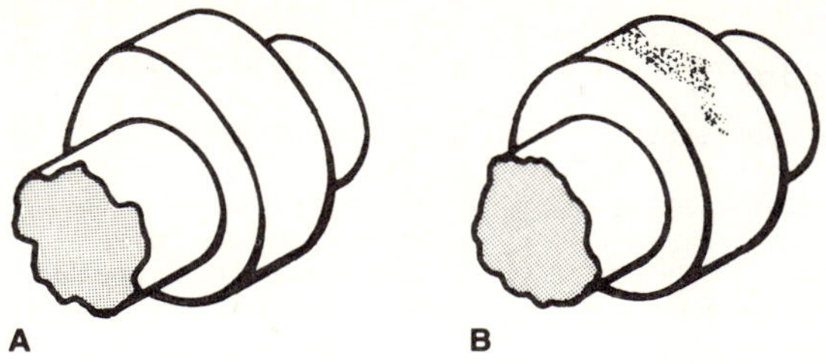

Figure 3-22. These drawings of a camshaft lobe show a new lobe with a close tolerance, Position A, and a worn lobe with increased tolerance, Position B.

engine wear, because this keeps the oil additives at top strength and removes dirt particles that can damage moving surfaces and interfere with hydraulic valve lifter operation.

In "old" oil, the additives have broken down. This allows the oil to be washed off the cylinder walls by the fuel. This, in turn, leads to rapid cylinder wear and the formation of gum and varnish on the walls.

The SAE oil viscosity rating will affect engine wear and performance as well. If too low a viscosity is used during high-speed or high-temperature operation, the thin oil film between moving parts will be forced out, and engine wear will increase. If too high a viscosity is used during low-temperature operation, part of the engine's power will be wasted in forcing engine parts to move through the thick oil. In extreme cases, a combination of cold temperature and high-viscosity oil will prevent the starting system from cranking the engine at all.

Economy

Economy is also partly determined by engine mechanical condition. Camshaft and valve assembly wear, for instance, will change valve timing. Other engine wear factors such as the amount of power wasted as blowby will also affect economy.

Some manufacturers of synthetic oils claim that the increased slipperiness of their oil will decrease engine friction and increase fuel economy. This has been proven in some testing, but the increased fuel mileage alone may not pay for the more expensive oil.

Emission Control

A small amount of oil is normally present in the combustion chamber when the air-fuel mixture is ignited. The oil must be there to lubricate the cylinder walls and to help to seal the chamber. Because oil does not burn as easily as gasoline, some of the oil leaves with the exhaust gases as an unburned hydrocarbon (HC), one of the gasoline engine's main pollution-causing emissions.

When the engine oil is old and diluted by blowby gases, or when the engine is worn and tolerances are large, more oil than normal will enter the combustion chamber. The oil can be coming down through the valve assembly or up past the piston rings. However the oil gets there, once it is in the chamber it is a source of HC emissions. If an older engine has not been properly maintained, it will probably emit more pollutants than if good care had been taken of its lubrication system.

SUMMARY

Motor oil has five major jobs in an engine: reducing friction, cooling, cleaning, sealing, and shock absorbing. Additives mixed in with the oil help it to do these jobs. Two oil rating systems are generally used: the API service classification and the SAE viscosity rating. The API number rates an oil's performance in a laboratory engine, and the SAE viscosity number rates the oil's thickness. An oil may have one or two API ratings. An oil's service classification and viscosity rating must be matched to an engine's requirements for best performance, economy, and emission control.

Synthetic motor oils are made from a petroleum base, but have chemical compounds added in for greater control over the motor oil's qualities and performance.

An engine lubrication system includes the pan, filter, pump, oil galleries or oil lines (or both), dipstick, and pressure warning lamp or pressure gauge. The oil is stored in the oil pan. The pump pulls it out of the pan, pressurizes it, and sends it up through the oil galleries or oil lines to lubricate the moving parts of the engine. To do its job right, the oil must be clean. The filter traps the dirt that the oil picks up, and so the oil and the filter must be changed periodically. A dipstick is used to measure the oil level in the pan when the engine is cold, and either a pressure warning lamp (for low oil pressure) or a pressure gauge (for a continuous pressure reading) is used when the engine is running.

The main effect of lubrication in an engine is the control of engine mechanical wear. Excessive wear will hurt performance, economy, and emission control. The use of the proper oil, and regular oil and filter changes, will keep engine wear to a minimum.

Tolerances: The clearances allowed between moving parts.

Review Questions

Select the single most correct answer.
Compare your answers with the correct answers on page 280.

1. Which of the following is *not* a primary job done by motor oil:
   a. Cooling the engine
   b. Reducing friction
   c. Reducing exhaust emissions
   d. Cleaning the engine

2. Additives in motor oil can:
   a. Replace qualities lost during refinement
   b. Strengthen natural qualities
   c. Add qualities not naturally present
   d. All of the above

3. Blowby gases come from the:
   a. Crankcase
   b. Combustion chamber
   c. Oil filter
   d. Carburetor

4. The oil used in a 1972 or later passenger-car gasoline engine should have an API service classification of:
   a. SD
   b. CD
   c. CC
   d. SE

5. The "W" in an SAE viscosity rating indicates that the oil:
   a. Has been tested at 0° F
   b. Is for winter use
   c. Has no additives
   d. None of the above

6. The highest viscosity number given to motor oils is:
   a. SAE 40
   b. SAE 50
   c. SAE 100
   d. SAE 70

7. The term "multigrade" means that an oil:
   a. Has been given two API service classifications
   b. Has many additives
   c. Has been tested for viscosity when hot and cold
   d. Can be used in gasoline or diesel engines

8. Synthetic motor oils:
   a. Are all partially petroleum based
   b. Cost less than conventional oils
   c. Reduce engine operating temperatures
   d. Can be safely mixed with conventional oils

9. If the crankcase is not ventilated:
   a. Exhaust emissions will increase
   b. The oil will foam and not protect the engine
   c. Too much oil will enter the combustion chambers
   d. Blowby gases will pressurize it and force oil out

10. The oil pump is usually driven by the:
    a. Crankshaft
    b. Camshaft
    c. Fuel pump
    d. Timing gears

11. The oil pressure relief valve:
    a. Is usually mounted in the oil pump housing
    b. Limits the maximum pressure in the oiling system
    c. Contains a spring-loaded ball or piston
    d. All of the above

12. When an oil filter becomes clogged:
    a. A bypass valve allows dirty oil to lubricate the engine
    b. The oil pump must slow down
    c. Engine operating temperature increases
    d. All of the above

13. To direct oil to critical points, the oiling system includes:
    a. Internal galleries
    b. Connecting rod squirt holes
    c. Holes drilled in the crankshaft
    d. All of the above

14. Most dipsticks will show you:
    a. What grade of oil to use
    b. If the oil is oxidized
    c. The maximum and minimum oil levels
    d. None of the above

15. Oil pressure warning lamps have a ___ installed in the engine:
    a. Sending unit
    b. Lamp bulb
    c. Oil tube
    d. Pressure gauge

16. Lubrication's greatest effect on performance, economy, and emission control is:
    a. Making the engine run cooler
    b. Keeping the compression ratio steady
    c. Controlling engine mechanical wear
    d. Increasing engine horsepower

Chapter

4

Cooling and Exhaust Systems

When the air-fuel mixture in a combustion chamber burns, the cylinder temperature can soar to 6,000° F (3,330° C). During the entire 4-stroke cycle, the average temperature is about 1,500° F (816° C). About one-third of this heat energy is turned into mechanical energy by the engine. The remaining two-thirds must be removed from the engine so that the engine can operate efficiently. If the heat energy remains, several things will happen:

- The engine oil temperature will increase, affecting engine lubrication.
- The incoming air-fuel mixture will be preheated too much, making it expand and reduce engine efficiency.
- Metal engine parts will heat, expand, and possibly be damaged.

Where does the extra heat energy from combustion go? About half of it is absorbed by the engine's metal, which is then cooled by the cooling system. The other half remains in the combustion gases, which are removed through the exhaust system. This chapter describes these two systems, and their relationship to performance, economy, and emission control.

COOLING SYSTEM FUNCTION

We have said that about one-third of the total heat energy from combustion is absorbed by the engine's metal. This is both good and bad:

- It is good because a too-cold engine will have poor fuel vaporization, poor lubrication, excessive acids in the blowby gases, and high HC emissions.
- It is bad because a too-hot engine will have weakened metal, poor volumetric efficiency, poor lubrication, and high oxides of nitrogen (NO_x) emissions.

Obviously, there must be a "just-right" engine operating temperature that will avoid these problems. This temperature is slightly different from engine to engine, depending on design and accompanying emission control devices. Earlier engines run at about 180° F (82° C). Late-model engines run as high as 212° F (100° C). These are average temperatures, and can be different under abnormal conditions.

Cooling System Operation

Most automotive engines use a liquid cooling system. The liquid, or coolant, is constantly circulated through the engine, absorbing heat from the engine metal. The coolant is then circulated outside of the engine and exposed to the stream of air passing the car. The air absorbs heat from the coolant, so that the coolant can go back to absorb more engine heat.

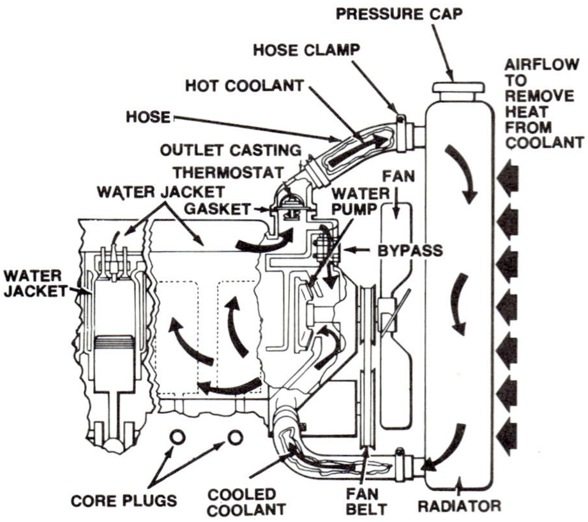

Figure 4-1. A typical liquid cooling system.

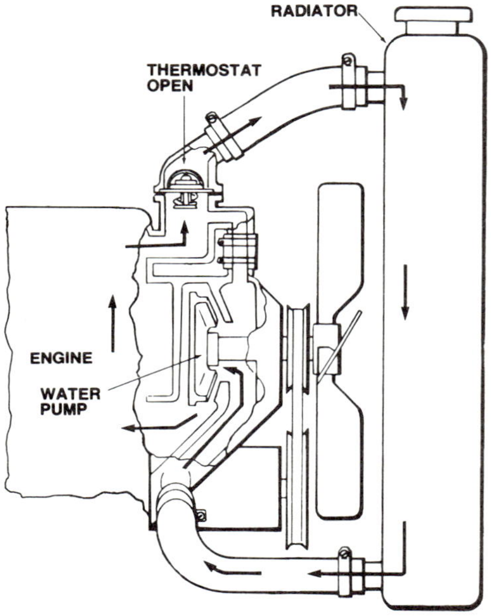

Figure 4-3. As the coolant warms, the thermostat opens and allows it to flow through the radiator.

Circulation Patterns

Within the engine, coolant circulates in passages called water jackets, figure 4-1. Outside the engine, the coolant circulates through hoses and the radiator. To keep the engine running at its ideal temperature, there are two patterns of coolant circulation. The two patterns are controlled by the thermostat, a temperature-sensitive valve that opens and closes a passage between the engine and the radiator.

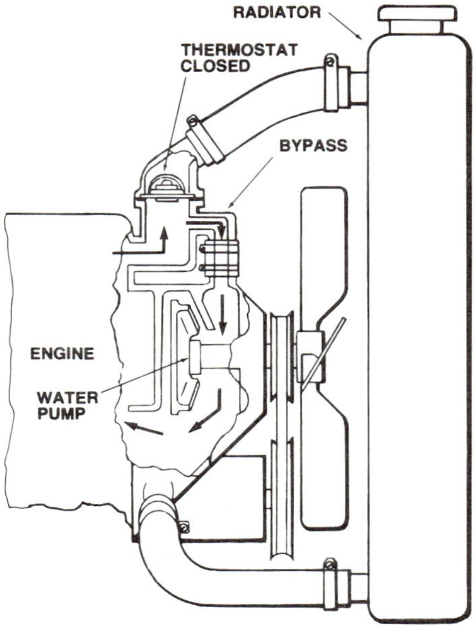

Figure 4-2. The thermostat prevents cold coolant from entering the radiator.

Circulation when cold

When the engine is cold, it must be quickly warmed to its ideal temperature. With the cooling system in full operation, it would take a long time for this to happen. To speed the warmup, the thermostat is closed when the coolant is cold, figure 4-2. This keeps the coolant in the engine from mixing with the coolant in the radiator. Because there is no heat transfer between the coolant in the engine and the air passing the radiator, the heat of combustion quickly warms the engine.

Circulation when warm

As the coolant within the engine gets warm, the thermostat opens, figure 4-3. This allows full coolant circulation between the engine and the radiator. The amount of coolant in the system, the size of the radiator, and the opening temperature of the thermostat help to determine the average operating temperature of the engine.

System Pressure

There are two reasons for pressurizing a cooling system: to increase water pump efficiency and to raise the boiling point of the coolant.

Pump efficiency is affected by pressure. Without a pressure cap installed on the system, a water pump is only about 85 percent efficient. With a 14-psi pressure cap, the pump becomes 100 percent efficient.

Coolant boiling-point control is equally important. Water boils at 212° F (100° C) at atmos-

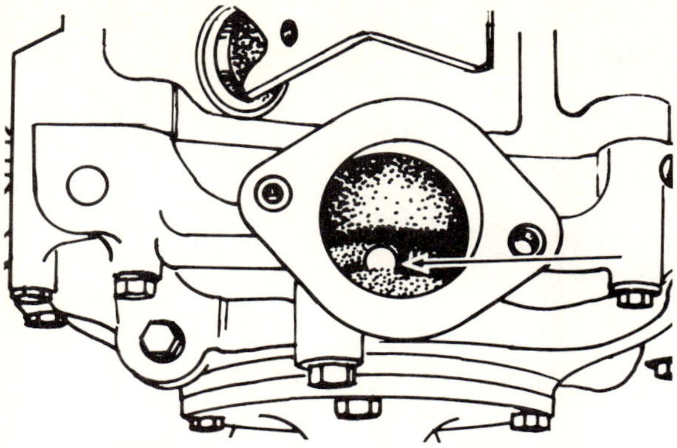

Figure 4-4. This internal passage in the thermostat housing directs cold coolant to the water pump.

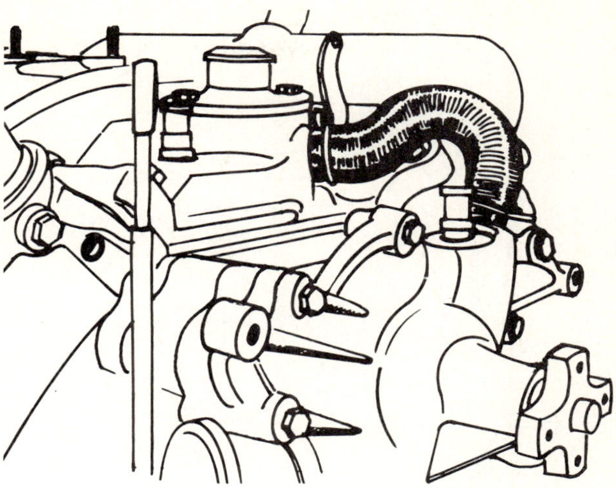

Figure 4-5. This external hose connects the thermostat housing to the water pump.

pheric pressure (14.7 psi) at sea level. An average solution of engine coolant boils at around 223° F (106° C) under the same conditions. Even this higher figure is dangerously close to the operating temperature of the engine. Pressurizing the system increases the boiling point of the coolant. Each pound increase in pressure raises the boiling point about 3° F (1.6° C). Thus, coolant under 15-psi pressure will boil at approximately 268° F, or 45° higher than without the cap. Most cooling systems are pressurized at 12 to 17 psi. A few imported cars are designed for lower pressures.

ENGINE COOLANT

Although pure water transfers heat very well, it has other disadvantages when used in a cooling system:

- Metal exposed to water will rust, and rust particles will clog water jackets.
- When the car is not running and the system is not pressurized, water will freeze at only 32° F (0° C), harming engine parts.

Early cooling systems used water blended with a rust inhibitor and alcohol, to lower the liquid's freezing point. Alcohol, however, boils at a lower temperature than does water, and tends to escape from the system.

Modern cooling systems use a mixture of water and antifreeze, an ethylene-glycol based solution. Glycol does not transfer heat as well as water, but its freezing point is much lower and its boiling point is much higher. Many modern cooling systems need the antifreeze in the coolant to activate the temperature warning lamp, which is set to register high temperatures considerably above the boiling point of water. When there is no antifreeze in the system, the water can boil away and cause severe damage to the engine block, yet the warning lamp will not light.

Different brands of antifreeze may contain corrosion preventives, rust preventives, and water pump lubricants. Some also contain small particles designed to seal minor leaks in the cooling system. A coolant mixture that contains from 35- to 50-percent ethylene-glycol antifreeze will protect the engine and cooling system while improving the system's high- and low-temperature operation.

COOLING SYSTEM COMPONENTS

A typical liquid cooling system, figure 4-1, includes the:

- Water jackets
- Thermostat-controlled bypass
- Core, or freeze, plugs
- Radiator
- Water pump
- Radiator fan
- Thermostat
- Radiator pressure cap
- Coolant hoses
- Drive belt
- Coolant recovery system.

The following paragraphs describe these parts.

Water Jackets, Bypass, And Core Plugs

Water jackets and coolant passages are cast into the cylinder block and head. They are designed so that the coolant will circulate freely and not stay in stationary pockets.

The thermostat-controlled bypass is a passage that can direct coolant back into the engine before it has circulated through the radiator. The bypass can be an internal passage in the block, figure 4-4, or an external connection, figure 4-5.

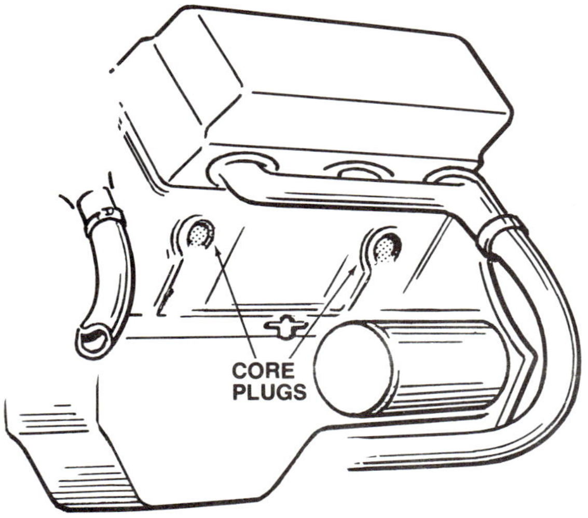

Figure 4-6. Core plugs are pressed into the cylinder block.

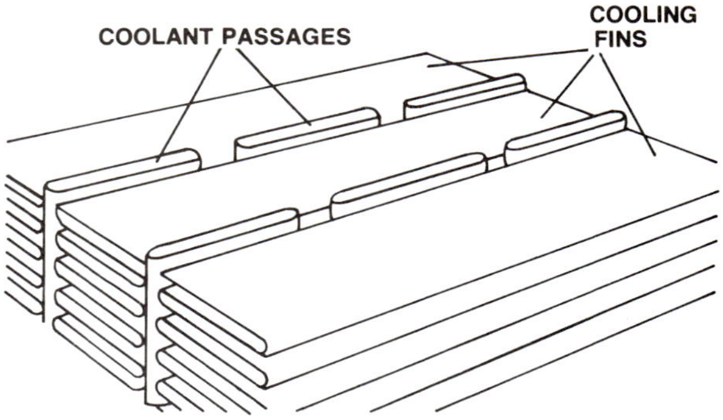

Figure 4-7. The tubes and cooling fins of the radiator core. (Chrysler)

Core plugs, also called freeze plugs, are soft metal plugs pressed into holes in the sides of the cylinder block, figure 4-6. The holes are used during the casting and machining of the block. If the coolant freezes and expands, it may push the core plugs out of their holes, relieving the pressure within the engine. This can prevent the rest of the engine from damage.

Radiators

The purpose of the radiator is to create a large amount of surface contact with the surrounding air. Radiators are made of thin metal tubes, usually copper or aluminum, with cooling fins attached, figure 4-7. Coolant flows through the tubes and transfers its heat to the radiator metal. The air flowing past the radiator tubes and fins absorbs this heat. The number of tubes and fins in a radiator determines the unit's heat-transferring capacity.

The assembly of tubes and fins is called the radiator core. To keep a steady stream of coolant flowing into and out of the core, two larger metal tanks are connected to opposite ends of the tubes. There are two patterns of radiator construction:

- Downflow radiators have vertical tubes and a tank at the top and the bottom of the core, figure 4-8, position A.
- Crossflow radiators have horizontal tubes and a tank on either side of the core, figure 4-8, position B.

Crossflow radiators allow a lower hood line on the car and are used on some late-model vehicles. Other than the flow direction, there is no difference in operation between the two types.

One of the radiator tanks, either the top tank or a side tank, has an open neck sealed with a pressure cap. On cars with automatic transmissions, one of the tanks contains a cooler for the transmission fluid, figure 4-9. The fluid and the coolant do not mix, but the extra heat of the transmission fluid adds to the work of the cooling system. Systems on automatic-transmission cars are built with a greater capacity than those on manual transmission cars.

Water Pumps

The water pump uses **centrifugal force** to circulate the coolant. It contains a fan-shaped impeller, figure 4-10, set in a round chamber with curved inlet and outlet passages. The chamber is called a scroll because of these curved areas. The impeller is driven by a belt from the crankshaft pulley, and spins within the scroll. Coolant from the radiator or the bypass enters the inlet and is picked up by the impeller blades. Centrifugal force flings the coolant outward, and the scroll walls direct it to the outlet passage and the engine. Because the water pump is driven by the engine, it keeps coolant circulating whenever the engine is running. Centrifugal pumps must turn fast to be efficient. Worn or loose belts will permit slippage which is not easily detected. Early water pumps had to be externally lubricated, but modern coolant mixtures contain a water pump lubricant. The pump bearings must be well sealed to prevent coolant leaks.

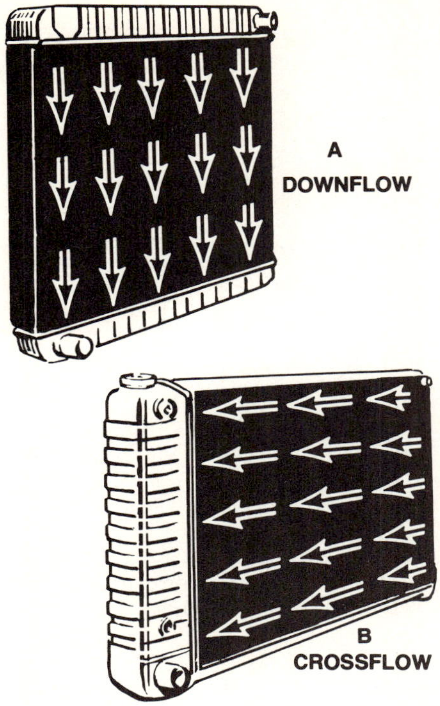

Figure 4-8. A radiator may be either a downflow type or a crossflow type.

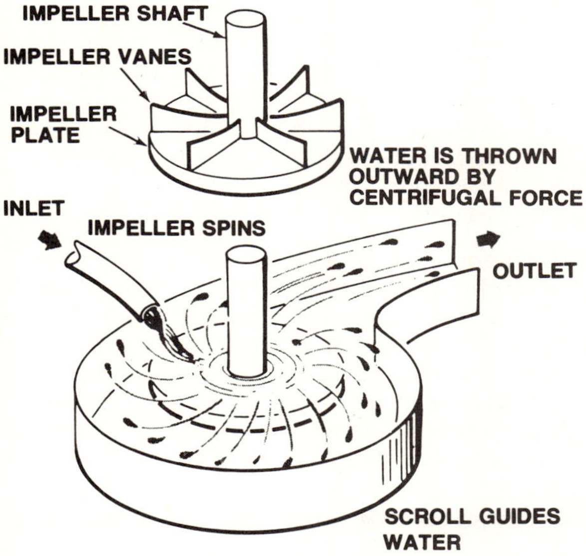

Figure 4-10. The water pump impeller is driven by the engine. Its spinning blades apply centrifugal force to the engine coolant.

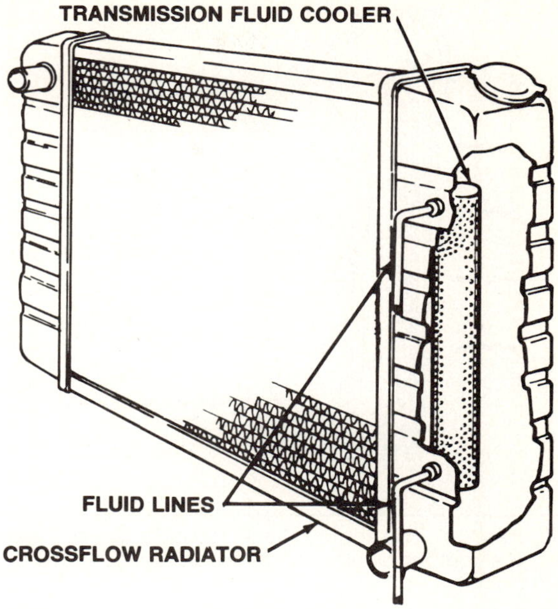

Figure 4-9. The transmission fluid cooler is installed in one of the radiator tanks. (Harrison)

Radiator Fan

When the car is traveling at highway speeds, airflow through the radiator core is great enough to absorb all excess heat. At low speeds or idle, however, there is not enough natural airflow. To increase the amount of air passing through the radiator, an engine-driven fan is mounted behind the radiator, figure 4-11. The fan is usually mounted on the same shaft as the water pump impeller, and is driven by the same belt. The fan blades pull extra air into the engine compartment. Many late-model cars have

Centrifugal Force: A force applied to a rotating object, tending to move the object toward the outer edge of the circle of rotation.

■ If A Little Is Good, A Lot Can Be Worse

It doesn't happen too often, but it is possible. Too much antifreeze in an engine's cooling system will *raise* the freezing point. The "right" amount of antifreeze varies from one part of the country to another and from one vehicle to another. However, in almost all cases, maximum freeze protection is provided by a coolant solution with about 67-percent antifreeze by volume. When the antifreeze concentration goes above this level, the coolant may freeze or become slushy at higher temperatures, and coolant heat transfer will be reduced.

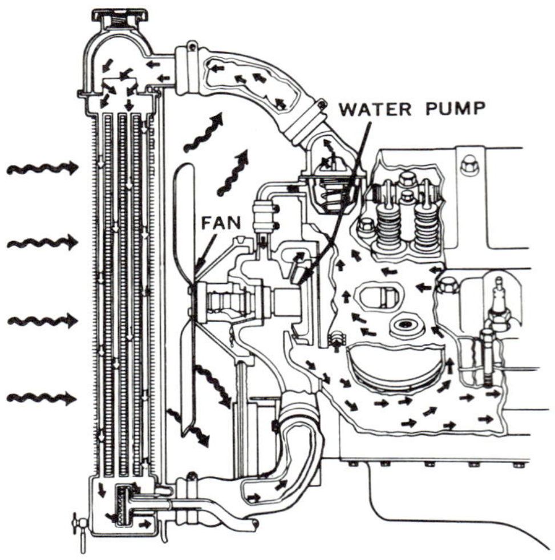

Figure 4-11. The engine-driven fan pulls air into the engine compartment through the radiator core. (Harrison)

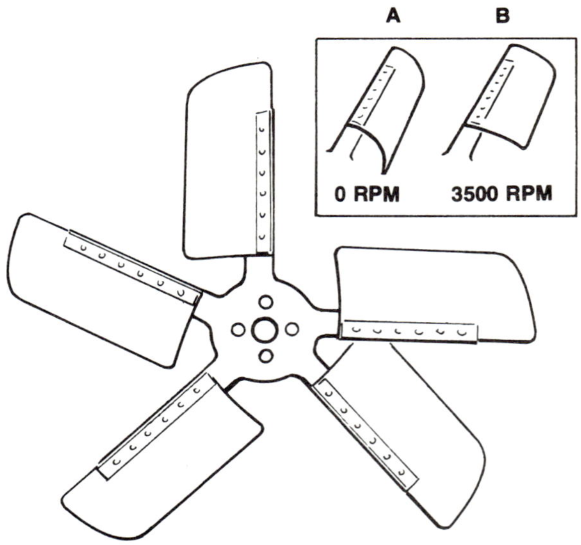

Figure 4-12. The flexible blades of this fan change shape as engine speed changes.

shrouds around the fan to concentrate the airflow where it is most needed.

Engine power drives the fan, but this power is wasted at high speeds when natural airflow will do the job. There are three ways to avoid this waste:

- Flexible fan blades
- Fluid fan drives
- Electric fans.

Flexible fan blades change their angle, or pitch, with changing engine speeds. At low speed, the blades are sharply angled, figure 4-12, position A. This pulls a lot of air through the radiator, but requires a lot of engine power. At high speeds, position B, centrifugal force flattens the fan blades. They pull less air and require less power. This also reduces fan noise at cruising speeds.

Fluid drive fans are heavier than conventional fans, but the power savings make them worthwhile. The drive, figure 4-13, is a clutch assembly that depends on a silicone fluid to transmit motion from the drive belt pulley to the fan itself. These are also called fluid clutch fans. When there is a lot of fluid present, the drive pulley and the fan rotate at the same speed. When there is little fluid present, the drive slips and the fan turns at less than drive pulley speed. When the drive is slipping, less engine power is used to drive the fan.

The amount of silicone fluid present is controlled by a **bimetal temperature sensor** spring exposed to the airflow from the radiator. When the airflow is cool, the spring lets very little silicone into the drive, and the fan turns slowly or idles. When the airflow is hot, the spring allows a lot of silicone into the drive, and the fan turns quickly.

Electric fans are used on some automobiles, usually to conserve horsepower. The fan is not driven by the engine but by an electric motor, figure 4-14. The motor is switched on and off by a temperature-sensitive switch that reacts to engine coolant temperature. When you work on a car that has such a fan, disconnect the battery ground cable or the fan power lead before getting near the fan. The fan can switch on at any time, whether or not the engine is running.

Thermostats

The thermostat regulates coolant flow between the engine and the radiator. When the coolant is cold, the thermostat is completely closed. As the coolant warms, the thermostat begins to open. The thermostat's position varies during normal operation, but it is rarely opened completely, unless the engine is working very hard or the weather is hot.

Bimetal Temperature Sensor: A sensor or switch that reacts to changes in temperature. Made of two strips of metal welded together that expand differently when heated or cooled, causing the strip to bend.

Aneroid Bellows: An accordian shaped bellows that responds to changes in coolant pressure or atmospheric pressure by expanding and contracting.

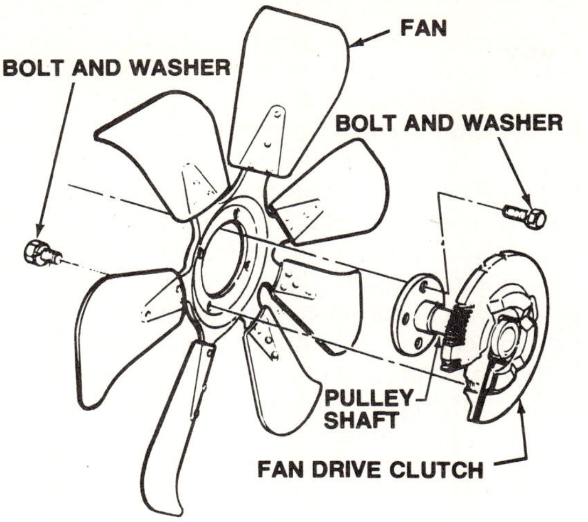

Figure 4-13. This fan has a fluid clutch controlling the speed of rotation. (Ford)

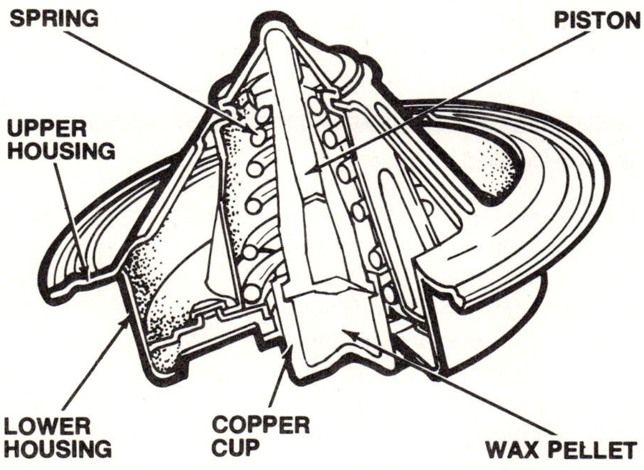

Figure 4-15. A cross section of a wax-actuated thermostat. (Stant)

All late-model thermostats are wax actuated, figure 4-15. A chamber in the thermostat is filled with wax that is solid when cold but melts and expands when heated. When the wax is cold, a spring holds the thermostat closed. As the coolant temperature increases, the wax expands and forces the thermostat to open against the spring tension. This type of thermostat will work reliably under normal system pressure. Earlier types that used an **aneroid bellows**, figure 4-16, could be affected by the coolant's pressure and not work reliably.

The thermostat's opening temperature is stamped on the outside. Replacement thermostats must have the same opening temperature for the system to work correctly. The thermostat is mounted in a metal housing at the top front of the engine, figure 4-17.

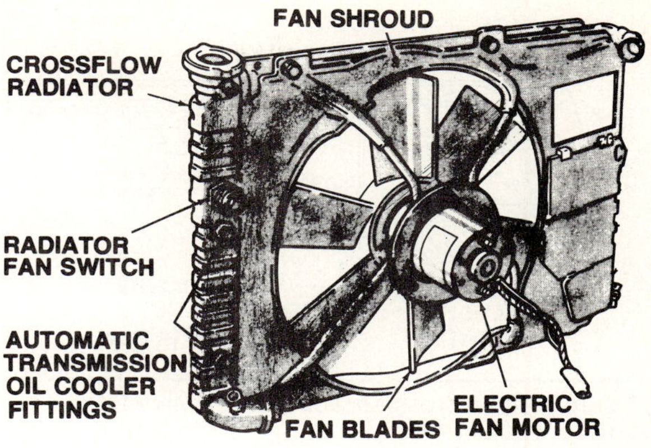

Figure 4-14. In this car, an electric motor drives the radiator fan. (Chrysler)

Figure 4-16. These bellows-actuated thermostats are not used in late-model cars because the bellows operation is affected by coolant pressure.

Figure 4-17. A typical thermostat location.

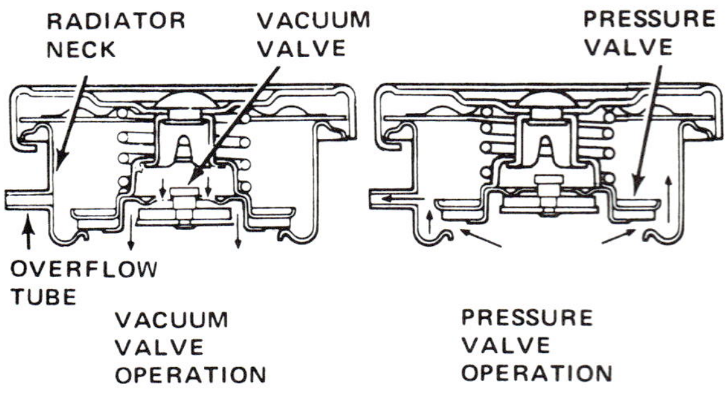

Figure 4-18. A cross-section of a typical radiator pressure cap. (AMC)

Radiator Pressure Cap

To improve the coolant's performance, the cooling system is sealed and pressurized. Part of the pressure comes from the water pump. The rest of the pressure comes from the expansion of the coolant as it warms. The radiator cap's job is to keep the system pressurized, vent any excess pressure, and allow atmospheric pressure to reenter the system as the coolant cools and contracts after the engine is shut off.

To do this, radiator caps have two valves, figure 4-18. A pressure valve relieves the excess pressure inside the system during engine operation. A vacuum valve allows outside air to enter the system when the engine is off. These pressure caps usually are on the top or side tank, figure 4-19, but occasionally can be found in unusual locations such as on a special hose section, figure 4-20.

Radiator Hoses And Drive Belts

Radiator hoses connect the cooling system passages to the radiator tanks, figure 4-1. The hoses are flexible and will absorb the motion between the vibrating engine and the stationary radiator. The hoses are made of rubber, and are often reinforced with a wire coil. The hoses can be molded in a specific shape or they can be ribbed, flexible lengths that adapt to different installations.

The hose ends fit over metal necks on the engine and radiator and are held in place with hose clamps, figure 4-21. Hose clamps can be held in place by spring tension or by a screw.

The drive belt that operates the radiator fan and water pump is a reinforced rubber V-belt, figure 4-22. It fits tightly over the crankshaft pulley, the fan pulley, and, usually, the alternator pulley. Belt tension is adjusted by moving the alternator in its mounting.

Coolant Recovery System

Most late-model cars have a coolant recovery system and a closed cooling system. Instead of a pressure cap that vents to the atmosphere, these systems use a pressure cap that is connected to a coolant tank or overflow reservoir, figure 4-20. When system temperature increases and the coolant expands, the extra coolant flows out of the radiator neck and through an overflow hose. Instead of being dumped on the ground, as in earlier cooling systems, this overflowing coolant is held in the coolant tank. When the system cools and a vacuum develops at the radiator cap, coolant is drawn back into the radiator through the same hose.

This system is important to late-model cars that tend to run hotter because of emission control equipment, automatic transmissions, air conditioning, and other power-consuming accessories. The coolant tank ensures that there is always enough coolant to fill the system because none is lost if the radiator overflows. Coolant recovery systems are said to increase cooling system efficiency by about 10 percent.

■ Air-Cooled Engines

All automotive engines are air-cooled. That is, even engines with liquid cooling systems rely on air to absorb the heat from the coolant. Some automotive engines, and many nonautomotive engines, do not use a liquid coolant but instead transfer engine heat directly from the cylinder's metal to the air. These types are called "air-cooled" engines to distinguish them from engines with liquid cooling systems.

Air-cooled engines can be found in everything from a fraction-of-a-horsepower model airplane to the 261-horsepower turbocharged Porsche Carrera. The Porsche and Volkswagen engines are the most common automotive air-cooled engines, along with General Motors air-cooled Corvair built in the 1960's.

The Volkswagen engine is typical of air-cooled construction. It has individual cylinder castings with cooling fins. The castings are bolted to a split crankcase in an opposing pattern. A cylinder head is attached to adjacent pairs of cylinders. A large fan directs air over the engine. A radiator-like oil cooler is sometimes added to help cool the engine.

Air-cooled engines are made largely from aluminum for quick heat transfer and light weight. Tolerances within an air-cooled engine are greater than those of a liquid-cooled engine, because the aluminum expands greatly during normal hot operation.

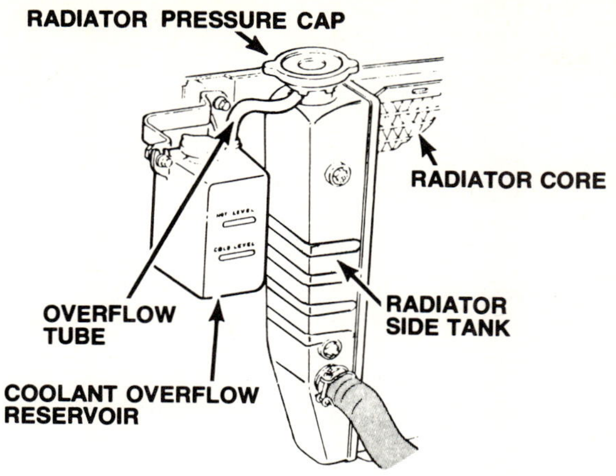

Figure 4-19. Most radiator caps are mounted on one of the radiator tanks. (Ford)

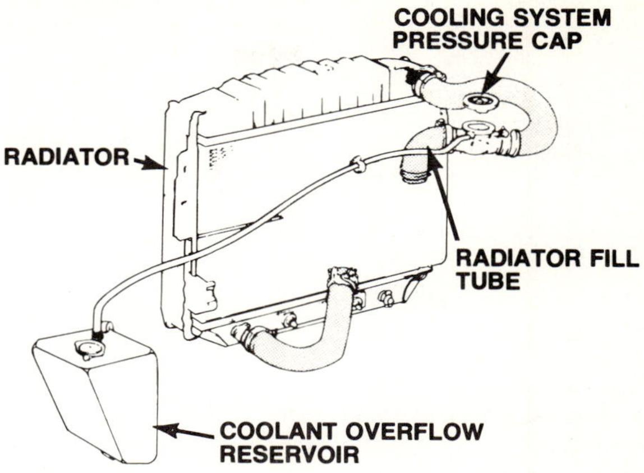

Figure 4-20. Some radiator caps are mounted away from the radiator to shorten the radiator profile. (Ford)

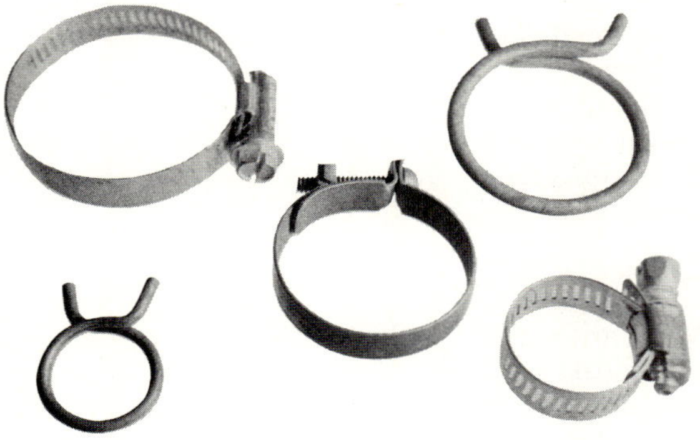

Figure 4-21. Various types of hose clamps.

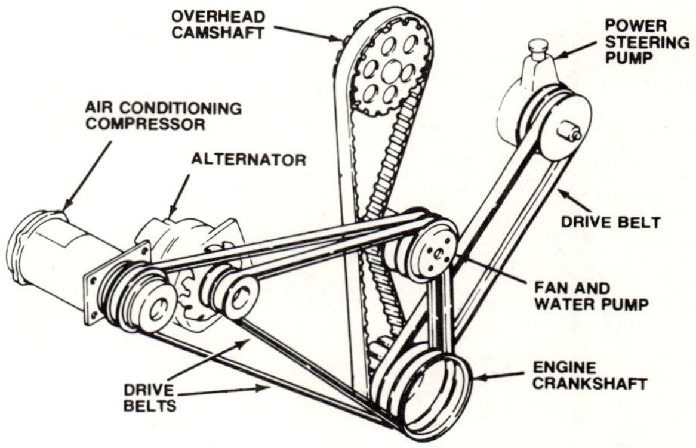

Figure 4-22. The fan and water pump drive belt also drives the alternator. (Harrison)

When more coolant must be added to a recovery system, it is added to the plastic tank, not directly to the radiator. The tank is marked with "hot" and "cold" fill levels.

ENGINE TEMPERATURE EFFECTS ON PERFORMANCE, ECONOMY, AND EMISSIONS

When the engine is too cold, performance is hurt because of poor fuel vaporization. The liquid fuel entering the cylinders also tends to wash oil off the walls, reducing the cylinder's lubrication. The liquid fuel runs past the piston rings into the crankcase, where it dilutes the oil and further affects lubrication.

A cold engine affects economy in several ways. Because the air-fuel mixture is not vaporizing and burning completely, less energy is

■ Hey, Look Me Over

The cooling system is often overlooked in periodic car service. According to the Car Care Council, in 1977 a major coolant manufacturer surveyed the condition of about 1,200 1-year-old to 3-year-old cars. The findings: 22 percent of the newest cars had evidence of corrosion in their coolant. Of the 2-year-olds, 32 percent had the problem; and 36 percent of the 3-year-olds were also afflicted. Perhaps even more significant was the finding that, when the three model years were averaged together, nearly *half* of the cars required some sort of cooling system service. New or old, the cooling system needs to be a part of periodic maintenance.

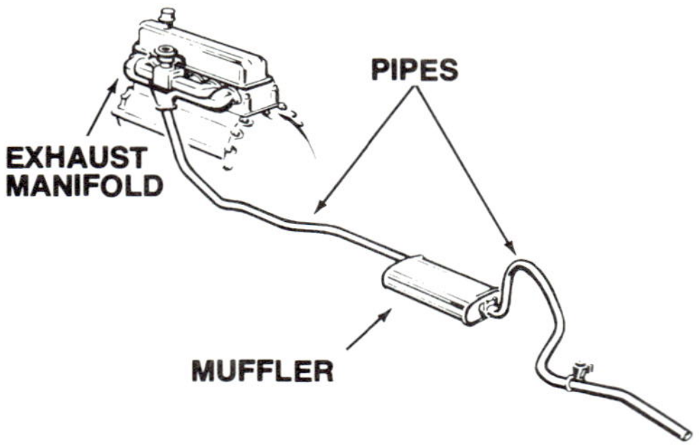

Figure 4-23. The exhaust system used with an inline engine.

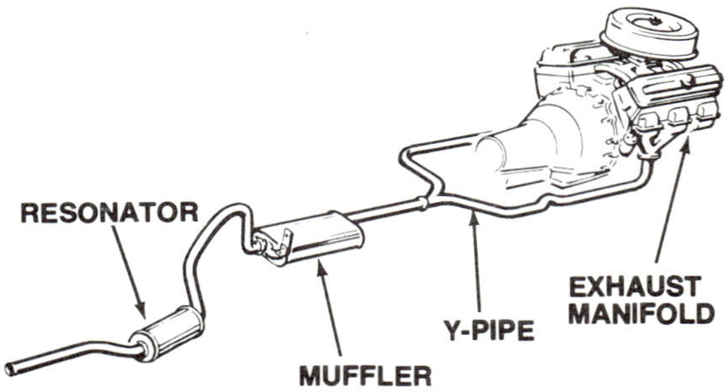

Figure 4-24. This V-type engine uses a Y-pipe to connect the two cylinder banks to a single exhaust pipe.

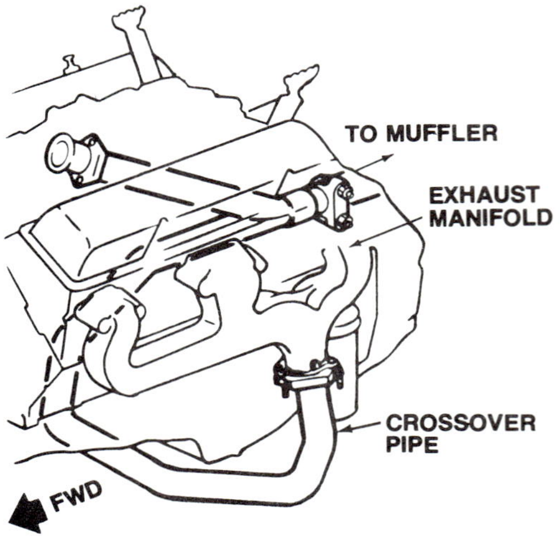

Figure 4-25. This V-type engine has a crossover pipe connecting the two cylinder banks to a single exhaust.

extracted from the fuel. Cold oil tends to form sludge, which increases engine friction, robs power, and increases mechanical wear.

Many emission control systems depend on coolant temperature to activate switches. If the coolant does not reach the proper temperatures, the emission control systems do not work properly. The liquid, unburned fuel in the combustion chambers also increases exhaust HC emissions.

A too-hot engine will lower volumetric efficiency and decrease performance. Hot oil will form carbon and varnish deposits that affect engine operation.

The increased mechanical wear allowed by hot, thinned oil will affect economy. The thin oil will also be drawn into the combustion chamber in greater amounts, fouling spark plugs and increasing both HC and CO emissions.

EXHAUST SYSTEM TYPES

The exhaust system routes engine exhaust gases to the rear of the car, quiets the exhaust noise, and, in some cases, reduces the pollutants in the exhaust. The system design varies according to engine design. The three major types are:

- Inline
- Single V-type
- Dual V-type.

These designs are described in the following paragraphs.

Inline

When an engine's cylinders are arranged in line, figure 4-23, all of the exhaust valves are on the same side of the engine. An exhaust system pipe connects to this side of the engine at the exhaust manifold. The single pipe connects to one or more units that may include a muffler, a resonator, and a catalytic converter.

Single V-Type

A V-type engine has exhaust valves on both cylinder banks. There are two ways in which these valves can be connected to a single exhaust pipe:

- There can be a Y-pipe behind the engine, figure 4-24.
- There can be a crossover pipe beneath the engine, figure 4-25.

Dual V-Type

Some V-type engines have two separate exhaust systems, one for each cylinder bank, figure 4-26. The systems are similar to the single systems used on an inline engine.

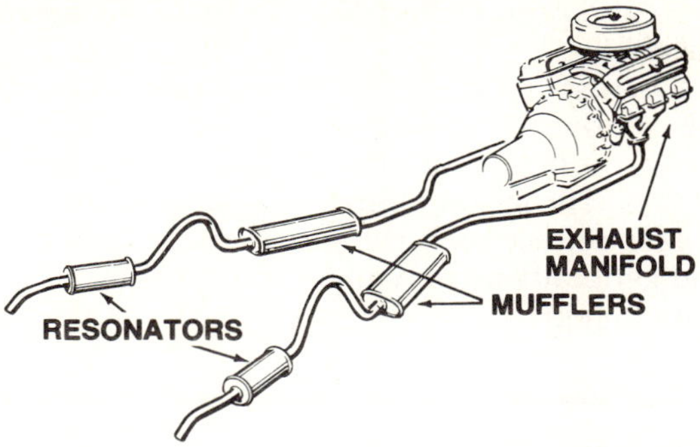

Figure 4-26. This V-type engine has two separate exhaust systems.

EXHAUST SYSTEM COMPONENTS

Major parts of the exhaust system include:
- Manifolds
- Pipes
- Mufflers
- Resonators
- Catalytic converters.

Manifolds and manifold heat control valves are covered in Chapter 13. We will study catalytic converters in detail in Chapter 19. The other exhaust system parts are described in the following paragraphs.

Pipes

The flow of exhaust gases from the engine should be as smooth as possible. If there are restrictions in the flow, there will be **backpressure** at the engine. Too much backpressure means that not all of the exhaust gases can leave a combustion chamber before the next cycle starts. This preheats and leans the incoming air-fuel mixture, hurting performance. It can also cause engine mechanical failures such as burned valves. Exhaust system pipes should be as straight as possible, avoiding sharp turns and restrictions. Pipes must also be able to withstand the constant presence of hot, corrosive exhaust gases and undercar hazards such as rocks.

The exhaust pipes on many late-model cars are formed with an inner and an outer skin. Occasionally, the inner skin will collapse and form a restriction in the system. From the outside, the exhaust pipe will look normal while the inside is partially or almost completely blocked. Close inspection, as explained in your *Shop Manual*, is necessary to locate such a defect.

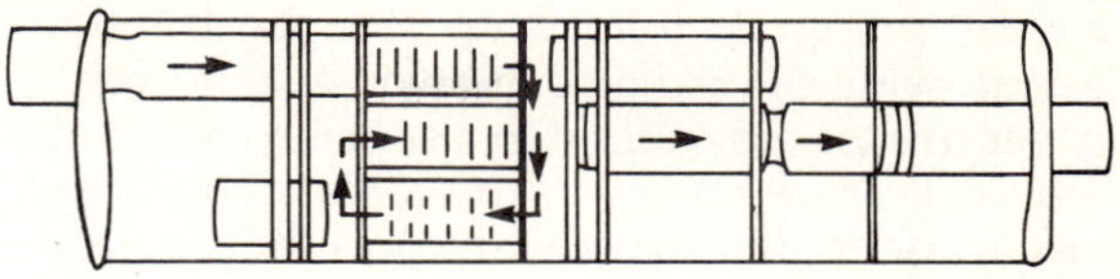

Figure 4-27. Exhaust gases must twist and turn to travel through this muffler.

Mufflers And Resonators

The muffler is an enclosed chamber that contains baffles, smaller chambers, and pipes to direct exhaust gas flow. The gas route through the muffler is full of twists and turns, figure 4-27. This quiets the exhaust flow but also creates backpressure. Mufflers must be carefully matched to the engine and exhaust system.

Some high-performance engines use less restrictive mufflers and add resonators to the system. A resonator contains a straight-through pipe that is surrounded by sound deadening material, often fiberglass. The resonator does not restrict exhaust gas flow very much, and it does reduce noise.

Exhaust System Effects On Performance, Economy, And Emissions

The main effect of the exhaust system is the backpressure it creates. Too much backpressure will restrict the flow of exhaust gases from the cylinders and hurt engine performance. The catalytic converter affects emissions, as we will learn in Chapter 19.

SUMMARY

The cooling and exhaust systems must remove about two-thirds of the heat produced by a car engine. Most cooling systems use coolant to remove this heat, by circulating it through the engine block where it picks up the heat, then routing it through the radiator, where the outside air flowing past takes away the heat. The coolant is partly water and partly a number of additives that allow a greater range of engine operating temperatures.

Backpressure: A pressure that tends to slow the exit of exhaust gases from the combustion chamber; usually caused by restrictions in the exhaust system.

The coolant is pushed through the system by a water pump. As it heats up, it expands, often overflowing the radiator. Newer systems recover this excess coolant in a separate reservoir, where it can be reused as the coolant cools. An electric or engine-driven fan cools the radiator by drawing more air through it at lower speeds or idle. A pressure cap helps keep the system pressurized, and a thermostat regulates the flow of coolant by sensing engine temperature. Engine operating temperatures that are either too hot or too cold will produce greater emissions and poor performance, and can damage the engine.

Exhaust systems carry the exhaust away from the combustion chambers through a series of pipes, mufflers, resonators, and catalytic converters. The system must be as straight as possible to reduce backpressure in the cylinders, and generally must dampen the noise from combustion.

Review Questions

Select the single most correct answer.
Compare your answers with the correct answers on page 280.

1. About ___ of the engine's heat energy is converted to mechanical energy.
   a. Two-thirds
   b. One-half
   c. One-third
   d. One-eighth

2. Late-model automobiles have an average operating temperature of approximately:
   a. 200° F
   b. 300° F
   c. 150° F
   d. 100° F

3. When the engine coolant is cold:
   a. The thermostat allows it to circulate through the radiator
   b. The thermostat is open
   c. The thermostat keeps it from flowing through the radiator
   d. None of the above

4. When coolant is pressurized:
   a. Its freezing point is raised
   b. Its boiling point is raised
   c. Its boiling point is lowered
   d. None of the above

5. Pure water is not used in cooling systems because:
   a. Its boiling point is too low
   b. Its freezing point is too high
   c. It can rust engine parts
   d. All of the above

6. Late-model cooling systems use a mixture of water and:
   a. Alcohol antifreeze
   b. Methanol antifreeze
   c. Ethylene-glycol antifreeze
   d. Polycarbonate antifreeze

7. The automobile's radiator:
   a. Creates a large surface contact with air
   b. Is made of tubes, fins, and tanks
   c. Can be a downflow or crossflow type
   d. All of the above

8. Which part of the water pump is driven by the engine?
   a. Impeller
   b. Bypass
   c. Scroll
   d. Inlet

9. The radiator fan must:
   a. Decrease airflow at high vehicle speeds
   b. Increase airflow at low vehicle speeds
   c. Decrease airflow at low vehicle speeds
   d. Increase airflow at high vehicle speeds

10. Flexible fans, fluid fan drives, and electric fans are all designed to:
    a. Increase the fan's pulling capacity
    b. Increase airflow at high vehicle speeds
    c. Keep engine power from being wasted by the fan
    d. None of the above

11. The radiator pressure cap must:
    a. Seal the system during normal operation
    b. Relieve excess pressure
    c. Relieve system vacuum
    d. All of the above

12. Radiator hoses are rubber so that they can:
    a. Withstand system pressure
    b. Absorb engine vibrations
    c. Contain boiling coolant
    d. Resist freezing coolant

13. The coolant recovery system:
    a. Stores excess coolant outside of the radiator
    b. Raises the coolant's boiling point
    c. Is used only on cars with air conditioning
    d. Is connected to the thermostat bypass

14. A malfunctioning cooling system can affect:
    a. Emission control systems
    b. The air-fuel mixture
    c. Lubrication
    d. All of the above

15. The exhaust system is designed to:
    a. Include a great amount of restriction
    b. Create excessive backpressure
    c. Have as little restriction as is necessary
    d. Increase combustion chamber termperatures

16. The many twists and turns in a muffler:
    a. Create backpressure
    b. Quiet the exhaust
    c. Must be carefully matched to the system
    d. All of the above

Chapter

5

Introduction to Emission Control

When emission control requirements were first introduced, automakers and car owners were able to comply by installing add-on or "hang-on" equipment, devices that were not an integral part of engine and vehicle design. As regulations became more strict, carmakers had to include emission control in basic engine design.

The first emission control regulation was adopted in California in 1961. Today, almost two decades later, emission control regulations are still being tightened and new control systems developed. Sophisticated computer-controlled systems appear on many cars, and "tamperproof" engines have been suggested as a way to guarantee emission control throughout the life of the car.

How did these great changes come about? What exactly *is* air pollution, and how does the automobile contribute to it? This chapter examines air pollution and automotive emissions, including the legislation controlling emissions and the ways in which carmakers have met the regulations.

AIR POLLUTION—A PERSPECTIVE

Air pollution is usually defined as the introduction of a contamination into the atmosphere in an amount large enough to injure human, animal, or plant life. There are many different types and causes of air pollution, but they all fall into two general groups: natural and manmade. Natural pollution is caused by such things as the plant life cycle, forest fires, volcanic eruptions, and dust storms and is often beyond our control. Manmade pollution from industrial plants and automobiles *can* be controlled.

Most urban and large industrial areas suffer periodic air pollution. During the late 1940's, a unique form of air pollution was identified in the Los Angeles area. This combination of irritating chemical compounds is called **photochemical smog**, figure 5-1. As smog became more of a problem, California became the first state to place controls on motor vehicles. When smog began to appear in other parts of the country, the federal government moved into the area of regulation. To understand why, we must first look at the elements produced by the automobile which form air pollution and smog.

MAJOR POLLUTANTS

An internal combustion engine emits three major gaseous pollutants into the air: **hydrocarbons** (HC), **carbon monoxide** (CO), and **oxides of nitrogen** (NO_x). In addition, an automobile engine gives off many small liquid or

Figure 5-1. Smog engulfs the Los Angeles Civic Center. When the base of the temperature inversion is 1,000 feet above ground, the inversion layer — a layer of warm air above a layer of cool air — prevents the natural dispersion of air contaminants into the upper atmosphere. (LA County)

solid particles, such as lead, carbon, sulfur and other **particulates**, which contribute to pollution. By themselves, all these emissions are not smog, but simply air pollutants.

Hydrocarbons (HC)

Gasoline is a hydrocarbon material. Unburned hydrocarbons given off by an automobile are largely unburned portions of fuel. For example, every 1,000 gallons of gasoline used by a car without emission controls produces about 200 pounds of hydrocarbon emissions.

Over 200 different varieties of hydrocarbon pollutants come from automotive sources. While most come from the fuel system and the engine exhaust, others are oil and gasoline fumes from the crankcase. Even a car's tires, paint, and upholstery give off tiny amounts of hydrocarbons. The three major sources of hydrocarbon emissions in an automobile are shown in figure 5-2:

1. Fuel system evaporation — 20 percent
2. Crankcase vapors — 20 percent
3. Engine exhaust — 60 percent

Hydrocarbons are the only major automotive air

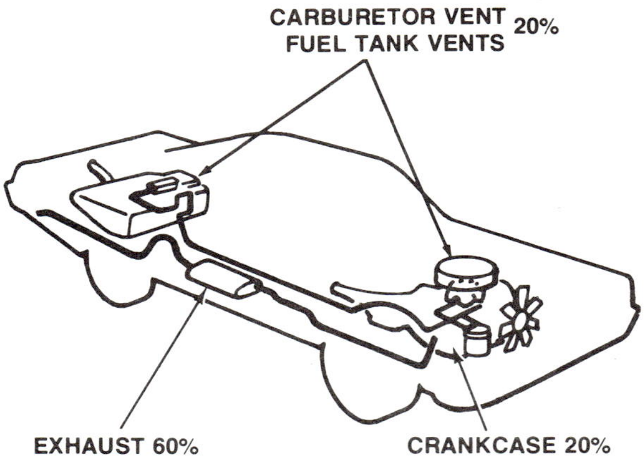

Figure 5-2. Sources of hydrocarbon emissions.

pollutant that come from sources *other than* the engine's exhaust.

Hydrocarbons of all types are destroyed by combustion. If an automobile engine burned gasoline completely, there would be no hydrocarbons in the exhaust, only water and carbon dioxide. But when the vaporized and compressed air-fuel mixture is ignited, combustion occurs so fast that gasoline near the sides of the combustion chamber does not get burned. This unburned fuel then passes out with the exhaust gases. The problem is worse with engines that misfire or are not properly tuned.

Carbon Monoxide (CO)

Although not part of photochemical smog, carbon monoxide is also found in automobile exhaust in large amounts. The same 1,000 gallons of gasoline which will produce 200 pounds of hydrocarbons when burned without emission controls will also produce 2,300 pounds of carbon monoxide. A deadly poison, carbon monoxide is both odorless and colorless. In a small quantity, it causes headaches and vision difficulties; in larger quantities, it is fatal. Large quantities of carbon monoxide that collect in the air push out the oxygen, which makes breathing difficult.

Because it is a product of incomplete combustion, the amount of carbon monoxide produced depends on the way in which hydrocarbons burn. When the air-fuel mixture burns, its hydrocarbons combine with oxygen. If the air-fuel ratio has too much fuel, there is not enough oxygen for this process to happen, and so carbon monoxide is formed. To make combustion more complete, an air-fuel mixture with less fuel is used. This increases the supply of oxygen, which reduces the formation of CO by producing harmless carbon dioxide (CO_2) instead.

Oxides of Nitrogen (NO_x)

Air is about 78 percent nitrogen, 21 percent oxygen, and 1 percent other gases. When the combustion chamber temperature reaches 2,500° F or greater, the nitrogen and oxygen in the air-fuel mixture combine to form large quantities of oxides of nitrogen (NO_x). NO_x is also formed at lower temperatures, but in far smaller amounts. By itself, NO_x is of no particular concern. But when the amount of hydrocarbons in the air reaches a certain level, and the ratio of NO_x to HC is correct, the two pollutants will combine chemically to form smog.

The amount of NO_x formed can be reduced by lowering the temperature of combustion in the engine, but this causes a problem which is difficult to solve. Lowering the combustion chamber temperature to reduce NO_x results in less efficient burning of the air-fuel mixture. This automatically means an increase in hydrocarbons and carbon monoxide, both of which are formed in large quantities at lower combustion chamber temperatures. Carmakers have used various emission control systems to combat this problem.

Particulates

Particulates are microscopic solid particles, such as dust and soot. They are fragments of solid or liquid matter which remain in the atmosphere for long periods of time. Because of this, particulates are prime causes of secondary pollution. For example, particulates such as lead and carbon tend to collect in the atmosphere. These are harmful substances and, in large amounts, can injure our health. Absorbed directly or indirectly into the body, particulates can damage the respiratory tract and lungs.

Particulates produced by automobiles are a small percentage of the total in the atmosphere. Most come from fixed sources such as factories, which spew industrial wastes into the air. But automobiles *do* produce particulates, and studies have shown that they can be reduced considerably. This is done by eliminating additives such as lead from gasoline, and by changing other characteristics of the fuel.

Sulfur Oxides

The sulfur created by processing and burning gasoline and other fossil fuels (coal and oil) gets into the atmosphere in the form of **sulfur oxides**. As this material breaks down, it will combine with water in the air to form corrosive sulfuric acid, which is a secondary pollutant. Recently, such pollution has become a major problem in the northeastern United States, as well as in other parts of the world. Clouds

Photochemical Smog: A combination of pollutants, primarily HC and NO_x. When acted upon by sunlight, they form chemical compounds that are harmful to human, animal, and plant life.

Hydrocarbon: A chemical compound made up of hydrogen and carbon. Abbreviated: HC. A major pollutant given off by an internal combustion engine. Gasoline, itself, is a hydrocarbon compound.

Carbon Monoxide: An odorless, colorless, tasteless poisonous gas. Abbreviated: CO. A major pollutant given off by an internal combustion engine.

Oxides of Nitrogen: Chemical compounds of nitrogen given off by an internal combustion engine. Abbreviated: NO_x. They combine with hydrocarbons to produce smog.

Particulates: Liquid or solid particles such as lead and carbon that are given off by an internal combustion engine as pollution.

Sulfur Oxides: Chemical compounds given off by processing and burning gasoline and other fossil fuels. As they decompose, they combine with water to form sulfuric acid.

■ The Case Of Incident Number 104

Manmade contaminants in the air we breathe have been a problem ever since we entered the industrial revolution a century and a half ago. Not all pollution is manmade, of course. Smoke from fires, dust, methane from decaying organic matter, vapors from pine trees, volcanic ash — all of these contribute to air pollution. But under normal conditions, Nature is able to cleanse herself of these contaminants.

With the arrival of heavy industry, man began to pump more junk into the air than Nature was able to clean out. Factories spewing out tons of particulates, cars emitting many more tons of dangerous and harmful emissions, agricultural spraying adding yet more tons of highly toxic poisons to the air — all of these can combine to make the air unhealthy to breathe in some areas.

Air pollution disasters have occurred around the world for two generations or so. Donora, Pennsylvania, achieved unwanted fame when pollution killed 17 and caused another 6,000 persons to become ill. London suffered through a black smog attack in 1952 that left 4,000 dead. New York City has gone through several periods of dangerously heavy air pollution. But in 1969, more than 20 Midwestern states were involved in a major air pollution attack for several days. It was close to being a fearful disaster, until the atmospheric conditions changed and the contaminants were cleared away. This near-disaster was officially known as Incident Number 104.

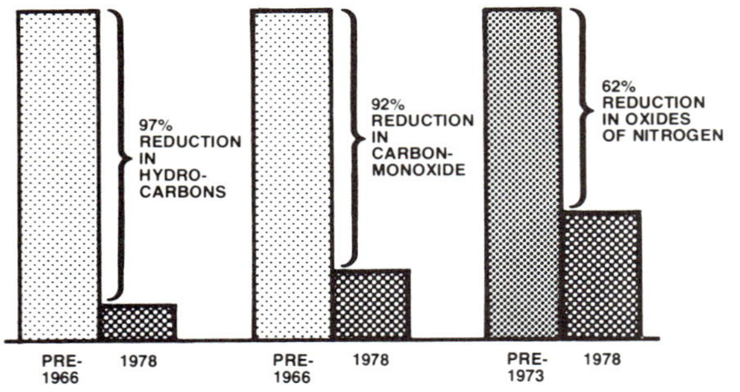

Figure 5-3. Automotive emission reductions.

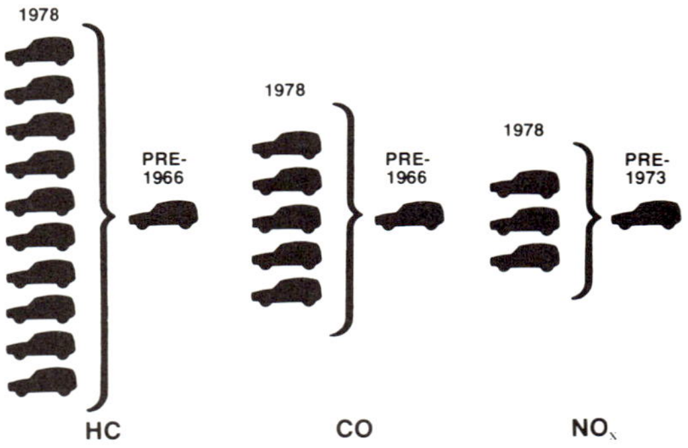

Figure 5-4. Vehicle emission comparisons.

containing sulfuric acid are carried by winds to the Northeast, where they deposit rain and snow with enough acid to kill wildlife.

POLLUTION AND THE AUTOMOBILE

While it is true that a single car gives off only a microscopic amount of these pollutants, remember that there are more than 117 million automobiles in use in the United States. Multiply each car's contribution toward air pollution by that figure, and you have the potential for a staggering amount of pollution — about 42 percent of the total problem in this country.

Great progress has been made in reducing — almost eliminating — automobile air pollution since 1966. Those HC emissions that come from the engine crankcase and fuel system — 40 percent of the total — have been almost totally eliminated. The other 60 percent of HC emissions — from the exhaust — has been lowered considerably. Total hydrocarbon emissions have been reduced by about 97 percent. Total carbon monoxide emissions have been reduced by a similar amount — about 92 percent — since 1966. Since 1973, total NO_x emissions have been reduced about 62 percent. These accomplishments are shown in figure 5-3. Figure 5-4 shows us that ten 1978 cars emit less HC than one pre-1966 model; five 1978 cars produce less CO than one pre-1966 car; and three 1978 cars emit less NO_x than one pre-1973 model.

SMOG — CLIMATIC REACTION WITH AIR POLLUTANTS

Smog and air pollution are not the same thing; smog is a form of air pollution, but air pollution is not necessarily smog. HC and NO_x are the two principal materials that combine in the atmosphere to form smog.

The Photochemical Reaction

Although smog can be created in a laboratory experiment, scientists still do not understand exactly what it is, or exactly how it is formed. But they do know that for smog to form in the atmosphere, three factors must be present — sunlight, relatively still air, and a high concentration of hydrocarbons and oxides of nitrogen. When these three factors are present at the same time, sunlight causes a chemical reaction between the HC and NO_x. The result is smog.

Temperature Inversion

Normally, air temperature becomes cooler at higher altitudes. Warm air near the ground rises, cooling itself by contact with the cooler air above it. When nature follows this normal pattern, smog and other pollutants are carried away. But some areas experience a natural weather pattern called a **temperature inversion**. When this happens, a layer of warm air prevents the upward movement of cooler air near the ground. This inversion acts as a "lid" over the stagnant air; the air cannot rise, so smog and pollution collect.

When the inversion layer is several thousand feet high, the smog will rise enough to provide reasonable visibility. But when the inversion layer is within a thousand feet of the ground, smog is trapped and visibility is reduced, figure 5-1. The distant landscape is impossible to see, and eye irritation, headaches, and difficulty in breathing follow. The Los Angeles area, where temperature inversion was first noted, provides a classic example of this phenomenon.

AIR POLLUTION LEGISLATION AND REGULATORY AGENCIES

Once the problem of air pollution was recognized, it became the subject of intense research and investigation. By the early 1950's it was

Year	Regulations	HC	CO	NO_x
Before Controls		850 ppm (16.8 g/mi)*	3.4% (125.0 g/mi)*	1000 ppm (4.0 g/m)*
1966-67	Calif.	275 ppm	1.5%	none
	U.S. Federal	none	none	none
1968-69	Calif.	275 ppm	1.5%	none
	U.S. Federal	275 ppm	1.5%	none
1970	Calif. & U.S. Federal	4.6 g/mi	46.0 g/mi	none
1971	Calif.	4.6 g/mi	46.0 g/mi	4.0 g/mi
	U.S. Federal	4.6 g/mi	46.0 g/mi	none
	Canadian	2.2 g/mi**	23.0 g/mi**	none
1972	Calif.	3.2 g/mi	39.0 g/mi	3.2 g/mi
	U.S. Federal	3.4 g/mi	39.0 g/mi	none
	Canadian	3.4 g/mi	39.0 g/mi	none
1973	Calif.	3.2 g/mi	39.0 g/mi	3.0 g/mi
	U.S. Federal	3.4 g/mi	39.0 g/mi	3.0 g/mi
	Canadian	3.4 g/mi	39.0 g/mi	3.0 g/mi
1974	Calif.	3.2 g/mi	39.0 g/mi	2.0 g/mi
	U.S. Federal	3.4 g/mi	39.0 g/mi	3.0 g/mi
	Canadian	3.4 g/mi	39.0 g/mi	3.0 g/mi
1975-76	Calif.	0.9 g/mi	9.0 g/mi	2.0 g/mi
	U.S. Federal	1.5 g/mi	15.0 g/mi	3.1 g/mi
	Canadian	2.0 g/mi	25.0 g/mi	3.0 g/mi
1977-79	Calif.	0.41 g/mi	9.0 g/mi	1.5 g/mi
	U.S. Federal	1.5 g/mi	15.0 g/mi	2.0 g/mi
	Canadian	2.0 g/mi	25.0 g/mi	3.0 g/mi

ppm = parts per million
g/mi = grams per mile
* approximate estimations of uncontrolled levels
** theoretical flow rate

Figure 5-5. Exhaust emission limits for new passenger cars by model years.

believed that smog in Los Angeles was caused by the photochemical process, and California, as we noted earlier, became the first state to pass laws to limit automotive emissions. Standards established by California one year have usually become U.S. Federal standards the next year.

Emission Control Legislation

Beginning with the 1961-model cars, California required control over crankcase emissions. This became standard for the rest of the United States with the 1963-model cars. That year, carmakers voluntarily equipped their cars with a blowby device which almost eliminated crankcase emissions on all cars.

The first Federal air pollution research program began in 1955. In 1963, Congress passed the Clean Air Act, giving money to states for development of air pollution control programs. This law was amended in 1965, giving the Federal government authority to set emission standards for new cars. These standards were first applied nationwide to 1968 models.

The Air Quality Act of 1967 was a new approach to air pollution. Major changes followed in 1970 — changes designed to turn piecemeal programs into a unified attack on pollution of all kinds. At this time, emission standards were established for 1973-74 cars and projected for 1975 and later models, figure 5-5.

Regulatory Agencies

The Environmental Protection Agency (EPA) is the U.S Federal agency responsible for enforcing the Air Quality Act. It is part of the Department of Health, Education, and Welfare. The EPA first set standards requiring that HC and CO emissions for 1975 passenger cars be reduced 90 percent from 1970 levels, with a 90-percent reduction in NO_x by 1976. These standards were later amended to those shown in figure 5-5. In addition, the EPA has established standards in other automotive areas, such as fuel consumption and fuel additives. California established its own Air Resources Board (ARB), whose authority roughly parallels that of the Federal EPA. The California ARB operates under permission from the EPA, but its authority extends only to those vehicles sold in or brought into California.

AUTOMOTIVE EMISSION CONTROLS

Early researchers dealing with automotive pollution and smog began work with the idea that *all* pollutants were carried into the atmosphere by the car's exhaust pipe. But auto manufacturers doing their own research soon discovered that pollutants were also given off from the fuel tank and the engine crankcase. The total automotive emission system, figure 5-6, contains three different types of controls. This picture shows that the emission controls on a modern automobile are not a separate system, but are part of an engine's fuel, ignition, and exhaust systems.

Automotive emission controls can be grouped into major families, as follows:

1. Crankcase emission controls — Positive crankcase ventilation (PCV) systems control HC emissions from the engine crankcase. These are described in detail in Chapter 16.
2. Evaporative emission controls — Evaporative emission control (EEC) systems control the evaporation of HC vapors from the fuel tank, pump, and carburetor. These are described in detail in Chapter 12.

Temperature Inversion: A weather pattern in which a layer or "lid" of warm air keeps the cooler air beneath it from rising.

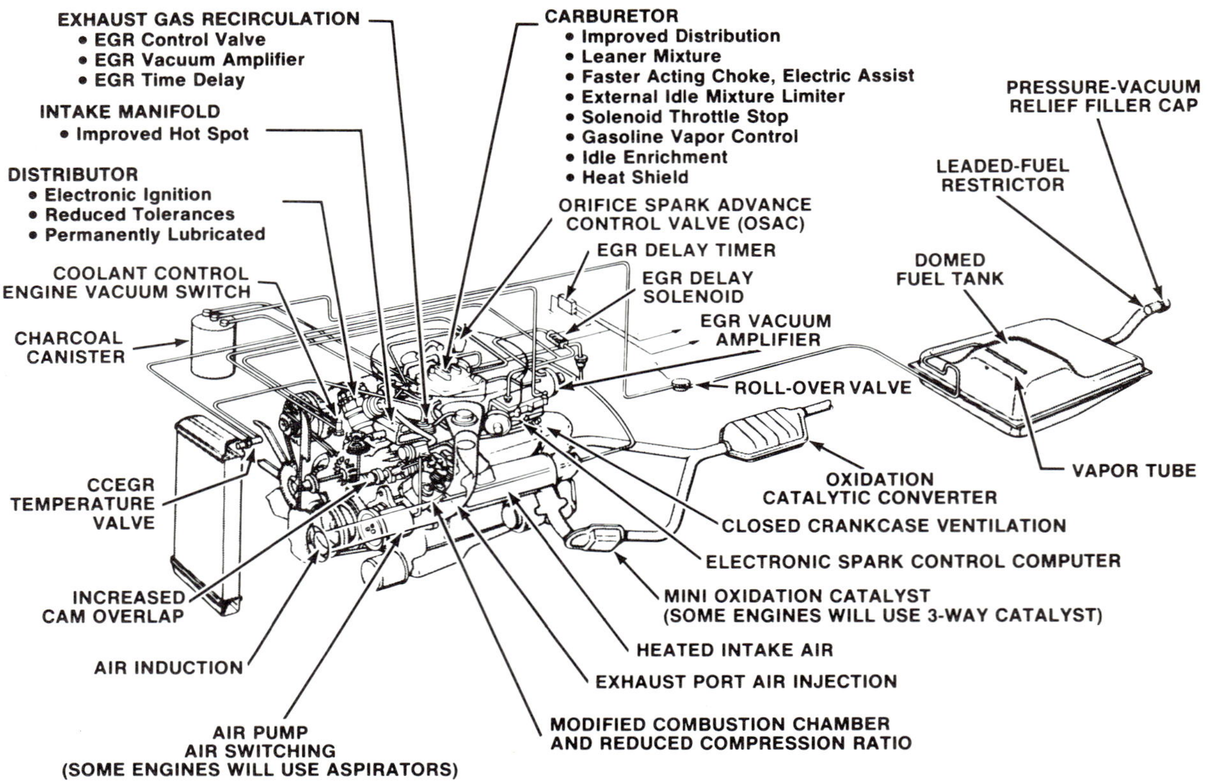

Figure 5-6. The complex emission controls on a modern automobile engine are an integral part of the fuel, ignition, and exhaust systems. (Chrysler)

3. Exhaust emission controls — Various systems and devices are used to control HC, CO, and NO_x emissions from the engine exhaust. These controls can be subdivided into the following general groups:

a. Air injection systems — These systems inject extra air into the exhaust system to help burn up HC and CO in the exhaust. They are described in Chapter 17.

b. Engine modifications — Various changes have been made in the designs of engines to help eliminate all three major pollutants.

c. Spark timing controls — Carmakers have used various systems to delay or retard ignition spark timing to control HC and NO_x emissions. Most of these systems modify the distributor vacuum advance, but many late-model cars have electronic spark controls. Ignition spark timing and its role in controlling emissions are discussed in Chapter 9.

d. Exhaust gas recirculation — The most effective way to control NO_x emissions is to recirculate a small amount of exhaust gas back to the intake manifold to dilute the incoming air-fuel mixture. Exhaust gas recirculation (EGR) systems are described in Chapter 18.

e. Catalytic converters — The first catalytic converters installed in the exhaust systems of 1975 and '76 cars helped the chemical oxidation, or burning, of HC and CO in the exhaust. Later catalytic converters, which began to appear on 1977 and '78 cars, also promote the chemical reduction of NO_x emissions. Catalytic converters are described in Chapter 19.

TUNE-UPS AND EMISSION CONTROL

Although a brand-new automobile must meet emission control standards when sold, the wear and tear of everyday operation soon can change this. Settings and tolerances change, engine parts wear, and conditions in other systems indirectly affect the engine's performance.

One of the major goals of a tune-up is to bring the car back in line with the emission control standards that it met when new. Here is a quick review of how changing engine conditions affect vehicle emissions, and how the corrections that can be made during a tune-up will reduce emissions.

How Air-Fuel Ratio Affects Emissions

The air-fuel ratio determines how much fuel is present in the cylinder when combustion begins. Different air-fuel ratios have different effects on exhaust emissions.

If there is too much fuel present, it will not burn completely. Unburned fuel in the exhaust is the principal source of HC emissions. The lack of oxygen in the mixture also increases the amount of CO in the exhaust.

If the air-fuel ratio is made leaner, these problems decrease. Combustion is more complete, reducing HC, and the excess oxygen allows the formation of harmless carbon dioxide (CO_2) instead of CO. These advantages of a lean air-fuel mixture disappear, however, if the mixture is made *too* lean.

A lean mixture is very hard to ignite, because the particles of fuel are far apart. If the mixture is so lean that the ignition spark does not cause combustion, a **misfire** occurs. The entire chamber full of vaporized, unburned fuel enters the exhaust system, and HC emissions skyrocket. Most engines with conventional ignition systems use air-fuel ratios for cruising of about 12:1 to 14:1, which allows some of the lean mixture advantages while avoiding misfire. Some computer-controlled ignition timing systems, such as Chrysler's Electronic Lean-Burn, are able to run an 18:1 mixture by precise control of the ignition spark. A mixture this lean burns at a very low temperature and NO_x formation is reduced.

How Ignition Timing Affects Emissions

The timing of the ignition spark affects both the quality and the temperature of combustion. The quality of combustion, or how completely the air-fuel mixture burns, directly affects HC and CO emissions. The temperature of combustion affects NO_x formation and, indirectly, HC and CO.

We have seen in previous chapters that the ignition spark timing must be set correctly at the engine idle speed and then must vary to match other engine speeds. Most ignition systems use mechanical devices to vary ignition timing. Some late-model systems have small computers, or microprocessors, that compute ignition timing.

If the spark timing is advanced until combustion is complete and HC and CO emissions decrease, then NO_x formation is greatly increased by the high combustion chamber temperatures. If the spark timing is retarded to lower combustion chamber temperatures and NO_x formation, the mixture does not have enough time to burn completely and HC and CO increase.

The different ignition timing settings that are specified for different engines are a carefully engineered compromise between these two extremes. Generally, initial timing is slightly retarded to reduce NO_x formation. Other emission control devices are added to reduce the resulting HC and CO emissions.

How Engine Mechanical Condition Affects Emissions

Engine mechanical wear cannot be corrected during a tune-up. This requires engine service, such as valve grinding or timing chain repair, or an overhaul. The tune-up mechanic, though, must be able to *recognize* engine wear and inform the car owner. Otherwise, the car owner may not realize that work is needed and may be disappointed that the tune-up did not completely restore the car's performance.

Misfire: When combustion does not occur during the power stroke.

■ How New Cars Are Emission-Certified

The availability of new cars each year depends on whether the carmakers successfully complete the emission certification process. The constant volume sampling test is performed by the Federal Environmental Protection Agency (EPA), although California's Air Resources Board tests some vehicles.

As each new model year approaches, automakers build prototype or emission data cars for EPA use. The automakers are responsible for conducting a 50,000 mile durability test of their emission control systems. The test cars are driven for 4,000 miles to stabilize the emission systems before testing.

The first step is to precondition a car. Then the car stands for 12 hours at an air temperature of 73° F to simulate a cold-start. The actual test is done on a chassis dynamometer, using a driving cycle which represents urban driving conditions. The car's exhaust is mixed with air to a constant volume and analyzed for harmful pollutants.

The entire test takes about 41 minutes. The first 23 minutes are a cold-start driving test. The next ten minutes are a waiting, or hot-soak, period. The final eight minutes are a hot-start test, representing a short trip in which the car is stopped and started several times while hot. If the emissions test results for all data cars are equal to or lower than all of the HC, CO, and NO_x standards, the manufacturer receives certification of the engine "family" and the car can be offered for sale to the public.

The engine mechanical wear that affects exhaust emissions includes items such as valve train wear, which will change the valve timing and affect combustion. Increased clearances from mechanical wear will also allow excessive amounts of oil into the cylinders, contributing to HC emissions.

How Lubrication, Cooling, And Exhaust Systems Affect Emissions

The lubrication system helps to keep engine mechanical wear to a minimum, reducing the effects of wear on emissions. If the lubrication system is neglected, engine wear will occur faster. Also, if the wrong viscosity oil is used or if the oil becomes diluted, it will be drawn into the combustion chamber and increase HC emissions.

The cooling system keeps the engine at an ideal operating temperature. If the engine is too cold, the fuel will not vaporize well and will not burn completely. Unburned gasoline can wash oil off the cylinder walls, which leads to rapid cylinder wear and increased blowby. High engine temperatures can thin the oil and lead to oil burning. The cooling system also helps to control the operation of some emission control systems. Their performance will suffer if the cooling system does not work correctly.

Exhaust backpressure prevents exhaust gases from being completely cleared from the cylinder. The incoming air-fuel mixture is diluted by these gases. This is done purposely in many cases to reduce emissions. Too much backpressure and too much dilution, however, can cause misfiring and increased emissions. The catalytic converter is a major part of many emission control systems. If it is physically damaged or if the catalyst is contaminated by gasoline additives, exhaust emissions will increase.

SUMMARY

Air pollution is not entirely manmade, and manmade pollution is not all caused by automobiles. The gasoline engine emits three major pollutants (HC, CO, and NO_x) as well as various particulates and sulfur oxides. Since the introduction of emission control laws, the major gaseous pollutants have been reduced by 62 to 97 percent in new cars.

The most visible and irritating form of air pollution is photochemical smog. This forms when HC and NO_x combine in the presence of sunlight.

California produced the first emission control legislation, and the Federal government followed. The Federal Environmental Protection Agency and California's Air Resources Board establish and enforce pollution control standards for automobiles. Carmakers have developed various ways of meeting control standards, including crankcase vapor control systems, evaporative emission controls, and exhaust emission controls. One major goal of a tune-up is to make the car meet its original emission standards and to recognize any problems that cannot be corrected by a tune-up.

Review Questions

Select the single most correct answer.
Compare your answers with the correct answers on page 280.

1. Smog:
   a. Is a natural pollutant
   b. Cannot be controlled
   c. Is created by a photochemical reaction
   d. Was first identified in New York City

2. The three *major* pollutants in automobile exhaust are:
   a. Sulfates, particulates, carbon dioxide
   b. Sulfates, carbon monoxide, nitrous oxide
   c. Carbon monoxide, oxides of nitrogen, hydrocarbons
   d. Hydrocarbons, carbon dioxide, nitrous oxide

3. Fuel evaporation accounts for what percentage of total HC emissions:
   a. 40%
   b. 60%
   c. 33%
   d. 20%

4. Carbon monoxide is a result of:
   a. Incomplete combustion
   b. A lean mixture
   c. Excess oxygen
   d. Impurities in the fuel

5. Which of the following is *not* true:
   a. High temperatures increase NO_x emission
   b. High temperatures reduce HC and CO emission
   c. Low temperatures reduce HC and CO emission
   d. Low temperatures reduce NO_x emission

6. Particulates are:
   a. Microscopic particles suspended in the atmosphere
   b. Produced in emissions by additives such as lead
   c. Harmful to our lungs
   d. All of the above

7. Sulfur oxides are harmful because:
   a. They damage catalytic converters
   b. They combine with water in the air to form sulfuric acid
   c. They are primary pollutants
   d. They are visible

8. Since 1966, total HC emissions have been reduced by about:
   a. 65%
   b. 47%
   c. 82%
   d. 97%

9. Smog is created by a combination of sunlight, still air, and:
   a. High levels of CO and HC
   b. High levels of CO and NO_x
   c. High levels of HC and NO_x
   d. High levels of HC and sulfur oxides

10. A temperature inversion increases air pollution by:
    a. Pushing cool air up
    b. Forming a "lid" over stagnant air
    c. Pushing warm air down
    d. None of the above

Chapter

6

What Is A Tune-Up?

The previous chapters have explained why engine maintenance is necessary for the best performance, economy, and emission control. Now we can look more closely at the combination of engine maintenance services commonly called a tune-up. This chapter describes:
- The difference between traditional and modern ideas of tune-up
- Carmakers' and equipment suppliers' tune-up recommendations
- The effects of emission control regulations on tune-up services.

TUNE-UP'S CHANGING MEANING

The automobile has undergone many changes since the first emission control regulations appeared in 1961. Engine equipment has changed, and the services necessary to maintain that equipment have changed. In relation to an engine tune-up, this means that:
- There are fewer items to be replaced
- Service intervals for replacement items have gradually lengthened
- There are more testing procedures called for, rather than replacement and adjustment procedures.

Later in this chapter we will discover why these changes have occurred. In the following paragraphs, we will see what the changes have done to the meaning of an engine tune-up.

Periodic Service — The Traditional Tune-Up

Before and during the infancy of emission control regulations, an engine tune-up was done at regular intervals based on items that had to be replaced at those intervals, figure 6-1. This is usually called preventive maintenance or periodic maintenance. The tune-up included:
- Replacing certain parts, such as filters and ignition system parts
- Adjusting tolerances, especially in the ignition and fuel systems
- Testing the engine's general mechanical condition to find any major problems.

Diagnosis And Correction — The Modern Tune-Up

As the number of serviceable items declined and service intervals lengthened, tune-up mechanics could no longer simply replace parts and make a few adjustments. They had to be aware of problems that could appear in systems where maintenance is not normally needed. The emphasis of a tune-up changed from routine services to a concentration on testing, figure 6-2. Extensive checks of overall performance

TUNE-UP RELATED SERVICES REQUIRED DURING 4 YEARS OR 50,000 MILES OF OPERATION — 1968 Dodge	
CHANGE crankcase oil	12 times
REPLACE crankcase oil filter	6 times
LUBRICATE manifold heat valve	12 times
Check PCV system	12 times
REPLACE PCV valve and CLEAN system	4 times
CLEAN air cleaner filter	12 times
REPLACE air cleaner filter	2 times
CLEAN carburetor choke shaft, fast-idle cam, and pivot pin	8 times
OIL distributor shaft	8 times
CLEAN and OIL the oil filler cap	8 times
CHANGE engine coolant	4 times
LUBRICATE distributor cam	4 times
Engine tune-up (REPLACE or ADJUST as needed):	
• Spark plugs	4 times
• Ignition points	4 times
• Distributor cap and rotor	4 times
• Ignition coil	4 times
• Ignition cables	4 times
• Carburetor choke	4 times
• Idle speed and fuel mixture	4 times
REPLACE fuel filter	2 times

Figure 6-1. These tune-up services concentrate on replacing or adjusting items at regular intervals.

TUNE-UP RELATED SERVICES REQUIRED DURING 4 YEARS OR 50,000 MILES OF OPERATION — 1978 Dodge with catalytic converter	
Change crankcase oil	6 times
Replace crankcase oil filter	3 times
CHECK manifold heat valve	1 time
CHECK PCV system	3 times
Replace PCV valve	1 time
Replace air cleaner filter	1 time
Clean carburetor choke shaft, fast-idle cam, and pivot pin	6 times
Clean and oil crankcase inlet air cleaner	1 time
Change engine coolant	2 times
Replace spark plugs	1 time
CHECK ignition cables	1 time
CHECK and adjust choke	1 time
CHECK and adjust idle speed and fuel mixture	1 time
Replace fuel filter	1 time
INSPECT and adjust drive belts	6 times
CHECK cooling system	3 times
INSPECT emission control system hoses	3 times
CHECK evaporative emission control system	1 time
Replace evaporative canister filter	1 time

Figure 6-2. These tune-up requirements emphasize checks and inspections over regular replacement of items.

and individual system performance are needed to pinpoint problems that could have been missed during a traditional replace-and-adjust tune-up. Testing, diagnosis, and correction, combined with those periodic services that remain, make up the modern tune-up procedure.

CARMAKERS' RECOMMENDATIONS

We have said that the changing tune-up is caused by changing service recommendations. Carmakers have always published servicing guidelines for their products. These guidelines let the owner and mechanic know how the car should be maintained and allow the carmaker to guarantee the car based on a specific program of maintenance. Today's maintenance schedules represent the minimum amount of service that the car will require when used in "normal" driving. Later, we will examine how well this "normal" service represents most cars' operating conditions. The following paragraphs explain how and why carmakers' maintenance recommendations have changed.

Services

Many tune-up related services have been eliminated from late-model cars. For example, Ford has announced that, between 1972 and 1976, ten routine maintenance items were dropped from its schedule. Not all of these were tune-up related, but as an example, one item listed is distributor cap and rotor replacement:

- The 1972 model requires a 24,000-mile replacement interval
- The 1976 model does not have a specified replacement interval.

Figure 6-3 shows the tune-up services recommended by Chevrolet for its 1968 and 1978 models. You can see that six services required in 1968 are not called for in 1978:

1. Ignition points replacement
2. Coolant replacement
3. Air cleaner filter rotation
4. Valve adjustment (on some engines)
5. Distributor cam lubricator rotation
6. Ignition condenser testing.

Some of these service parts are not on 1978 models, such as ignition points, while other service requirements were simply eliminated. You can also see that six items appear in the 1978 recommendations that did not appear in 1968. These are mainly *tests* of emission control systems that did not exist in 1968.

	CHEVROLET			
	1968 SERVICE INTERVALS		1978 SERVICE INTERVALS	
	MONTHS	MILES	MONTHS	MILES
REPLACE:				
Crankcase oil	4	6,000	12	7,500
Crankcase oil filter	8	12,000	24	15,000
PCV valve	12	12,000	12	15,000
Ignition points		12,000	NONE	
Spark plugs		12,000		22,500
Fuel filter	12	12,000	12	15,000
Coolant	24		NOT REQUIRED	
Air cleaner filter				
rotate		12,000	NOT REQUIRED	
replace		24,000		30,000
ADJUST:				
Valves		12,000	NOT REQUIRED	
Idle speed		12,000	18	22,500
Rotate distributor				
cam lubricator		12,000	NONE	
Ignition timing		12,000		22,500
CHECK:				
Ignition condenser		12,000	NONE	
PCV system	4	6,000	12	15,000
Ignition cables		12,000		22,500
Cooling system	NOT REQUIRED		12	15,000
Carburetor choke	NOT REQUIRED		18	22,500
Vacuum advance		12,000	18	22,500
Thermostatic air cleaner	NONE		18	22,500
Idle stop solenoid	NONE			22,500
Early fuel evap. system	NONE		18	22,500
Evap. control system	NONE		24	30,000

Figure 6-3. A comparison of the tune-up services and intervals for a 1968 Chevrolet and a 1978 Chevrolet.

Intervals

The Ford announcement in 1976 pointed out that service intervals had been extended for more than 25 items, including:

- Spark plug replacement — from every 12,000 miles in 1972 to every 15,000 or 20,000 miles in 1976, depending on the engine
- Fuel filter replacement — from every 12,000 miles in 1972 to once only at 15,000 or 20,000 miles in 1976
- Air cleaner filter replacement — from every 12,000 miles in 1972 to every 20,000 or 30,000 miles in 1976
- Engine coolant replacement — from every 24 months in 1972 to every 36 months in 1976.

The Chevrolet servicing table in figure 6-3 shows that *every* 1978 tune-up service interval has been lengthened compared to 1968 intervals.

Legal Requirements

There are two main reasons behind these changes in carmakers' tune-up services and intervals:

1. The 1972 Federal Clean Air Act requires that carmakers provide a 5-year, 50,000-mile warranty on their products' emission control devices.
2. During the EPA's new-car certification procedures, recommended maintenance schedules *must* be followed. These schedules are then given to the new-car buyer.

The extended warranty requirement has caused carmakers to add many testing requirements to their maintenance schedules, as we saw in figure 6-3. This makes the car owner responsible for maintaining the emission control devices so that they will operate properly for 5 years or 50,000 miles.

The rule that maintenance schedules be followed during EPA certification tests had two results. Low maintenance, solid-state ignition systems became standard equipment because they are more precise than breaker-point systems, making it easier for emissions to be controlled over a long period of operation. We will learn more about this in Chapter 11. Also, the complete maintenance schedules that must be published for consumers have become a sales issue. Manufacturers compare their schedules to those of the competition and also to the schedules recommended in earlier years. For example, Ford's 1976 sales literature pointed out that the changes in maintenance requirements meant a change in maintenance costs. In a specific case, the maintenance costs were figured for comparable 1972 and 1976 Ford cars. At 1976 labor rates and parts prices, the 1972 model would cost $136 per year to maintain. The 1976 model would cost only $75 per year.

When such comparisons are made between competing car models, it is obviously to a carmaker's advantage to have a cheaper maintenance schedule. This is one of the reasons that they have worked to eliminate service items and to extend service intervals.

PARTS AND EQUIPMENT SUPPLIERS' RECOMMENDATIONS

Carmakers supply specific maintenance schedules for their products. More general recommendations are made by:

- Manufacturers of replacement parts
- Manufacturers of tune-up testing equipment.

Their recommendations differ from carmakers' schedules for several reasons.

Carmakers tend to base their schedules on the least demanding vehicle operating conditions in order to reduce services and lengthen intervals. Parts and equipment suppliers base their recommendations on more severe operating conditions, which are closer to what the average car encounters. This tends to shorten their recommended service intervals. Of course, they are also interested in selling as many parts and as much equipment as possible, so they recommend more frequent replacement and testing.

SEVERE SERVICE EFFECTS ON TUNE-UP REQUIREMENTS

If the term "severe service" makes you think of police cars, taxicabs, and other special-duty vehicles, you may be surprised at today's severe service definition. A car is in severe service if:

- Most of its trips are less than 10 miles
- It is used in stop-and-go driving
- Its engine is idled for long periods
- It is driven in dusty or sandy conditions
- It is driven in very hot or very cold weather
- It is used to pull a trailer.

Other severe service factors include individual driving habits such as the use of air conditioning, extended high-speed driving, and rapid acceleration.

Today's common definition of normal driving is actually an *ideal* condition: long-distance driving at cruising speeds on dust-free highways. Almost any other driving condition qualifies as severe service.

Carmakers include special severe service recommendations in their maintenance schedules. These usually call for cutting certain service intervals in half. A few tune-up items are affected by severe service recommendations, including:

- Air cleaner filters (affected by dusty or sandy conditions)
- Crankcase oil and filter
- Spark plugs.

Transmission and rear-end services are also affected by severe service operation.

EMISSION INSPECTION EFFECTS ON TUNE-UP REQUIREMENTS

Periodic motor vehicle inspection (PMVI) programs exist in a few states and are under study by the Federal government and other agencies. In New Jersey, Arizona, and the city of Chicago, cars must be examined at yearly intervals for general vehicle condition. This includes exhaust emissions as well as suspension, lighting, brakes, and other safety-related areas.

Car owners may also have their cars tested at other locations, such as some service stations, auto clubs, car dealers, or locations specializing in tune-ups. Figure 6-4 shows a typical checklist from an inspection station. If PMVI programs become widespread, more vehicle tune-ups will be done as car owners are required to keep their cars within emission control standards.

TUNE-UP MERCHANDISING

Not many car owners know how important tune-ups are to performance, economy, and emission control. Because of this, tune-up mechanics cannot just *offer* their services, but they must actively *sell* their services. Average motorists must be made aware of:

- Their cars' need for tune-ups
- What the tune-up should include.

Many companies supply promotional items that can help a tune-up shop sell its services. These companies include carmakers, replacement parts suppliers, test equipment makers, and magazines aimed at the automotive service trade. The promotional products range from eye-catching signs and banners to ideas and suggestions for bringing tune-ups to the car owner's attention. Some ideas include:

- Displaying new and used parts side-by-side to emphasize the wear caused by everyday use
- Asking the car owners what fuel mileage their cars get, then showing them the appropriate EPA mileage estimate, to emphasize what the car *might* be getting if properly tuned
- Setting up specific appointments for the work to be done
- After the tune-up, showing the customer the parts that were replaced and explaining *why* they were replaced
- Showing the customer before-and-after test results in writing.

Our definitions of an engine tune-up do not tell you what specific services should be included. The services contained in your tune-up package should be clearly explained to the car owner. First, of course, you must decide for yourself what items will be serviced. Many shops offer a one-price service that covers an overall engine test and the replacement of a few common parts. If any unusual problems are found, they will be fixed only at additional cost. A similar tune-up package charges a fixed labor rate for engine testing, then adds the costs of any replacement items needed. These methods are often used in high-volume tune-up specialty shops. Other shops usually charge for the actual amount of time spent working on the car plus the cost of the parts replaced.

Again, large companies help the tune-up mechanic by publishing lists of items that should be checked during a complete tune-up. The items are usually listed in the order in which they should be done. Figures 6-5, 6-6, 6-7, and 6-8 show typical service checklists created by various companies.

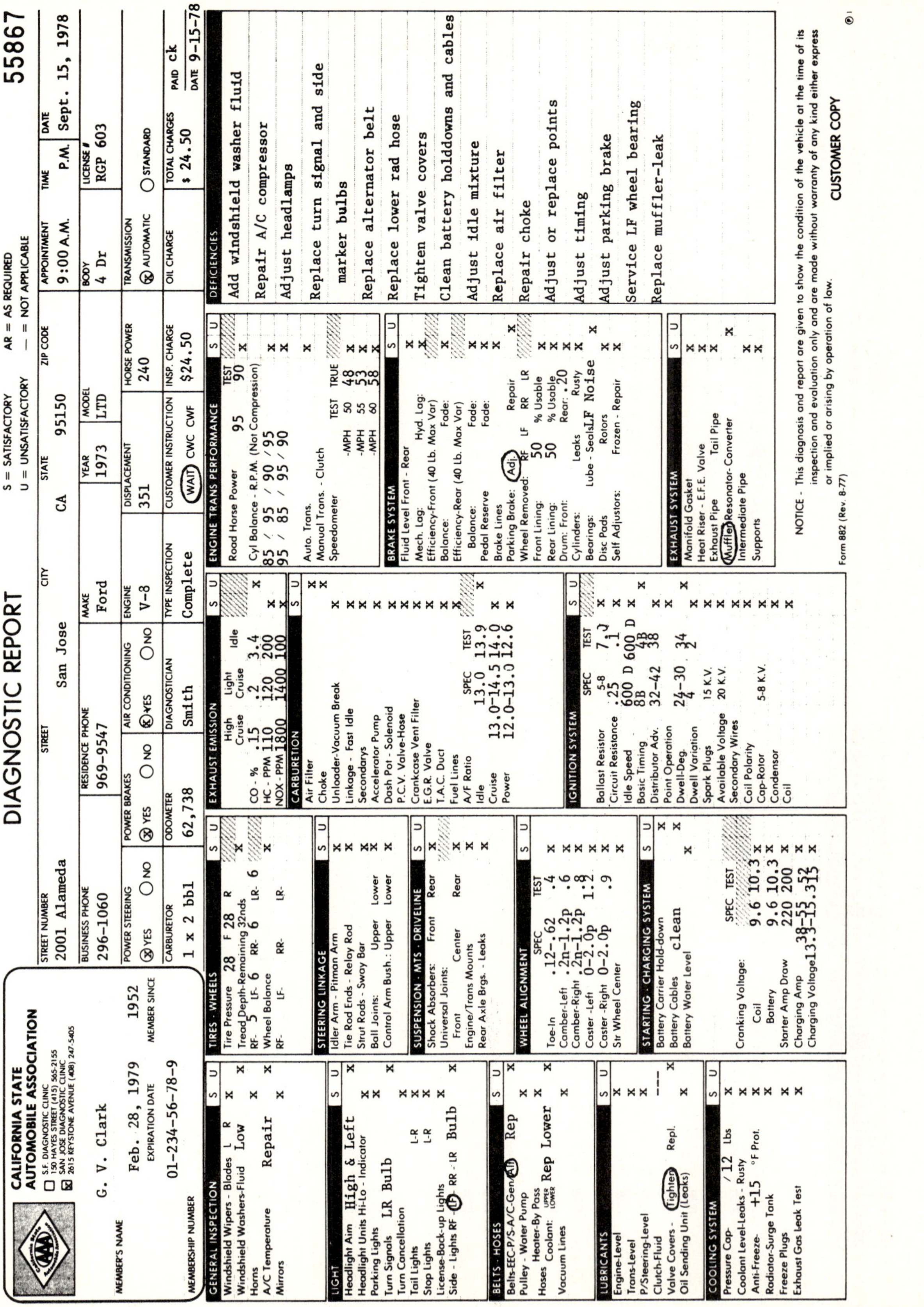

CALIFORNIA STATE AUTOMOBILE ASSOCIATION
☐ S.F. DIAGNOSTIC CLINIC 150 HAYES STREET (415) 565-2155
☒ SAN JOSE DIAGNOSTIC CLINIC 2615 KEYSTONE AVENUE (408) 247-5405

DIAGNOSTIC REPORT

S = SATISFACTORY U = UNSATISFACTORY AR = AS REQUIRED — = NOT APPLICABLE

55867

MEMBER'S NAME: G. V. Clark	STREET NUMBER: 2001 Alameda	STREET	CITY: San Jose	STATE: CA	ZIP CODE: 95150	APPOINTMENT: 9:00 A.M. · TIME: P.M. · DATE: Sept. 15, 1978
EXPIRATION DATE: Feb. 28, 1979 · MEMBER SINCE: 1952	BUSINESS PHONE: 296-1060	RESIDENCE PHONE: 969-9547	MAKE: Ford	YEAR: 1973 · MODEL: LTD		BODY: 4 Dr · LICENSE #: RGP 603
	POWER STEERING: ⊗ YES ○ NO	POWER BRAKES: ⊗ YES ○ NO	AIR CONDITIONING: ⊗ YES ○ NO · ENGINE: V-8	DISPLACEMENT: 351	HORSE POWER: 240	TRANSMISSION: ⊗ AUTOMATIC ○ STANDARD
MEMBERSHIP NUMBER: 01-234-56-78-9	CARBURETOR: 1 x 2 bbl	ODOMETER: 62,738	DIAGNOSTICIAN: Smith · TYPE INSPECTION: Complete	CUSTOMER INSTRUCTION: (WAIT) CWC CWF	INSP. CHARGE: $24.50	OIL CHARGE · TOTAL CHARGES: $ 24.50 · PAID ck · DATE 9-15-78

GENERAL INSPECTION (S U)
Windshield Wipers - Blades L R — X
Windshield Washers-Fluid Low — X
Horns — X
A/C Temperature Repair — X
Mirrors — X

LIGHT (S U)
Headlight Aim High & Left — X
Headlight Units Hi-Lo - Indicator — X
Parking Lights — X
Turn Signals LR Bulb — X
Turn Cancellation — X
Tail Lights L-R — X
Stop Lights L-R — X
License-Back-up Lights — X
Side - Lights RF-(LF)- RR - LR Bulb — X

BELTS - HOSES (S U)
Belts-EEC-P/S-A/C-Gen(Alt) Rep — X
Pulley - Water Pump — X
Hoses Heater-By Pass; Coolant: Upper/Lower Rep Lower — X
Vacuum Lines — X

LUBRICANTS (S U)
Engine-Level — X
Trans-Level — X
P/Steering-Level — X
Clutch-Fluid — ---
Valve Covers - (Tighten) Repl. — X
Oil Sending Unit (Leaks) — X

COOLING SYSTEM (S U)
Pressure Cap- / 12 Lbs — X
Coolant Level-Leaks - Rusty — X
Anti-Freeze- +15 °F Prot. — X
Radiator-Surge Tank — X
Freeze Plugs — X
Exhaust Gas Leak Test — X

TIRES - WHEELS (S U)
Tire Pressure 28 F 28 R — X
Tread Depth-Remaining 32nds: RF- 5 LF- 6 RR- 6 LR- 6
Wheel Balance RF- LF- RR- LR- — X

STEERING LINKAGE (S U)
Idler Arm - Pitman Arm — X
Tie Rod Ends - Relay Rod — X
Strut Rods - Sway Bar — X
Ball Joints: Upper Lower — X
Control Arm Bush.: Upper Lower — X

SUSPENSION - MTS - DRIVELINE (S U)
Shock Absorbers: Front Rear — X
Universal Joints: Front Center Rear — X
Engine/Trans Mounts — X
Rear Axle Brgs. - Leaks — X

WHEEL ALIGNMENT (S U) — SPEC / TEST
Toe-In .12-.62 / .4 — X
Camber-Left .2n-1.2p / .6 — X
Camber-Right .2n-1.2p / .8 — X
Caster-Left 0-2.0p / 1:2 — X
Caster-Right 0-2.0p / .9 — X
Str Wheel Center — X

STARTING - CHARGING SYSTEM (S U)
Battery Carrier Hold-down — X
Battery Cables clean — X
Battery Water Level — X
Cranking Voltage: SPEC / TEST
Coil 9.6 / 10.3 — X
Battery 9.6 / 10.3 — X
Starter Amp Draw 220 / 200 — X
Charging Amp 38-55 / 52 — X
Charging Voltage 13.3-15.3 / 13 — X

EXHAUST EMISSION (S U) — High Cruise / Light Cruise / Idle
CO - % .15 / .2 / 3.4 — X
HC - PPM 110 / 120 / 200 — X
NOX - PPM 1800 / 1400 / 100 — X

CARBURETION (S U)
Air Filter — X
Choke — X
Unloader-Vacuum Break — X
Linkage - Fast Idle — X
Secondarys — X
Accelerator Pump — X
Dash Pot - Solenoid — X
P.C.V. Valve-Hose — X
Crankcase Vent Filter — X
E.G.R. Valve — X
T.A.C. Duct — X
Fuel Lines — X
A/F Ratio: SPEC / TEST
Idle 13.0 / 13.9 — X
Cruise 13.0-14.5 / 14.0 — X
Power 12.0-13.0 / 12.6 — X

IGNITION SYSTEM (S U) — SPEC / TEST
Ballast Resistor 5-8 / 7.0 — X
Circuit Resistance .25 / .1 — X
Idle Speed 600 D / 600 D — X
Basic Timing 8B / 4B — X
Distributor Adv. 32-42 / 38 — X
Point Operation — X
Dwell-Deg. 24-30 / 34 — X
Dwell Variation 4 / 2 — X
Spark Plugs — X
Available Voltage 15 K.V. — X
Secondary Wires 20 K.V. — X
Coil Polarity — X
Cap-Rotor 5-8 K.V. — X
Condensor — X
Coil — X

ENGINE TRANS PERFORMANCE (S U)
Road Horse Power 95 / TEST 90 — X
Cyl Balance - R.P.M. (Not Compression)
85 / 95 / 90 / 95
95 / 85 / 95 / 90 — X
Auto. Trans. — X
Manual Trans. - Clutch
Speedometer: TEST / TRUE
-MPH 50 / 48 — X
-MPH 55 / 53 — X
-MPH 60 / 58 — X

BRAKE SYSTEM (S U)
Fluid Level Front - Rear — X
Mech. Lag: Hyd. Lag: — X
Efficiency-Front (40 Lb. Max Var) — X
Balance: Fade:
Efficiency-Rear (40 Lb. Max Var) — X
Balance: Fade: — X
Pedal Reserve — X
Brake Lines — X
Parking Brake: (Adj) Repair — X
Wheel Removed: RF LF RR LR
Front Lining: 50 % Usable — X
Rear Lining: 50 % Usable — X
Drum: Front: Rear: .20 — X
Cylinders: Leaks Rusty — X
Bearings: Lube - Seals LF Noise — X
Disc Pads Rotors — X
Self Adjustors: Frozen - Repair — X

EXHAUST SYSTEM (S U)
Manifold Gasket — X
Heat Riser - E.F.E. Valve — X
Exhaust Pipe Tail Pipe — X
(Muffler) Resonator-Converter — X
Intermediate Pipe — X
Supports — X

DEFICIENCIES
Add windshield washer fluid
Repair A/C compressor
Adjust headlamps
Replace turn signal and side marker bulbs
Replace alternator belt
Replace lower rad hose
Tighten valve covers
Clean battery holddowns and cables
Adjust idle mixture
Replace air filter
Repair choke
Adjust or replace points
Adjust timing
Adjust parking brake
Service LF wheel bearing
Replace muffler-leak

NOTICE - This diagnosis and report are given to show the condition of the vehicle at the time of its inspection and evaluation only and are made without warranty of any kind either express or implied or arising by operation of law.

CUSTOMER COPY

Form 882 (Rev. 8-77)

Figure 6-4. This diagnostic checklist is used by an automobile association's diagnostic service center. (CSAA)

Step by Step Tune-Up Procedure

ignition manufacturers institute

1. Diagnose – Get the Story
2. Clean and inspect battery
3. Test battery
4. Test cranking voltage
5. Test compression
6. Replace or service spark plugs
7. Inspect coil
8. Inspect distributor cap and rotor
9. Inspect primary wires and secondary cables
10. Replace breaker points
11. Replace condenser
12. Lubricate distributor cam
13. Set breaker point dwell angle
14. Set ignition timing
15. Check centrifugal advance action
16. Check vacuum advance (and retard) action
17. Inspect drive belt condition and test belt tension
18. Test generator or alternator output
19. Test voltage regulator
20. Test charging system insulated and ground circuits
21. Test charging voltage
22. Service manifold heat control valve
23. Service carburetor air cleaner (check action of hot air damper)
24. Replace positive crankcase ventilation system valve
25. Test emission control and EGR Systems, as required
26. Replace fuel filter
27. Test fuel pump
28. Adjust carburetor (use "Lean roll" procedure on E.E.C. engines)
29. Adjust valve clearances (if necessary)

Figure 6-5. Tune-up checklist prepared by the Ignition Manufacturers Institute. (IMI)

NAPA ECHLIN®

TUNE UP PROCEDURE

- CLEAN BATTERY & CABLES
  Hydrometer & Cranking Voltage tests.
- COMPRESSION TEST OR CYLINDER BALANCE TEST
- RESISTANCE IN DISTRIBUTOR CIRCUIT
  Dwellmeter Test
- VACUUM & MECHANICAL ADVANCE CHECK
- CONTACTS & CONDENSER
  Lubricate Rubbing Block.
  Adjust Spacing.
- CAP & ROTOR INSPECTION
- SPARK PLUGS (Gap to Specs.)
  Inspection, Clean Or Replace
- SPARK PLUG WIRES
  Inspection & Test.
- COIL TEST
- IGNITION TIMING
- FUEL PUMP & FILTER
- P.C.V. VALVE, HOSES & FILTER
- MANIFOLD HEAT CONTROL VALVE
- CHECK EMISSION CONTROLS FOR PROPER OPERATION
- CARBURETOR
  Adjust, Clean Or Repair.
- AIR CLEANER
  Clean Or Replace Gasket & Element.
- CHARGING SYSTEM
  Voltmeter Check - Regulator
- ROAD TEST

Printed in U.S.A. Form No. 1-162-73

Figure 6-6. This checklist was prepared by a replacement parts manufacturer. (Echlin)

ATLAS®

10-Step Tune-up Procedure

This procedure is a method of diagnosis and service that allows you to pinpoint engine problems quickly and make the necessary adjustments. Following this 10-step procedure will conserve time by avoiding unnecessary tests and adjustments.

1 Ignition Tests
Use the oscilloscope motor analyzer to check the entire ignition system. Check the primary circuit first, then the secondary circuit. Step 1 will indicate the need for a cylinder power balance test.

2 Distributor Service
Perform routine distributor service in this Step: visual inspection for damage, replacement of any defective parts, dwell adjustment on breaker-point distributors, and lubrication as necessary.

3 Spark Plug Service
Check the condition of the plugs. Clean, regap, or replace the plugs as necessary.

4 Ignition Timing
Set the initial timing to specifications. Check spark advance and adjust as necessary.

5 Carburetor Adjustment
Check hose connections and the operation of the throttle linkage, choke, and PCV system. Set the idle speed and mixture to specifications.

6 Battery, Charging System, and Starting System
Check the battery and clean or service as necessary. Test the battery; if required, run a complete charging system test. If the battery is OK but cranking is slow, check the starting system.

7 Fuel System and Manifold Heat Control Valve
Visually inspect all fuel system components and conduct pump pressure and flow or volume tests. Replace fuel filter as required. Check the operation of the manifold heat control valve.

8 Cylinder Power Balance Test
Perform this test if the previous steps have not achieved smooth engine operation. A compression test may also be desirable.

9 Air Filters
Check the air cleaner element, the evaporative control filter, and the PCV filter and replace as necessary.

10 Road Test
Drive the vehicle and check for indications of spark knock, improper idling, hesitation, surge, or stalling. Check overall performance and advise the car owner of any possible safety-related defects.

Figure 6-7. Atlas Supply Company supplies both replacement parts, test equipment, and the 10-step tune-up procedure. (Atlas)

CUSTOMER ________ PHONE ________ DATE ________

MAKE-YEAR-MODEL ________ MILEAGE ________

SIGNATURE ________

Engine Performance Test Procedure

Using Sun Tune-Up Tester, Model TUT-915

TEST PROCEDURE	TEST CONDITION OF:	SPECS.	RESULTS	GO	NO-GO
1. Cranking Battery Voltage	Battery, Starting System				
2. Dwell	Breaker Point Setting				
3. Dwell Variation	Distributor Mechanical Condition				
4. Idle Speed	Idle Speed Adjustment				
5. Initial Timing	Spark Timing Setting				
6. Timing Advance	Dist. Mech. and Vac. Advance				
7. Coil Polarity	Coil Installation, System Polarity				
8. Maximum Coil Output	Coil, Condenser Ignition Primary				
9. Secondary Circuit Insulation	High Tension Wires, Cap, Rotor				
10. Spark Plug Firing Voltage	Plug & Rotor Gaps, Wires				
11. Spark Plugs Under Load	Worn or Fouled Spark Plugs				
12. Secondary Circuit Condition	Plugs, Wires, Cap, Rotor Resistance				
13. Coil and Condenser Condition	Coil Windings, Cond. Leakage				
14. Breaker Point Condition	Point close, bounce, open, arc.				
15. Cam Lobe Accuracy	Cyl. to Cyl. Timing, Distributor Breaker Cam				
16. Cylinder Power Balance	Individual Cyl. Comp. Factors				
Record RPM	1 2 3	4	5 6	7	8
17. Charging Voltage	Generator-Alternator, Volt. Reg.				

RECOMMENDATIONS:

692-744-1

Sun SERVICE CONTROL SYSTEM

Figure 6-8. This checklist is intended to help you pinpoint problems through testing. (Sun)

THE TOTAL TUNE-UP

According to the Ignition Manufacturers Institute (IMI), a trade association of major parts manufacturers, a quality tune-up must include these five elements:

- A trained mechanic with
- Dependable test equipment and
- Accurate specifications, performing a
- Specific test procedure and using
- Quality replacement parts.

This definition of an engine tune-up agrees with our earlier explanation. When tuning a late-model car, the mechanic must be able to understand the information obtained from the car's condition and from test equipment. The equipment must be dependable and accurate, or the test results are meaningless. A vital part of diagnosis is having accurate specifications, so that you can compare how the car *should* be operating to how it is *actually* operating. A specific test procedure ensures that all engine performance areas are thoroughly checked in the proper order. Using quality replacement parts reduces the risk of installing new parts that are already defective or might wear out in less than the recommended service interval.

The remaining chapters in this *Classroom Manual* cover the operation of the various engine systems that are involved in tune-up testing and service. The relationship of each of these systems to engine performance, economy, and emission control is also explained.

SUMMARY

The changes in the automobile caused by emission control regulations have in turn caused changes in engine tune-up ideas and procedures. The traditional tune-up's replace-adjust-test structure has shifted to the modern emphasis on comprehensive testing and fault diagnosis. In general, today's service recommendations contain fewer replacement items, longer service intervals, and more testing of engine systems.

Carmakers' recommendations are based on ideal driving conditions which very few cars actually meet. Shorter servicing intervals are often recommended by replacement parts manufacturers and test equipment suppliers. Carmakers recognize this difference and include severe service recommendations in their schedules.

Periodic motor vehicle inspections are not widespread, but they may be an answer to enforcing emission control standards on cars in use. Tune-up services will still have to be sold, however. Large companies supply advertising aids and tune-up procedure checklists to help the tune-up mechanic attract and keep business. The total tune-up is an important idea if late-model cars are to be serviced properly.

Review Questions

Select the single most correct answer.
Compare your answers with the correct answers on page 280.

1. Emission control regulations for cars first appeared in:
   a. 1958
   b. 1961
   c. 1964
   d. 1969

2. Which of the following is generally *not true* for a modern tune-up:
   a. There are fewer items to be replaced
   b. The service intervals have gradually lengthened
   c. Tune-ups are no longer as important as they used to be
   d. There are more testing procedures now than before

3. Which of the following is *true* of the carmakers' service recommendations:
   a. They represent the minimum service under normal driving
   b. They represent the maximum service under normal driving
   c. They represent the maximum service under severe driving
   d. They represent the minimum service under severe driving

4. The 1972 Federal Clean Air Act requires cars to have a ___ warranty on their emission control devices.
   a. 2-year, 20,000-mile
   b. 3-year, 30,000-mile
   c. 4-year, 40,000-mile
   d. 5-year, 50,000-mile

5. Which of the following is *not* a part of severe service driving:
   a. Most of the trips are 10 miles or less
   b. The car frequently is used for trailer towing
   c. The car is driven at moderate speeds for extended periods of time on highways.
   d. The car is driven in dusty or sandy conditions

6. Which of the following *are* included as normal service driving:
   a. Long distance driving
   b. Driving at cruising speeds
   c. Driving on dust-free highways
   d. All of the above

PART TWO

Electrical Systems

Chapter

7

The Battery, Starting, and Charging Systems

At a glance, you may think that the ignition system is the only automotive electrical system that needs to be considered during a tune-up. If you ignore other systems, however, you may miss the causes of engine performance problems. This chapter contains a brief explanation of the battery, the charging system, and the starting system, and shows how they relate to engine performance.

To examine the car's electrical systems we should be familiar with some electrical terms and ideas. Electricity can be defined as the flow of electrical current through a **conductor**, which is a material that allows easy current flow. **Insulators** are materials that do not allow current flow. The amount of current flow through a conductor is called the **amperage**. It is measured in **amperes**, commonly called amps. The force that causes current flow is called **voltage**, and is measured in **volts**. In order for current to flow, there must be a complete path of conductors to carry the current. These paths are called **circuits**. Although there can be no current flow if the circuit is not complete, there *can* be voltage present in the incomplete circuit. All circuits contain some amount of **resistance** to current flow. This resistance is measured in **ohms**.

The voltage for a car's electrical circuits comes from the battery. The circuits are made up of:

- Wiring
- The engine, frame, and body of the car.

The engine, frame, and body are called the **ground**. The cable from one battery terminal is bolted to the engine or frame, figure 7-1. This is called the **ground cable**. The cable from the other battery terminal is connected to the wiring in the circuits. This is called the **hot, or insulated, cable**. The insulated side of every circuit in the car is the wiring running from the battery to the devices in the circuit. The ground side of every circuit is the vehicle chassis, figure 7-1. This is called a single-wire, or a ground-return, system.

We will be showing you pictures of circuits that are called circuit diagrams. These show the wires and devices that are in the circuit. They do not show the car engine, frame, or body. Instead, the ground symbol, figure 7-1, is used to show that the wire or device is bolted to the car chassis. You can think of the ground symbol as showing a direct connection back to the battery ground cable.

BATTERY OPERATION AND CONSTRUCTION

The automobile battery does not actually store electricity, as is often believed. It converts electrical energy to chemical energy, which is stored until the battery is connected to an external

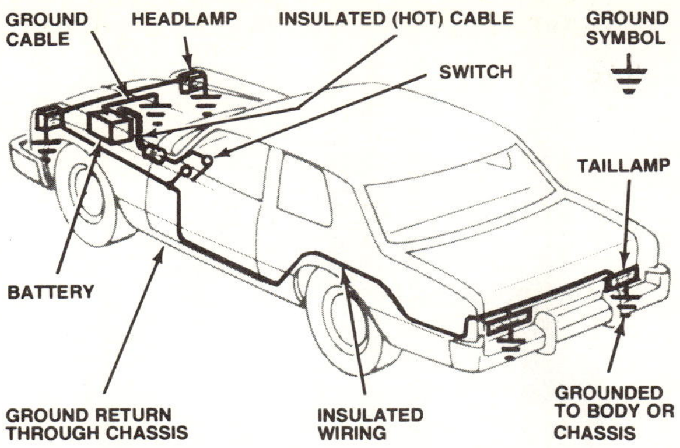

Figure 7-1. Half of the automotive electrical system is the ground path through the vehicle chassis.

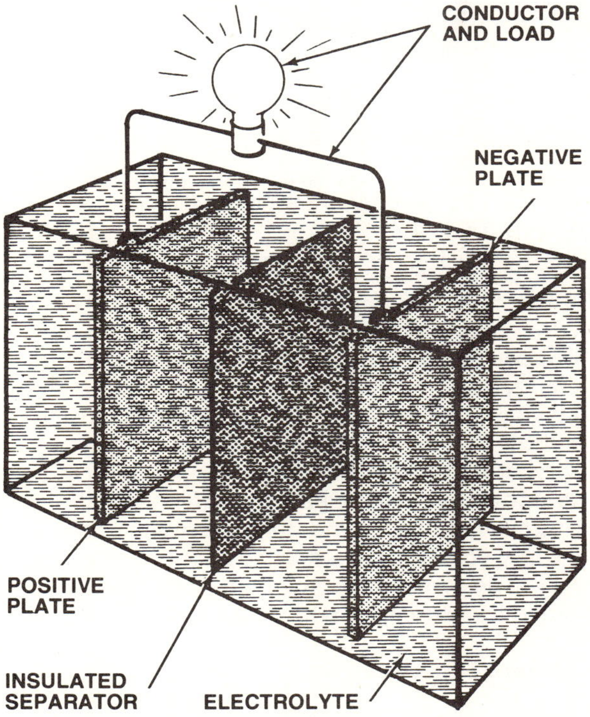

Figure 7-2. The potential difference between the two plates of a battery can cause current to flow in an outside circuit. (Chevrolet)

circuit. The stored chemical energy is then converted back to electrical energy, which flows from one battery terminal, through the circuit, and back to the other battery terminal.

Battery Electrochemical Action

All automotive batteries operate because of the chemical action of two dissimilar metals in the presence of a conductive and reactive solution called an **electrolyte**. Because this chemical action produces electricity, it is called electrochemical action. The chemical action of the electrolyte causes one metal plate to be positively charged and the other metal plate to be negatively charged. This difference between the charged plates is the battery's voltage.

We can see from the simple battery shown in figure 7-2 that when a positive plate and a negative plate are put in a container of electrolyte, and a conductor is connected between them, the battery's voltage will cause electrical current to flow through the conductor.

Conductors: Materials that allow easy flow of electricity.

Insulators: Materials that will not conduct electricity.

Amperage: The amount of current flow through a conductor.

Amperes: Units for measuring the rate of electrical current flow.

Voltage: The electromotive force that causes current flow. The potential difference in electrical force between two points when one is negatively charged and the other is positively charged.

Volts: Units for measuring the amount of electrical force.

Circuit: A circle or unbroken path through which an electric current can flow.

Resistance: Opposition to electrical current flow.

Ohms: Units for measuring electrical resistance.

Ground: The connection of an electrical circuit or unit to the engine or chassis to return electrical current to the battery.

Ground Cable: The battery cable that provides a ground connection from the vehicle chassis to the battery.

Hot (Insulated) Cable: The battery cable that conducts battery current to the automotive electrical system.

Electrolyte: The chemical solution in a battery that conducts electricity and reacts with the plate materials.

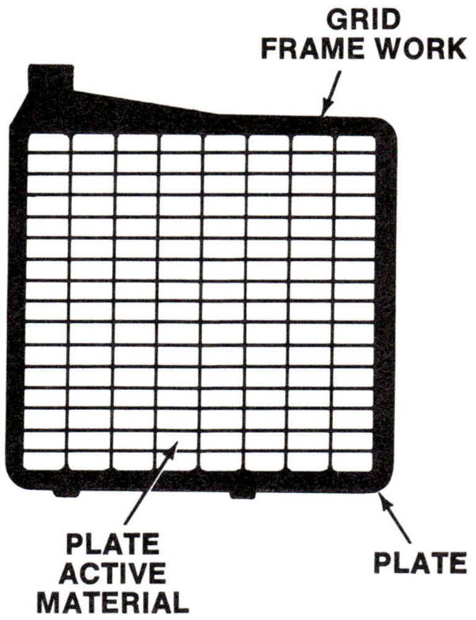

Figure 7-3. The grid provides a support for the plate active material.

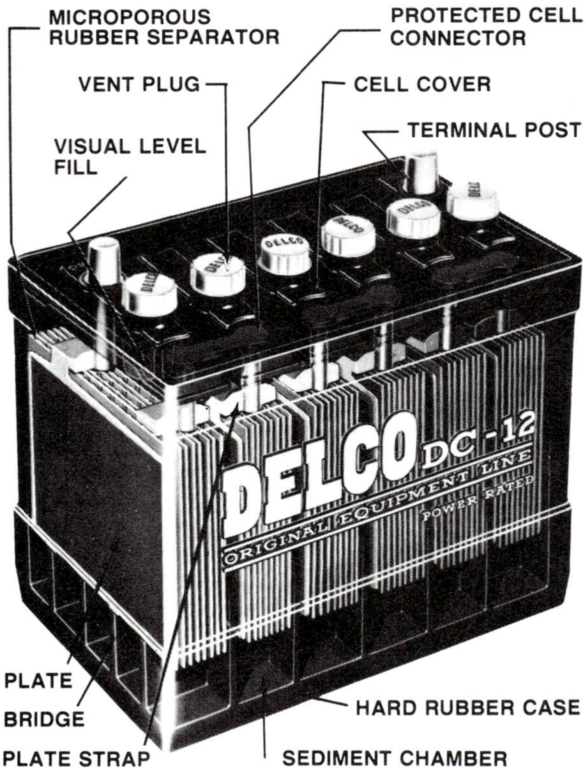

Figure 7-5. A cutaway view of an assembled battery. (Delco-Remy)

As the current flows, the chemical structure of the plates and the electrolyte changes. This is called discharging. When a battery is totally

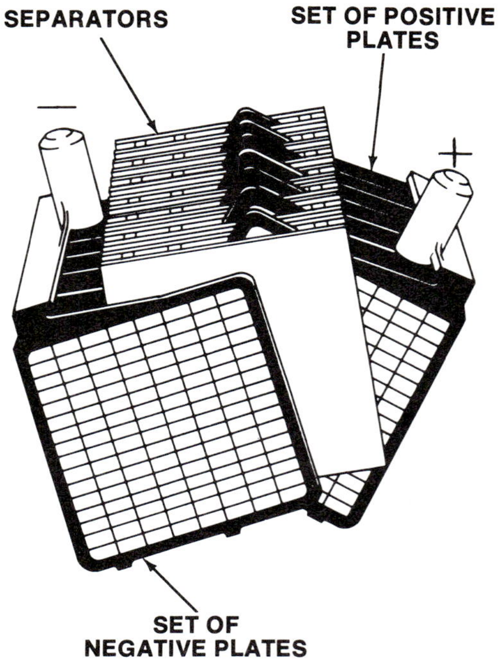

Figure 7-4. Two groups are interlaced to form a battery element.

discharged, the chemical energy has all been converted to electrical energy and used in the outside circuit. Automotive batteries can be recharged, however, by applying an outside voltage to the battery terminals and forcing current to flow through the battery in the opposite direction. This current flow causes chemical action that restores the battery materials to their original condition, and the battery can again supply voltage. This is called charging the battery. The condition of the battery materials is called the battery's state of charge.

Battery Construction

Battery construction begins with the positive and negative plates. The plates have a grid framework of conductive metal, figure 7-3. The dissimilar metals, called the active materials of the battery, are pasted onto the grids. The active materials in an automotive battery are two different lead compounds, sponge lead and lead peroxide.

A number of similar plates, all positive or all negative, are connected together into a plate group. A positive and a negative plate group are then interlaced so that their plates alternate, figure 7-4. To keep the plates from touching each other, they are separated by chemically inactive separators made from paper, plastic, or fiberglass. This assembly of two plate groups and separators is called a battery element.

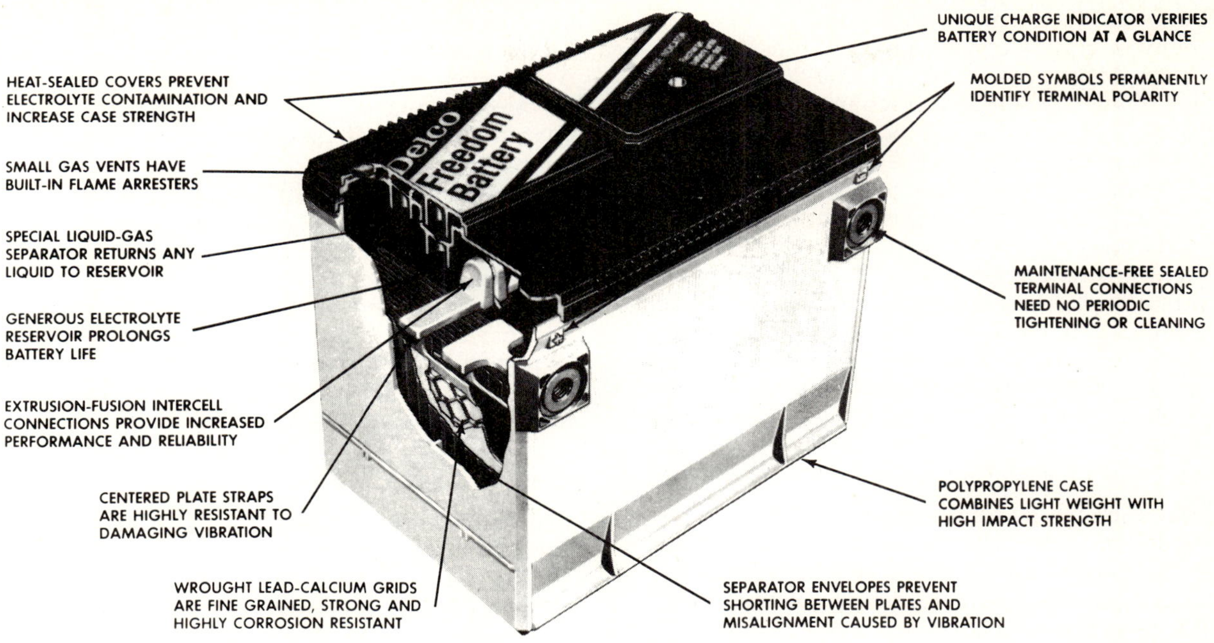

Figure 7-6. A cutaway view of a maintenance-free battery. Notice the internally threaded connections that form the battery terminals.

The element is placed in the cell of a battery case, figure 7-5. No matter how many plates a cell has, or what size the plates are, one cell of an automotive battery can produce only about 2.1 volts. Six-volt batteries have three cells; 12-volt batteries have six cells.

The cases of most modern automotive batteries are made of plastic materials, which are lighter, stronger, and more corrosion resistant than the rubber used in older batteries. The battery cover has vent caps to provide an opening for adding electrolyte and for the escape of gases that form during charging and discharging. Each battery has two terminals for connection to an external circuit. The terminals of automotive batteries are either two tapered posts on top of the case, as shown in figure 7-5, or two internally threaded connectors on the side of the case, figure 7-6. The positive terminal, marked (+), is connected to all of the positive plates in the battery. The negative terminal, marked (−), is connected to all of the negative plates in the battery.

The electrolyte in an automobile battery is a solution of sulfuric acid and water. In a fully charged battery, the solution is about 35-percent acid and 65-percent water. The state of charge of a battery can be measured by checking the **specific gravity** of the electrolyte.

Specific gravity is the weight of a given volume of liquid divided by the weight of an equal volume of water. Pure water has a specific gravity of 1.000. Battery electrolyte should have a specific gravity of about 1.260 to 1.280, at 80° F(26.7° C). That is, battery electrolyte is 1.260 to 1.280 times as heavy as water. As the battery discharges and the chemical structure of the battery materials change, the electrolyte's acid is changed to water. This means that the electrolyte's specific gravity will decrease, because it is mostly water. The specific gravity of the electrolyte can tell you approximately how discharged the battery has become:

1.265 specific gravity	100% charged
1.225 specific gravity	75% charged
1.190 specific gravity	50% charged
1.155 specific gravity	25% charged
1.120 specific gravity or lower	discharged

These values may vary slightly, according to the design factors of a particular battery.

Specific Gravity: The weight of a volume of liquid divided by the weight of the same volume of water at a given temperature and pressure. Water has a specific gravity of 1.000.

Wet- and dry-charged batteries
Batteries may be manufactured and sold as either wet-charged or dry-charged types. A wet-charged battery is completely filled with electrolyte when it is built. Even when a wet-charged battery is not in use, a slow reaction takes place between the plates and the electrolyte. This is a self-discharging reaction, and will eventually discharge the battery almost completely.

A dry-charged battery is shipped from the factory without electrolyte. During manufacture, the plates are charged and then completely washed and dried. The battery is then assembled and sealed to keep moisture out. It will remain charged as long as it is sealed, and it can be stored for a long time. A dry-charged battery is put into service by adding electrolyte, checking the state of charge, and charging if required.

Maintenance-free batteries
Most battery manufacturers are now producing sealed, maintenance-free batteries, figure 7-6. This means that the battery will not need to have water added to its electrolyte during its lifetime.

The main reason for battery water loss is the heat of the charging current, which causes water evaporation. Maintenance-free batteries have plate grids made of a special lead-calcium alloy that reduces the amount of charging current needed and so reduces the battery's internal heat. With less heat, there is less water loss. Maintenance-free batteries also have greater electrolyte capacity. This combination means that the battery will have enough electrolyte for a long time.

Maintenance-free batteries are not completely sealed. They all have small vents in the top or case to relieve the gases produced during charging and discharging. However, the amount of gassing is reduced by the battery construction. Because gassing is a major cause of battery terminal corrosion, a maintenance-free battery tends to have less terminal corrosion than a conventional battery.

■ Early Starting Systems

The earliest cars generally used a hand crank to turn the engine over until the compression built up enough to ignite the air-fuel mixture and the engine began to run on its own. Much later, electric starters were used. In the meantime, several other systems saved the driver from having to crank his engine by hand.

One of these was the Prest-O-Lite gas starter, which injected a small air-fuel charge into the cylinders, instead of mechanically turning the crankshaft. The driver operated a small hand pump to charge the cylinders, then turned on the ignition switch to fire the charge in proper firing order. This would turn over the engine and generate the current necessary for continued operation.

Another system used compressed air or compressed exhaust to drive the pistons. The exhaust was stored in a tank when the engine was still running, and would presumably still be compressed when the driver needed to start the car again. The compressed air would drive a "starter motor" attached to the front of the engine where the crank used to be. This would turn over the engine until it began firing by itself.

BATTERY RATINGS

The battery must provide good cranking power for the car's engine and adequate reserve power for the electrical system. Test standards and rating methods developed by the Battery Council International (BCI) and the SAE are designed to test a battery's ability to meet these requirements and do its job. The most common battery rating methods are explained in the following paragraphs.

Ampere-Hour Rating

The oldest battery rating method, and the one with which many people are most familiar, is the **ampere-hour rating**. This rating method was the industry standard for decades. In recent years, however, it has been replaced by the cold-cranking and reserve capacity ratings, which give better indications of a battery's performance.

The ampere-hour method is also called the 20-hour discharge rating method. This rating represents the steady current flow which the battery will deliver for 20 hours at a temperature of 80° F(26.7° C) without cell voltage falling below 1.75 volts (a total of 10.5 volts for a 12-volt battery). For example, a battery that will continuously deliver 3 amperes for 20 hours is rated as a 60 ampere-hour battery (3 amperes × 20 hours = 60 ampere-hours).

Cold-Cranking Rating

The **cold-cranking rating** indicates the power a battery can supply for engine cranking under low temperature conditions. The ampere figures (at 0° F or −17.8° C) for this rating may be listed, for example, as 305, 315, 380, or 440. These figures represent the current flow in amperes that a battery can deliver for 30 seconds at 0° F(−17.8° C) while maintaining a voltage of 1.2 volts per cell (a total of 7.2 volts for a 12-volt battery).

Reserve Capacity Rating

The **reserve capacity rating** indicates the power available from a battery. Reserve capability is listed in minutes, such as 68, 81, 101, or 105. It is a measurement of the length of time a fully charged battery can supply 25 amperes and maintain a voltage of no less than 1.75 volts per cell, or 10.5 volts for a 12-volt battery. The reserve capacity defines the length of time a vehicle can be driven after the charging system fails.

Watts Rating

A battery may also be rated by watts, which represent the available cranking power. The **watts rating** is determined by multiplying the current available from the battery by the battery voltage at 0° F(−17.8° C). Wattage ratings for most 12-volt batteries range from 2,000 to 4,000 watts.

BATTERY SELECTION

The BCI publishes application charts that list the correct battery for any car. Optional heavy-duty batteries are normally used in cars with air conditioning or several major electrical accessories or in cars operated in cold climates. To ensure adequate cranking power and to meet all other electrical needs, a replacement battery may have a higher rating, but never a lower rating, than the original unit.

The battery must also be the correct physical size for the car, and have the correct type of terminals. BCI standards include a coding system, called the group number, that identifies battery length, width, height, terminal design, and other special features.

BATTERY CHARGING VOLTAGE AND LIFE

A battery is charged by forcing current to flow through it in the direction opposite to discharge current. The battery offers some resistance to this charging current, because of the battery's chemical voltage and the resistance of the battery's internal parts. For charging current to enter the battery, the charging voltage must be great enough to overcome this resistance.

Charging Voltage Levels

When a battery is fully charged, its internal resistance is very high. Very little charging current can flow through it. When the battery is discharged, its internal resistance is low, and charging current flows freely.

The temperature of the battery also affects the charging voltage because temperature affects the resistance of the electrolyte. Cold electrolyte has higher resistance than warm electrolyte, so a colder battery is harder to charge.

Battery Life

All batteries have a limited life, but some conditions can shorten that life even more. Important factors that affect battery life are discussed in the following paragraphs.

Overcharging

When charging current is forced through a fully charged battery, there is a violent chemical reaction within the battery. The water in the electrolyte is rapidly broken down into hydrogen and oxygen gases. The gas bubbles can wash active materials off the plates, as well as lower the level of the electrolyte. Overcharging also causes excessive heat, which can buckle the plates.

Undercharging and sulfation

When a battery plate is discharged, it is covered with lead sulfate. If the battery is constantly undercharged, the lead sulfate will build up and crystallize. This is called battery **sulfation**. The crystals are difficult to break down by normal recharging, and the battery may become useless.

Electrolyte level

If the electrolyte level drops below the top of the plates and active material is exposed to the air, the material will harden and resist chemical

Ampere-Hour Rating: A battery rating based on the amperes of current that a battery can supply steadily for 20 hours, with no battery cell falling below 1.75 volts. Also called a 20-hour discharge rating.

Cold-Cranking Rating: A battery rating based on the amperes of current that a battery can supply for 30 seconds at 0° F, with no battery cell falling below 1.2 volts.

Reserve-Capacity Rating: A battery rating based on the number of minutes a battery at 80° F can supply 25 amperes, with no battery cell falling below 1.75 volts.

Watts Rating: A battery rating based on the available cranking power of the battery. The rating can be found by multiplying the current available from the battery by the battery voltage at 0° F.

Sulfation: The crystalization of lead sulfate on the plates of a constantly discharged battery.

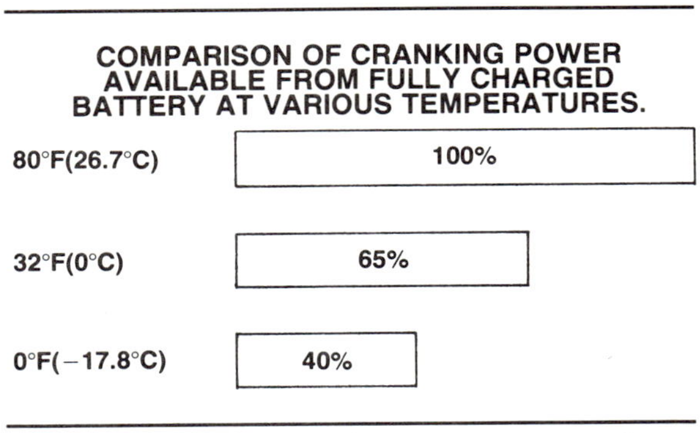

Figure 7-7. Battery power decreases as temperature decreases.

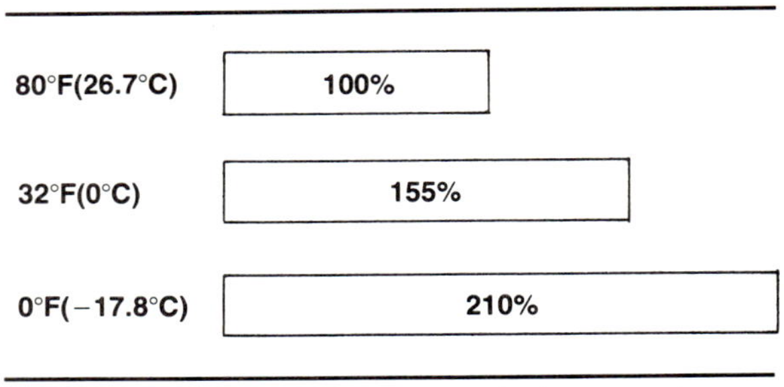

Figure 7-8. This graph represents the increased power required to crank an engine at low temperatures.

action. Also, the remaining electrolyte will have a high concentration of acid. This can cause the plates to deteriorate rapidly. Overfilling the battery can be just as harmful. The electrolyte will be weakened, reducing battery efficiency. Any spillage of electrolyte will cause corrosion around the battery terminals.

Corrosion
Battery corrosion is caused by spilled electrolyte and by electrolyte condensation from gassing. The sulfuric acid attacks and can destroy connectors, terminals, and the battery carriers. Corroded connectors reduce the amount of voltage that the battery can supply to the car. Corrosion on the battery cover can also form a current leakage path, which can allow the battery to discharge.

■ Repairing Storage Batteries

Car mechanics of two generations ago found battery repairs a frequent part of their service routine. Those old batteries were usually in wooden or hard rubber cases, and the plate groups were in "jars" sealed into the case with a sealing compound. The mechanic could easily loosen this sealant to pull the plates out for inspection and repairs.

Early repair manuals give detailed instructions on how to straighten the plates, how to replace the wooden spacers between the plate groups, how to clean the sediment off the inside of the case, or how to take the elements apart.

Even though the old batteries could be easily serviced and today's batteries cannot be, they all look about the same on the outside. In fact, the biggest advance in batteries in the last generation is the development of the maintenance-free battery.

Cycling
The operation of a battery from charged to discharged and back to charged is called **cycling**. If a battery is repeatedly cycled, the active material on the positive plates may shed and fall into the bottom of the case. This will reduce the capacity of the battery and shorten its useful life.

Temperature
High temperatures caused by overcharging or engine heat can shorten battery life. Low temperatures in winter can also harm a battery. If the electrolyte freezes, it can expand and break the case, ruining the battery. The freezing point of electrolyte depends on the state of charge of the battery. A fully charged battery, with a proper concentration of sulfuric acid, will not freeze at −50° F (−45.5° C). A discharged battery, with an electrolyte which is mostly water, can freeze at 18° F (−7.7° C). As we saw before, cold temperatures make it harder to keep the battery fully charged, but this is when a full charge is most important. Figure 7-7 compares the energy levels available from a fully charged battery at various temperatures. Figure 7-8 compares the energy required to crank an engine at those temperatures. As you can see, the colder a battery is, the less energy it can supply. Yet the colder an engine gets, the more energy is required to crank it. This is why battery care is especially important in cold weather.

Vibration
A battery must be mounted securely to protect it from vibration. Vibration can shake the active materials off the plates and shorten the battery's life. Severe vibration can even crack a battery case or loosen cable connections.

CHARGING SYSTEM OPERATION

The charging system converts some of the engine's mechanical energy into electrical energy. This electrical energy is used to maintain the battery's state of charge and to operate the car's electrical system while the engine is running.

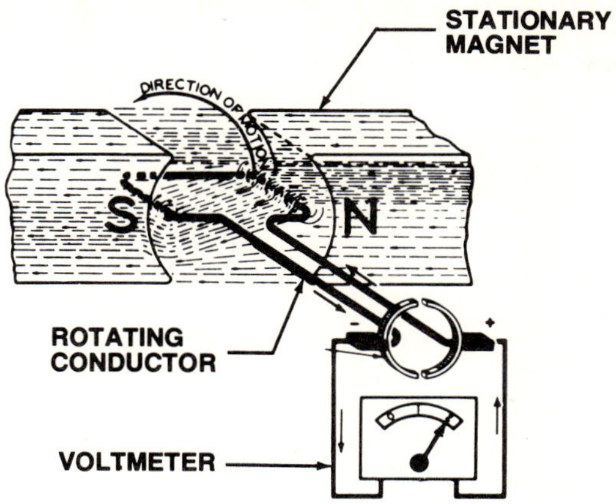

Figure 7-9. A simple generator. (Prestolite)

During cranking, all electrical energy for the car is supplied by the battery. Once the engine starts and runs, the charging system must produce enough electrical energy both to recharge the battery and to supply the demands of other parts of the electrical system, such as the ignition system, lighting systems, and accessories. If the starting system is in poor condition and uses too much energy, or if the charging system cannot both recharge the battery and supply the other parts, more energy must be drawn from the battery for short periods.

Voltage Sources

The two devices that convert mechanical energy to electrical energy are the generator and the alternator. Both of these units use the principle of **induction** to create an electrical voltage. When the lines of force of a magnetic field cut through a conductor (wire), a voltage will appear in the wire. The voltage will cause current flow in a complete circuit such as the automobile electrical system.

The generator is constructed with a stationary magnet, figure 7-9, and loops of wire that rotate within the magnet's field. The loops of wire are turned by a drive belt from the engine. When the loops turn, the lines of force from the magnet cut through them, and a voltage is induced within the loops. This voltage is used to recharge the battery and operate the rest of the car's electrical system.

The current that recharges the battery must be **direct current** (d.c.), that is, current that always flows in the same direction. Induction as used in generators and alternators will cause **alternating current** (a.c.), a current that flows first in one direction and then in the other. This alternating current must be **rectified**, or changed to direct current, before it can be used in the car's electrical system. The generator is constructed so that the current is rectified mechanically. The alternator cannot use a mechanical rectifier. Until a practical way was developed to rectify alternator output, generators were used in cars.

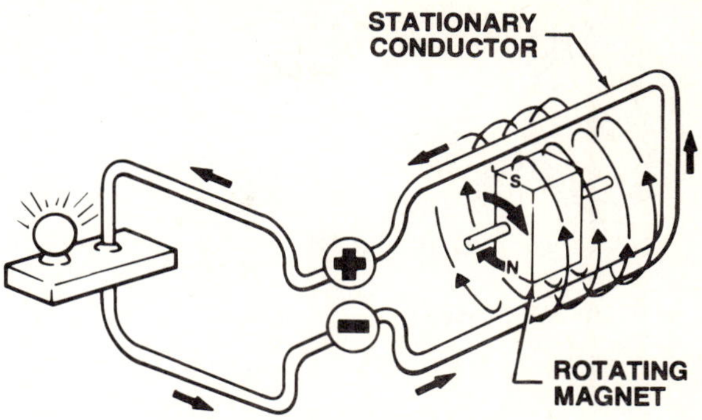

Figure 7-10. A simple alternator.

Today, alternators use **solid-state** electronic parts as rectifiers. These are called **diodes**, and they act as one-way electrical check valves. Current can flow through a diode in one direction but not in the other.

The alternator, figure 7-10, works on the same principle of induction as does the generator. A magnet, turned by the engine, rotates within stationary conductors. As the lines of

Cycling: Battery electrochemical action. One complete cycle is the operation from fully charged to discharged and back to fully charged.

Induction: The production of an electrical voltage in a conductor or coil by moving the conductor or coil through a magnetic field, or by moving the magnetic field past the conductor or coil.

Direct Current: A flow of electricity in one direction through a conductor.

Alternating Current: A flow of electricity through a conductor, first in one direction, then in the opposite direction.

Rectified: Electrical current changed from alternating (a.c.) to direct (d.c.).

Solid-State: A method of controlling electrical current flow, in which the parts are primarily made of semiconductor materials.

Diodes: Electronic devices made of semiconductor material that allow current flow in one direction but block it in the other.

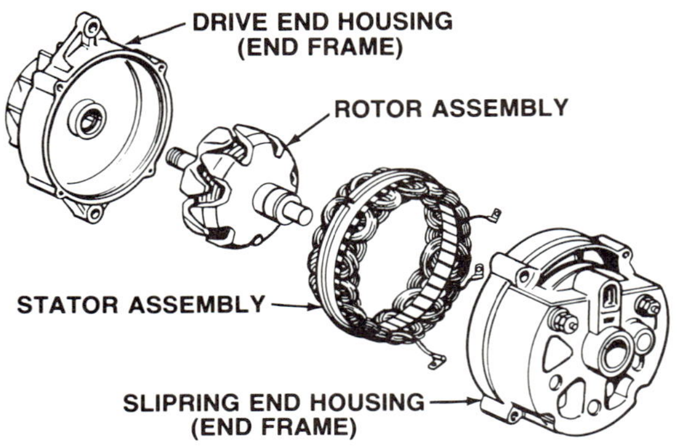

Figure 7-11. The alternator housing encloses the rotor and stator.

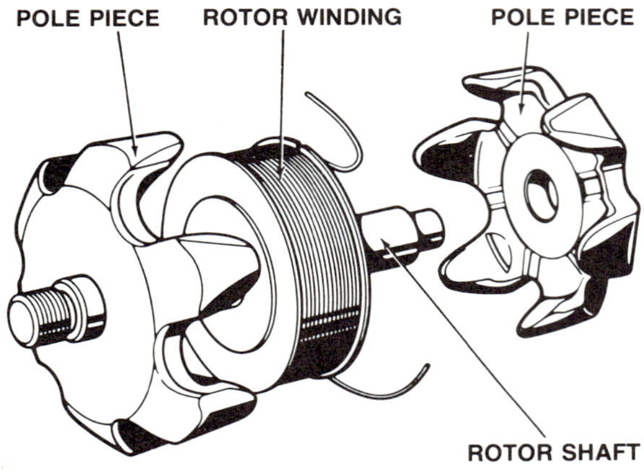

Figure 7-13. The magnetic field of the rotor is created by current flow through the rotor winding. (Bosch)

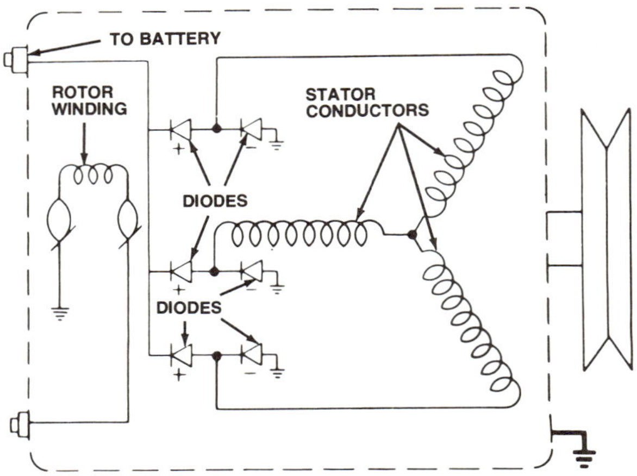

Figure 7-12. The circuit diagram of a typical automotive alternator. (Prestolite)

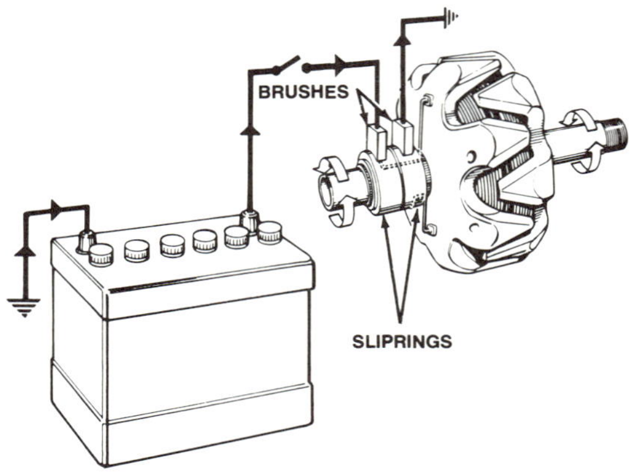

Figure 7-14. The sliprings and brushes carry current to the rotor winding.

force from the magnet cut through the conductors, voltage is induced. The following paragraphs describe how alternators work in actual automotive charging systems.

CHARGING SYSTEM COMPONENTS

The charging system used on all late-model cars consists of:

- An alternator
- A regulator to control the alternator's voltage
- An indicator to monitor the system's performance.

The following paragraphs describe these units and show the designs used by major manufacturers.

Alternator

A disassembled alternator is shown in figure 7-11. The rotor is the magnet that is turned by the engine. The stator is the collection of three looped conductors within which voltage is induced. The two end frames enclose the rotor and stator. The end frames hold the bearings that support the rotor shaft and provide the terminals where the stator conductors and rotor winding are connected to the rest of the electrical system.

In most alternators, each of the three stator conductors is connected to two diodes, figure 7-12. This allows all of the alternator's output to be rectified and used in the car's electrical system. Some heavy-duty alternators have more than six diodes. The diodes are mounted on one alternator end housing.

Whenever current flows through a conductor, a magnetic field is created around the

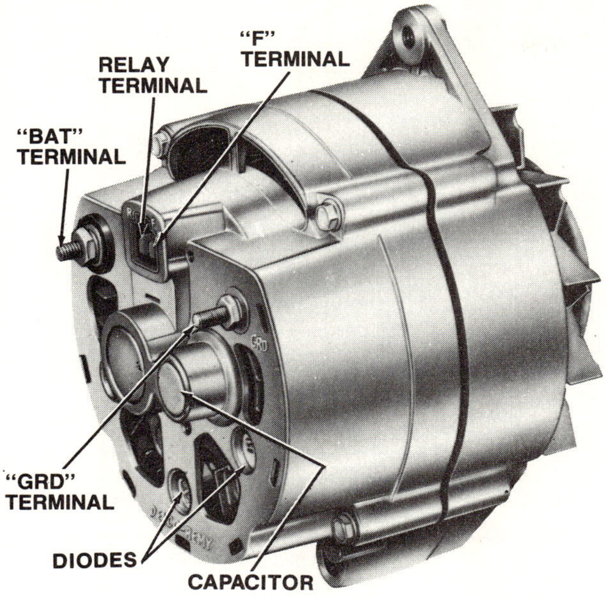

Figure 7-15. The Delcotron 10-DN. (Delco-Remy)

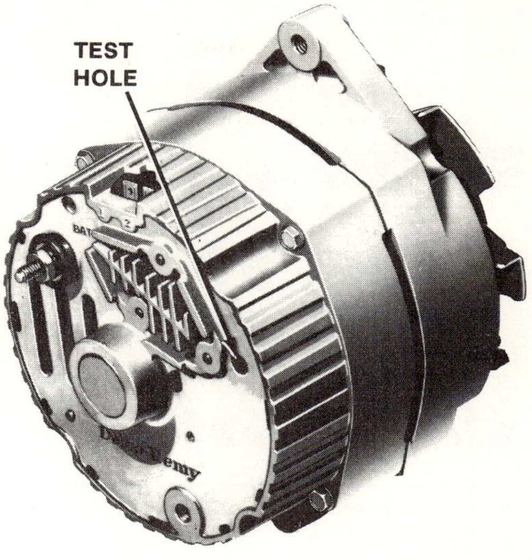

Figure 7-16. A 10-SI series alternator. (Delco-Remy)

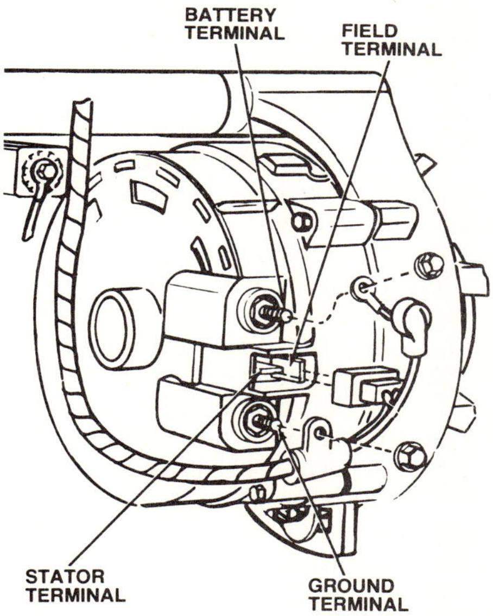

Figure 7-17. The Motorcraft side-terminal alternator. (Ford)

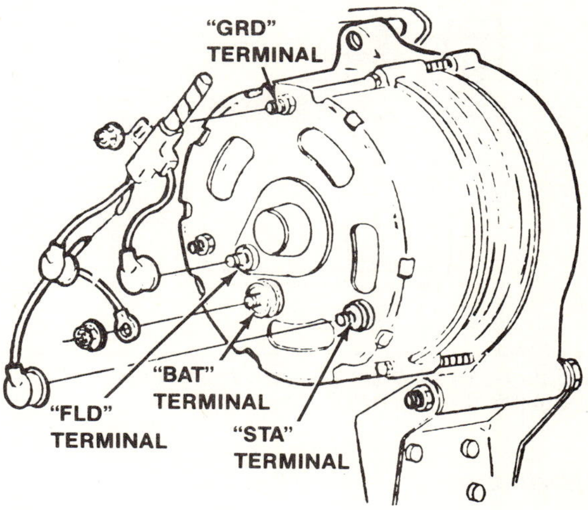

Figure 7-18. A Motorcraft rear-terminal alternator. (Ford)

conductor. The magnetic field of the rotor is created by the flow of electrical current through a coil of wire in the rotor assembly, figure 7-13. This coil is called the rotor winding, and the current flow is called the field current. Field current reaches the rotor through sliprings and brushes, figure 7-14. Each slipring is attached to one end of the rotor winding. The sliprings are mounted on the rotor shaft and rotate with the shaft. The brushes are held in place within the alternator end housing. Spring tension holds each brush against the surface of one of the sliprings. Field current flows through one brush and its slipring, through the rotor winding, and back out through the other slipring and brush.

Specific alternators

All General Motors (GM) cars have alternators made by Delco-Remy, called Delcotrons. Until the early 1970's, the most common model was the 10-DN, figure 7-15. The latest Delcotron model is the 10-SI, figure 7-16. Ford Motor Company products use Motorcraft alternators. These have terminals either on the side, figure 7-17, or on the rear, figure 7-18. Most Chrysler products use the alternator shown in figure 7-19.

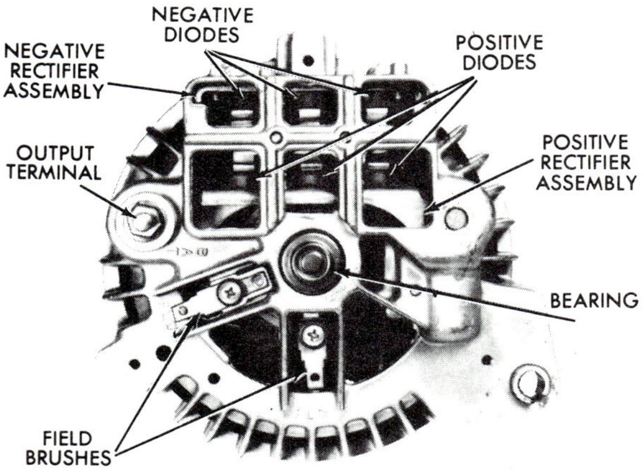

Figure 7-19. The terminals on a Chrysler standard-duty alternator. (Chrysler)

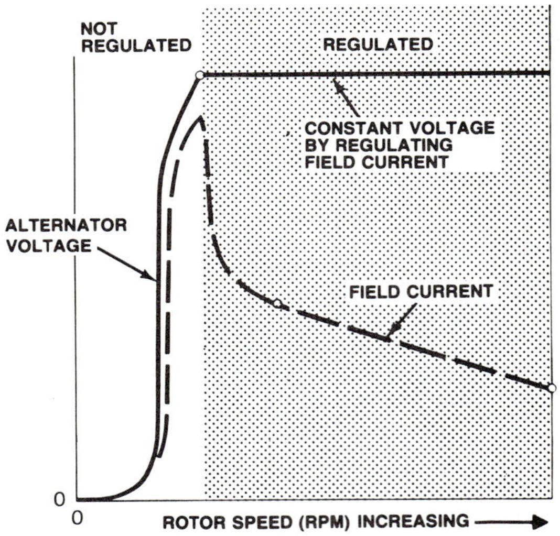

Figure 7-20. Field current is decreased as rotor speed increases to keep alternator output voltage at a constant level. (Bosch)

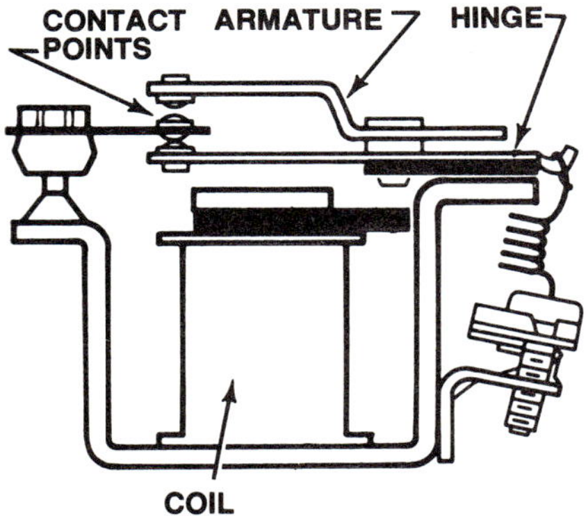

Figure 7-21. An alternator's electromagnetic voltage regulator. (Delco-Remy)

Regulator

The regulator used with an alternator has only one job: to limit the voltage strength of the alternator's output. The car's electrical system is designed to work on about 12 volts. If a greater voltage is applied to the battery and electrical circuits, they could be damaged. The regulator limits the alternator output to about 12 volts.

The regulators used with generators have two other jobs:

- Limiting the amperage of the generator's output
- Controlling the circuit between the battery and the generator so that the battery does not discharge through the generator windings when the engine is not running.

The alternator's regulator does not need to do these jobs. The battery cannot discharge through the alternator stator because the diodes do not allow current to flow in that direction. The alternator uses induction to limit its current output. When an induced voltage causes current flow in a conductor, that conductor creates its own magnetic field. As current flow increases, the magnetic field expands. This expanding magnetic field induces another voltage within the same conductor. The second voltage is a countervoltage which tends to oppose any change in the original current flow.

When induced voltage causes current flow in the conductors of the stator, a countervoltage is also formed. The countervoltage opposes any increased current flow in the stator. The more current the alternator puts out, the greater this countervoltage becomes. The alternator reaches its maximum current output when the countervoltage is great enough to totally stop any further increase in the *current*. However, the two *voltages* will continue to increase as alternator speed increases, so a method of voltage regulation is needed. This self-regulation of current does not happen in generators because of the way the conductors are arranged on the rotating center.

Voltage regulation

Alternator output voltage is directly related to the strength of the rotor's magnetic field and the speed of rotation. An increase in either of these will increase voltage output. Similarly, a decrease in either of these will decrease voltage output. Rotor speed is controlled by engine speed, and cannot be changed simply to control

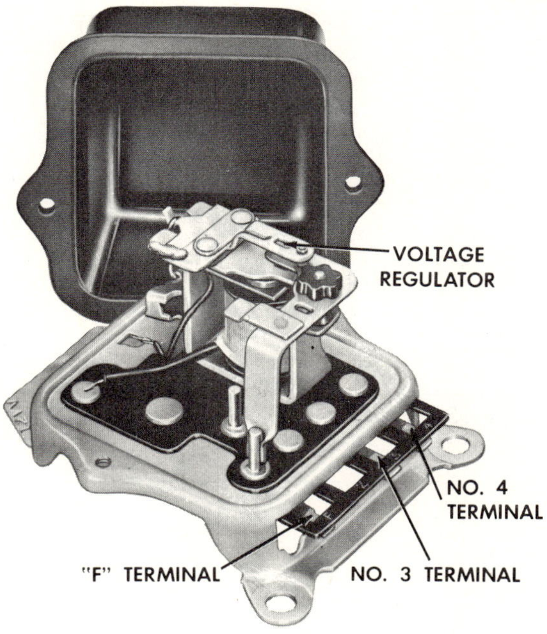

Figure 7-22. A single-unit Delco voltage regulator. (Delco-Remy)

Figure 7-24. The Ford electromagnetic regulator.

the alternator. The field strength can be controlled by changing the field current flow in the rotor winding. This is how both alternator and generator voltage regulators work. Figure 7-20 shows how the field current (dashed line) is reduced to keep alternator voltage output (solid

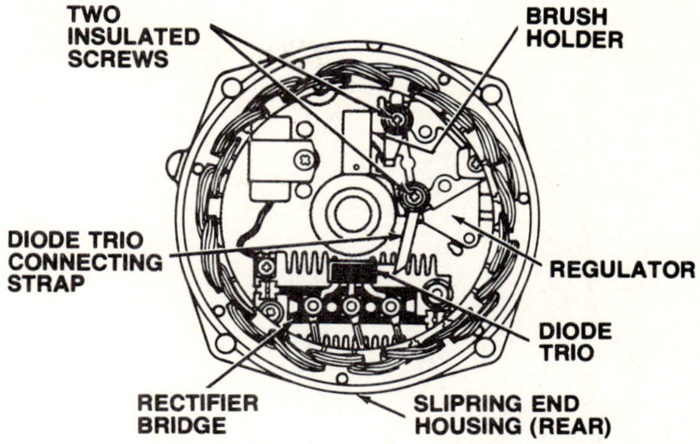

Figure 7-23. The rear end housing of the later model 10-SI Delcotron.

line) at a constant maximum, even when the rotor speed increases.

At low rotor speeds, the field current is allowed to flow at full strength for relatively long periods, and is reduced only for short periods. At high rotor speeds, the field current is reduced for long periods and flows at full strength only for short periods. The regulator that controls this flow can be an electromagnetic switch or a solid-state device.

An electromagnetic regulator is a mechanical switch that is opened and closed by magnetism. It has a coil, figure 7-21, and a hinged armature. Current flow through the coil creates a magnetic field that moves the armature. The current flow through the coil is caused by the alternator's voltage. As the alternator's voltage varies, the current flow through the regulator coil varies. Field current for the alternator must flow through contact points on the regulator armature. The position of the armature determines how much field current can reach the alternator. As the alternator's voltage varies, and the coil's magnetic field varies, the position of the armature changes and the field current flow is affected.

Solid-state voltage regulators have no moving parts. Their solid-state devices are able to act as switches. When the regulator is connected to the alternator's output circuit, the solid-state devices react to the output voltage level and control the field current flow.

Specific regulators

A mechanical regulator was used with the Delcotron 10-DN, figure 7-22. The 10-SI has a small solid-state regulator mounted within the alternator housing, figure 7-23. Ford has used a

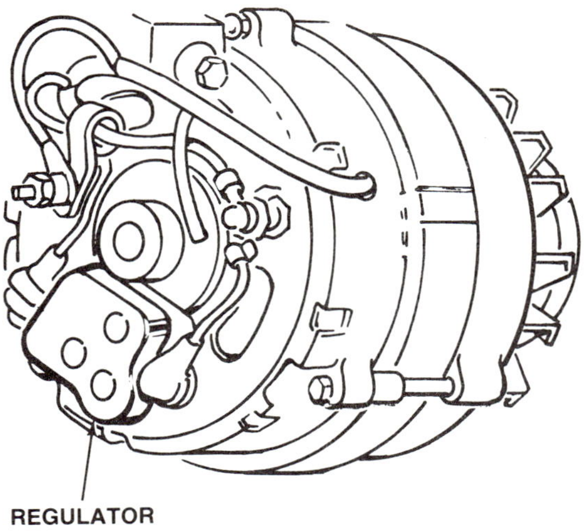

Figure 7-25. The Motorcraft 55-amp alternator has a solid-state regulator mounted on the rear housing.

Figure 7-27. A Chrysler electromagnetic voltage regulator.

mechanical regulator, figure 7-24, and a solid-state regulator mounted on the alternator, figure 7-25. Many late-model Ford products use a new solid-state regulator, figure 7-26. Chrysler has used a mechanical regulator, figure 7-27, but all late-model Chrysler products use a solid-state unit, figure 7-28.

Indicators

A charging system indicator lets the driver check on the charging system. It can be:
- A warning lamp that lights when the alternator is not supplying voltage greater than battery voltage, figure 7-29

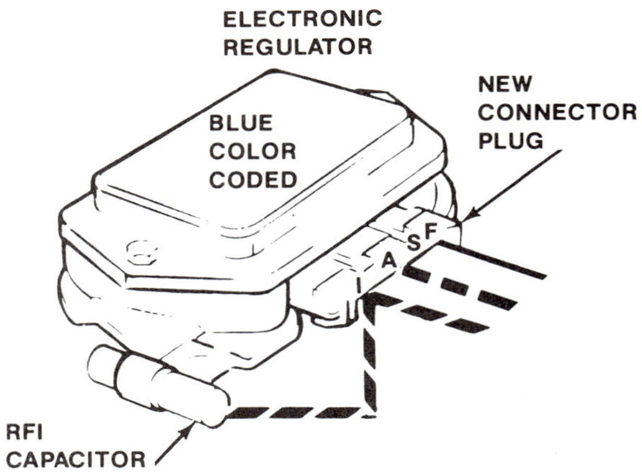

Figure 7-26. Ford's totally solid-state regulator used in a system with an ammeter. (Ford)

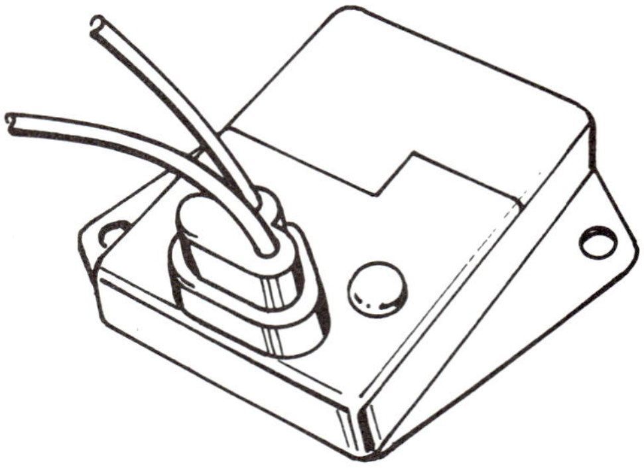

Figure 7-28. The Chrysler solid-state regulator is mounted on the firewall.

- An ammeter that shows current flow into and out of the battery, figure 7-30
- A voltmeter that shows the battery or electrical system voltage, figure 7-31.

STARTING SYSTEM OPERATION

The automotive starting system can also be called the cranking system. The only job of the starting system is to crank the engine fast enough for the engine to fire and run. The ignition and fuel systems must supply the spark and fuel for the engine to start and run, but the starting system cranks the engine to get it running.

System Circuits

The starter motor draws a great deal of energy from the battery. This current flows through large-diameter cables between the battery and

Figure 7-29. A warning lamp.

Figure 7-30. An ammeter.

Figure 7-31. A voltmeter.

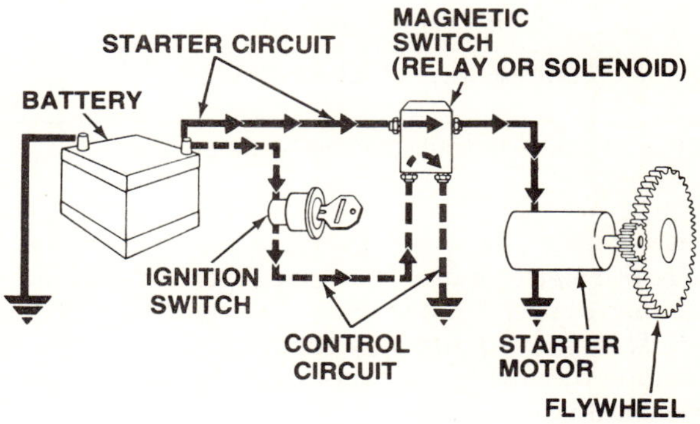

Figure 7-32. In this diagram of the starting system, the starter circuit is shown as a solid line and the control circuit is shown as a dashed line. (Delco-Remy)

the motor. The driver must be able to control the current flow through these cables with the ignition switch. If the large cables were routed from the battery to the instrument panel and then to the motor, there would be too much voltage lost from resistance (**voltage drop**). To avoid such a voltage drop, the starting system has two connected circuits. One circuit, the starter circuit, consists of the large cables between the battery and the motor. The other circuit, the control circuit, includes the ignition switch and a magnetic switch that allows the driver to open and close the starter circuit.

The starter circuit, shown as the solid lines of figure 7-32, consists of the:

1. Battery
2. Magnetic switch coil
3. Starter motor
4. Large cables.

The control circuit, shown as the dashed lines of figure 7-32, consists of the:

1. Ignition switch
2. Magnetic switch contact points
3. Normal size wiring.

When the ignition switch is closed by the driver, a small amount of current flows through a coil in the magnetic switch. This creates a magnetic field in the switch that closes a set of contact points, which are part of the starter circuit. When they are closed, current can flow directly from the battery to the starter motor. When the motor has cranked the engine and the engine is running, the driver opens the ignition switch. This stops the current flow through the magnetic switch and the magnetic field disappears. The contact points open, opening the starter circuit and stopping current flow to the motor.

Starter Motor Drives

The starter motor has a pinion gear that engages with a gear on the engine flywheel. When current flow reaches the motor, the pinion gear spins and the engine is cranked. If the starter motor were always connected to the engine, it would be turned by the engine once the engine

Voltage Drop: The measurement of the loss of voltage caused by the resistance of a conductor or a circuit device.

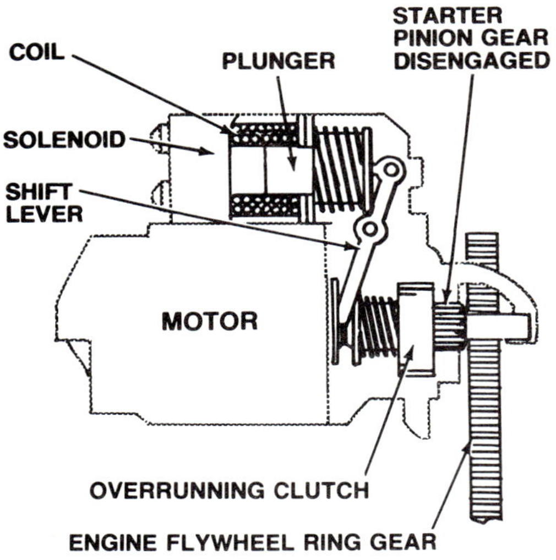

Figure 7-33. A typical solenoid-actuated drive.

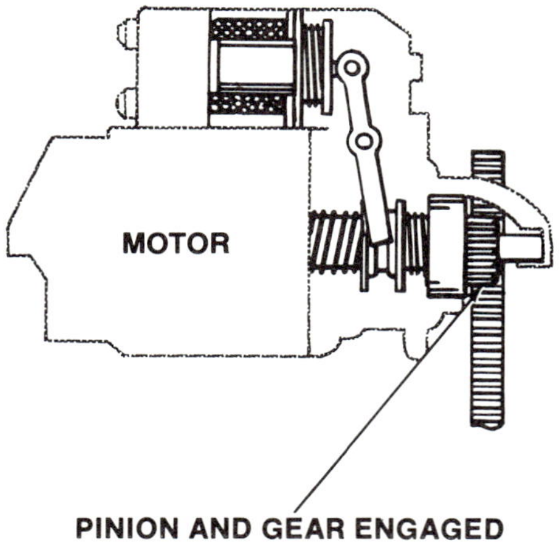

Figure 7-34. The movement of the solenoid plunger meshes the pinion gear and the flywheel ring gear.

had started and was running. This would damage the starter motor, because it is not designed to turn at such high speeds. The pinion gear must be moved into and out of mesh with the engine flywheel gear every time the engine is to be cranked.

This is the job of the starter motor drive. On most automotive starters, the drive is operated by magnetism. The most common type is a solenoid-actuated drive, figure 7-33. When current flow reaches the motor through the starter circuit, it flows through the coils of the solenoid and creates a magnetic field. The solenoid's plunger is pulled into the coils by the field force. This motion is transferred through a shift lever to the pinion gear. The gear is moved into mesh with the flywheel, figure 7-34. When the engine has started and the driver stops current flow to the starter motor, the solenoid's magnetic field disappears. A return spring forces the solenoid plunger out of the coil and the pinion gear is moved out of mesh with the engine flywheel.

To protect the motor from being turned at high speed before the driver releases the ignition switch, starter drives include an overrunning clutch, figure 7-33. This is a mechanical device, mounted on the armature shaft, that transfers the rotation of the shaft to the pinion gear so that the engine can be cranked. The clutch works through an arrangement of rollers and notches. When the motor armature is turning faster than the flywheel, the clutch transfers the motion. When the flywheel is turning faster than the motor armature, the clutch will not transfer the motion. The clutch simply spins by itself, protecting the armature.

STARTING SYSTEM COMPONENTS

We have already studied the battery, which is an important part of the starting system. The other circuit parts are the:

1. Ignition switch
2. Starting safety switch (on some systems)
3. Magnetic switch (relays or solenoids)
4. Starter motor.

Ignition Switch

The ignition switch has jobs other than controlling the starting system. The ignition switch normally has at least four positions:

- Accessories
- Off
- On (Run)
- Start.

Switches on late-model cars also have a Lock position to lock the steering wheel. All positions except Start are **detented**; that is, the switch will remain in that position until moved. When the ignition key is turned to Start and released, it will return to the On (Run) position. The Start position is the actual starter switch part of the ignition switch. It sends battery current to the magnetic switch.

Starting Safety Switch

The starting safety switch is also called a neutral start switch. It is an additional switch in the control circuit that prevents the system from operating when the automobile's transmission is in gear. Safety switches are most commonly

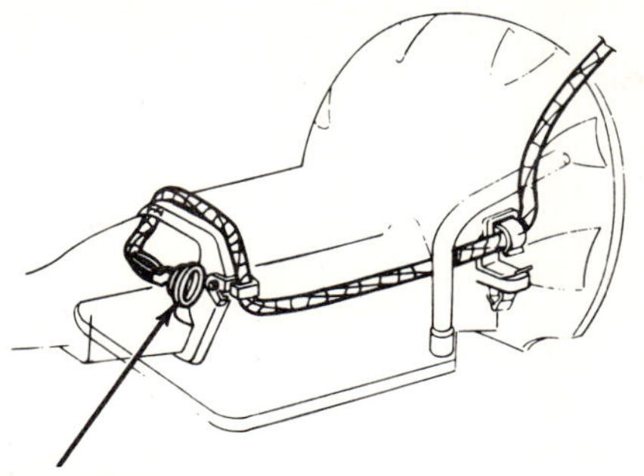

Figure 7-35. An electrical safety switch mounted on the transmission housing. (Chrysler)

Figure 7-37. The Ford starter relay or magnetic switch.

used with automatic transmissions but can also be used with manual transmissions. The safety switch can be an electrical switch, figure 7-35, that opens the control circuit if the car is in gear. It also can be a mechanical block, figure 7-36, that will not let the ignition switch turn to Start if the car is in gear.

Magnetic Switch (Relays And Solenoids)

A magnetic switch in the starting system allows the control circuit to open and close the starter circuit. The switch can be a:

- **Relay**, which uses the magnetic field to attract a hinged armature, closing the contact points
- **Solenoid**, which uses the magnetic field to pull a plunger into its coil, closing the contact points. The plunger's movement can also be used to do a mechanical job, such as shifting the starter motor gear in a solenoid-actuated drive.

GM cars use the solenoid that is mounted on the starter as the system magnetic switch. Ford

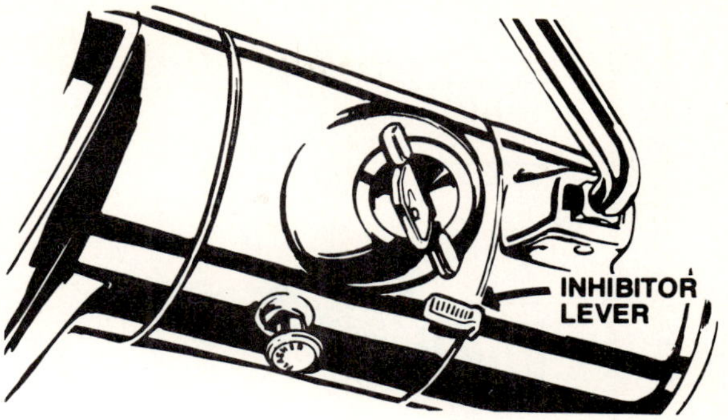

Figure 7-36. A lever on the steering column blocks the movement of the ignition key when the transmission is in gear.

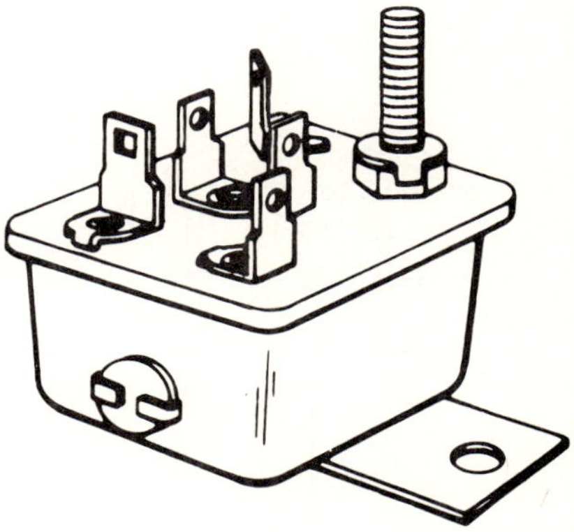

Figure 7-38. The Chrysler starter relay.

products usually have a remotely mounted solenoid, called a starter relay, figure 7-37, or they use a starter-mounted solenoid. Chrysler products have a starter relay, figure 7-38.

Detented: Positions in a switch that allow the switch to stay in that position. In an ignition switch, the On, Off, Lock, and Accessory positions are detented.

Relay: An electromagnetic switch. A relay uses a small amount of current flow to control the flow of a larger amount of current through a separate circuit.

Solenoid: An iron core with a wire coil surrounding it. The core moves when electrical current is applied to the coil. Used to convert electrical energy to mechanical energy.

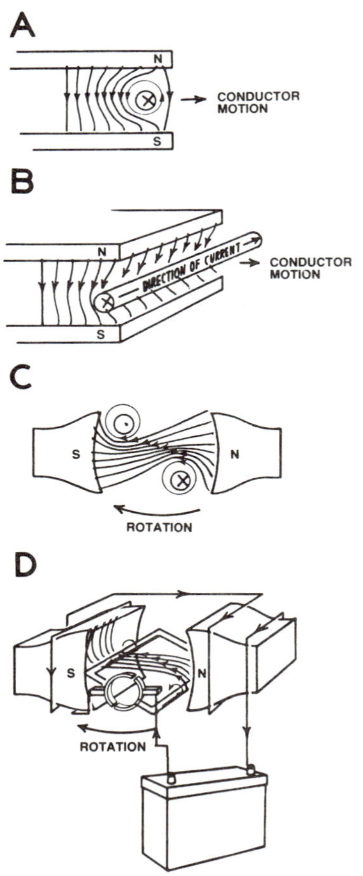

Figure 7-39. The motor principle. (Prestolite)

Starter Motor

The starter motor converts electrical energy from the battery into mechanical energy to turn the engine. It does this through the interaction of magnetic fields, figure 7-39. When current flows through a conductor, a magnetic field is formed around the conductor. If the conductor is placed in another magnetic field, the two fields will be weakened at one side and strengthened at the other side, position A. The conductor will tend to move from the strong field into the weak field, position B. Positions C and D show how a simple motor can use this movement to make the conductors rotate. An automotive starter motor has many conductors and uses a lot of current to create enough rotational force to crank the engine.

Figure 7-40 shows a cutaway view of a starter motor. The armature is the collection of conductors that will spin to crank the engine. The starter drive gear is mounted on the armature shaft. The pole pieces are the stationary magnetic fields. The motor housing encloses the armature and pole pieces, holds the bearings that support the armature shaft, and provides the terminals for connecting the motor to the rest of the starting system.

All GM starters are solenoid-actuated, figure 7-41. Chrysler uses two solenoid-actuated starters. One type has a reduction gear within the motor drive assembly, figure 7-42.

Ford uses both a solenoid-actuated starter and a movable pole shoe starter, figure 7-43. This has one pole piece that is hinged and held up out of the motor housing by spring tension. When the ignition switch is turned to Start, a small amount of current flows to the movable pole shoe. The magnetic field that is created pulls the pole piece down into the starter housing. This motion does two things:

- It shifts the pinion gear into mesh with the flywheel
- It closes contact points to connect the starter motor directly to the battery.

When the engine has started and the driver releases the ignition switch, there is no magnetic field to hold the pole piece down. The spring pulls the pole piece back up. This disengages the pinion gear and opens the contact points to stop current flow to the motor.

SUMMARY — EFFECTS OF BATTERY, CHARGING, AND STARTING SYSTEMS ON ENGINE PERFORMANCE

The symptoms caused by defects in these systems are often very similar. To make the relationships between these systems more clear, we have organized the possible defects into groups according to the engine performance symptoms that they may cause:

- No cranking or slow cranking
- Normal cranking — no starting or hard starting
- Engine quits under electrical load.

Before using these paragraphs to identify defects, make sure all other systems have been checked out first.

No Cranking Or Slow Cranking

If the engine does not turn over, or if it cranks too slowly, the starting system may be at fault. However, this could also be caused by a discharged battery, which in turn can be caused by a battery defect or a weak charging system.

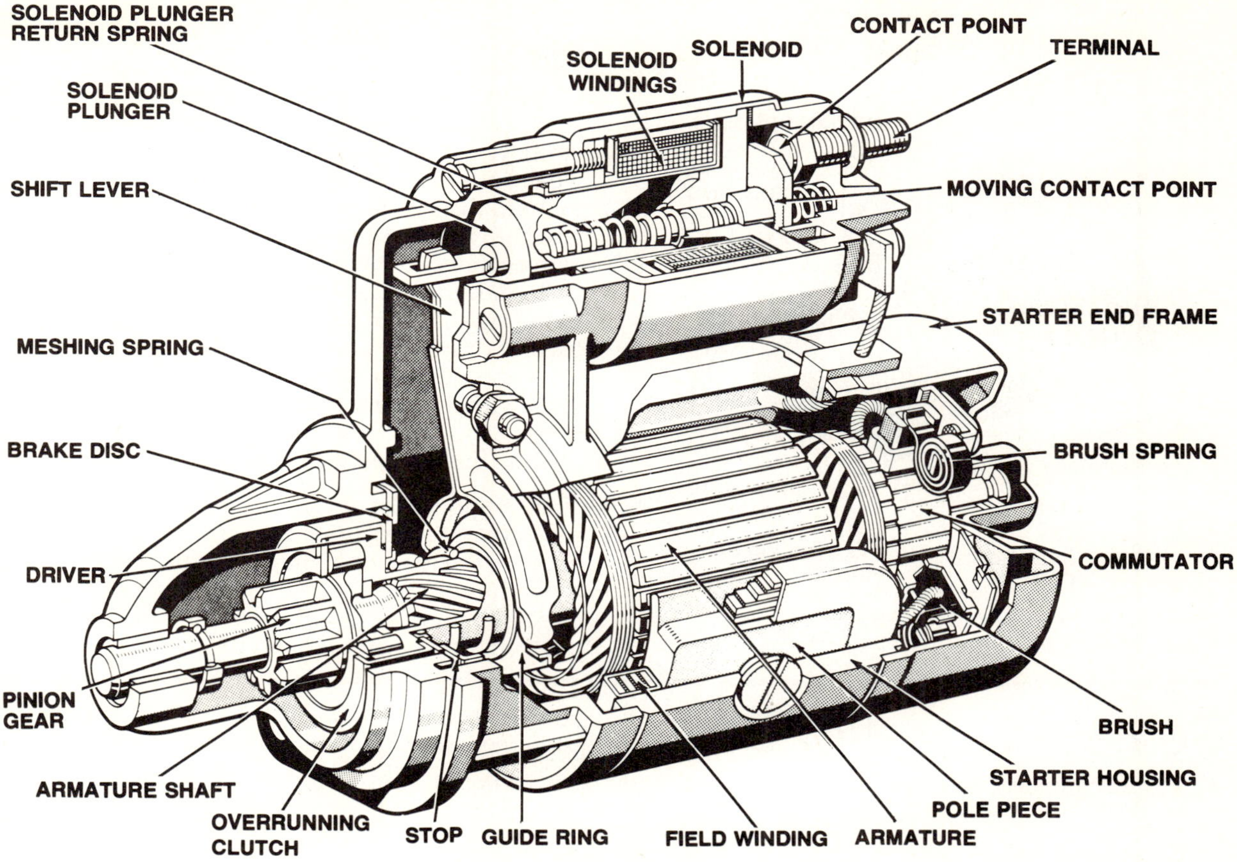

Figure 7-40. A cutaway view of a typical starter motor. (Bosch)

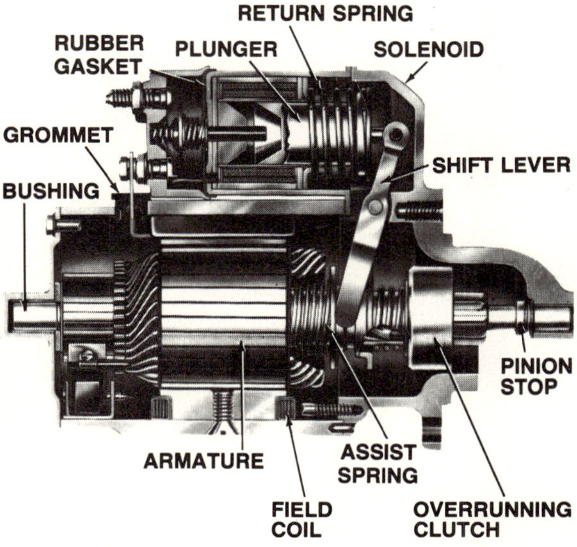

Figure 7-41. The Delco-Remy solenoid-actuated drive motor. (Delco-Remy)

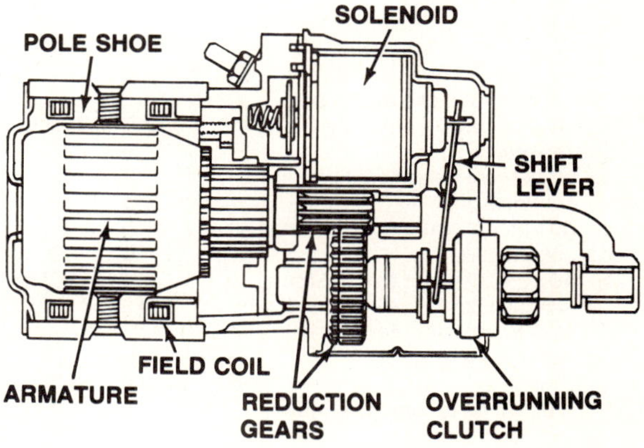

Figure 7-42. The Chrysler reduction gear starter motor.

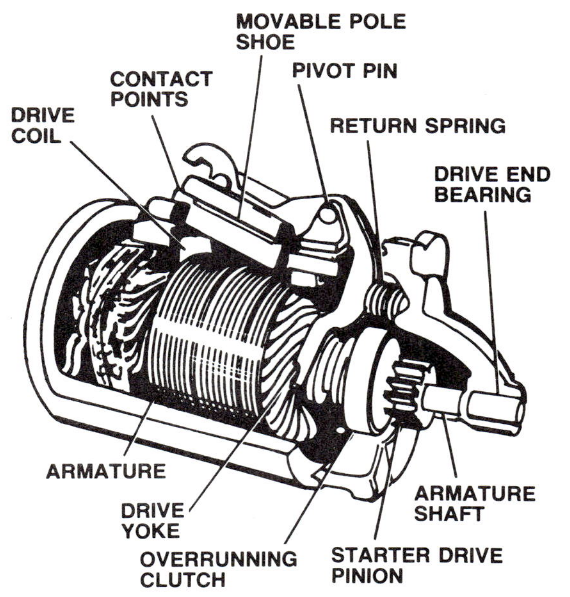

Figure 7-43. The Ford movable pole shoe starter

Normal Cranking — No Starting Or Hard Starting

When the engine is being cranked at normal speed but the engine will not start or is hard to start, the engine is probably not getting enough ignition system voltage. This can be caused by a weak battery that does not have enough energy for both the starting and the ignition systems. The battery's condition is caused either by a battery defect or by a problem in the charging system. If the battery and the charging system are all right, the problem may be excessive current draw through the starting circuit, not leaving enough energy for the ignition system even with a healthy battery.

Quits Under Electrical Load

If the engine quits at idle or under acceleration when many accessories are being used, it is because the ignition voltage has dropped below the minimum requirement. This is caused by either a battery defect or a charging system problem that has weakened the battery.

Review Questions

Select the single most correct answer.
Compare your answers with the correct answers on page 280.

1. Which of the following statements is *not* true of a replacement battery?
   a. It may have the same rating as the original battery
   b. It may have a higher rating than the original battery
   c. It may have a lower rating than the original battery
   d. It should be selected according to an application chart

2. The correct ratio of water to sulfuric acid in battery electrolyte is *approximately*:
   a. 80 percent water to 20 percent sulfuric acid
   b. 60 percent water to 40 percent sulfuric acid
   c. 40 percent water to 60 percent sulfuric acid
   d. 20 percent water to 80 percent sulfuric acid

3. Each cell of an automobile battery can produce about ______ volts.
   a. 1.2
   b. 2.1
   c. 4.2
   d. 6

4. Batteries may be rated in terms of:
   a. Ampere-hours
   b. Reserve capacity
   c. Cranking performance at 0° F
   d. Any of the above

5. Battery electrolyte is a mixture of water and:
   a. Lead peroxide
   b. Sulfuric acid
   c. Lead sulfate
   d. All of the above

6. The electrolyte in a fully charged battery will generally not freeze until the temperature drops to:
   a. 32° F
   b. 0° F
   c. −20° to −30° F
   d. −50° to −60° F

7. Wattage ratings for most 12-volt batteries range from:
   a. 200 to 400 watts
   b. 1,000 to 2,000 watts
   c. 2,000 to 4,000 watts
   d. 4,000 to 6,000 watts

8. The principal cause of battery water loss is:
   a. Spillage from the vent caps
   b. Leakage through the battery case
   c. Conversion of water to sulfuric acid
   d. Evaporation due to heat of the charging current

9. Alternator output voltage is directly related to:
   a. Field strength
   b. Rotor speed
   c. Both field strength and rotor speed
   d. Neither field strength nor rotor speed

10. Which of the following is *not* used to indicate charging system performance?
    a. Ammeter
    b. Ohmmeter
    c. Voltmeter
    d. Warning lamp

11. Which of the following statements is true?
    a. The generator rotates a looped conductor inside a fixed magnetic field
    b. The generator rotates a fixed magnet inside a fixed, looped conductor
    c. The alternator rotates a looped conductor inside a fixed magnetic field
    d. None of the above

12. During cranking, all electrical energy for the automobile system is supplied by the:
    a. Alternator
    b. Battery
    c. Generator
    d. None of the above

13. Which of the following does *not* convert mechanical energy to electrical energy?
    a. Alternator
    b. Battery
    c. Generator
    d. None of the above

14. Which of the following is true of solenoids?
    a. They use the electromagnetic field of a coil to pull a plunger into the coil
    b. They are generally used to engage the starter motor with the engine flywheel
    c. They operate with a movable plunger and usually do a mechanical job
    d. All of the above

15. Which of the following is *not* part of the starter control circuit?
    a. The ignition switch
    b. The starting safety switch
    c. The starter relay
    d. The starter motor

16. The ignition switch will *not* remain in which of the following positions?
    a. Accessories
    b. Off
    c. On (run)
    d. Start

Chapter

8

The Ignition System

The ignition system must provide a high-voltage spark in the correct cylinder of the engine at the correct time for the most efficient combustion. If any one of these is missing, the engine will not run right. This chapter shows you the complete ignition system, figure 8-1. Chapter 9 discusses basic ignition timing and timing control. Chapter 10 describes breaker-point distributor operation, ignition coils, and condensers. Chapter 11 explains the solid-state electronic ignition systems used as standard equipment by domestic carmakers.

NEED FOR HIGH VOLTAGE

Energy is supplied to the automotive electrical system by the battery. The battery can supply about 12 volts, but the voltage required to ignite the air-fuel mixture can range from 5,000 to 25,000 volts or more, depending on engine operating conditions.

This high voltage is needed to force an arc across the spark plug air gap. Required voltage levels increase when the:

1. Spark plug air gap increases
2. Engine operating temperature increases (resistance increases with greater pressure)
3. Air-fuel mixture contains less fuel (fewer volatile fuel particles)
4. Air-fuel mixture is at a greater pressure (resistance increases with greater pressure).

Since part of the ignition system's job is to provide a high-voltage spark, battery voltage must be greatly increased to meet the needs of the ignition system. This can be done by using electromagnetic induction, which we discussed in Chapter 7.

HIGH VOLTAGE THROUGH INDUCTION

A current-carrying conductor or coil is surrounded by a magnetic field. As current flow in the coil is increased or decreased, the magnetic field will expand or contract. If a second coiled conductor is placed within this magnetic field, figure 8-2, the expanding or contracting magnetic flux lines will cut the second coil. A voltage will be induced in the second coil. This transfer of energy between two unconnected conductors is called **mutual induction**.

Induction In The Ignition Coil

The ignition coil uses the principle of mutual induction to step up, or transform, low battery voltage to high ignition voltage. The ignition coil, figure 8-2, contains two windings of copper wire around a soft iron core. The primary winding is made of a few hundred turns of heavy

Figure 8-1. The components of a typical ignition system as they are installed in a car.

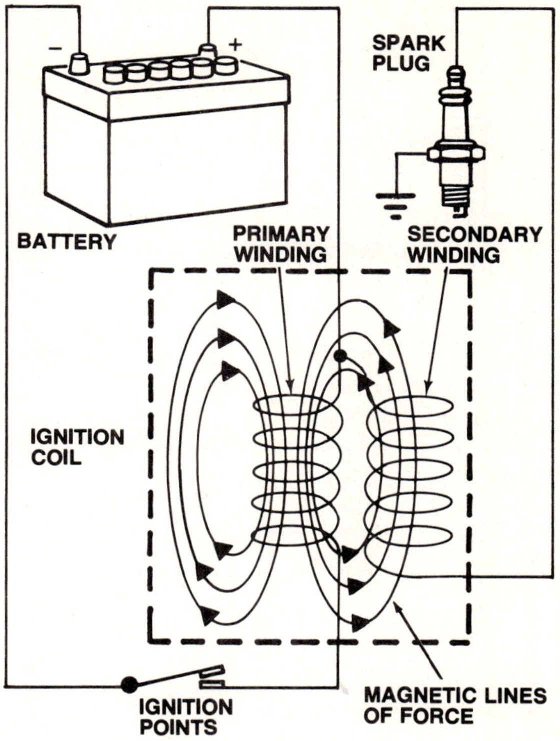

Figure 8-2. Mutual induction in the ignition coil supplies voltage to fire the spark plugs.

wire. It is connected to the battery so that current flows through it. The secondary winding is made of many thousand turns of fine wire. When current flow in the primary winding increases or decreases, a voltage is induced in the secondary winding.

The ratio of the number of turns in the secondary winding to the number of turns in the primary winding can be between 100 to 1 and 200 to 1. This ratio acts as a voltage multiplier. That is, any voltage induced in the secondary winding will be 100 to 200 times the voltage present in the primary winding.

Several factors govern the coil's induction of voltage. Only two of these factors can be controlled easily in an ignition system. Induced voltage will increase with:

- More magnetic flux lines (a stronger magnetic field caused by greater current flow)
- More rapid movement of flux lines (faster collapse of the field caused by an abrupt end to current flow).

Normal voltage applied to the coil primary winding is about 9 or 10 volts. (At high speeds, voltage may rise to 12 volts or more.) This voltage causes from 1 to 4 amperes of current to flow in the primary winding. When this current is flowing, a magnetic field builds up around the windings. Building up a complete magnetic field is called **magnetic saturation**, or coil saturation. When this current flow is stopped, the primary winding's magnetic field collapses. A greater voltage is self-induced in the primary winding by the collapse of its own magnetic field. This self-induction creates from 250 to 400 volts in the primary winding. If it develops 250 volts, and the ratio of turns multiplies this by 100, then 25,000 volts will be induced in the secondary winding. This is enough voltage to ignite the air-fuel mixture under almost all operating conditions.

This kind of ignition system, based on the induction of a high voltage in a coil, is called an **inductive-discharge ignition** system. An inductive-discharge system, using a battery as the source of low-voltage current, has been the standard automotive ignition system for 70 years.

Mutual Induction: The transfer of energy between two unconnected conductors, caused by the expanding or contracting magnetic flux lines of the current-carrying conductor.

Magnetic Saturation: The condition when a magnetic field reaches full strength and maximum flux density.

Inductive-Discharge Ignition: A method of igniting the air-fuel mixture in an engine cylinder. It is based on the induction of a high voltage in the secondary winding of the coil.

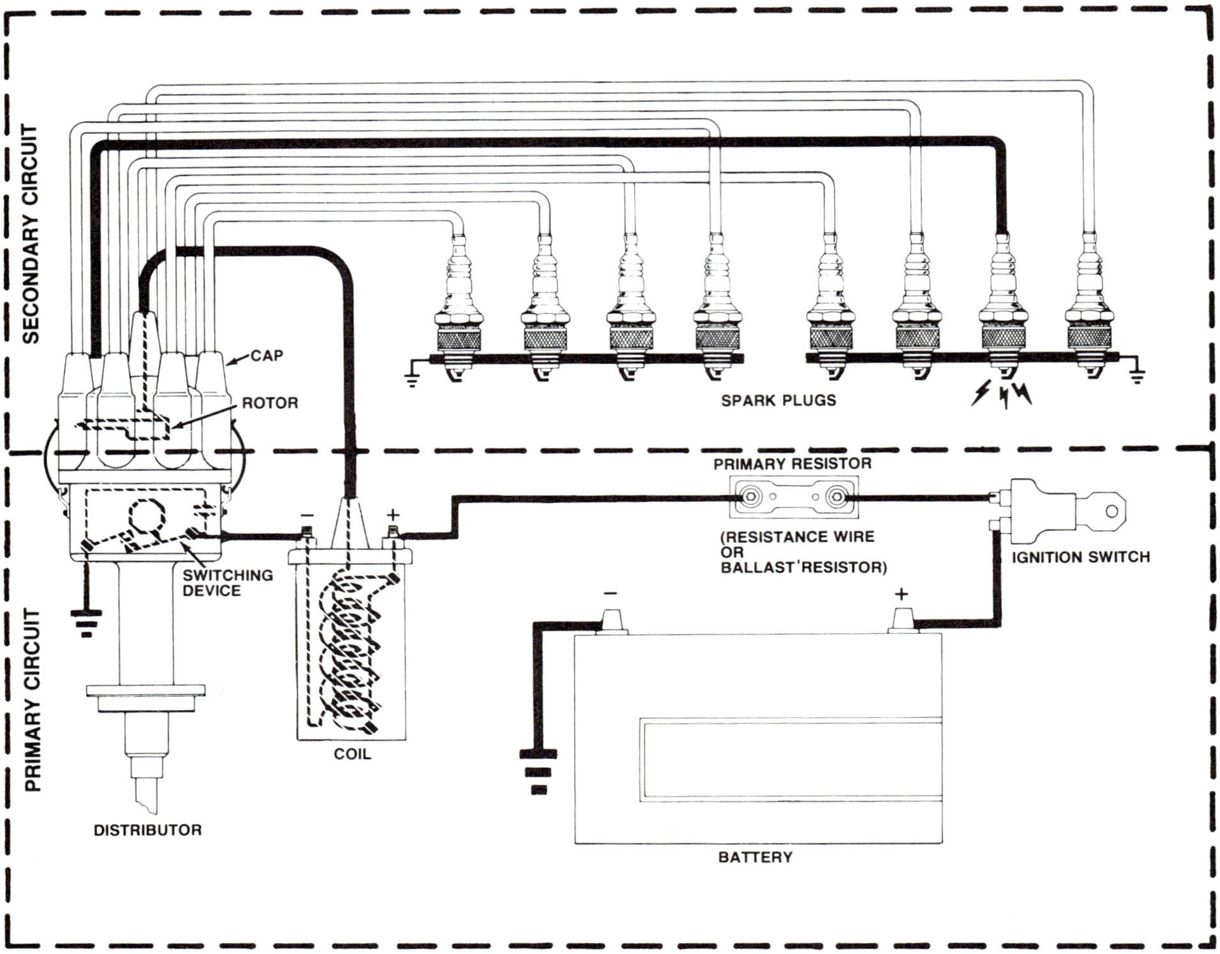

Figure 8-3. The ignition system is divided into the primary circuit and the secondary circuit. (Prestolite)

We have seen that the ignition system can transform low battery voltage into high ignition voltage. Now we can look at the ignition system circuitry to see how the system works.

BASIC CIRCUITS AND CURRENT FLOW

The ignition system, figure 8-3, consists of two interconnected circuits:

1. The primary (low-voltage) circuit
2. The secondary (high-voltage) circuit.

When the ignition switch is turned on, battery current flows:

- Through the ignition switch and the primary resistor
- To and through the coil primary winding
- Through a switching device (such as breaker points)
- To ground and the grounded terminal of the battery.

Low-voltage current flow in the coil primary winding creates a magnetic field. When the switching device interrupts this current flow:

- A high-voltage surge is induced in the coil secondary winding.
- Current flows through an ignition cable from the coil to the distributor.
- Current passes through the distributor cap, rotor, rotor air gap, and another ignition cable.
- Current flows to the proper spark plug, where it arcs to ground.

The inductive-discharge battery ignition system was invented by Charles F. Kettering in 1908. He used a set of contact points as a mechanical switch to open and close the circuit to the primary winding of the ignition coil. These contacts are opened by the rotation of a cam on the distributor shaft. They are called **breaker points** because they continually break the primary circuit.

Kettering's ignition system was quickly adopted as the standard for the automobile industry and was used virtually unchanged for over 60 years. In the early 1970's, solid-state electronic components began to replace breaker points as switching devices for the primary circuit. The electronic ignitions on late-model cars still use the inductive-discharge principles to produce a high-voltage spark. However, the primary circuits are controlled by electronic, rather than mechanical, switching devices.

PRIMARY CIRCUIT COMPONENTS

The primary circuit, figure 8-3, contains the:
1. Battery
2. Ignition switch
3. Primary (ballast) resistor (in some systems)
4. Starting bypass (in some systems)
5. Coil primary winding
6. Switching device in the distributor.

Battery

The battery supplies low-voltage current to the ignition primary circuit. This current flows when the ignition switch is in the Start or the Run position.

Ignition Switch

The ignition switch controls low-voltage current flow through the primary circuit. This current can flow when the ignition switch is in the Start or the Run position. Other switch positions route current flow to accessory circuits and lock the steering wheel in position.

Various manufacturers use different ignition switch circuitry. The differences lie in how battery current is routed to the switch. Regardless of variations, full system voltage is always present at the switch, as if it were connected directly to the battery. The following descriptions apply to late-model domestic automobiles.

General Motors automobiles draw ignition current from a terminal on the Delco-Remy starter motor solenoid, figure 8-4. Ford Motor Company systems draw ignition current from a terminal on the Motorcraft starter relay.

In Chrysler Corporation cars, ignition current comes through a wiring splice installed between the battery and the alternator. American Motors cars use the Motorcraft starter relay with one of two different alternators. Ignition current is drawn through a wiring splice between the battery and the starter relay to simplify the circuitry. Older foreign and domestic automobiles may use different connections, such as drawing ignition current from a terminal on the voltage regulator.

Ballast (Primary) Resistor

For an ignition coil to have uniform secondary voltage capabilities over a wide range of engine speeds, complete saturation of its magnetic fields must be developed at these varying speeds. Magnetic saturation depends on the amount of voltage applied to the coil and the amount of current flowing in the windings. It also depends on the length of time during which current flows.

An automobile ignition system does not operate with uniform current and voltage or current flow time. When an engine is cranking, the high current draw of the starter motor drops system voltage to about 10 volts. Therefore, the ignition coil must be able to produce enough secondary

Breaker Points: The metal contact points that act as an electrical switch in a distributor. They open and close the ignition primary circuit.

■ Charles Franklin Kettering (1876-1958)

Charles F. Kettering was a leading inventor and automotive engineer. After graduating from Ohio State University, he became chief of the inventions department at the National Cash Register Company. While there he designed a motor used in the first electrically operated cash register.

In 1909, he helped form a company called Dayton Engineering Laboratories Company, later to be known as Delco. In 1917, he became the president and general manager of the General Motors Research Corporation.

Kettering invented both the automobile self-starter and the inductive-discharge battery ignition system. The accompanying illustration is an early sketch by Kettering of his design for an ignition system. He was involved in the invention and perfection of high-octane gasoline; improvements for engines, especially diesel engines; electric refrigeration; and much more. He also helped establish the Sloan-Kettering Institute For Cancer Research in New York City.

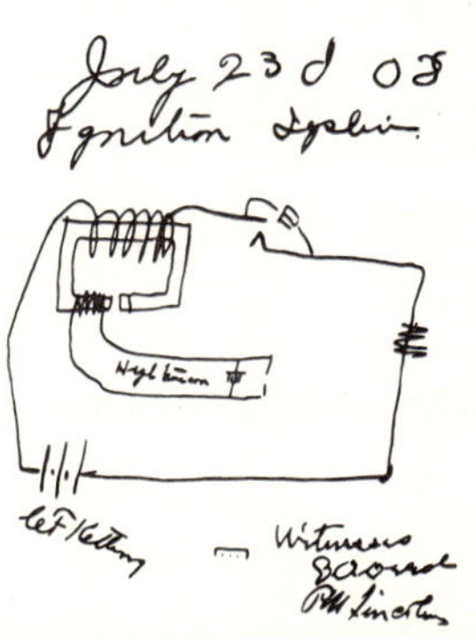

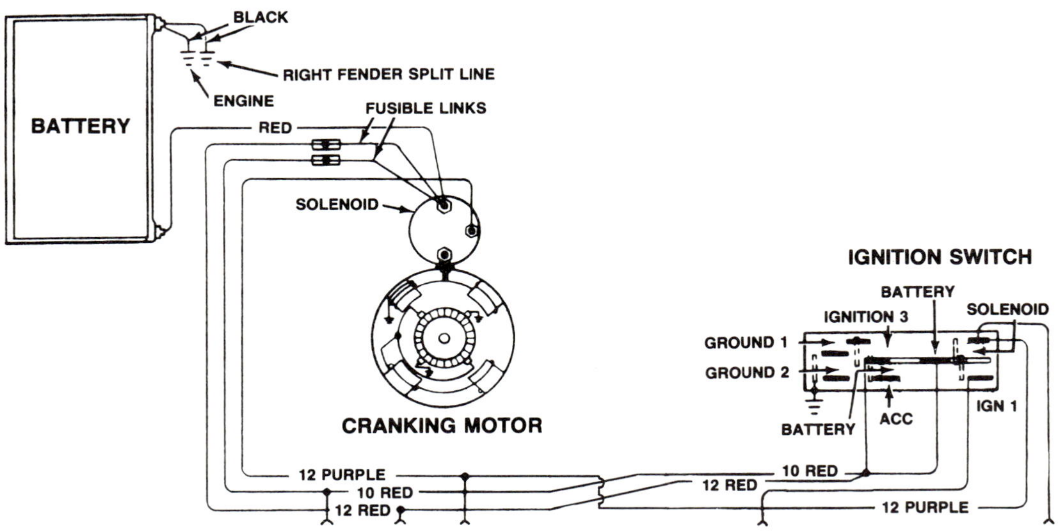

Figure 8-4. General Motors products draw ignition current from a terminal on the starter solenoid. (Buick)

The Lincoln-Zephyr V-12 Ignition System

The ignition system for the Lincoln-Zephyr V-12 engines used two coils contained in a single housing mounted on top of a distributor with two distributor caps, two sets of breaker points, a rotor with two contacts, and two condensers. The distributor was mounted on the front of the engine and was connected directly to the camshaft.

Looking at the distributor from the driver's seat position, the right-hand coil and a fixed set of breaker points fired the right bank of cylinders, numbers 2-4-6-8-10-12. The left-hand coil and an adjustable set of breaker points fired the left bank of cylinders, numbers 1-3-5-7-9-11.

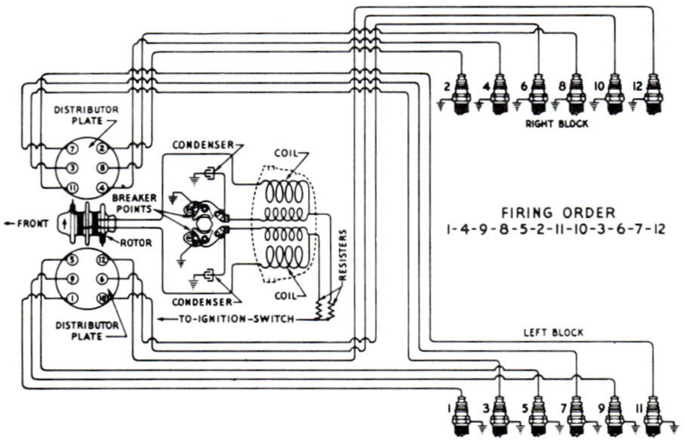

voltage to fire the engine with only 10 volts of primary voltage applied. Also, at high engine speeds, primary current flow time is reduced to only a few milliseconds. Therefore, uniform coil saturation must develop under extremes of low voltage and short current flow time. To achieve this, most 12-volt ignition coils are designed to operate on 9 or 10 volts under most conditions.

As we have mentioned, the starter motor drops ignition system voltage to about 10 volts when the engine is cranking. As soon as the engine starts, however, system voltage rises to 12 volts or more. To keep the ignition primary voltage at the desired level, a resistor is installed in the primary circuit. This is called the **ballast (primary) resistor**, figure 8-5.

The ballast resistor compensates for changes in voltage and current flow caused by engine speed and temperature changes. The resistor provides about one-half of the total primary circuit resistance (the coil is the other half).

At low speeds, current flows through the circuit for relatively long periods of time. The current flow heats the resistor, and its resistance increases. This drops the applied voltage at the coil. At higher speeds, the breaker points open more often and current flows for shorter periods of time. The ballast resistor cools, and its resistance drops. Higher voltage is applied to the coil, but the shorter current flow duration results in about the same magnetic saturation of the coil.

The ballast resistor simply evens out the voltage and current flow of the primary circuit. In doing so, it reduces peak voltage at the coil and thus reduces current flow that would burn the

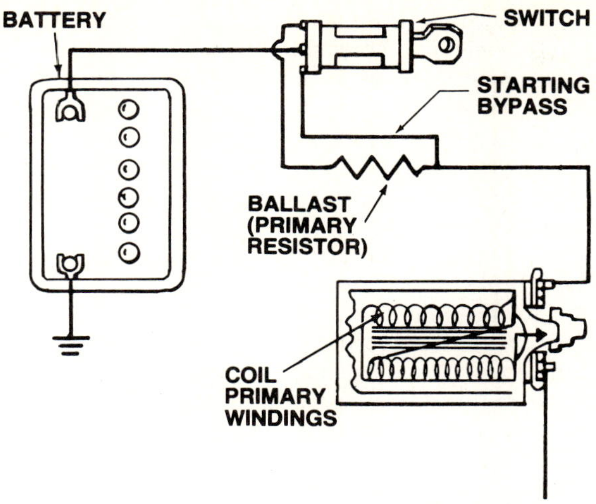

Figure 8-5. A ballast (primary) resistor can protect the primary circuit from excessive voltage.

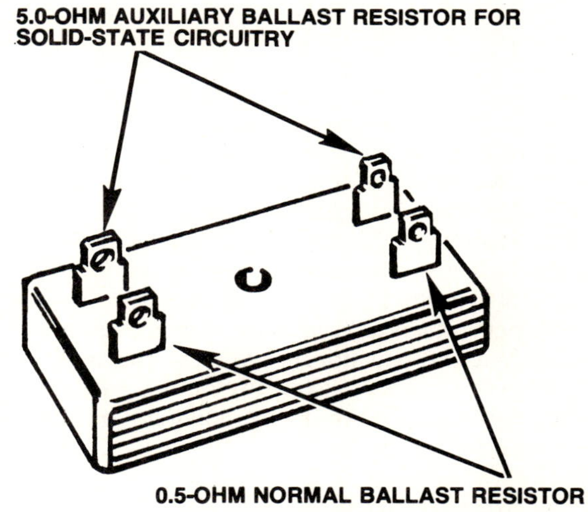

Figure 8-7. Chrysler's dual ballast resistor protects both the primary circuit and the solid-state ignition circuitry.

breaker points faster. This is its most noticeable effect.

During cranking the ballast resistor is bypassed because the starter motor already drops the system voltage to the desired level. This is done with a low-resistance starting bypass circuit in parallel with the ballast resistor, figure 8-5. When the engine starts, the bypass circuit is opened, and primary current flows through the resistor. Ignition primary voltage is reduced to the desired level. Bypassing the ballast resistor at times other than cranking would cause rapid burning of the breaker points and could damage the coil primary winding.

Although late-model domestic cars use electronic ignition with no breaker points, a ballast

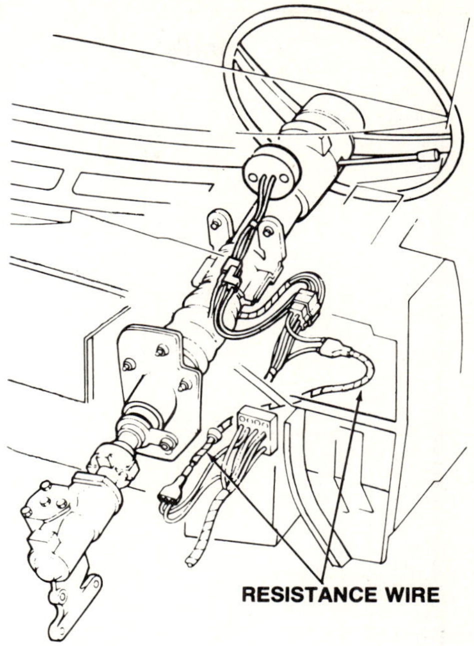

Figure 8-6. Ford products have a length of resistance wire installed near the ignition switch. (Ford)

resistor is sometimes installed to stabilize primary voltage and current flow. The following paragraphs describe the installation in late-model domestic cars, with both breaker-point and electronic ignitions.

Most Ford products (except with Dura-Spark I electronic ignitions) have a length of resistance wire installed near the ignition switch, figure 8-6. Chrysler products with electronic ignitions have two ballast resistors in one ceramic holder mounted on the firewall, figure 8-7. A 0.5-ohm resistor is connected between the ignition switch and the coil; a 5-ohm resistor is part of the electronic ignition control module circuitry. Chrysler products with breaker-point ignition and 1977 318 V-8 and all 1978 and later Lean-Burn systems have only the 0.5-ohm ballast resistor. General Motors and 1975-77 American Motors cars do not use ballast resistors with their electronic ignition systems. However, GM and AMC breaker-point ignitions have primary resistors in the form of resistance wires in the car wiring harnesses. AMC 1978 ignitions have a resistance wire.

Ballast (Primary) Resistor: A resistor in the primary circuit that stabilizes ignition system voltage and current flow.

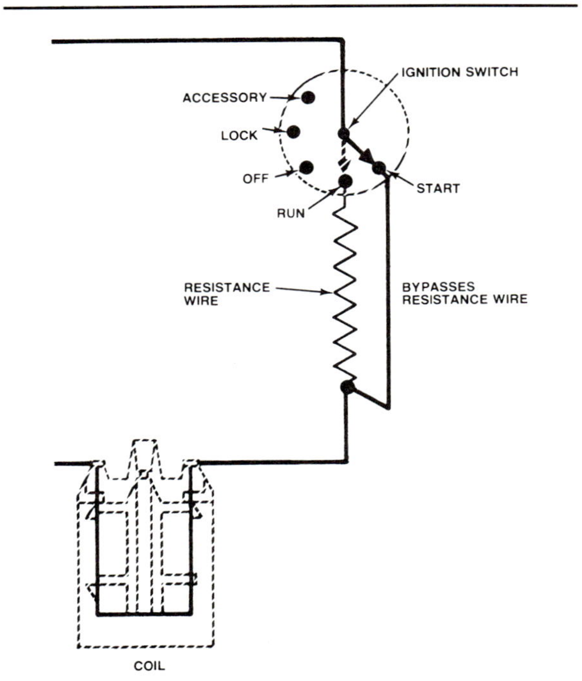

Figure 8-8. The ignition switch can control the starting bypass circuit.

During normal engine operation, the ballast resistor reduces coil primary voltage to about 9 or 10 volts from the charging system's 14-volt level. During cranking, battery voltage decreases to about 10 volts because of the great current demand of the starter motor. When the cranking system is operating, the ballast resistor is bypassed to provide full available battery voltage to the primary circuit. The ignition breaker points and coil are not damaged by this, because:

- The ballast resistor is bypassed for a very short period.
- The battery voltage available during cranking is already reduced to a safe level.

Bypassing the ballast resistor at times other than cranking would cause very rapid burning of the ignition breaker points and could damage the coil primary winding.

Starting Bypass

When the ballast resistor is bypassed during cranking, battery current flows to the primary circuit through a parallel circuit branch called the **starting bypass**. Current flow through this parallel branch can be controlled either by the ignition switch or by the starter relay or solenoid.

When the starting bypass is controlled directly by the ignition switch, figure 8-8, the ballast resistor is connected between the Run position contacts and the coil primary winding. When the ignition switch is in the Start position, full battery voltage is applied to the coil primary winding. Larger Ford automobiles and pre-1977 Chrysler products use this starting bypass method.

When the starting bypass is controlled by a starter relay or solenoid, figure 8-9, the ballast resistor is again connected between the Run ignition switch contacts and the coil. When the ignition switch is turned to Start, current flow through the starter relay or solenoid closes a set of contact points. Battery current will flow through the relay or solenoid to the coil primary winding through a parallel circuit.

Small Ford automobiles and 1977 and later Chrysler products control the starting bypass through the starter relay. The starting bypass on GM cars is controlled through the starter solenoid.

Coil Primary Winding

The coil primary winding, figure 8-10, is made of about 100 to 150 turns of a relatively heavy copper wire. The coil turns are insulated from each other by a thin coat of enamel. The two ends of the winding are connected to two terminals on the top of the coil. With a **negative-ground electrical system**, the coil positive (+) terminal is connected to the positive terminal of the battery; the coil negative (−) terminal is connected to the ignition breaker points and through the points to ground. With a **positive-ground electrical system**, the coil primary connections are reversed.

Switching Devices

The magnetic field of the coil primary winding must collapse totally in order to induce a high voltage in the secondary winding. For the field to collapse, current flow through the primary winding must stop very rapidly. Current flow must then start and stop again, to induce the next high-voltage discharge. The primary circuit needs a switching device that *rapidly* breaks and completes the circuit to start and stop current flow. For more than 60 years, ignition systems have used breaker points as a mechanical switch. In recent years, solid-state electronic devices have replaced breaker points.

Breaker points

The ignition breaker point assembly, figure 8-11, includes the:

1. Fixed contact
2. Movable contact
3. Movable arm

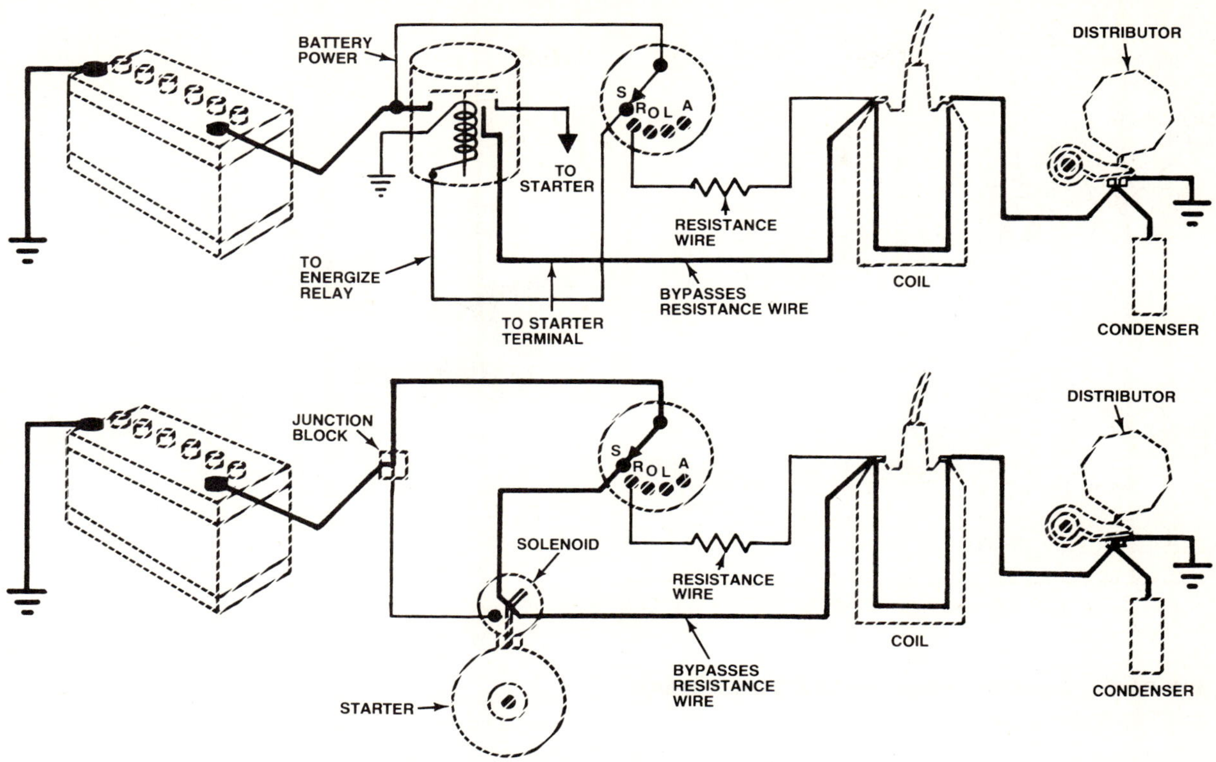

Figure 8-9. The starting bypass circuit can be controlled by the starter relay or the solenoid.

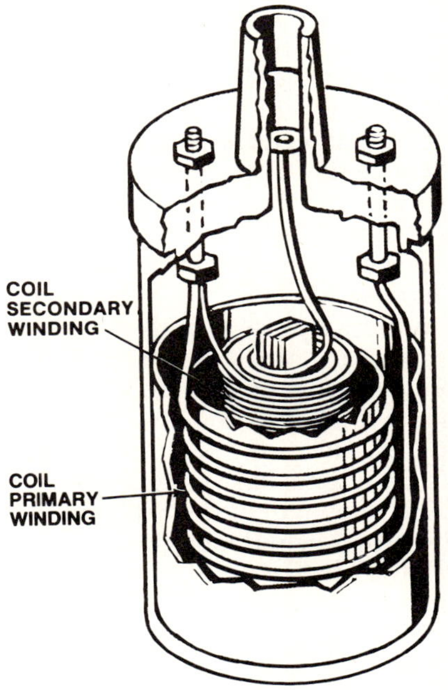

Figure 8-10. An ignition coil, cut away to show the primary and secondary windings.

4. Rubbing block
5. Pivot
6. Spring
7. Breaker plate.

Both breaker points are made from tungsten, a metal that is extremely hard and has a high melting point. The fixed contact is grounded through the distributor housing. The movable contact is insulated from the distributor housing and is connected to the negative terminal of the coil primary winding. Because current flows from the movable to the fixed contact, the movable contact can be labeled + and the fixed contact can be labeled − in a negative-ground system.

Starting Bypass: A parallel circuit branch that bypasses the primary ballast resistor during cranking.

Negative-Ground Electrical System: An automotive electrical system in which the battery negative terminal is connected to ground.

Positive-Ground Electrical System: An automotive electrical system in which the battery positive terminal is connected to ground.

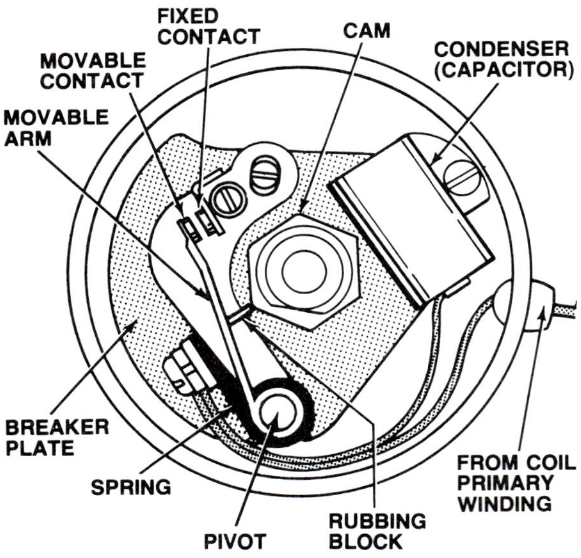

Figure 8-11. Ignition breaker (contact) point assembly.

■ Dual-Point, Dual-Coil, Dual-Plug Ignition

Some early Nash automobiles used a "Twin Ignition" system. This system had two ignition coils, two sets of spark plugs and cables, a distributor with 16 plug terminals and two coil wire terminals, a rotor with offset tips, two sets of breaker points and two condensers.

The engines were of an inline design with the spark plugs installed on each side of a cylinder. The overhead valves were located directly over each cylinder in a vertical position.

The breaker points were synchronized by using a dual-bulb test lamp. With the ignition switch on, the distributor cam was turned to just break a stationary set of points, thereby lighting one bulb. A movable set of points was then adjusted to break contact, and light the second bulb, at the same instant. This adjustment assured that both sparks would occur in a cylinder at the same time.

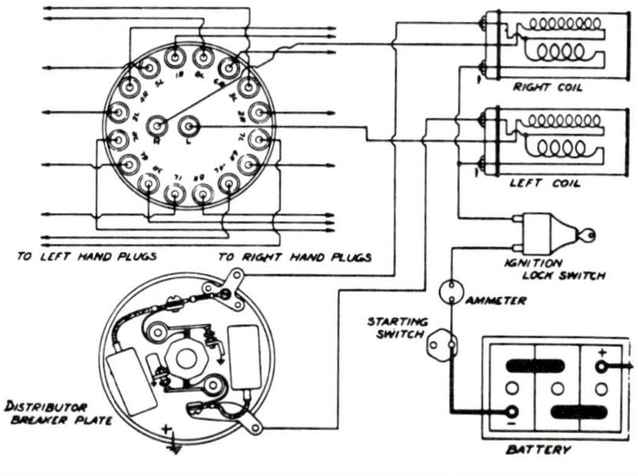

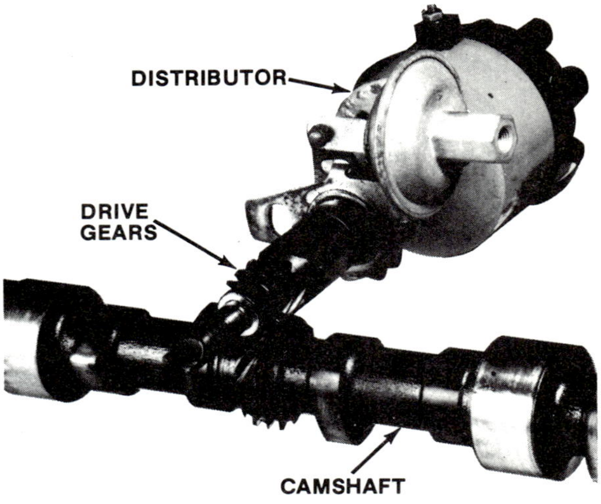

Figure 8-12. The gear on the distributor shaft is driven by another gear on the engine camshaft.

The movable contact is mounted on a movable arm. The arm also holds a rubbing block, a small piece of plastic or other synthetic nonconductive material that rides on the surface of the distributor cam. As the cam rotates, the raised portions, or lobes, push the arm to open the breaker points. The spring closes the points when the cam lobes move away from the rubbing block.

The pivot and spring control the movement of the arm. The entire assembly is mounted on the breaker plate, which is attached to the distributor housing with screws.

The distributor cam is mounted on the centrifugal advance shaft, or driven assembly, which is driven through the centrifugal advance weights by the distributor shaft. The distributor shaft is driven by the engine camshaft through gears, figure 8-12. Every opening of the breaker points induces a pulse of high-voltage current in the secondary circuit, and a spark plug fires. The breaker points must open as many times during one rotation of the cam as the engine has spark plugs. The cam, therefore, has as many lobes as the engine has spark plugs. The cam and points shown in figure 8-11 would be installed with a 6-cylinder engine.

When the breaker points open, voltage is still present at the movable arm. This voltage can cause current to arc across the opening point gap, figure 8-13. This would damage the points and affect the ignition system's operation. To avoid this, a **condenser**, a type of **capacitor**, is connected to the movable arm. The condenser consists of two conducting surfaces separated by an insulator. A condenser can store voltage without using any of the voltage. The voltage at

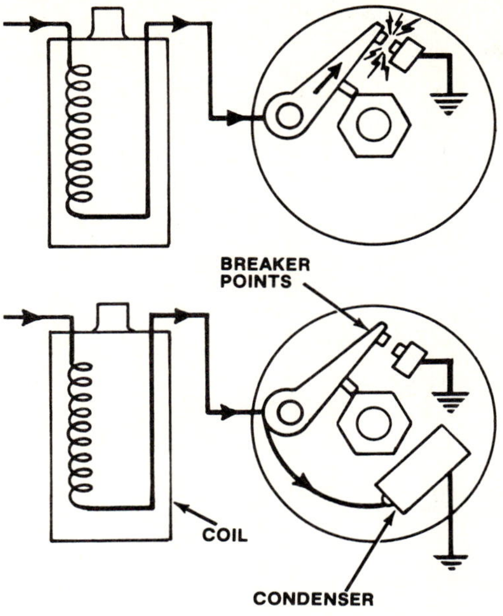

Figure 8-13. The condenser prevents arcing as the breaker points open.

the movable point arm is stored in the condenser instead of causing an arc across the point gap. We will study condensers in more detail in Chapter 10.

Solid-state switching devices

Although there has been continuing development and improvement of the breaker-point ignition system, it has difficulty meeting today's exhaust emission control standards.

Emission control devices not only require maximum performance from an ignition system, but they also require *consistent* performance. Since breaker points wear during normal operation, ignition system settings and performance change. However, solid-state switching devices do not wear. Therefore, ignition performance remains consistent, and emissions can be effectively controlled for longer periods.

Major domestic manufacturers now install solid-state ignition systems in their automobiles. We will study the more common solid-state ignition systems in detail in Chapter 11. The following paragraphs explain the general operation of the most common type of solid-state switching device used by domestic manufacturers.

A solid-state control module is responsible for switching the primary current flow on and off. The module must be signalled *when* to turn the flow off. The signal is a very slight electrical current produced by a magnetic pulse generator. This pulse generator is usually installed in the distributor housing, where breaker points

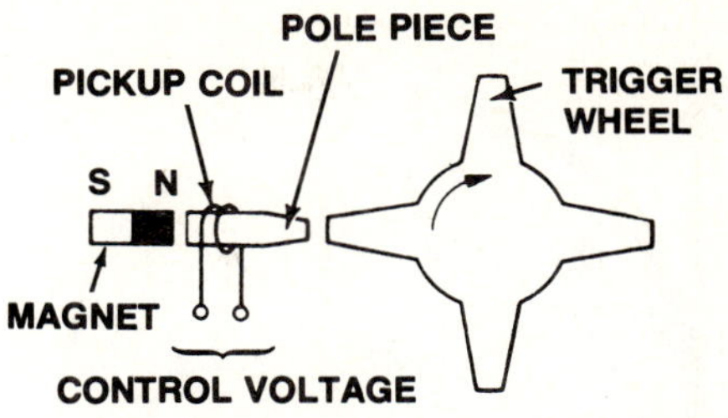

Figure 8-14. A simple magnetic pulse generator. (Bosch)

would have been. The pulse generator, figure 8-14, consists of a trigger wheel, a permanent magnet, a pole piece affected by the permanent magnet, and a pickup coil wound around the pole piece. The only moving part is the trigger wheel, which rotates as a distributor cam would.

The trigger wheel is made of steel that cannot be permanently magnetized, but which has a low **reluctance**. That is, it provides a low-resistance path for magnetic force lines. The trigger wheel has as many teeth as the engine has cylinders, except for the Buick uneven-firing V-6 engine with high energy ignition (HEI).

As the trigger wheel rotates, its teeth come near the pole piece. Magnetic lines from the pole piece concentrate in the low-reluctance wheel, increasing the strength of the magnetic field and inducing a voltage in the pickup coil. The pickup coil is connected to the electronic control module, which senses this voltage and switches the primary current off. Each time a tooth of the trigger wheel comes near the pole piece, the control module is signalled to switch the primary current off. Solid-state circuitry in the module determines when the primary current will be turned on again.

Condenser: A capacitor used in breaker-point ignition systems to prevent arcing across the points.

Capacitor: A device that can store voltage without affecting the voltage in any way. Formed by bringing two conductive surfaces close together, separated only by an insulator. Also called condenser.

Reluctance: An object's resistance to magnetic lines of force. Magnetic lines will concentrate in areas of low reluctance, and avoid areas of high reluctance.

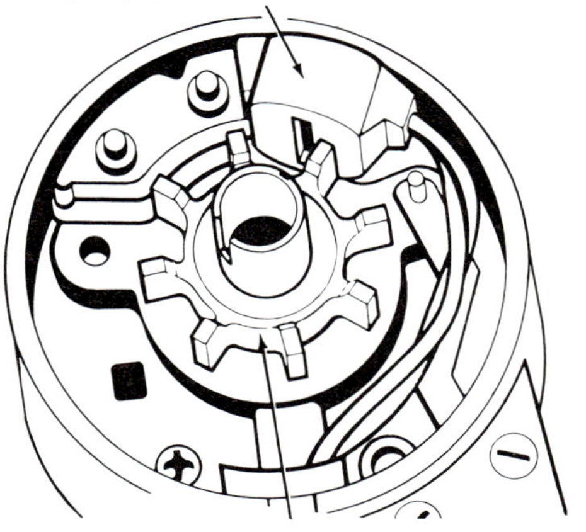

Figure 8-15. A magnetic pulse generator installed in the distributor housing. (Ford)

The simple pulse generator shown in figure 8-14 would be installed in a 4-cylinder engine. The typical construction of a pulse generator for an 8-cylinder engine is shown in figure 8-15.

We have used the terms "trigger wheel" and "pickup coil" in describing the magnetic pulse generator. Various manufacturers have different names for these components, but all serve the same purpose:

Manufacturer	*Stationary Pickup Coil*	*Rotating Trigger Wheel*
AMC	Sensor	Trigger Wheel
Chrysler	Pickup Coil	Reluctor
Ford	Stator	Armature
General Motors	Pole Piece and Magnetic Pickup	Timer Core

SECONDARY CIRCUIT COMPONENTS

The secondary circuit must conduct surges of high voltage. To do this, it has large conductors and terminals and heavy-duty insulation. The secondary circuit consists of the:

1. Coil secondary winding
2. Distributor cap and rotor
3. Ignition cables
4. Spark plugs.

The secondary circuit has the same components and function whether the primary circuit uses breaker points or a solid-state switching device.

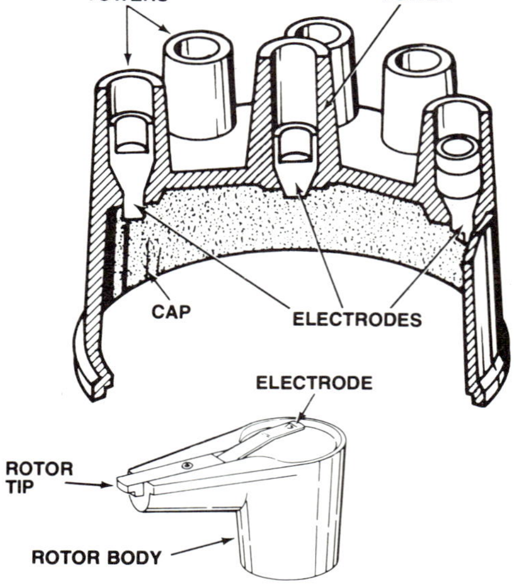

Figure 8-16. A distributor rotor and a cutaway view of the distributor cap.

Coil Secondary Winding

The coil secondary winding, figure 8-10, is made from 15,000 to 30,000 turns of a very fine copper wire, which are insulated from each other by a thin coat of enamel. One end of the secondary winding is connected to the primary winding, and the other end is connected to the high-voltage terminal on the top of the coil.

Distributor Cap And Rotor

The distributor cap and rotor, figure 8-16, receive high-voltage current from the coil secondary winding. Current enters the distributor cap through the central terminal, called the coil tower. The rotor carries the current from the coil tower to the spark plug electrodes in the rim of the cap. The rotor is mounted on the distributor shaft and rotates with it, so that the rotor electrode moves from one spark plug electrode to another in the cap to follow the specified firing order.

Distributor rotor

The rotor is keyed to the distributor shaft to maintain its correct relationship with the shaft and the spark plug electrodes of the cap. The key may be a flat section or a slot in the top of the shaft. Most rotors are hand-pressed onto the shaft. Many Delco-Remy distributors, figure 8-17, have the rotor keyed to the shaft by two locators and secured by two screws.

The rotor is made of silicone plastic, **bakelite**,

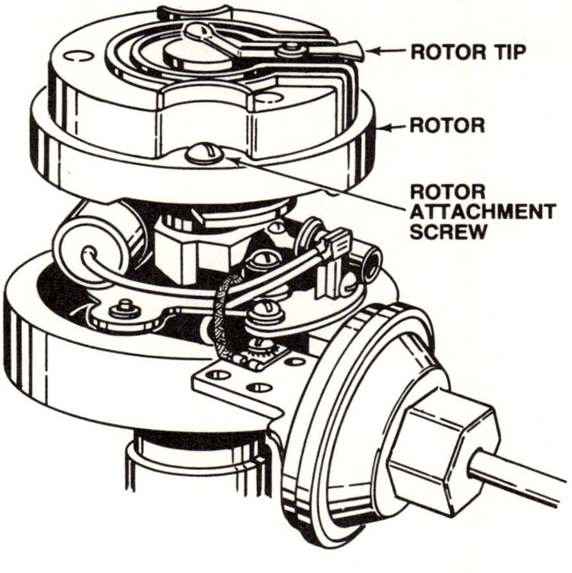

Figure 8-17. A Delco-Remy V-6 or V-8 rotor.

or a similar synthetic material that is a very good insulator. A metal electrode on top of the rotor conducts current from the carbon terminal of the coil tower.

Rotor air gap
An air gap of a few thousandths of an inch exists between the tip of the rotor electrode and the spark plug electrode of the cap. If they actually touched, both would wear very quickly. Because the gap cannot be measured when the distributor is assembled, it is usually described in terms of the voltage required to create an arc between the electrodes. Only about 3,000 volts are required to create an arc across a normal air gap, but some Delco-Remy distributors require as much as 9,000 volts. This adds additional resistance to the secondary circuit.

Distributor cap
The distributor cap is also made of silicone plastic, bakelite, or a similar material, which resists chemical attack and protects other distributor parts. Metal electrodes in the spark plug towers and a carbon insert in the coil tower provide electrical connections with the ignition cables and the rotor electrode, figure 8-18. The cap is keyed to the distributor housing, and is held on by two or four spring-loaded clips or by screws.

Ignition Cables

Ignition cables, also called spark plug cables, conduct high-voltage current. They are de-

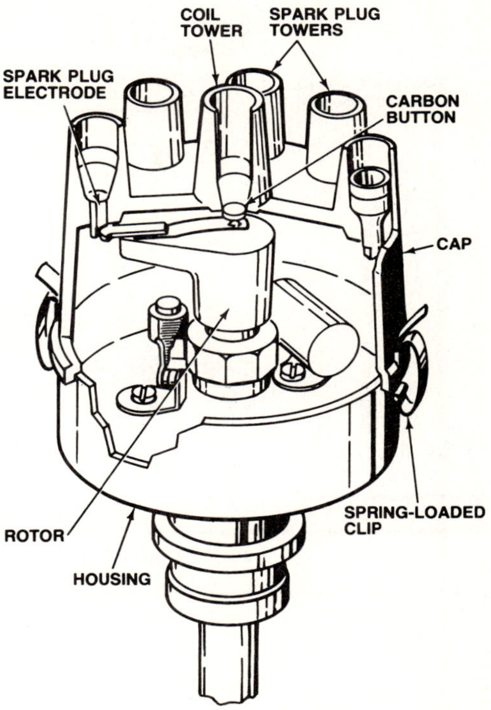

Figure 8-18. The distributor cap and rotor assembled with the distributor housing.

signed to offer a high resistance to avoid interference with radio and television signals. The cable terminals, figure 8-19, are designed to make a physically strong contact with the coil and distributor electodes.

The ignition cables must have special connectors, often called spark plug boots, figure 8-20. The boots provide a strong and well-insulated contact between the cable and the spark plug.

Spark Plugs

Spark plugs allow the high-voltage current to arc across a small air gap. The three basic parts of a spark plug, figure 8-21, are:
1. A ceramic core, or insulator, which insulates the center electrode and acts as a heat conductor
2. Two electrodes, one insulated in the core and the other grounded on the shell
3. A steel shell that holds the insulator and electrodes in a gas-tight assembly and which has threads for plug installation in the engine.

Bakelite: A plastic material that is a good insulator. Distributor caps are often made of bakelite.

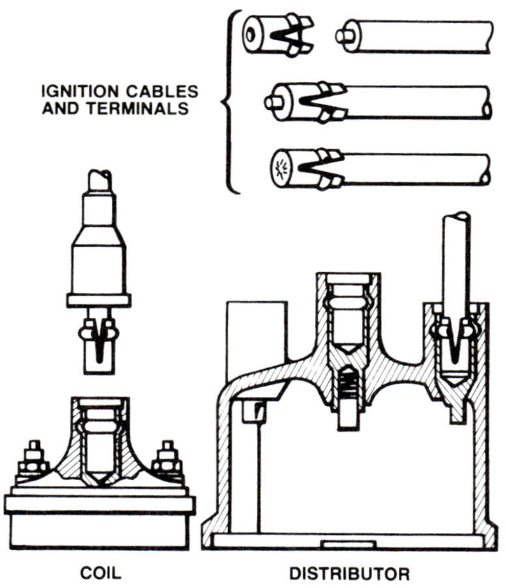

Figure 8-19. Ignition cables and terminals.

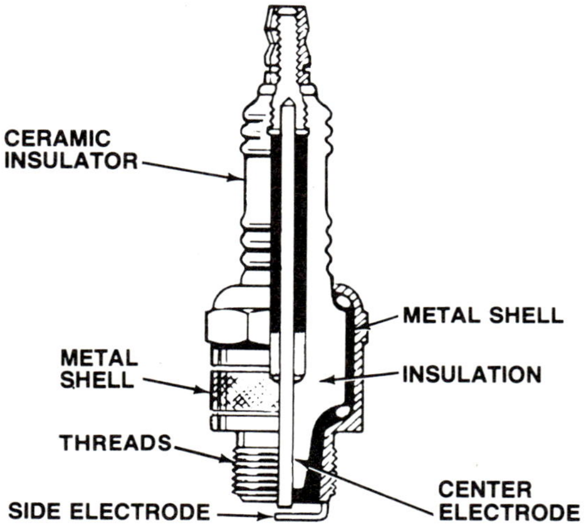

Figure 8-21. A cutaway view of the spark plug.

The metal shell of the side electrode is grounded against the engine. The other electrode is encased in the ceramic insulator. A spark plug boot and cable are attached to the top of the plug. High-voltage current flows through the center of the plug and arcs from the tip of the insulated electrode to the side electrode and ground. This spark ignites the air-fuel mixture in the combustion chamber, providing power to the engine.

The burning gases in the engine can corrode and wear the spark plug electrodes. Electrodes

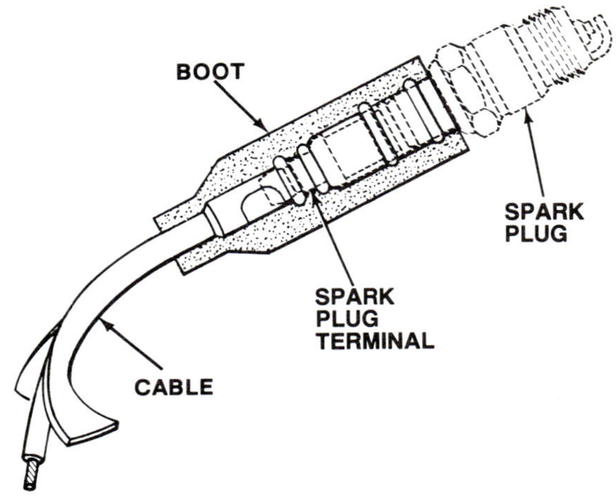

Figure 8-20. The spark plug boot fits over the top portion of the spark plug.

are made of metals that resist this attack. Most electrodes are made of high-nickel steel, but platinum and silver alloys have also been used.

Spark plug firing action

The arc of current across a spark plug air gap provides two types of discharge:

- Capacitive
- Inductive.

When a high-voltage surge is first delivered to the spark plug center electrode, the air-fuel mixture in the air gap cannot conduct an arc. The spark plug acts as a capacitor, with the center electrode storing a negative charge and the grounded side electrode storing a positive charge. The air gap between the electrodes acts as an insulator. This is the opposite of the normal negative-ground polarity, and results from the connection of the coil secondary winding, as shown in figure 8-22.

Secondary voltage increases, and the charges in the spark plug strengthen, until the different charges at the electrodes are great enough to **ionize** the spark plug air gap. That is, the air-fuel mixture in the gap is changed from a nonconductor to a conductor by the positive and negative charges of the two electrodes. The air gap is no longer an insulator, and current flows between the electrodes. The voltage level at this instant is called ionization voltage. The current that flows across the spark plug air gap at the instant of ionization is the capacitive portion of the spark. It flows from negative to positive, and uses the energy stored in the plug itself when the plug was acting as a capacitor, before ionization. This is the portion of the spark that starts the combustion process within the engine.

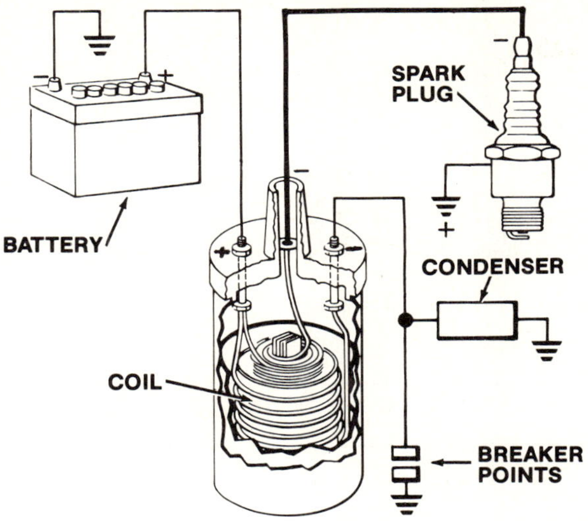

Figure 8-22. The spark plug should have a negative charge at the center electrode and a positive charge at the side electrode.

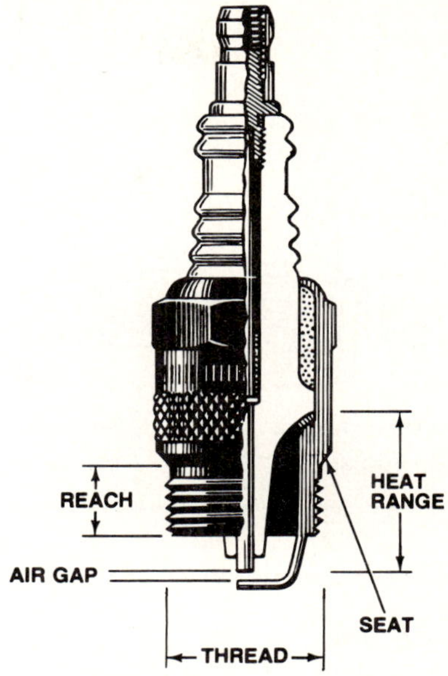

Figure 8-23. The design features of a spark plug.

The ionization voltage level is usually less than the total voltage produced in the coil secondary winding. The remainder of the secondary voltage (that amount of voltage not needed to force ionization) is dissipated as current flow across the spark plug air gap. This is the inductive portion of the spark discharge, which causes the *visible* flash or arc at the plug. It contributes nothing to the combustion of the air-fuel mixture, but is the cause of electrical interference and severe electrode erosion. High-resistance cables and spark plugs suppress this inductive portion of the spark discharge.

SPARK PLUG CONSTRUCTION

Spark Plug Design Features

Spark plugs are made in many different sizes and types to fit different engines. The most important differences among plugs are:

1. Reach
2. Heat range
3. Thread and seat
4. Air gap.

These are illustrated in figure 8-23.

Reach

The **reach** of a spark plug is the length of the shell from the seat to the bottom of the shell, including threaded and unthreaded portions. If incorrect plugs are installed and the reach is too short, the electrodes will be in a pocket and the spark will not ignite the air-fuel mixture, figure 8-24. Also, the exposed threads on the engine will accumulate carbon deposits which must be removed before the correct plugs can be installed.

If the spark plug reach is too long, the exposed plug threads could get hot enough to ignite the air-fuel mixture at the wrong time. Plug removal may be difficult because of carbon deposits on the plug threads. Also, engine damage could result from interference between moving parts and the exposed plug threads.

Heat range

The **heat range** of a spark plug determines its ability to dissipate heat from the firing end. The design of the insulator core controls the plug's heat dissipation ability, figure 8-25. A cold spark plug has a short insulator tip that provides a short path for heat to travel, and permits

Ionize: To break up molecules into two or more oppositely charged ions. The air gap between the spark plug electrodes is ionized when the air-fuel mixture is changed from a nonconductor to a conductor.

Reach: The length of the spark plug shell from the seat to the bottom of the shell.

Heat Range: The measure of a spark plug's ability to dissipate heat from its firing end.

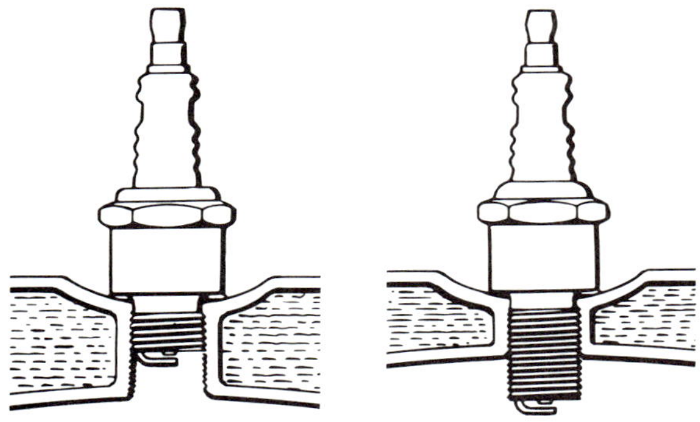

Figure 8-24. Spark plug reach.

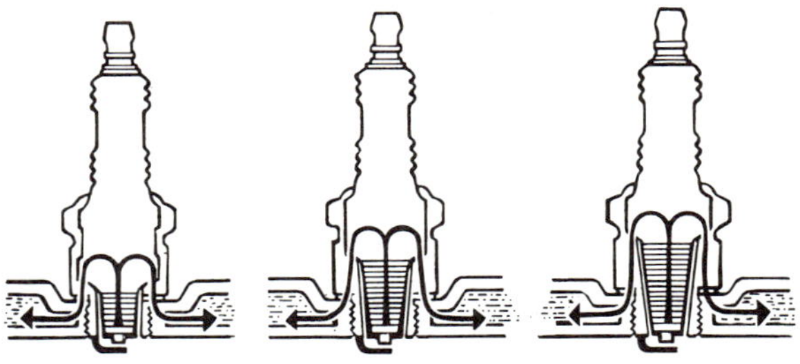

Figure 8-25. Spark plug heat range.

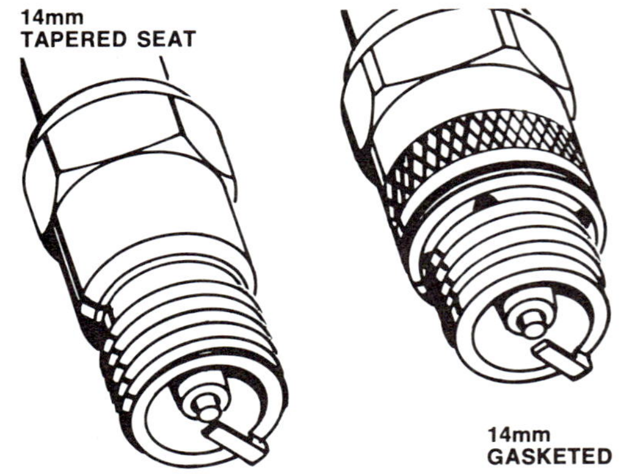

Figure 8-26. Spark plug thread and seat types.

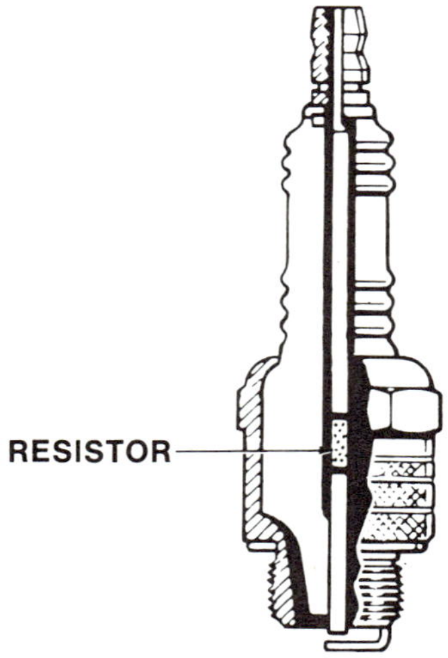

Figure 8-27. A resistor spark plug.

the heat to dissipate rapidly to maintain lower electrode temperature. A hot spark plug has a long insulator tip that creates a longer path for heat to travel. This slower heat transfer maintains higher electrode temperature.

Thread and seat

Automotive spark plugs are made with two different thread diameters: 14 millimeters and 18 millimeters, figure 8-26. All 18-mm plugs have tapered seats that match the tapered seats in the engine. No gaskets are used. Fourteen-millimeter plugs are made either with a flat seat that requires a gasket or with a tapered seat that does not. The gasket-type 14-mm plugs are the most common.

The steel shell of a spark plug is hex-shaped so that a wrench will fit it. Fourteen-millimeter, tapered-seat plugs have shells with a 5/8-inch hex; 14-mm gasketed and 18-mm tapered-seat plugs have shells with a 13/16-inch hex.

Air gap

The correct spark plug air gap is important to engine performance and plug life. A gap that is too narrow will increase current flow, which may lead to burnt electrodes. A gap that is too wide will require higher voltage to create an arc. If the required voltage is too great, the plug will not fire. Typical air gaps with breaker-point ignitions are in the .030- to .040-inch range (.75 to 1.0 mm). Several of the electronic ignition systems use gaps from .045 to .080 (1.0 to 2.0 mm). These wider gaps are necessary to fire the leaner air-fuel mixtures of late-model engines. Special spark plugs are manufactured for these electronic ignitions, because trying to stretch a normal spark plug to this greater gap could damage it.

Special-Purpose Spark Plugs

Besides the basic spark plug described previously, there are several special-purpose plugs

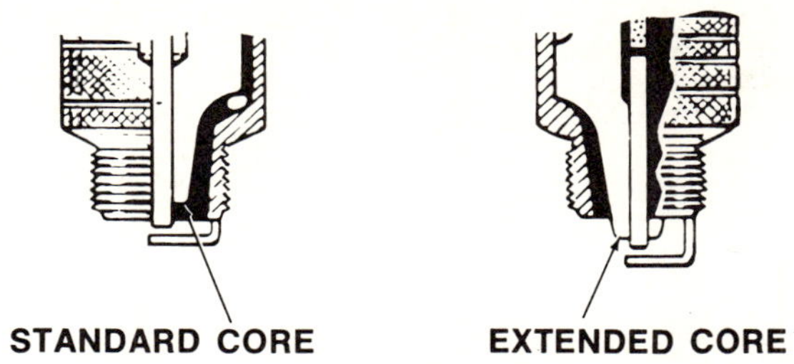

Figure 8-28. A comparison of a standard and an extended core spark plug.

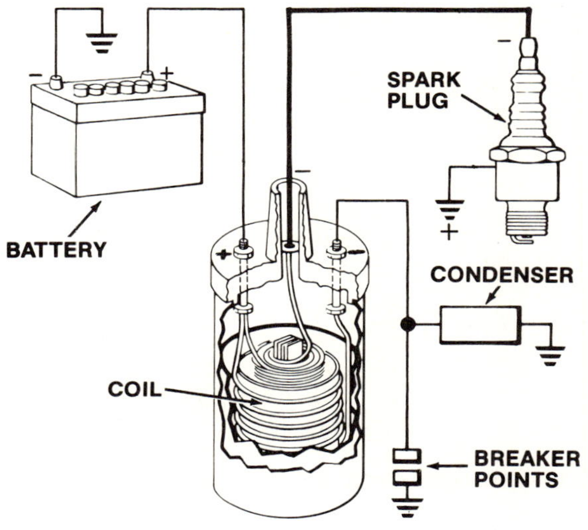

Figure 8-30. Negative, or ground, ignition polarity.

that are made for motorcycle, racing, and marine engines.

Resistor-type spark plugs have a resistor located in the center electrode, figure 8-27. This resistor, of about 10,000 ohms, reduces the inductive portion of the spark discharge. This reduces radio interference and increases plug electrode life.

In an **extended-core spark plug**, figure 8-28, the insulator core and electrodes extend farther into the combustion chamber. The extended core operates hotter under slow-speed driving conditions to burn off combustion deposits and cooler at high speed to prevent spark plug overheating. These are also called extended-tip or power-tip plugs. They are manufactured with or without a resistor.

A **series-gap, or booster, spark plug** has an extra gap in the center electrode, figure 8-29. This creates a slightly higher ionization voltage to reduce electrode erosion and spark plug fouling.

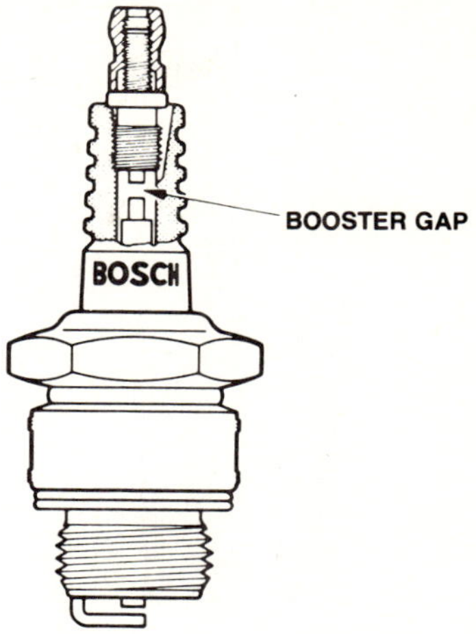

Figure 8-29. A booster, or series-gap spark plug. (Bosch)

IGNITION POLARITY

If the ignition coil is correctly connected, figure 8-30, voltage will be applied to the spark plug so that the center electrode is negatively charged and the grounded electrode is positively charged. The center electrode is hotter than the side electrode because the side electrode is in contact with the engine block and the cooling system. Electrical current will flow more easily from a hot object to a cold object. This means that less ignition voltage is required to form a spark when the spark is going from the hot center electrode to the cool side electrode. This correct ignition connection is called **negative polarity**. If the coil is incorrectly connected,

Resistor-Type Spark Plug: A plug that has a resistor in the center electrode to reduce the inductive portion of the spark discharge.

Extended-Core Spark Plug: Also called power tip. The insulator core and the electrodes in this type of spark plug extend further into the combustion chamber than they do on other types.

Series-Gap Spark Plug: Also called booster. A spark plug in which the center electrode has an extra gap to increase ionization voltage.

Negative Polarity: Also called ground polarity. A correct polarity of the ignition coil connections. Coil voltage is delivered to the spark plugs so that the center electrode of the plug is negatively charged and the grounded electrode is positively charged.

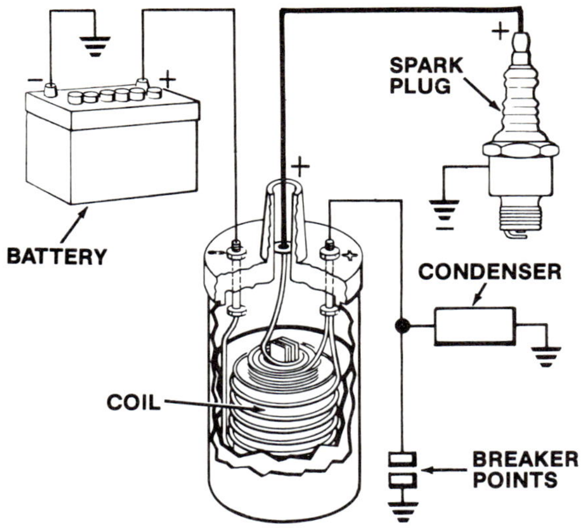

Figure 8-31. Reverse, or positive, ignition polarity.

Positive Polarity: Also called reverse polarity. An incorrect polarity of the ignition coil connections. Coil voltage is delivered to the spark plug so that the center electrode of the plug is positively charged and the grounded electrode is negatively charged.

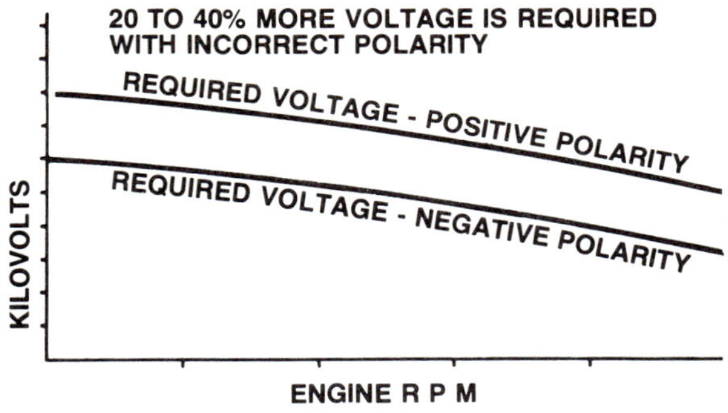

Figure 8-32. More voltage is required to fire a spark plug that has reverse polarity. (Prestolite)

figure 8-31, the spark will be moving from the cooler side electrode to the hotter center electrode. This is called **positive polarity**, or reverse polarity. Figure 8-32 shows how much more ignition voltage is required to fire a spark plug with positive polarity than one with negative polarity.

SUMMARY

Through electromagnetic induction, the ignition system transforms the low voltage of the battery into the high voltage needed to fire the spark

■ Spark Plug Design

Many people have tried to redesign the spark plug. Not all of the "new" designs have worked out. For example, a plug manufactured before World War I had an insulated handle at the top. By pulling this handle up, an auxiliary gap was opened, presumably to create a hotter spark and stop oil fouling. A window in the side of the plug showed whether the gap was open or closed.

Another "revolutionary" type of plug had a screw connector that allowed the inner core assembly to be removed and cleaned quickly.

Still another design had threads and electrodes at each end of the plug. The plug could be removed, the terminal cap installed on the other end, and then reinstalled upside down. All of the photos were provided by Champion Spark Plug Company.

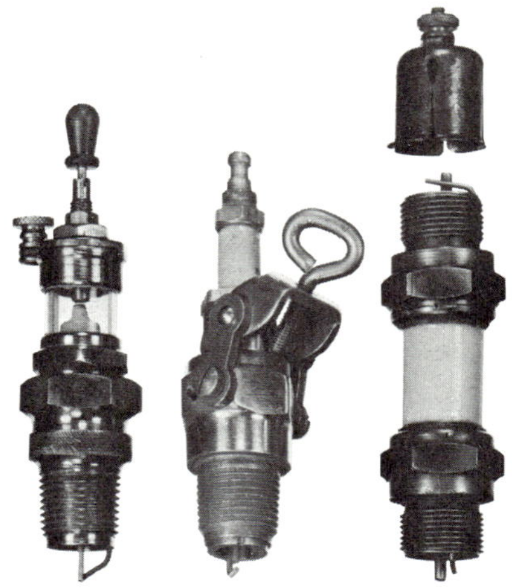

plugs. Induction occurs in the ignition coil where current flows through the primary winding to build up a magnetic field. When the field collapses rapidly, high voltage is induced in the coil secondary winding. All domestic original-equipment ignitions are the battery-powered, inductive discharge type.

The ignition system is divided into two circuits: the primary and the secondary. The primary circuit contains the battery, the ignition switch, the ballast (primary) resistor, the starting bypass, the coil primary winding, and a switching device in the distributor. For over 60 years, mechanical breaker points were used as the primary circuit switching device. In the last decade, solid-state electronic components have replaced breaker points as primary circuit switching devices.

The ignition secondary circuit generates the high voltage and distributes it to the engine's spark plugs. This circuit contains the coil secondary winding, the distributor cap and rotor, the ignition cables, and the spark plugs. The spark plugs allow the high voltage to arc across an air gap and ignite the air-fuel mixture in the combustion chamber. Important design features of a spark plug are its reach, heat range, thread and seat size, and air gap. For efficient spark plug firing, ignition polarity must be established so that the center electrode of the plug is negative and the ground electrode is positive.

Review Questions

Select the single most correct answer.
Compare your answers with the correct answers on page 280.

1. The voltage required to ignite the air-fuel mixture can range from ______ volts.
   a. 5 to 25
   b. 50 to 250
   c. 500 to 2,500
   d. 5,000 to 25,000

2. Which of the following does *not* require higher voltage levels to cause an arc across the spark plug gap?
   a. Increased spark plug gap
   b. Increased engine operating temperature
   c. Increased fuel in air-fuel mixture
   d. Increased pressure of air-fuel mixture

3. The coil transforms low voltage from the primary circuit to high voltage for the secondary circuit through:
   a. Magnetic induction
   b. Capacitive discharge
   c. Series resistance
   d. Parallel capacitance

4. Voltage induced in the secondary winding of the ignition coil is how many times greater than the self-induced primary voltage?
   a. 1 to 2
   b. 10 to 20
   c. 100 to 200
   c. 1,000 to 2,000

5. The two circuits of the ignition system are:
   a. The "Start" and "Run" circuits
   b. The point circuit and the coil circuit
   c. The primary circuit and the secondary circuit
   d. The insulated circuit and the ground circuit

6. Which of the following components is part of both the primary and the secondary circuits?
   a. The ignition switch
   b. The distributor rotor
   c. The condenser
   d. The coil

7. Which of the following is *not* contained in the primary circuit of an ignition system?
   a. Battery
   b. Spark plugs
   c. Ignition switch
   d. Coil primary winding

8. When the cranking system is operating, the ballast resistor:
   a. Reduces coil primary voltage to about 7 volts
   b. Heats up, increasing resistance and reducing voltage
   c. Is bypassed to provide full available voltage
   d. Cools, and increases primary current flow

9. Which of the following is true of the coil primary windings in a negative-ground system?
   a. They consist of 100 to 150 turns of very fine wire
   b. The turns are insulated by a coat of enamel
   c. The negative terminal is connected directly to the battery
   d. The positive terminal is connected to the breaker points and to ground

10. Current flow through the starting bypass can be controlled by the:
    a. Ignition switch
    b. Starter relay
    c. Solenoid
    d. Any of the above

11. In order to collapse the magnetic field of the coil, the primary circuit requires a:
    a. Ballast resistor
    b. Switching device
    c. Condenser
    d. Starting bypass circuit

12. Breaker points are usually made of:
    a. Silicon
    b. Tungsten
    c. Aluminum
    d. Copper

13. Which of the following statements is true?
    a. Breaker points wear during operation, changing settings and performance
    b. A breaker-point ignition system has difficulty meeting exhaust emission control standards
    c. Solid-state switching devices do not wear, keeping ignition performance consistent
    d. All of the above

14. In the solid-state ignitions used as original equipment on late-model domestic cars, the breaker points and distributor cam have been replaced by:
    a. RFI filter capacitors
    b. Auxiliary ballast resistors
    c. Magnetic pickup triggering devices
    d. Integrated coil and distributor cap assemblies

15. Bakelite is a synthetic material used in distributors because of its good:
    a. Permeability
    b. Conductance
    c. Insulation
    d. Capacitance

16. Which of the following are basic parts of a spark plug?
    a. Ceramic core
    b. Two electrodes
    c. Steel shell
    d. All of the above

17. That part of the energy that flows through a spark plug and causes the visible flash or arc:
    a. Contributes nothing to the combustion of fuel mix
    b. Causes electrical interference and electrode corrosion
    c. Is called the inductive portion of the spark
    d. All of the above

18. Which of the following is *not* an important design feature among types of spark plugs?
    a. Reach
    b. Heat range
    c. Polarity
    d. Air gap

19. In the illustration at right, the dimension arrows indicate the:
    a. Heat range
    b. Resistor portion of the electrode
    c. Extended core length
    d. Reach

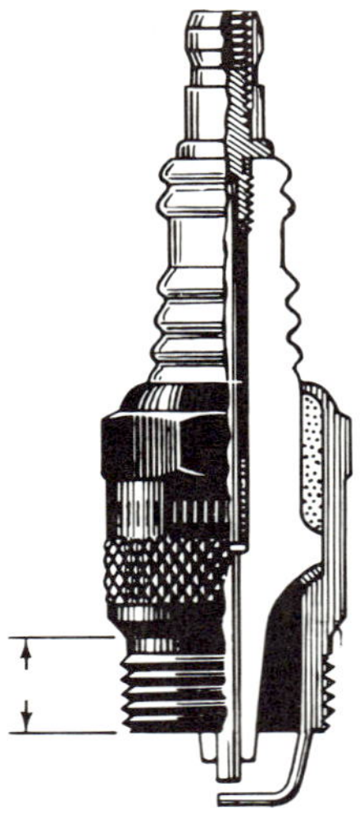

Chapter

9

Ignition Timing, Spark Advance, and Spark Timing Controls

In Chapter 1 we discussed an engine's initial timing requirements. We also saw that, because the air-fuel mixture's burn time does not change greatly, the engine's ignition timing must be advanced for efficient operation at varying speeds. This chapter reviews these basic ideas and then explains:

- How ignition timing is changed to match engine operating conditions
- The effects of ignition timing on performance, economy, and emission control
- How spark timing emission control systems affect ignition timing to reduce exhaust pollutants
- The new computer-controlled systems that electronically determine ignition timing.

INITIAL TIMING

Initial timing is the engine's correct timing setting at or near idle speed. This is normally within a few degrees of top dead center. The initial timing is set so that combustion will be complete by 10° atdc. In figure 9-1, position A, when the engine turns at 625 rpm, the crankshaft rotates about 11 degrees during the air-fuel mixture's burn time of 0.003 second. Timing should be set at 1° btdc for efficient combustion.

SPARK ADVANCE

When the engine in figure 9-1 speeds up, the ignition timing must be advanced to match the speed:

- At 1,000 rpm, position B, the crankshaft rotates 18 degrees during the burn time. The spark should occur at 8° btdc for combustion to be completed at 10° atdc.
- At 2,000 rpm, position C, the crankshaft rotates 36 degrees in 0.003 second. Spark timing must be advanced to 26° btdc for efficient combustion.

Several other factors, besides engine speed, affect ignition timing requirements:

- Combustion chamber design
- Valve timing
- Ignition system condition
- Compression ratio
- Type of fuel used
- Engine load (power demanded from the engine)
- Combustion chamber temperature.

Only two of these factors, engine speed and engine load, change quickly or often enough during operation to create a need for rapid automatic changes in ignition timing. Most automotive distributors have two spark advance mechanisms that react to engine operating changes and alter ignition timing:

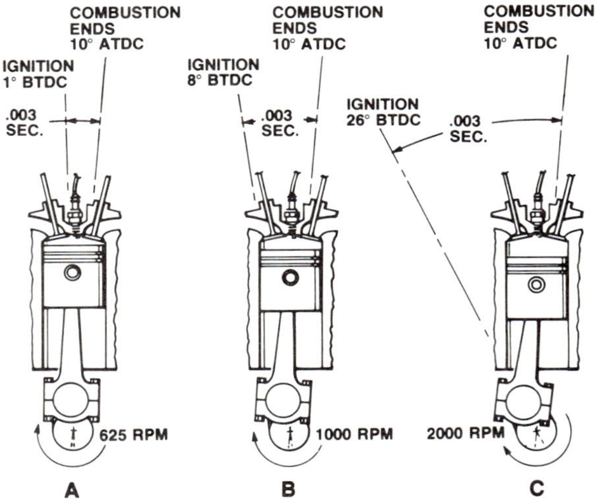

Figure 9-1. Initial ignition timing and timing advance with increasing engine speed.

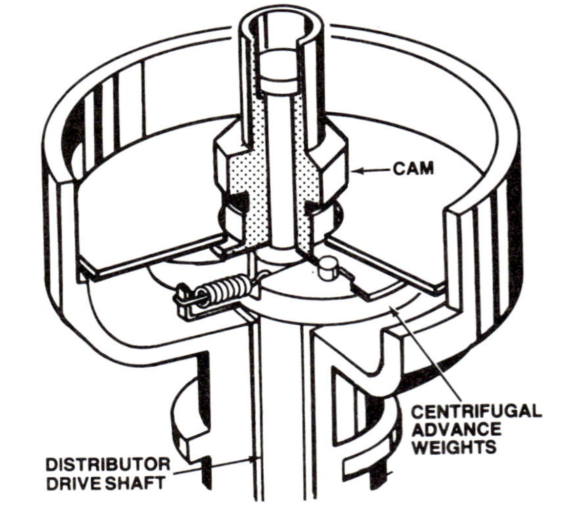

Figure 9-2. The centrifugal advance weights transfer the rotation of the distributor drive shaft to the cam and rotor.

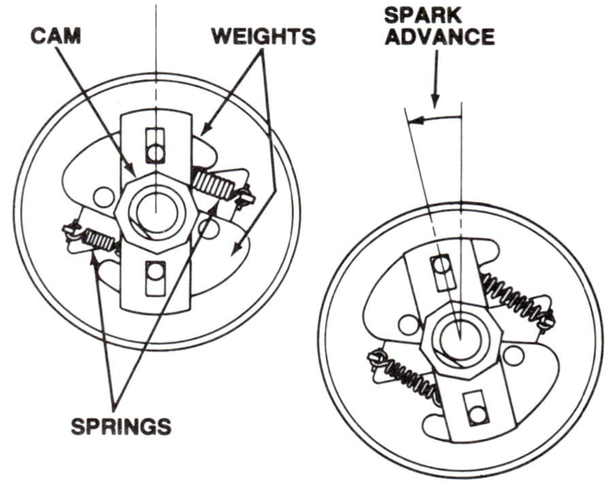

Figure 9-3. When the centrifugal advance weights move, the position of the cam and rotor changes.

1. The centrifugal advance changes ignition timing to match engine speed.
2. The vacuum advance changes ignition timing to match engine load.

In most breaker-point and electronic ignition systems, these two spark advance mechanisms work as follows:

- The centrifugal advance alters the position of the distributor cam or trigger wheel on the distributor shaft.
- The vacuum advance alters the position of the breaker points or the magnetic pickup coil.

These changes in position alter the time, relative to crankshaft position, at which the primary circuit is opened.

Centrifugal Advance

The **centrifugal**, (or **mechanical**) **advance** mechanism consists of two weights connected to the distributor drive shaft by two springs, figure 9-2. The distributor cam, or electronic trigger wheel, and the distributor rotor are mounted on another shaft. The second shaft fits over the drive shaft like a sleeve. Drive shaft motion is transmitted to the second shaft through the centrifugal advance weights. When the weights move, the relative position of the drive shaft and the second shaft is changed.

As engine speed increases, distributor shaft rotation speed increases. The advance weights move outward because of centrifugal force. The outward movement of the weights shifts the second shaft and cam, or trigger wheel, and rotor ahead of distributor shaft rotation, figure 9-3. The primary circuit is opened earlier in the compression stroke, and the spark occurs earlier.

Each advance weight is connected to the distributor drive shaft by a control spring. These springs are selected to allow the correct amount

Centrifugal (Mechanical) Advance: A method of advancing the ignition spark, using weights in the distributor that react to centrifugal force.

Vacuum Advance: The use of engine vacuum to advance or retard ignition spark timing by moving the distributor breaker plate.

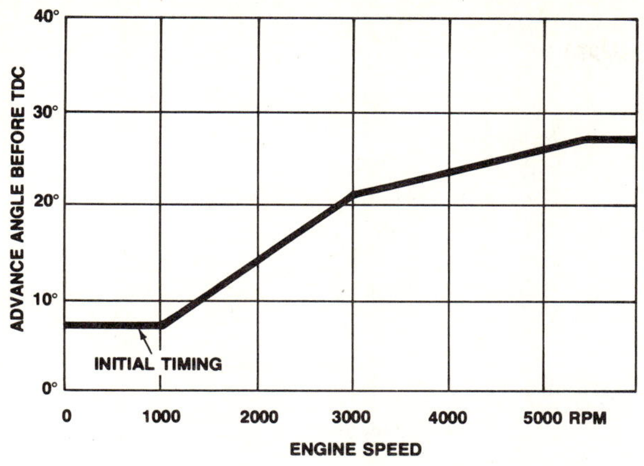

Figure 9-4. A typical centrifugal advance curve.

of weight movement and ignition advance for a particular engine.

At low engine speeds, spring tension holds the weights in, so that initial timing is maintained. As engine speed increases, centrifugal force overcomes spring tension and the weights move outward. The advance is not a large, rapid change, but rather a slow, gradual shift. Figure 9-4 shows a typical centrifugal advance curve. The advance curve can be changed by changing the tension of the control springs. Remember that centrifugal advance responds to engine speed.

In most distributors, the centrifugal advance mechanism is mounted below the cam and breaker points, figure 9-2, or below the trigger wheel and pickup coil. Most Delco-Remy distributors have the advance mechanism above the cam or trigger wheel, just below the rotor, figure 9-5.

Vacuum Advance

The **vacuum advance** mechanism allows efficient engine performance within a range of air-fuel ratios. These ratios are important, since there are limits to how rich or how lean they can be and still remain fully combustible. The air-fuel mixtures with which an engine can operate efficiently range from 8:1 to 18.5:1 by weight. These ratios are generally stated as eight parts of air combined with one part of gasoline (8:1), which is the richest mixture that an engine can tolerate and still fire regularly. A ratio of 18.5 parts of air mixed with one part of gasoline (18.5:1) is the leanest that an engine can tolerate without misfiring.

An average air-fuel ratio is about 15:1. This mixture takes about 3 milliseconds (0.003 second) to burn. A lean mixture, with more air and less fuel, will take more time to burn. The

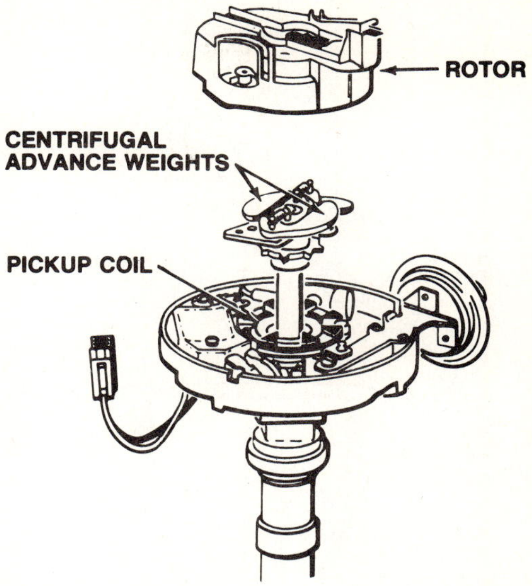

Figure 9-5. Many Delco-Remy distributors have the centrifugal advance mechanism above the cam and rotor, or above the pickup coil and trigger wheel, as shown.

■ Are Spark Timing Controls Confusing?

As you study this chapter, you may be confused by the various spark control systems. If so, here is a summary which may help you understand this form of emission control.

The greatest amount of exhaust emissions are produced at idle and during low-speed operation. *ALL* spark control systems are designed to reduce these emissions by preventing vacuum advance at idle, and when the car is in the lower gears. This means that a spark control system should work *ONLY* when the engine is at normal operating temperature, and the car is operating at low speeds or in the low gears. When the engine is at normal operating temperature and the car is at cruising speed or in high gear, vacuum advance takes place just as it would without a spark control system.

ALL spark control systems have some form of temperature control to prevent the system from operating if the engine is too cold or too hot.

Here are two troubleshooting tips:

1. If vacuum advance *is* present when the engine is at idle or in the low gears, or is *not* present when the engine is in high gear, then a speed sensor, transmission switch, control module, or solenoid is defective.
2. If vacuum advance is *not* present when the engine is cold or overheated, then a temperature sensor or switch is defective.

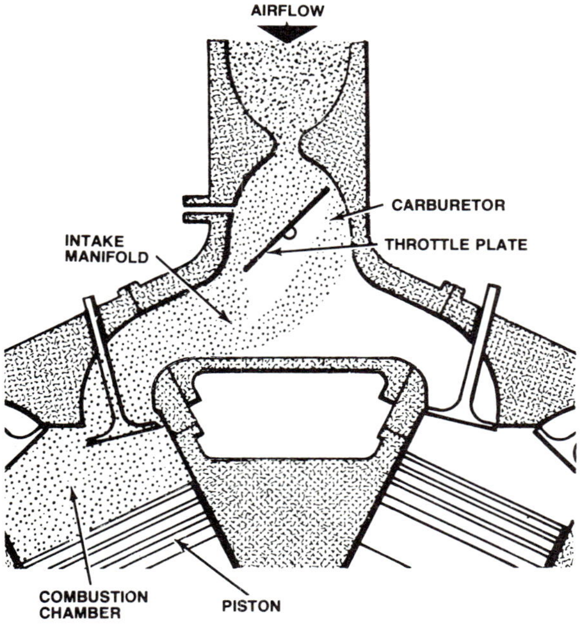

Figure 9-6. Air flows through the carburetor and intake manifold to reach the combustion chambers.

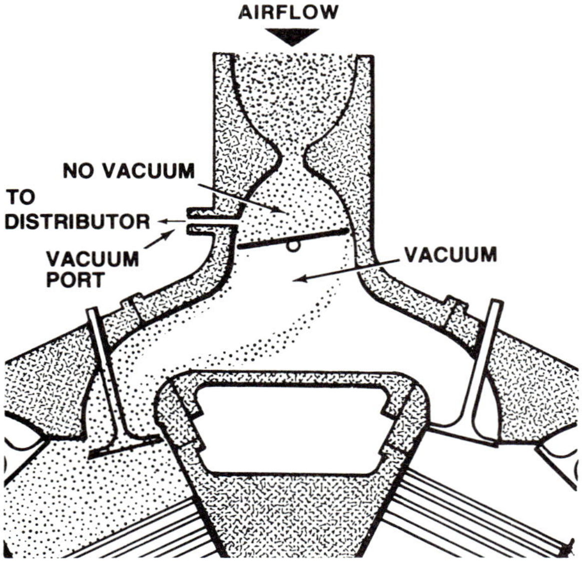

Figure 9-8. When the throttle is closed during idle or deceleration, there is no vacuum at the port.

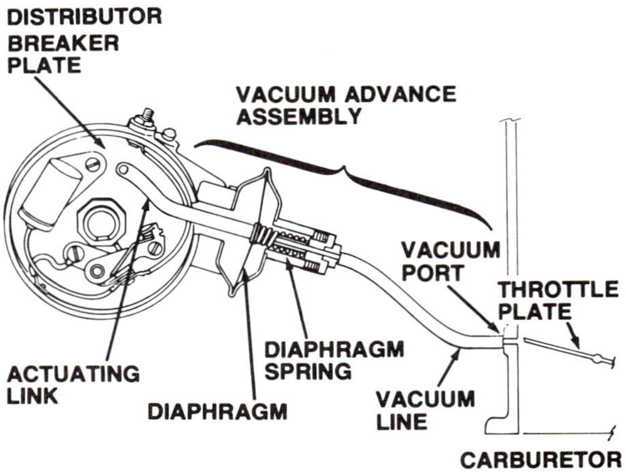

Figure 9-7. The vacuum advance assembly is connected to a port in the carburetor.

ignition timing must be advanced to provide maximum combustion pressure at the correct piston position. A rich air-fuel mixture, with more fuel and less air, will burn more quickly and emit more exhaust pollutants. Ignition timing should be retarded for efficient combustion and emission control.

Engine Vacuum

The reciprocating engine can be considered as an air pump. As a piston moves downward, air pressure in the cylinder decreases. Air from the atmosphere rushes in to fill the void.

The fuel delivery system uses this air movement to carry fuel to the cylinders. On most cars, the air must travel through a carburetor, figure 9-6, to reach the cylinders. Under several operating conditions, air movement caused by the downstroke of a piston is not great enough to draw fuel into the cylinder. The carburetor forces the air to flow through a restriction, called a **venturi**. This speeds the airflow and creates a low pressure (vacuum) area. Fuel is drawn into the airflow by the vacuum, and the resulting air-fuel mixture enters the cylinder combustion chamber. The air-fuel ratio will change as the vacuum in the carburetor changes.

The ignition timing must be changed as the air-fuel ratio changes, so that the mixture has the correct amount of time to burn. The vacuum advance mechanism is connected to a small hole, or port, in the carburetor, figure 9-7, just above the throttle plate. This is called **ported vacuum**. When a vacuum exists at the port, timing is advanced. When no vacuum exists at

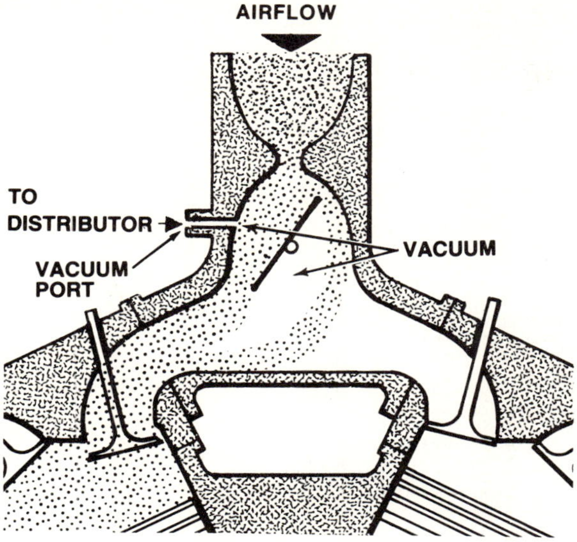

Figure 9-9. When the throttle is partially open, the port is exposed to manifold vacuum.

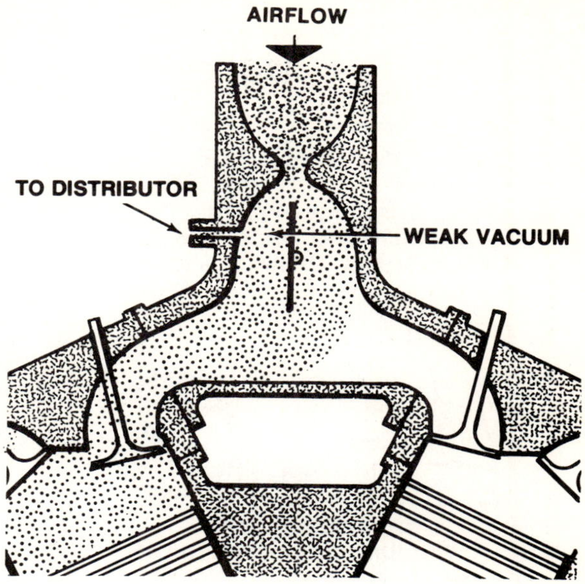

Figure 9-10. When the throttle is fully open, the vacuum at the port is too weak to cause any vacuum advance.

the port, the timing remains at a basic setting or is affected by only the centrifugal advance. The vacuum advance mechanism will be explained in more detail later; the following paragraphs explain the relationship between carburetor vacuum and the need for advanced timing.

The automobile driver controls engine load and carburetor vacuum through the action of the throttle plate, which is a variable restriction in the carburetor airflow. When the engine is at idle, figure 9-8, the throttle plate is almost closed. Very little air flows through the carburetor to mix with the fuel. With a rich air-fuel mixture, no spark advance is needed. High vacuum exists in the intake manifold, but there is no vacuum at the port because it is above the closed throttle plate. There is no vacuum-controlled spark advance with a closed throttle. The large amount of vacuum *below* the throttle plate exists because of the small amount of air entering the engine. This is called **manifold vacuum**.

When the throttle plate is partially open, figure 9-9, more air can flow through the carburetor. The air-fuel mixture becomes lean and requires an advanced spark. Because the port is now exposed to vacuum, ignition timing is advanced.

At medium cruising speeds, the engine operates with a partly open throttle and an air-fuel ratio of 15:1 or 16:1. Airflow speed is high, and the vacuum signal at the carburetor port is strong enough to provide enough vacuum advance for these relatively lean air-fuel mixtures.

A lean mixture is not the only factor that affects ignition timing requirements. During part-throttle operation, the cylinders are only partially filled with the air-fuel mixture. Because there is less to compress, the compression pressure is less. A less highly compressed mixture takes a longer time to burn.

At wide-open throttle, figure 9-10, the power circuit in the carburetor provides a richer mixture than at part-throttle cruising. Also, the cylinders are more completely filled because airflow volume increases. These factors cause a faster burn that requires the ignition timing to retard slightly for proper efficiency. Manifold vacuum and ported vacuum drop at wide-open throttle and become approximately equal. Vacuum may drop to as little as 5 inches of mercury, and typically 7 inches of vacuum are required to operate a vacuum advance mechanism.

Venturi: A restriction in an airflow, such as in a carburetor, that speeds the airflow and creates a vacuum.

Ported Vacuum: Vacuum immediately above the throttle plate in a carburetor.

Manifold Vacuum: Low pressure in an engine's intake manifold, below the carburetor throttle.

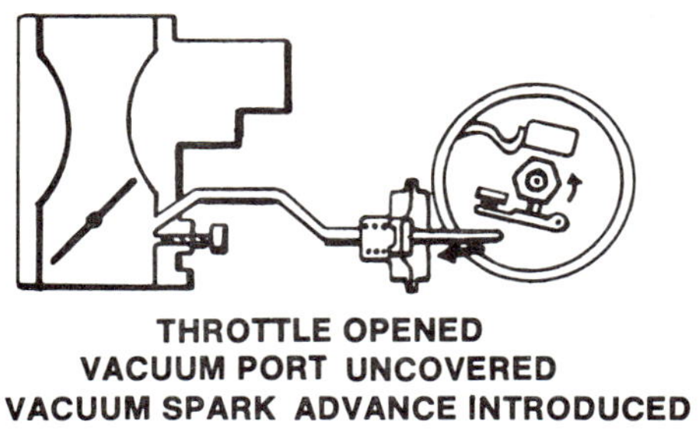

Figure 9-11. Vacuum at the carburetor port causes the position of the breaker plate to be shifted.

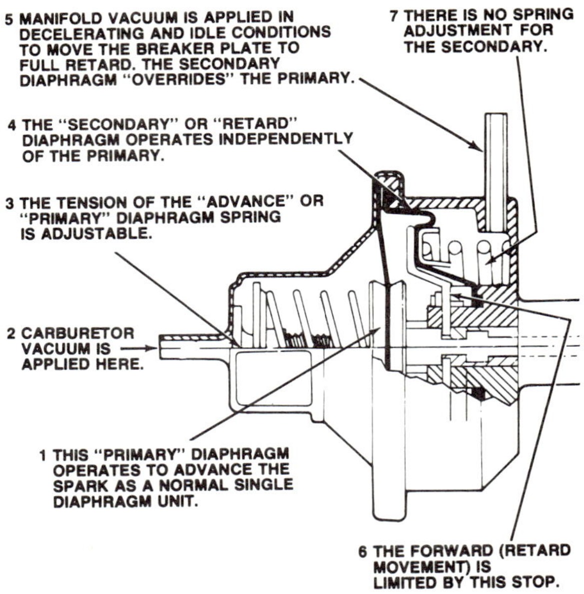

Figure 9-12. A typical dual-diaphragm vacuum advance assembly. (Ford)

Vacuum advance also retards when the throttle is opened quickly from idle or a part throttle position. This occurs because airflow velocity will lag behind throttle opening. This reduced vacuum will not draw enough fuel through the main metering circuit of the carburetor but the accelerator pump supplies extra fuel. This momentarily rich mixture does not need as much spark advance for complete combustion. Because the vacuum at the vacuum port of the carburetor is low when the throttle is first opened quickly, the vacuum advance decreases to meet the needs of the momentarily rich mixture. Remember that vacuum advance responds to engine load.

Vacuum advance mechanism

The vacuum advance mechanism at the distributor, figure 9-7, consists of the:

- Movable breaker plate on which the points or electronic pickup coil is mounted
- Vacuum assembly, a housing with a flexible **diaphragm** and a spring
- Actuating link that connects the vacuum diaphragm to the breaker plate
- Tubing to connect the vacuum unit to the vacuum source.

The diaphragm is pulled toward a vacuum, pulling the actuating link, figure 9-11. This rotates the breaker plate to change the position of the point rubbing block in relation to the cam. Ignition timing is advanced. In an electronic system, the pickup is shifted in relation to the trigger wheel, advancing timing.

Some vacuum advance units can be adjusted by changing spring tension against the diaphragm.

Some Ford, Volkswagen, and AMC models have a dual-diaphragm vacuum unit to retard timing during certain engine operating conditions. In these systems, figure 9-12, one diaphragm works the way we just described to advance ignition timing. The second diaphragm is exposed to intake manifold vacuum (between the throttle plate and the combustion chamber). Here, vacuum is great during idle and closed-throttle deceleration. This vacuum acts upon the second diaphragm to shift the breaker plate in the retard direction.

For exhaust emission control, the retard diaphragm can override the advance diaphragm, causing retarded timing even when the advance diaphragm is at an advanced position, figure 9-13.

Total Ignition Advance

The two types of advance mechanisms work on different parts of the distributor to advance timing. Therefore, their effects are additive: the total ignition advance will be the sum of the centrifugal advance plus the vacuum advance. This total advance is added to the initial ignition

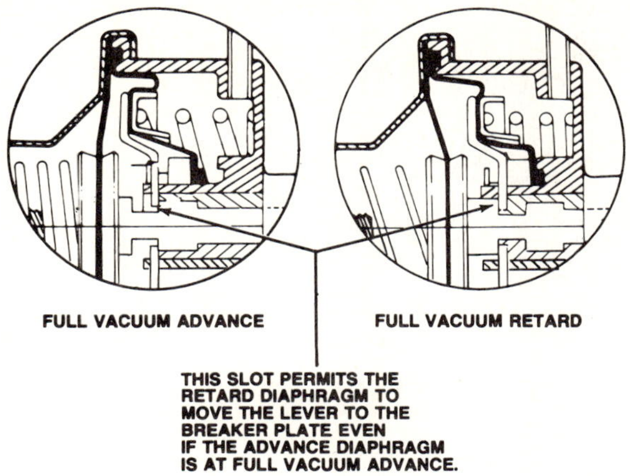

Figure 9-13. In a dual-diaphragm unit, the retard diaphragm can override the action of the advance diaphragm. (Ford)

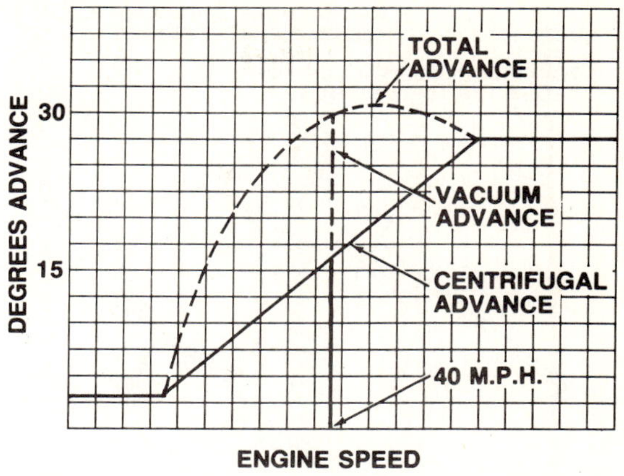

Figure 9-14. This typical advance curve shows that total advance is the total of vacuum plus centrifugal advances.

timing to determine actual ignition timing under any operating condition. Figure 9-14 shows the advance curves of a typical ignition system.

EFFECTS OF TIMING ON PERFORMANCE, ECONOMY, AND EMISSIONS

Igniting the air-fuel mixture at exactly the right instant will create the highest possible amount of heat and pressure. This means that the engine is getting the most energy possible from the fuel, leading to good performance and fuel economy. This efficient combustion process uses advanced ignition timing. High-temperature combustion occurs at an early point during the process. Unfortunately, NO_x is created during high-temperature combustion. Later in the combustion process, the exhaust gases are cooler. They do not heat up the exhaust manifold as much. These cool exhaust gases will carry whatever fuel was not burned in the cylinder into the atmosphere as HC.

If ignition timing is retarded, combustion is less efficient. Performance and economy are worse, too, but NO_x emissions decrease. This is because the peak temperature of combustion is reduced, lowering NO_x formation. At the same time, the time of peak temperature is shifted toward the end of the combustion process. This increases the temperature of the exhaust gases. HC emissions are also reduced because any unburned fuel from the cylinder is burned in the exhaust system.

SPARK TIMING EMISSION CONTROLS

Spark timing controls were widely used during the early 1970's to reduce exhaust emissions. All of the systems described in the following paragraphs work with the distributor vacuum advance. Early controls concentrated on advancing or retarding the ignition timing under certain operating conditions. Later systems simply delay vacuum advance at low and intermediate speeds, and allow it during high-speed cruising.

When catalytic converters were introduced in 1975, many spark timing control systems were discontinued. Converters reduce HC and CO emissions, and exhaust gas recirculation (EGR) provides more effective control of NO_x emissions.

Early Distributor Controls

Early emission-control equipment advanced or retarded ignition timing under certain engine operating conditions, usually during starting, deceleration, and idle.

The deceleration vacuum advance valve, figure 9-15, was used in the middle to late 1960's on Chrysler, Ford, AMC, and Pontiac products with manual transmissions. In a car with a manual transmission, the air-fuel mixture becomes extremely rich upon deceleration or gear shifting.

This valve momentarily switches the vacuum for the vacuum advance from a low vacuum source at the carburetor to a high manifold

Diaphragm: A thin, flexible wall to separate two cavities, such as the diaphragm in a mechanical fuel pump or distributor vacuum advance unit.

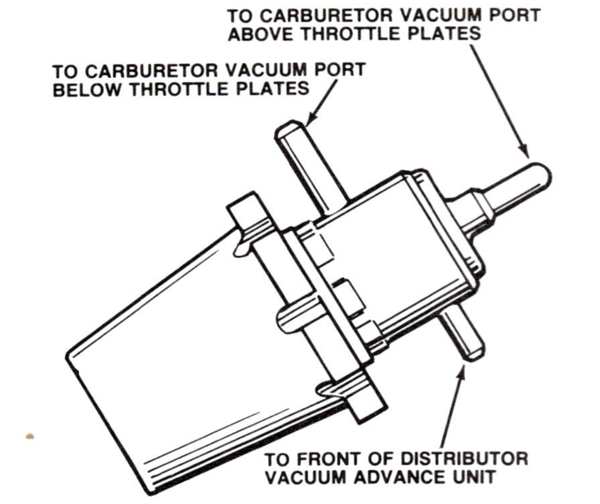

Figure 9-15. A deceleration vacuum advance valve.

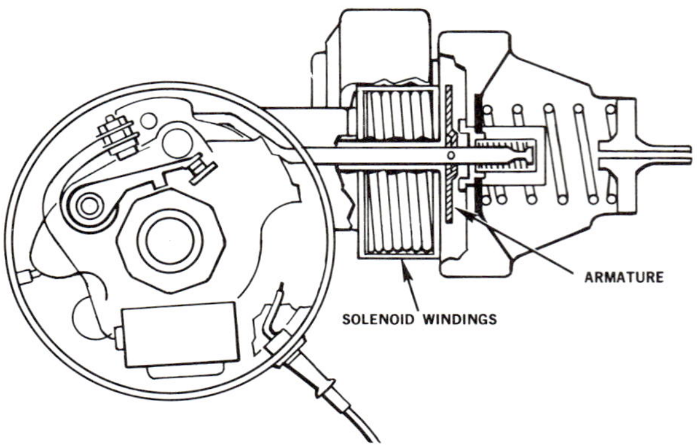

Figure 9-17. The installation of a distributor advance solenoid.

vacuum source during deceleration, then back to the low carburetor vacuum source. This prevents over-retard during deceleration or gear shifting which could cause high CO emissions from some engines.

The deceleration vacuum advance valve was discontinued in the early 1970's because it was most effective against CO emissions, but not against HC and NO_x emissions. As emission limits for HC and NO_x became stricter, manufacturers developed other devices which worked against all three major pollutants.

The distributor retard solenoid was used in 1970 and 1971 on some Chrysler products with V-8 engines and automatic transmissions. This electric solenoid is attached to, and controls the action of, the distributor vacuum advance unit, figure 9-16.

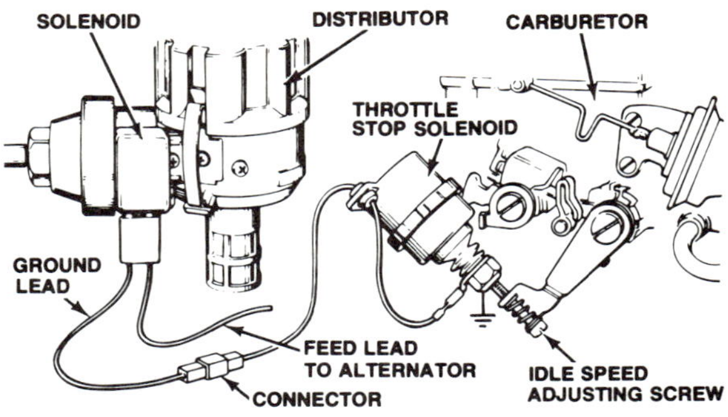

Figure 9-16. A distributor retard solenoid.

The solenoid is energized by contacts mounted on a carburetor throttle stop solenoid. When the throttle is closed, the carburetor solenoid is contacted by the idle adjusting screw to complete the ground circuit. The contacts in the carburetor solenoid carry current to the distributor solenoid windings. Since the distributor solenoid plunger is connected to the vacuum diaphragm, movement of the solenoid shifts the breaker plate in the retard direction.

When the engine speed increases, the idle adjusting screw breaks contact with the carburetor solenoid. Current flow to the distributor solenoid is stopped, and normal vacuum advance is allowed.

The distributors on some 1972 and 1973 Chrysler V-8 engines have a spark timing advance solenoid that helps the engine start by providing a 7½ degree spark advance. The solenoid is mounted in the distributor vacuum unit, figure 9-17. It is activated by power from the starter relay at the same terminal that sends power to the starter solenoid. The solenoid is activated only while the engine is cranking.

The starting advance solenoid is not an emission control device by itself, but it allows lower basic timing settings, which can help in emission control while providing advanced timing for quicker starting.

Principles Of Recent Systems

The most common types of control systems used in the early to mid-1970's are:

- Vacuum delay valves
- Speed- and transmission-controlled timing.

Vacuum delay valves

The vacuum delay valve "filters" the carburetor vacuum, slowing its application to the distributor advance unit. Generally, vacuum must be present in the system for 15 to 30 seconds before it is allowed to affect the advance mechanism.

One method of vacuum delay is used in Ford's spark delay valve (SDV) system, figure

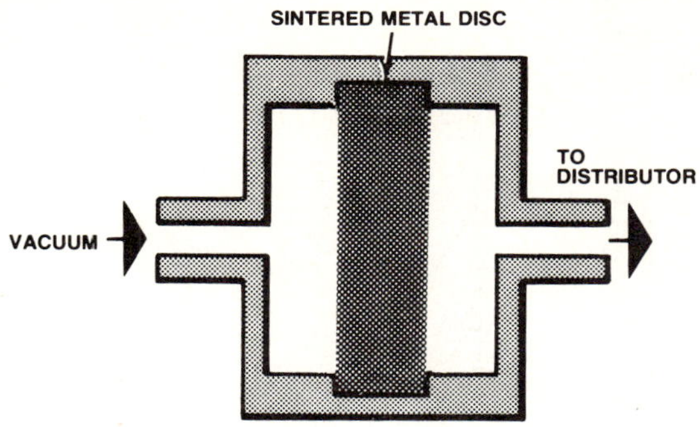

Figure 9-18. A cross-section view of a vacuum delay valve that uses sintered metal to slow the application of vacuum.

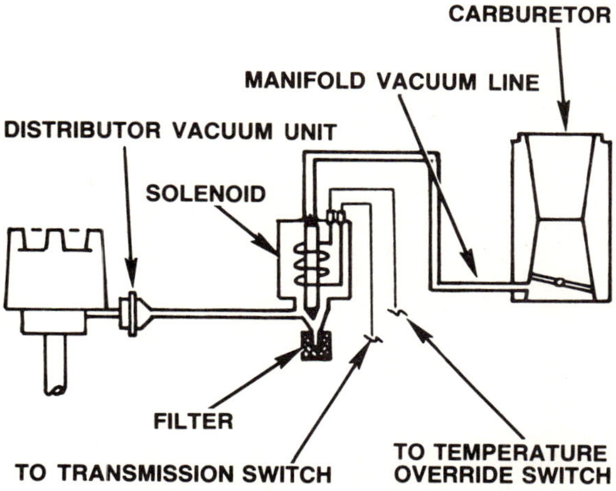

Figure 9-20. A simplified transmission-controlled spark system.

9-18. Here, vacuum must work its way through a **sintered**, or sponge-like, metal disc to reach the distributor. Many GM engines also use this type of spark delay valve.

Another method of vacuum delay is used in Chrysler's orifice spark advance control (OSAC) system, figure 9-19. A small **orifice** is placed in the vacuum line to delay vacuum buildup.

Manufacturers often use vacuum delay valves in combination with other emission control systems. All valves operate on one of the two principles described above.

Speed- and transmission-controlled timing
These systems prevent any distributor vacuum advance when the car is in a low gear or is traveling slowly. A solenoid controls the application of vacuum to the advance mechanism, figure 9-20. Current flow through the solenoid is controlled by a switch that reacts to vehicle operating conditions.

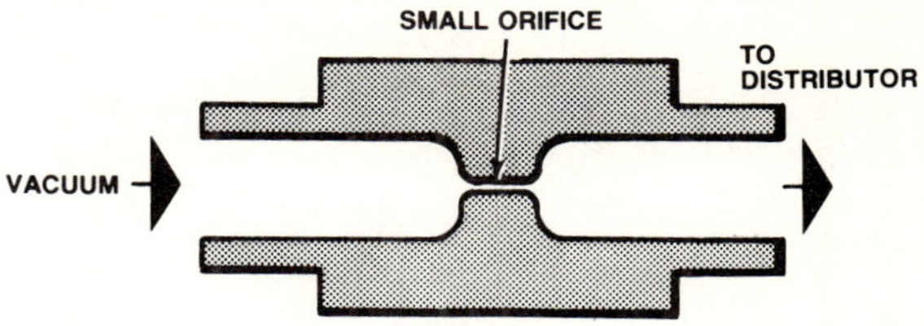

Figure 9-19. A cross-section view of a vacuum delay valve that uses a small orifice to delay the application of vacuum.

Sintered: Welded together without using heat to form a porous material, such as the metal disc used in some vacuum delay valves.

Orifice: A small opening in a tube, pipe, or valve.

■ Starting The Locomobile

If you own a 1927 or '28 Locomobile, you had better know how to-adjust the spark advance lever when you start your car. The spark lever allows you to advance or retard spark timing for all conditions. Here are the owner's manual instructions for positioning the lever when starting your Locomobile:

1. Be sure gear shift lever is in neutral position and the hand brake is applied.
2. Retard the spark lever, which is at right center of steering wheel, about one half way down from the full advance position.
3. The throttle control lever is to the left of center of the steering wheel. Move lever up an inch or more from the closed position.
4. Turn ignition switch arm to the "on" position.
5. Pull choke button as far as it will go, then place the right foot on the starting switch button and press down. On pushing down on the starting switch button, the starter gear will engage itself in the teeth of the flywheel and the motor will start to turn over. As soon as the motor starts, release the starting button and push the choke button part way in. As the motor warms up a little, push the choke button all the way in. Do this as soon as possible to avoid excessive choking, as any excess gasoline thrown into the cylinders will pass by the rings into the oil in the crankcase, thus diluting and partly destroying its lubricating value.
6. Advance spark by pushing lever up as far as it will go and bring the throttle lever toward the closed position to slow the motor down.

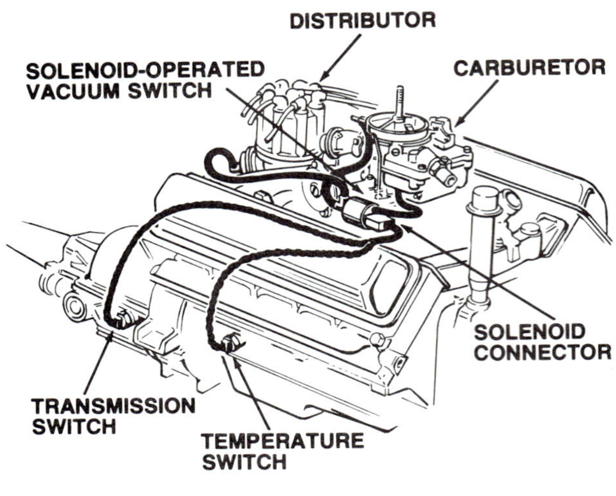

Figure 9-21. A typical General Motors transmission-controlled spark system.

A control switch used with a manual transmission reacts to shift lever position. A control switch used with an automatic transmission will usually react to hydraulic fluid pressure. Both systems prevent any vacuum advance when the car is in a low or intermediate gear.

A speed-sensing switch may be connected to the vehicle speedometer cable. The switch signals an electronic control module when the vehicle speed is below a certain specified level. The module triggers a solenoid that controls engine vacuum at the distributor.

Both vacuum-delay systems and speed- and transmission-controlled systems usually have an engine temperature bypass. This allows normal vacuum advance at high and low engine temperatures. Before March, 1973, some systems had an ambient temperature override switch. Most of these switches were discontinued at the direction of the Environmental Protection Agency (EPA). Later temperature override systems sense coolant temperature or under-hood temperature.

The systems using these principles are known by many different trade names, and are all somewhat different. The following paragraphs describe the major systems used by domestic manufacturers.

General Motors

Transmission-controlled spark (TCS)

The TCS system was introduced by GM on its 1970 automobiles. A solenoid-operated vacuum switch near the carburetor, figure 9-21, prevents vacuum advance in low and intermediate gears. Vacuum is blocked, and the vacuum advance unit is vented to the atmosphere.

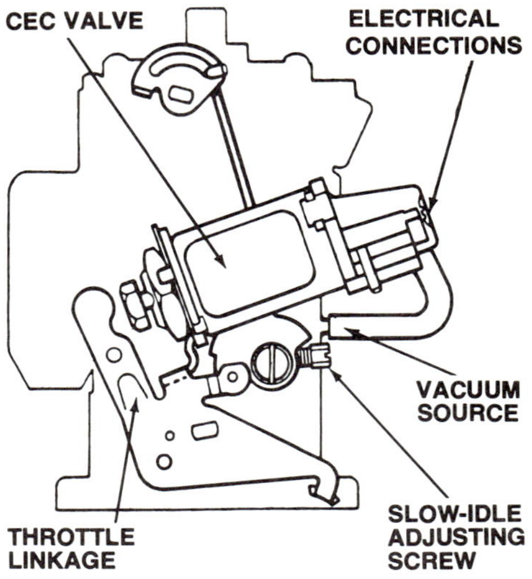

Figure 9-22. The General Motors combination emission control valve.

In high gear, the solenoid closes the vent and opens the vacuum port. This applies vacuum to the distributor advance unit. The solenoid is controlled by the transmission switch, which senses shift lever position on manual transmissions and fluid pressure on automatics.

Transmission control switch and solenoid operation on GM systems varies for different models and years. In some systems, the transmission switch is closed in low and intermediate to activate the solenoid and deny vacuum advance. In other systems, the switch is open and the solenoid deactivated to deny vacuum advance. In either case, switch and solenoid positions change from their low-gear positions to allow vacuum advance when the transmission is shifted into high gear.

A coolant temperature switch provides manifold vacuum to the distributor at low coolant temperatures. Some GM systems also have a hot coolant override switch for vacuum advance in all gears at high coolant temperatures. A time delay relay opens the solenoid vacuum switch during starting and for the first few seconds after the engine is started. This provides vacuum advance for easier starting and warmup.

Combination emission control (CEC) valve

The CEC valve is part of the transmission-controlled spark system on some 1971 through 1973 GM cars. This valve provides vacuum spark advance control and deceleration throttle position control in high gear. The CEC valve is mounted on the side of the carburetor, figure 9-22, and is a simple solenoid with a vacuum

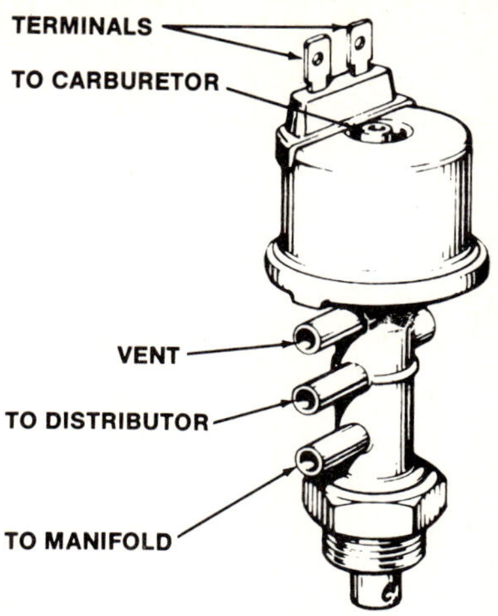

Figure 9-23. A distributor vacuum control switch.

valve at one end and a throttle check rod extending from the other.

When the transmission is in low and intermediate gears, the solenoid is deenergized and vacuum to the distributor advance unit is shut off. The vacuum is vented to the atmosphere through a filter on the CEC valve.

When the transmission is in high gear, the TCS switch energizes the CEC solenoid to allow vacuum to the distributor advance unit. At the same time, the CEC throttle rod extends to hold the throttle open by a predetermined amount during high-gear deceleration. When the transmission is shifted to a lower gear, the CEC valve is deenergized and the throttle rod retracts, allowing a normal idle and cutting off vacuum advance.

Distributor vacuum control switch
The distributor vacuum control switch, figure 9-23, is used on some Oldsmobiles along with a normally closed TCS switch. The vacuum control switch does the dual jobs of a TCS solenoid and a thermal vacuum switch. When the engine is at normal operating temperatures, it permits vacuum spark advance only when the transmission is in high gear. When engine coolant temperature increases, the switch applies full manifold vacuum to the distributor, regardless of transmission gear position.

When the transmission is in low or intermediate gears, the TCS switch is closed to energize the solenoid. The solenoid seals off the carburetor port and vents the distributor vacuum unit to the atmosphere. During high-gear operation, the deenergized solenoid seals off the vent port and applies carburetor vacuum to the distributor.

In case of engine overheating, due to prolonged idling with a retarded spark, the thermal portion of the switch seals off the vent port and applies manifold vacuum to the distributor.

Some Buick engines use a combination thermostatic vacuum switch and TCS solenoid that functions in the same way as the distributor vacuum control switch described above.

Speed control switch (SCS) system
The SCS system, used on 1972 Cadillacs and Pontiacs, controls distributor vacuum advance by vehicle speed rather than by transmission gear selection.

A speed sensor switch in the speedometer cable or in the transmission speedometer drive gear assembly controls a normally deenergized vacuum solenoid. At low speeds, the vacuum solenoid is energized because the speed switch is closed. The distributor vacuum unit is vented to the atmosphere, denying vacuum spark advance. At higher speeds, the speed switch opens and the solenoid is deenergized. Normal vacuum advance is allowed. If the engine begins to overheat, a thermostatic vacuum switch applies manifold vacuum to the distributor regardless of vehicle speed.

American Motors

Before 1971, AMC products used dual-diaphragm vacuum units and deceleration valves to control exhaust emissions. Some models have a thermostatic vacuum valve, figure 9-24, that allows normal vacuum advance at low engine temperatures. As engine temperature increases, the valve connects manifold vacuum to the distributor advance unit.

After 1971, many AMC products use a speed- or transmission-controlled spark timing system similar to General Motors' TCS. A control switch reacts to vehicle operating conditions and controls current flow to a solenoid vacuum valve, figure 9-25.

Transmission control switches on AMC systems are closed in low and intermediate gears to energize the solenoid and block vacuum advance. In high gear, the switch opens, the solenoid deenergizes, and vacuum advance is applied.

Before March 1973, cars with automatic transmissions used a speed-sensing control switch. On later cars, fluid pressure operates the automatic transmission control switch. On cars with manual transmissions, gear lever position activates the control switch.

Some AMC cars have an ambient temperature override switch that allows normal vacuum advance at low temperatures. Later products may have a coolant temperature override switch that applies manifold vacuum to the distributor

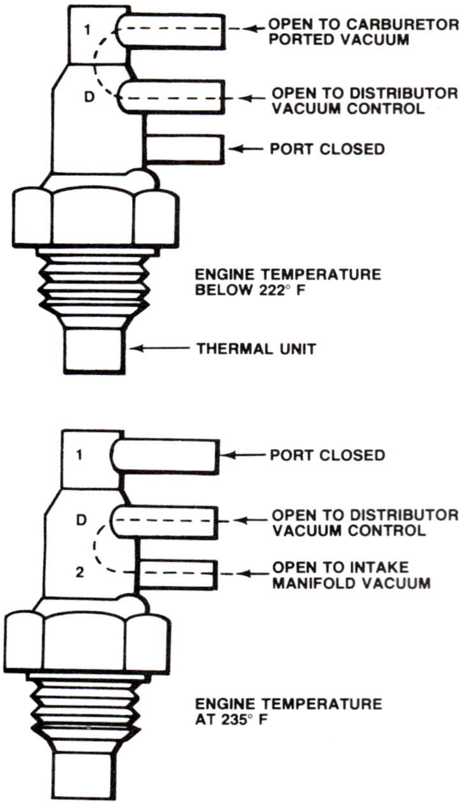

Figure 9-24. American Motors' thermostatic vacuum valve.

when coolant temperature is low, and carburetor ported vacuum when coolant temperature is high. When this coolant switch is used with a transmission control switch, the carburetor vacuum application is controlled but manifold vacuum is *always* applied when coolant temperature is low.

Chrysler

NO_x spark control

Chrysler products for 1971 and 1972 sold in California have a transmission-controlled spark system similar to those of GM and AMC. The Chrysler NO_x control system for cars with manual transmissions has a normally open transmission switch and a deenergized solenoid vacuum valve. When the transmission is in low and intermediate gears, the switch is closed and the solenoid is energized to cut off vacuum advance. When the transmission is in high gear, the switch opens and the solenoid is deenergized to provide vacuum advance.

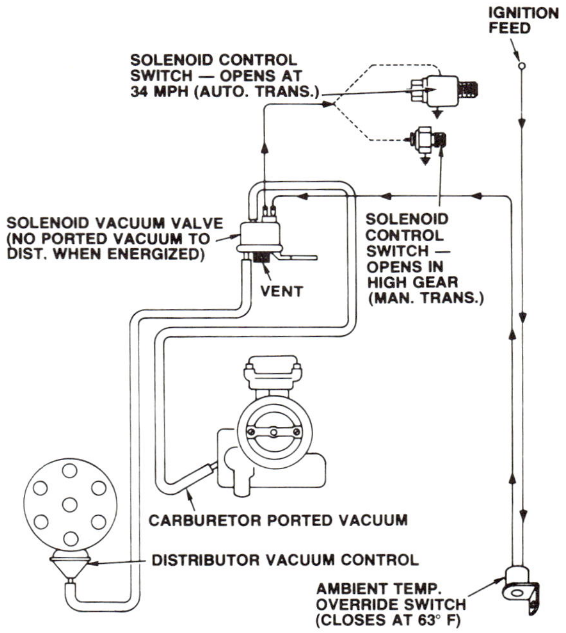

Figure 9-25. The AMC transmission-controlled spark system. (AMC)

An ambient temperature switch was used on 1971 cars to allow vacuum advance at all times when the temperature is low. This switch was discontinued in 1972.

The NO_x control system on cars with automatic transmissions uses an electric control unit, figure 9-26, to actuate the solenoid vacuum valve. The control unit is a reversing relay that receives signals from grounding switches: a temperature switch, a speed switch, and a vacuum switch (the vacuum switch was removed in 1972). All switches must be open to activate the control unit, energize the solenoid, and cut off vacuum advance.

Orifice spark advance control (OSAC) system

The OSAC system is used on 1973 and later models. The OSAC valve is installed in the vacuum line between the carburetor and the distributor, figure 9-27. It contains a small orifice that delays the rate of vacuum buildup in the distributor vacuum unit during acceleration from idle to part throttle. When the throttle is closed, vacuum is cut off to the OSAC valve. A check valve within the OSAC valve opens to relieve vacuum in the line to the distributor.

The first OSAC valves used in 1973 had a temperature control device that bypassed OSAC operation and allowed immediate vacuum advance at low ambient temperature. The temperature control device was eliminated in March of 1973, and the OSAC valve was moved

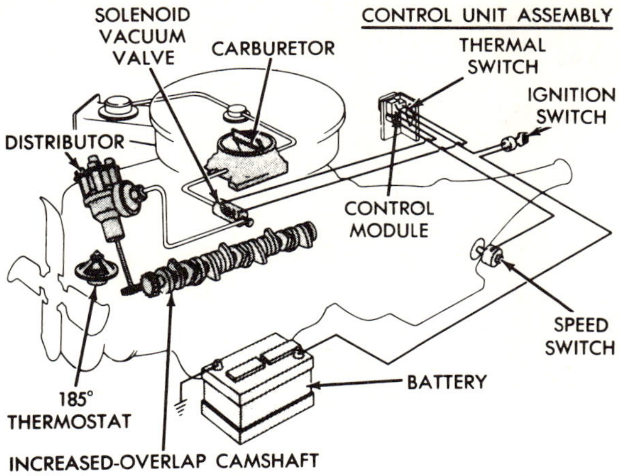

Figure 9-26. Chrysler's NO_x control system. (Chrysler)

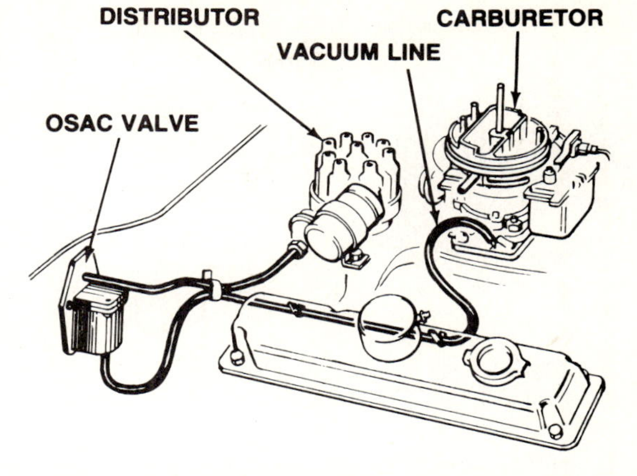

Figure 9-27. The Chrysler orifice spark advance control system.

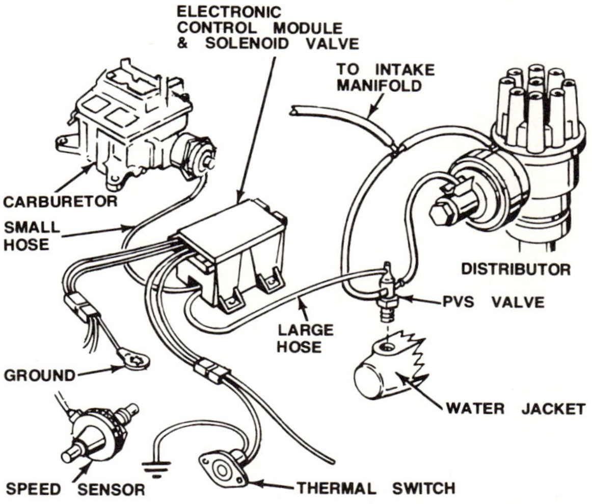

Figure 9-28. Ford's electronic distributor modulator system.

from its original location on the firewall to the air cleaner cover.

Some OSAC installations include a thermal ignition control valve, which is another type of coolant-sensing valve. It is installed in the engine cooling system, and the vacuum lines from the OSAC valve to the distributor run through it. A manifold vacuum line also runs to the thermal valve. When coolant temperature increases, the thermal valve applies full manifold vacuum to the distributor.

When the catalytic converter was introduced in 1975 cars, the OSAC valve was eliminated from many models. However, it continues on some 1975 and later Chrysler products.

Ford

Electronic distributor modulator (EDM) system

The EDM system was used on some 1970 and 1971 Ford vehicles. The system allows vacuum spark advance only during high-gear operation.

The system consists of a speed sensor, an ambient temperature switch, a coolant temperature switch, an electronic control module, and a 3-way solenoid valve, figure 9-28.

The speed sensor is connected between two sections of the speedometer cable. A rotating magnet, driven by the cable, turns inside a stationary winding. Above a selected speed, the voltage induced by the magnet is impressed on the electronic control module.

The ambient temperature switch is mounted near either front door hinge and senses outside air temperature. It will override the speed sensor signal when air temperature is low, to permit the distributor vacuum advance unit to operate normally.

The coolant temperature switch is also called the distributor vacuum control valve or the ported vacuum switch (PVS). It overrides the module to apply intake manifold vacuum to the distributor when coolant temperature is high.

The electronic control module contains the 3-way solenoid valve and is mounted in the passenger compartment under the dash. The module receives signals from the speed sensor and ambient temperature switch. It determines whether or not the solenoid will be energized.

If the ambient temperature is great enough, the solenoid will be energized at speeds above 23 mph. Carburetor port vacuum is then applied to the distributor vacuum unit. On deceleration, the reduced voltage signal from the speed sensor deactivates the solenoid. Carburetor vacuum is then shut off from the advance unit.

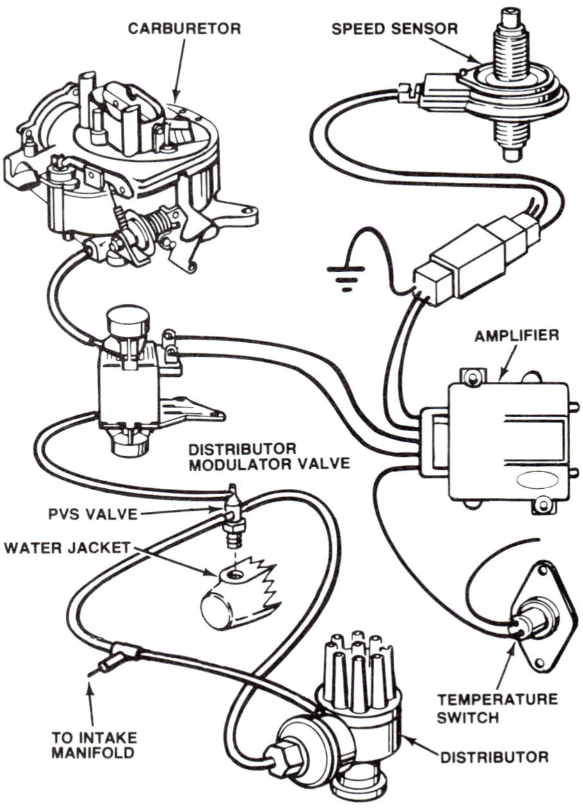

Figure 9-29. The Ford electronic spark control system.

Electronic spark control (ESC) system
The ESC system, used on some 1972 Ford products, is very similar to the EDM system just described. The main difference is that the ESC module is separate from the solenoid valve, figure 9-29, while the earlier EDM module contained the solenoid valve.

Transmission-regulated spark (TRS) system
The TRS system was introduced by Ford in 1972 and, like the EDM and ESC systems, denies vacuum spark advance at low speed. The system consists of a 2-way distributor solenoid valve, a transmission switch, and an ambient temperature switch, figure 9-30. In some applications, a spark delay valve is also used.

The distributor solenoid valve is installed between the carburetor and the distributor. In low and intermediate gears, the transmission switch is closed to energize the solenoid and cut off the vacuum to the distributor.

When the transmission is shifted into high gear, the transmission switch opens, the solenoid is deenergized, and vacuum is applied to the distributor.

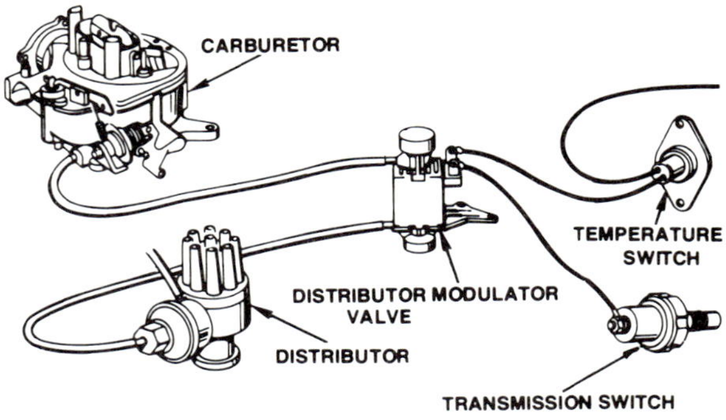

Figure 9-30. The Ford transmission-regulated spark system.

At low temperatures, the ambient temperature switch contacts open to deenergize the solenoid and allow normal vacuum advance in all gears. As temperature increases, the contacts close, and the solenoid is controlled by the transmission switch.

On some applications, the vacuum lines are routed through a ported vacuum switch like that used in the EDM and ESC systems. When coolant temperature increases, manifold vacuum is applied to the distributor by the ported vacuum switch.

When exhaust gas recirculation was introduced in 1973, the TRS system on some cars was modified to become the TRS + 1 system. The "+ 1" part of the system controls EGR valve operation.

The TRS and TRS + 1 systems were discontinued in 1974.

Spark delay valve (SDV) system
This system, figure 9-31, uses a spark delay valve to restrict the vacuum applied to the carburetor during acceleration. The valve is installed between the carburetor and the distributor, and vacuum must work its way through a sintered metal disc to reach the advance diaphragm. When carburetor vacuum decreases, a check valve in the SDV opens to relieve vacuum from the advance unit.

On engines built before March 1973, a delay valve bypass is used to allow full advance at low ambient temperature. An ambient temperature switch controls a solenoid vacuum valve. At low temperature, the switch closes to energize the solenoid and apply vacuum to the distributor, bypassing the SDV.

Temperature-activated vacuum (TAV) system
The TAV system is used on some Ford products to match the vacuum spark advance to engine requirements by switching between two vacuum sources, figure 9-32. The 3-way vacuum solenoid valve is connected to the distributor,

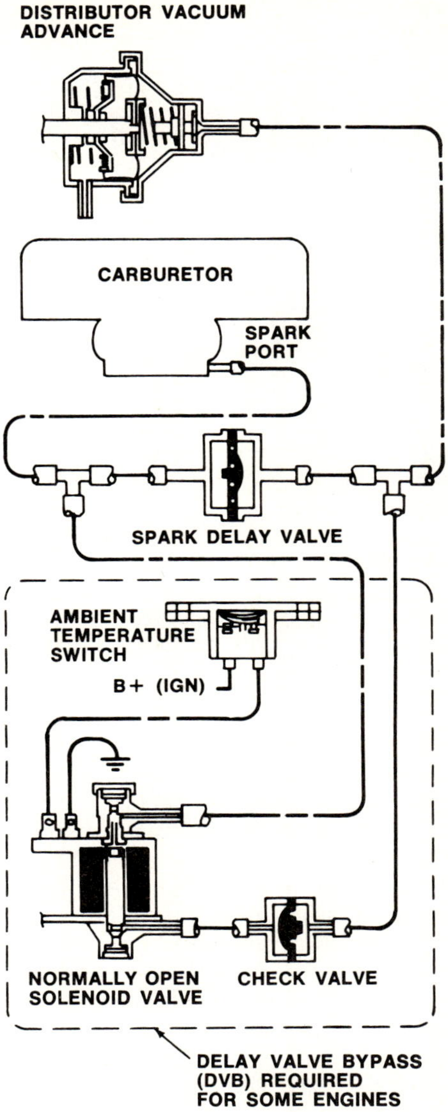

Figure 9-31. Ford's spark-delay valve system.

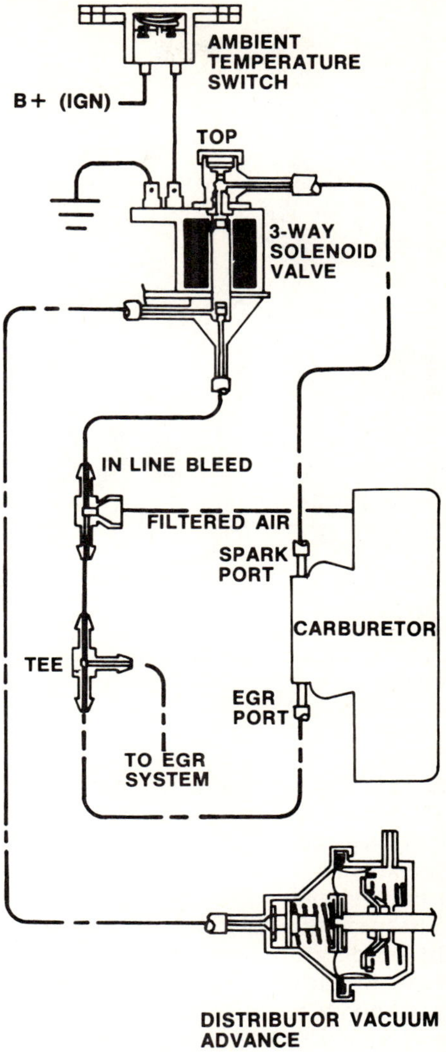

Figure 9-32. Ford's temperature-activated vacuum system.

the carburetor spark port, and the carburetor EGR port.

An **ambient temperature** switch controls the 3-way valve. At high temperatures, the switch is closed, the solenoid is energized, and EGR vacuum controls the spark advance. At lower temperature, the switch is open, the solenoid is deenergized, and carburetor spark port vacuum controls the vacuum advance.

A variation of the TAV system is the cold-temperature-activated vacuum (CTAV) system. In this system, the ambient temperature switch is located in the air cleaner. The temperature required to energize the solenoid is greater than in the TAV system, and a latching relay is used to allow only one cycle of the temperature switch each time the ignition is turned on.

ELECTRONICALLY CONTROLLED TIMING

Extremely accurate ignition system performance is needed to meet emission standards and fuel mileage requirements. Centrifugal and vacuum advance mechanisms often cannot respond quickly enough to changes in engine operating

Ambient Temperature: The temperature of the air surrounding a particular device or location.

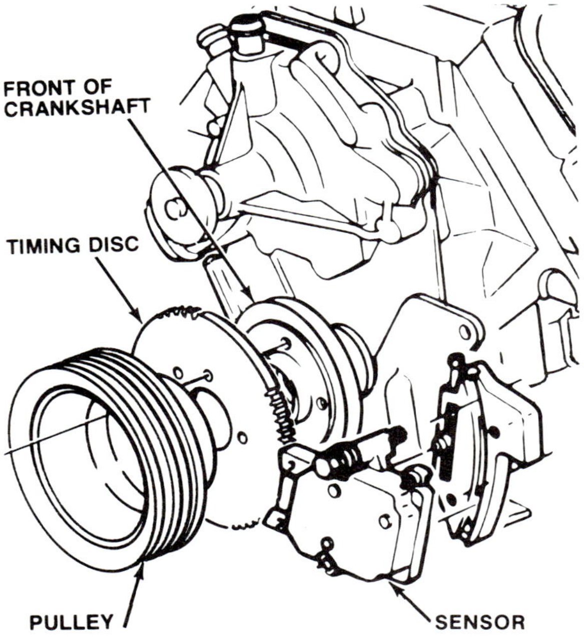

Figure 9-33. Crankshaft position signals can be taken directly from the crankshaft.

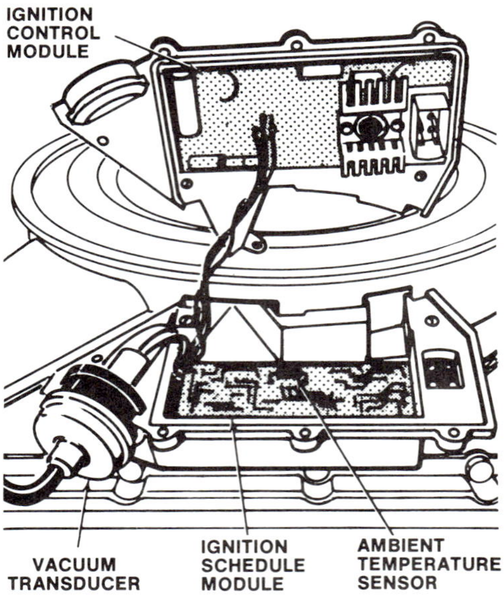

Figure 9-34. The early Electronic Lean-Burn computer contains two separate modules. Later models have one module performing all functions.

conditions. Major manufacturers are developing computer-controlled ignition systems to gain this necessary accuracy.

In these systems, an electronic control module receives signals from various system sensors. These signals may include information on coolant temperature, atmospheric pressure and temperature, throttle position and rate of change of position, and crankshaft position. Integrated circuits in the control module are programmed to interpret this information and calculate the proper ignition timing for each individual spark.

The systems currently used work with the manufacturers' standard solid-state ignition systems. Some changes are made to the standard ignition, because it no longer has to control spark timing, but many components remain the same. We will study these ignition systems in detail in Chapter 11. The following material is concerned only with the spark timing function of these systems.

Two types of computer-controlled ignition systems are used by domestic manufacturers. One type depends upon distributor shaft rotation to send a crankshaft position signal to the control module. The other type of system receives crankshaft position information from a sensor mounted near the crankshaft, figure 9-33. The sensor reacts to the rotation of a special disc attached to the crankshaft itself.

Signals taken directly from the crankshaft are much more accurate than those taken from the distributor shaft. The gears or chain driving the camshaft and the gears driving the distributor shaft have tolerances, or looseness. While small in actual measurements, these tolerances can combine to cause a significant difference between crankshaft position and ignition timing.

These high-technology systems are still being developed, studied, and evaluated. It may be a number of years before they are accepted for industry-wide use. Chrysler, GM, and Ford are using such systems. They are called:

- Chrysler Electronic Lean-Burn (ELB)
- GM Microprocessed Sensing and Automatic Regulation (MISAR)
- Ford Electronic Engine Control (EEC).

The electronic timing-regulation function of all the systems is similar, but the electronics are fundamentally different. The Lean-Burn system uses an **analog** computer, while MISAR and EEC use **digital** microprocessors.

The practical difference between analog and digital electronics in this kind of application is that a digital computer can instantly alter timing, for example, from 1 to 65 degrees. An analog computer must calculate through all the points on a theoretical curve to make such an adjustment. Since an electronic spark advance adjustment takes only a few milliseconds, this fact is not really significant to the driver or serviceman. However, a digital system is more

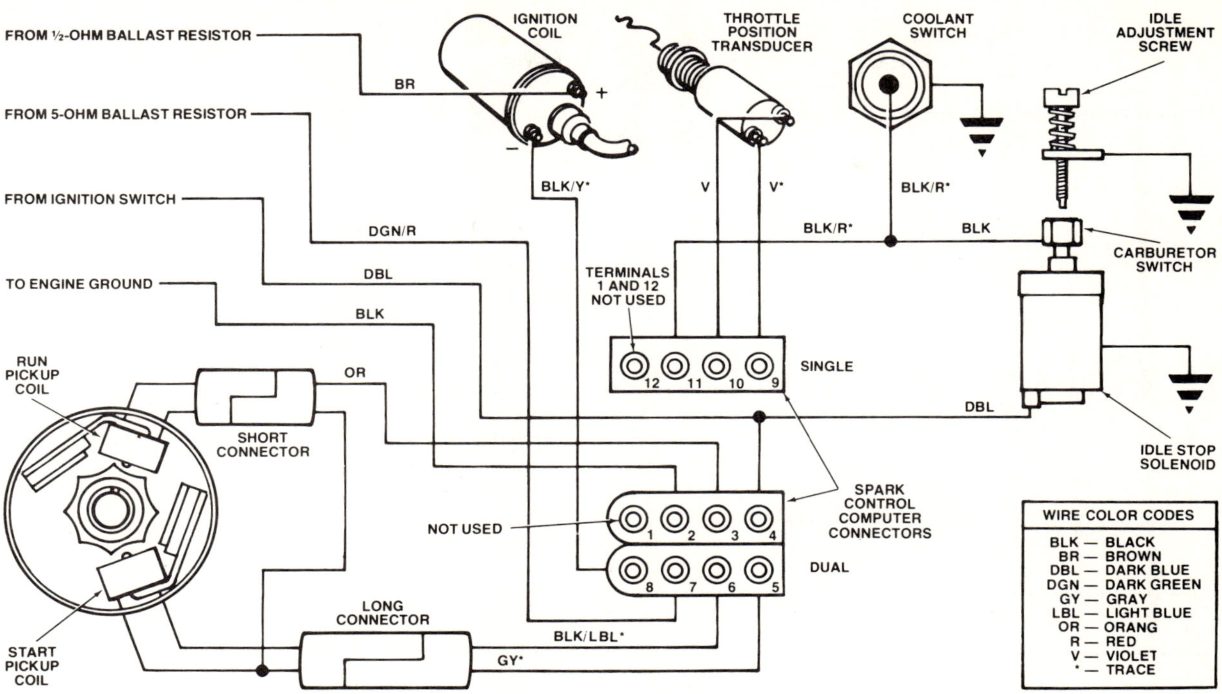

Figure 9-35. 1976-77 (except 318 V-8) Lean-Burn system wiring diagram.

flexible and, at this time, more economical to build than an analog system.

Chrysler

Chrysler's ELB has been used on some V-8 engines since 1976. Its use has steadily expanded since then.

The spark control computer used with V-8 engines is mounted on the air cleaner, figure 9-34. With the 4-cylinder engine, it is mounted on an inner fender panel. Early model computers have two circuit boards: the ignition schedule module and the ignition control module. The ignition schedule module receives signals from various engine sensors and computes them to determine the exact spark timing required. It then directs the ignition module to advance or retard the timing accordingly. Computers for 1977 318-cid engines and all 1978 and later systems have only one circuit board that does both jobs.

Six or seven sensors, figures 9-35, 9-36, and 9-37, feed information to the computer:

- The Start pickup coil in the distributor provides a fixed amount of advance during cranking.
- The Run pickup coil supplies a basic timing signal and allows the computer to determine engine speed. The system for 1977 318-cid engines and all 1978 and later V-8 systems use only one pickup coil to provide all timing signals to the computer. The 4-cylinder engine's distributor contains a Hall-effect pickup assembly. This unit is explained in Chapter 11.
- The coolant temperature sensor signals the computer when coolant temperature is low. On V-8 engines, the sensor is on the water pump housing. On 4-cylinder engines, the sensor is on the thermostat housing.
- The air temperature sensor is located inside the early model computer. It is a **thermistor** that provides a varying amount of electrical resistance with changing air temperature. The 318-cid engine and all 1978 and later systems have no air temperature sensor.

Analog: Describing a computer that uses similar (analogous) electrical signals to make its calculations.

Digital: Describing a computer that makes calculations with quantities represented electronically as digits.

Thermistor: An electrical conductor that changes its resistance as the surrounding temperature changes.

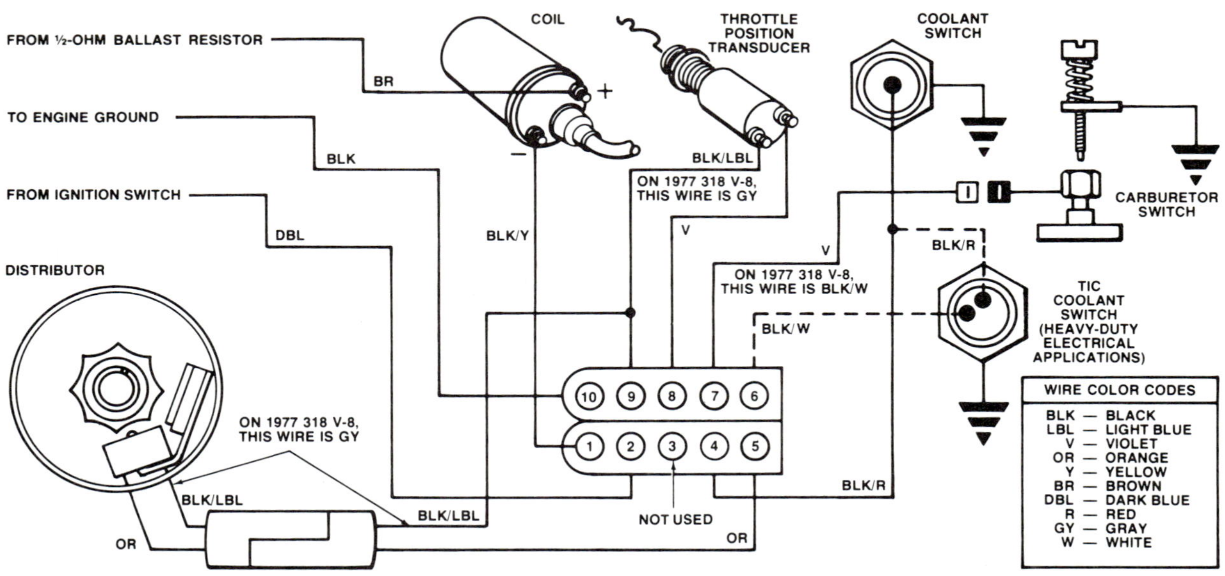

Figure 9-36. 1977 318 V-8 and all 1978 and later V-8 Lean-Burn system wiring diagram.

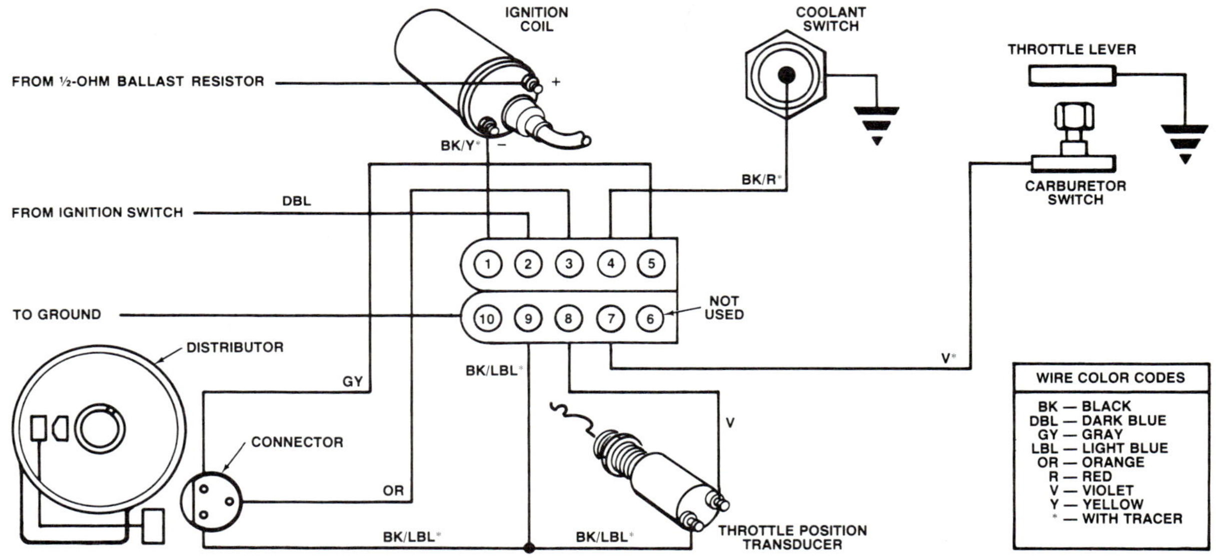

Figure 9-37. The 4-cylinder engine Lean-Burn system wiring diagram.

• The carburetor switch sensor tells the computer if the engine is at idle or off idle.

The remaining two sensors are **transducers**, devices that change mechanical movement to an electrical signal of varying voltage. The varying voltage signal is interpreted by the computer:

• The throttle position transducer is connected to the throttle lever. The varying voltage signal tells the computer the position and rate of change of position of the throttle plates.

• The vacuum transducer contains a diaphragm that is exposed to engine vacuum. Movement of the diaphragm signals the computer of changes in engine vacuum.

General Motors

The MISAR system, figure 9-38, was introduced on 1977 Oldsmobile Toronados. The spark control microprocessor is contained in a control module that is mounted under the instrument panel in the passenger compartment. The mod-

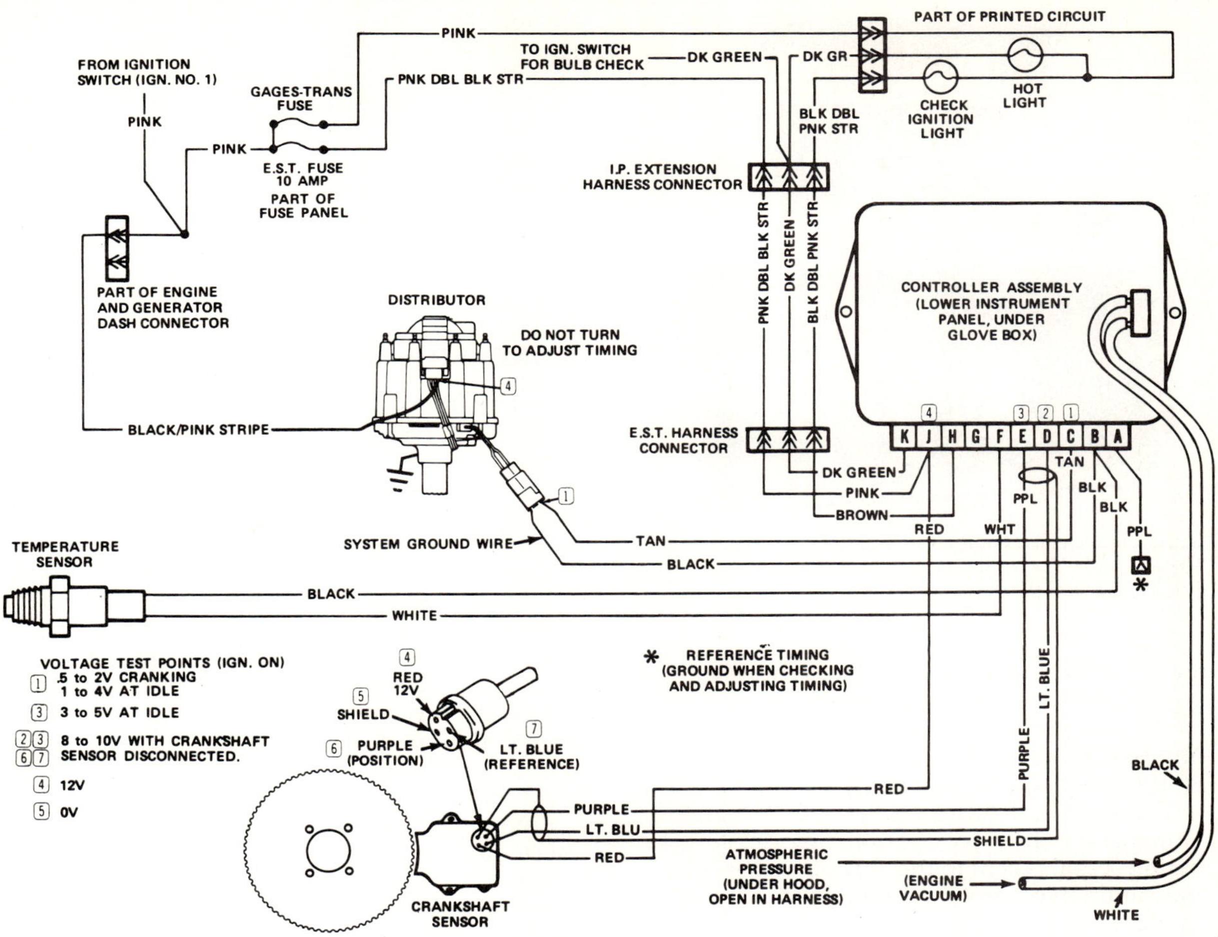

Figure 9-38. A circuit diagram of the General Motors MISAR system, with the early model crankshaft position sensor. Later systems have the sensor mounted in the distributor. (Oldsmobile)

ule monitors signals from these engine sensors:

- Coolant temperature
- Manifold vacuum
- Atmospheric pressure
- Crankshaft speed and position.

The coolant sensor is not a simple on-off switch, but rather a thermistor, supplying varying amounts of resistance with changes in temperature.

The vacuum sensor is located in the control module. The sensor is a solid-state unit and is connected to the intake manifold by a vacuum line. A second line connected to the module is open in the engine compartment, to provide an atmospheric pressure signal.

On 1977 systems, crankshaft speed and position signals are provided by a rotating disc and a stationary sensor on the front of the engine, shown in figure 9-33. These components replace the pickup coil and trigger wheel of the distributor. In 1978, the crankshaft speed and position sensor was moved into the distributor.

Ford

In 1978, Ford introduced its first generation electronic engine control (EEC-I) system on the Lincoln Versailles. The system controls both spark timing and EGR valve operation. A digital microprocessor installed in the passenger compartment, figure 9-39, receives signals from various sensors.

An electromagnetic pickup at the flywheel end of the crankshaft signals crankshaft position and speed. It also replaces the pickup coil and trigger wheel of the distributor. Other infor-

Transducer: A device that changes one form of energy into another. In an ignition system, it may sense a mechanical movement and change it to an electrical signal.

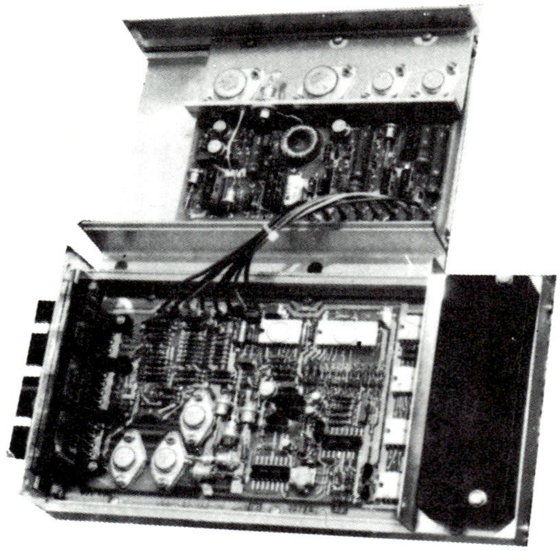

Figure 9-39. The control module of Ford's Electronic Engine Control system.

mation is fed to the computer by a throttle position sensor, a coolant temperature sensor, a barometric pressure sensor, an inlet temperature sensor, a manifold pressure sensor, and an EGR valve position sensor, figure 9-40.

The module determines the optimum spark timing and EGR valve operating mode. A spark timing signal is sent to the Dura-Spark solid-state ignition control module. This module is similar to that used with the standard Ford Dura-Spark ignition, as we will see in Chapter 11.

The EEC-I distributor is different from the standard distributor, however. It contains only a rotor, a coil electrode, and spark plug electrodes because its only function is to distribute the ignition spark. The distributor cap and rotor are designed to minimize the chances of **cross-firing**, when the spark jumps to the incorrect spark plug electrode.

Cross-Firing: Ignition voltage jumping from the distributor rotor to the wrong spark plug electrode inside the distributor cap. Also, ignition voltage jumping from one spark plug cable to another due to worn insulation.

EEC-II was introduced in 1979 on some 351 V-8 engines. The new system has all of EEC-I's features, plus it controls the engine air-fuel ratio through an electronically-adjustable carburetor. We will learn more about this part of EEC-II in Chapter 19. The ignition system portion of EEC-II is the same as that used in EEC-I. The EEC-I system is still used on the Lincoln Versailles' 302 V-8 engine.

SUMMARY

Most ignition distributors have two devices to advance or retard spark timing in response to changing conditions of engine operation. The centrifugal advance alters the position of the distributor cam or trigger wheel on the distributor shaft. It changes timing to match engine speed. The vacuum advance alters the position of the breaker points or magnetic pickup coil with respect to the cam or trigger wheel. It changes timing to match engine load.

Ignition timing affects NO_x formation in the cylinders and the presence of HC in the exhaust. Retarded ignition timing reduces these emissions but also hurts performance and fuel economy.

Many different spark timing emission control systems have been used. Regardless of the variation of their design and operation, these systems all regulate distributor vacuum advance. Vacuum advance is generally allowed only at cold startup, during high gear operation, and if the engine overheats.

Auxiliary spark timing control systems are not used on all engines. Development of the exhaust gas recirculation (EGR) systems, catalytic converters, and the expanded use of air injection systems have gradually reduced the use of TCS and SCS systems on late-model engines.

Electronic spark timing systems use computers to regulate all ignition timing for the best combination of emission control, fuel economy, and performance.

Review Questions

Select the single most correct answer.
Compare your answers with the correct answers on page 280.

1. Which of the following affect ignition timing?
   a. Valve timing
   b. Compression ratio
   c. Engine load
   d. All of the above

2. Distributors commonly have two automatic spark advance mechanisms, which are:
   a. Initial advance and mechanical advance
   b. Initial advance and dynamic advance
   c. Dynamic advance and static advance
   d. Centrifugal advance and vacuum advance

3. Centrifugal advance responds directly to engine:

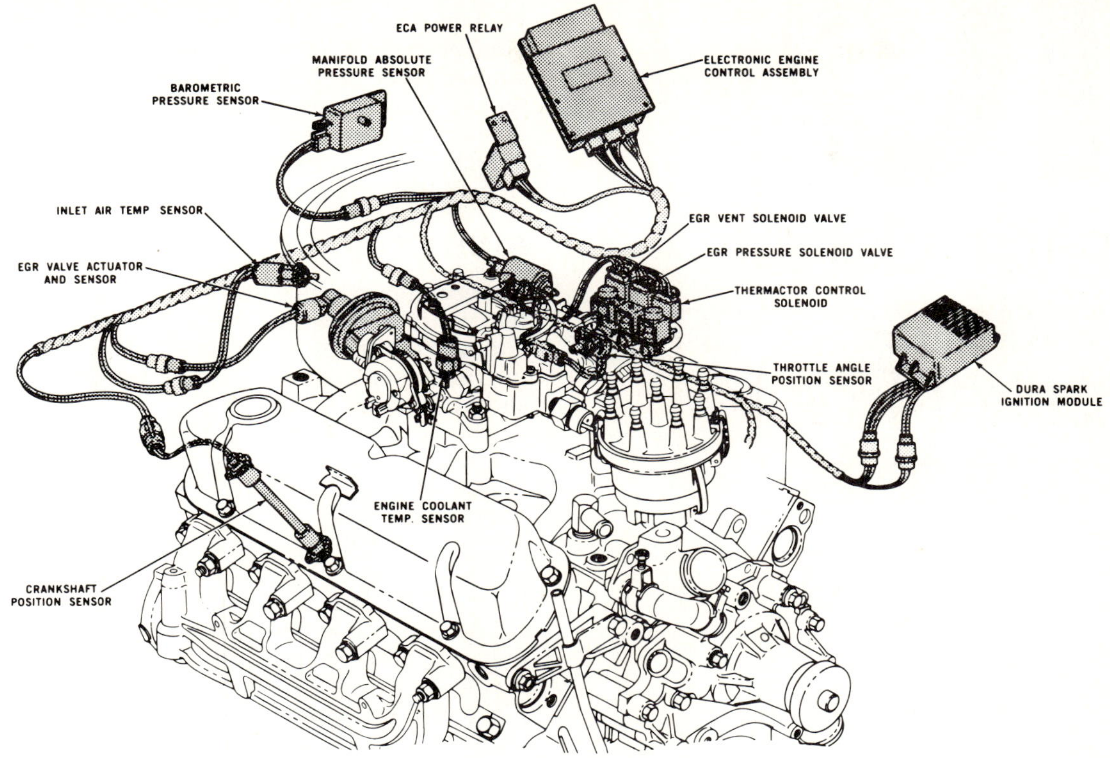

Figure 9-40. Ford's Electronic Engine Control system. (Ford)

a. Load
b. Horsepower
c. Speed
d. Torque

4. The purpose of the venturi in the carburetor is to:
   a. Mix the air and fuel mixture
   b. Speed the airflow and create a low-pressure area
   c. Carry fuel to mix with air
   d. Raise the air pressure sufficiently

5. Vacuum advance responds directly to engine:
   a. Load
   b. Horsepower
   c. Speed
   d. Compression ratio

6. Which of the following is *not* a component of the vacuum advance mechanism?
   a. Movable breaker plate
   b. Weight springs
   c. Diaphragm
   d. Actuating link

7. Automotive emission control equipment is designed to reduce exhaust emissions of which pollutants?
   a. Hydrocarbons
   b. Oxides of nitrogen
   c. Carbon monoxide
   d. All of the above

8. The most common types of recent spark timing control systems do *not* include:
   a. Decelerator vacuum advance valves
   b. Vacuum delay valves
   c. Speed-controlled timing
   d. Transmission-controlled timing

9. The illustration below shows a:

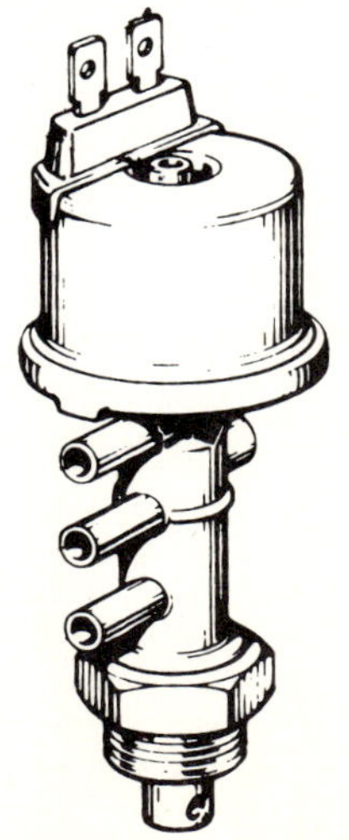

   a. Spark delay valve
   b. OSAC valve
   c. Deceleration vacuum advance valve
   d. Distributor vacuum control switch

10. Which of the following is *not* part of the General Motors TCS system?
    a. A CEC valve
    b. A spark delay valve
    c. A solenoid-operated vacuum switch
    d. A transmission control switch

11. Almost all spark timing emission control systems work with the vacuum advance units to:
    a. Increase vacuum advance at low and intermediate vehicle speeds
    b. Cut off vacuum advance during high-speed acceleration
    c. Retard the timing during high-speed deceleration
    d. Cut off vacuum advance at low and intermediate vehicle speeds

12. The Ford EDM and ESC spark timing control systems have vehicle speed sensors mounted in the:
    a. Transmission
    b. Distributor housing
    c. Speedometer cable
    d. None of the above

10

Ignition Coils, Condensers, and Breaker-Point Distributors

We have explained the general operation of a breaker-point ignition system, and we have studied some of the system parts. In this chapter, we will examine ignition coils and condensers in more detail. Then, we will look at breaker-point distributors, including the specific units used by major manufacturers.

IGNITION COILS

The ignition coil, as we have seen, is made of two windings of copper wire. To increase the strength of the coil's magnetic field, the windings are installed around a laminated core of soft iron, figure 10-1.

The primary winding is 100 to 200 turns of a heavy wire. The secondary winding is 15,000 to 30,000 turns of a thin wire. The ratio of secondary turns to primary turns is usually between 100 to 1 and 200 to 1, as we learned in Chapter 8.

The coil must be protected from the underhood environment, or its efficiency could decrease. The coil windings are insulated by layers of paper, and the entire case is filled with oil or wax for more insulation. The top of the coil is molded from an insulating material, often bakelite. Metal inserts for the winding terminals are installed in the cap. The entire unit is sealed, to keep out dirt and moisture. An ignition coil cannot be repaired, but must be replaced if defective.

Coil Installation

Ignition coils are usually mounted with a bracket on a fender panel in the engine com-

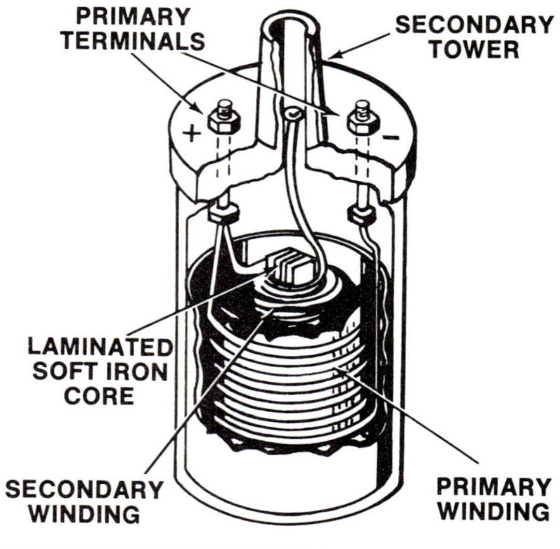

Figure 10-1. The laminated iron core within the coil strengthens the coil's magnetic field.

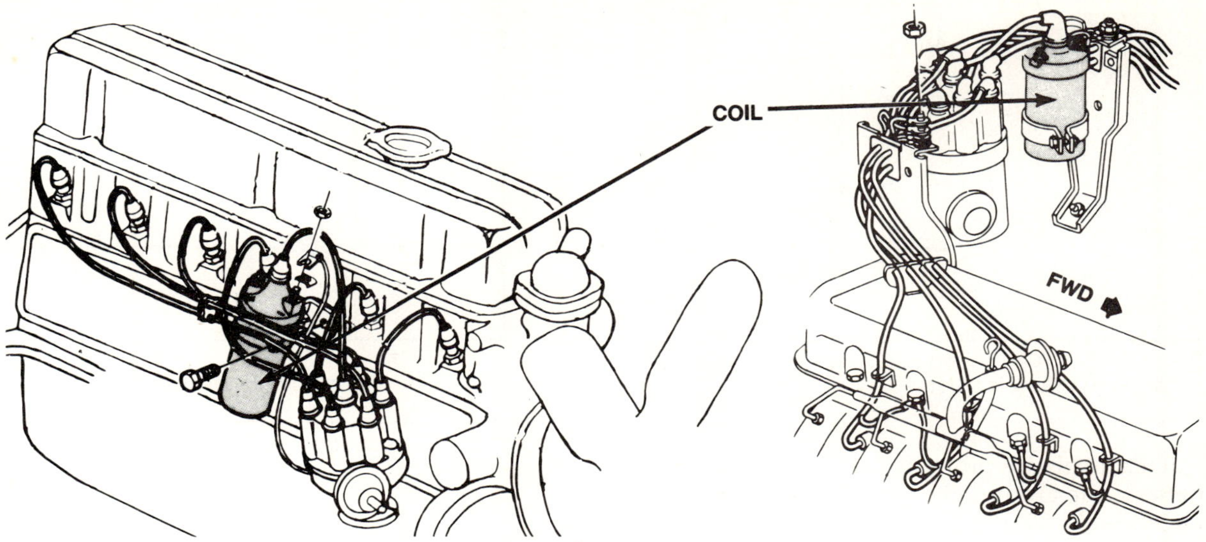

Figure 10-2. Ignition coils are commonly mounted on the engine (left) or on a fender panel (right). (Chevrolet)

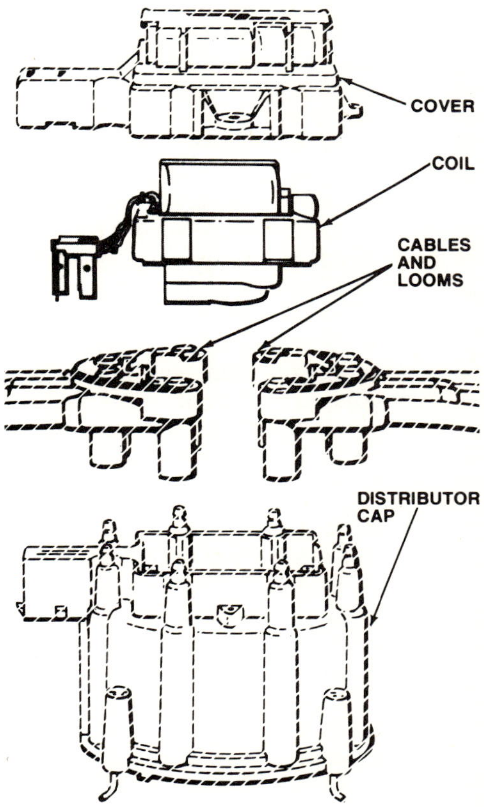

Figure 10-3. Most of Delco-Remy's HEI solid-state ignition systems have a coil mounted on the distributor cap.

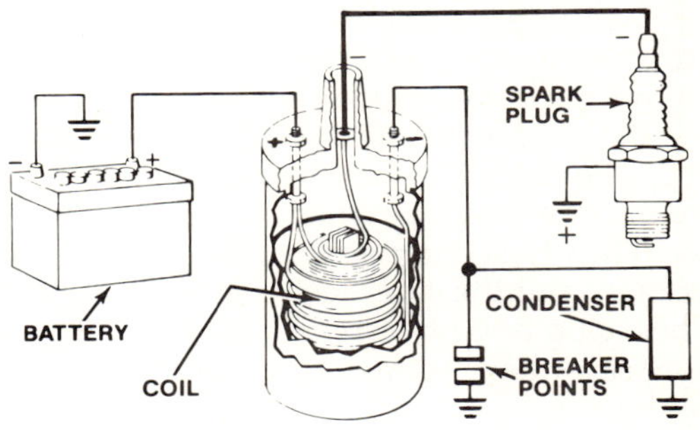

Figure 10-4. When coil connections are made properly, the spark plug center electrode is electrically negative.

partment or on the engine, figure 10-2. One common ignition coil has an unusual design and location. The Delco-Remy High Energy Ignition (HEI) solid-state ignition system has a coil mounted on the distributor cap, figure 10-3. The coil output terminal is connected directly to the center electrode of the distributor cap. The connections to the primary winding are made through a multiple-plug connector.

In any system, the connections to the primary winding must be made correctly. We learned in Chapter 8 that if spark plug polarity is reversed, greater voltage is required to fire the plug. Plug polarity is established by the ignition coil connections.

One end of the coil secondary winding is connected to the primary winding, figure 10-4, so that the secondary circuit is grounded through the ignition primary circuit. When the

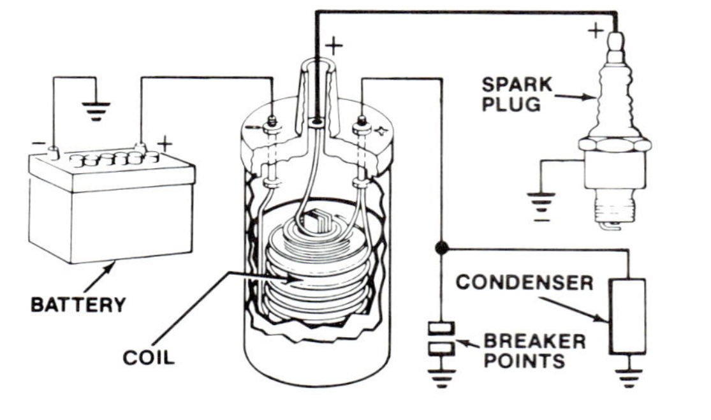

Figure 10-5. When coil connections are reversed, spark plug polarity is reversed.

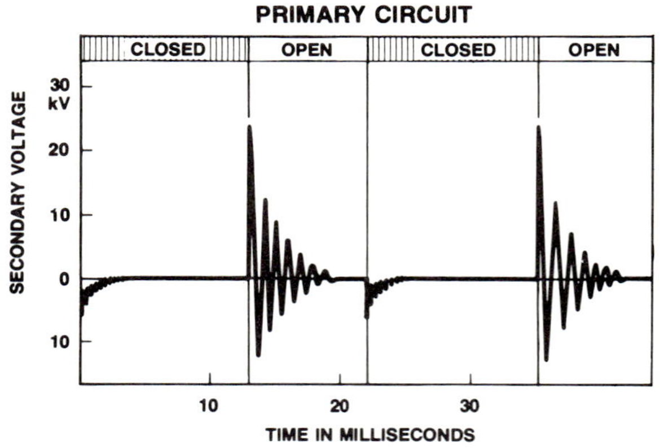

Figure 10-6. A secondary circuit no-load voltage trace. (Bosch)

coil terminals are properly connected to the battery, the grounded end of the secondary circuit is electrically positive. The other end of the secondary circuit, which is the center electrode of the spark plug, is electrically negative. The plug's grounded side electrode is positive, and plug polarity is correct. Whether the secondary winding is grounded to the primary + or − terminal depends on whether the windings are wound clockwise or counterclockwise.

If the coil connections are reversed, figure 10-5, plug polarity is reversed. The grounded end of the secondary circuit is electrically negative. The plug center electrode is electrically positive, and the side electrode is negative. When plug polarity is reversed, 20 to 40 percent more secondary voltage is required to fire the spark plug.

Coil terminals are usually marked + and − To establish the correct plug polarity with a negative-ground electrical system, the + terminal must be connected to the positive terminal of the battery (through the ignition switch, starter relay, and other circuitry). The − coil terminal must be connected to the distributor breaker points and condenser.

Coil Voltage

Coils must supply the correct amount of voltage for any system. When a coil is replaced, the new unit must be of the correct capacity. The following paragraphs describe coil voltage production.

Available voltage

The ignition coil is capable of supplying much more secondary voltage than the average engine requires. The peak voltage that a coil can produce is called its **available voltage**.

Three important coil design factors determine available voltage level:

- Secondary-to-primary turns ratio
- Primary voltage level
- Primary circuit resistance.

The turns ratio is a multiplier that creates high secondary voltage output. The primary voltage level that is applied to a coil is determined by the ignition circuit's design and condition. Installing a ballast resistor of the wrong value will affect this voltage level, as will loose or corroded connections. Generally, a primary circuit voltage loss of 1 volt can decrease available voltage by 10,000 volts.

If there was no spark plug in the secondary circuit (that is, if the circuit was open), the coil secondary voltage would have nowhere to discharge quickly. The voltage would oscillate in the secondary circuit, dissipating as heat. The voltage would be totally gone in just a few milliseconds. Figure 10-6 shows the trace of this no-load, open-circuit voltage. This is called secondary voltage **no-load oscillation**. The first peak of the voltage trace represents the maximum available voltage from that particular coil. Available voltage is usually between 20,000 and 50,000 volts.

Required voltage

When there is a spark plug in the secondary circuit, the coil voltage creates an arc across the plug air gap. Figure 10-7 compares a typical no-load oscillation to a typical secondary firing voltage oscillation. At about 15,000 volts, the spark plug air gap ionizes and becomes conductive. This is the ionization voltage level, as we learned in Chapter 8. This voltage level is also called the **firing voltage**, or **required voltage**.

As soon as a spark has formed, the energy demands of the spark cause the secondary voltage to drop to the much lower **spark voltage** level. This is the inductive portion of the spark. Spark voltage is usually about one-quarter of the firing voltage level.

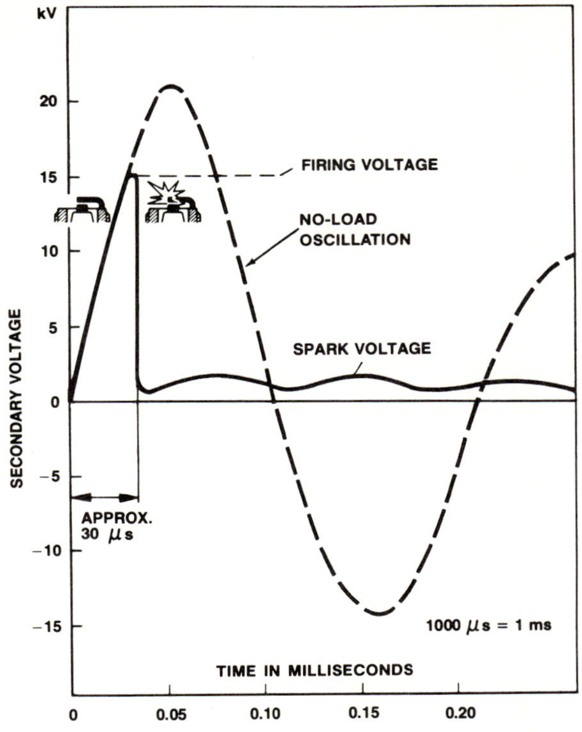

Figure 10-7. The dashed line shows no-load voltage; the solid line shows the voltage trace of firing voltage and spark voltage. (Bosch)

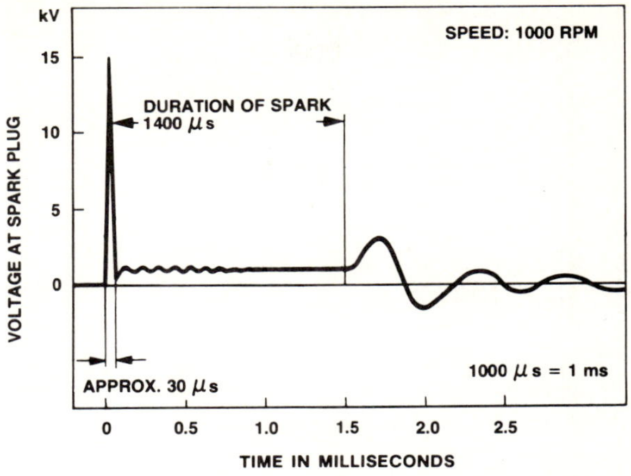

Figure 10-8. The voltage trace of an entire secondary ignition pulse. (Bosch)

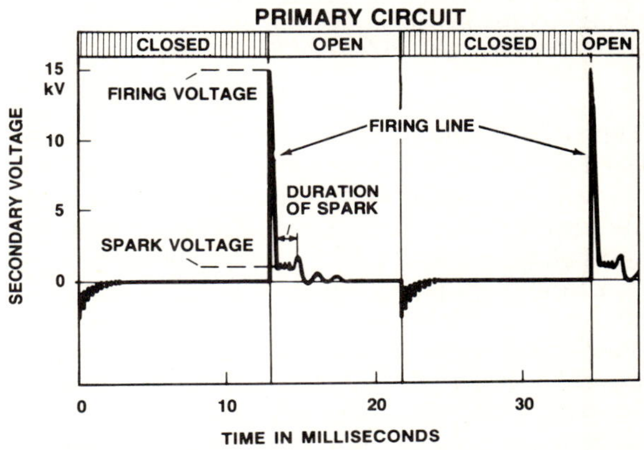

Figure 10-9. As the primary circuit opens and closes, the ignition cycle repeats. (Bosch)

Figure 10-8 shows the entire trace of the spark. When the secondary voltage falls below the inductive air-gap voltage level, the spark can no longer be maintained. The spark gap becomes nonconductive. The remaining secondary voltage oscillates in the secondary circuit, dissipating as heat. This is called secondary **voltage decay.** At this time, the primary circuit closes and the cycle repeats, figure 10-9. The traces shown in figures 10-8 and 10-9 are similar to the secondary circuit traces you will see on an oscilloscope screen.

Some conditions that cause required voltage levels to increase are:

- Eroded electrodes in the distributor cap, rotor, or spark plug
- Damaged ignition cables
- Reversed plug polarity
- High compression pressures
- A lean air-fuel mixture that is more difficult to ionize.

Voltage reserve

The physical condition of the automobile engine and ignition system can affect both available and required voltage levels, as we have seen.

Available Voltage: The peak voltage that a coil can produce.

No-Load Oscillation: The rapid, back-and-forth, peak-to-peak oscillation of voltage in the ignition secondary circuit when the circuit is open.

Firing Voltage (Required Voltage): The voltage level that must be reached to ionize and create a spark in the air gap between the spark plug electrodes.

Spark Voltage: The inductive portion of a spark that maintains the spark in the air gap between a spark plug's electrodes. Usually about one-quarter of the firing voltage level.

Voltage Decay: The rapid oscillation and dissipation of secondary voltage after the spark in a spark plug air gap has stopped.

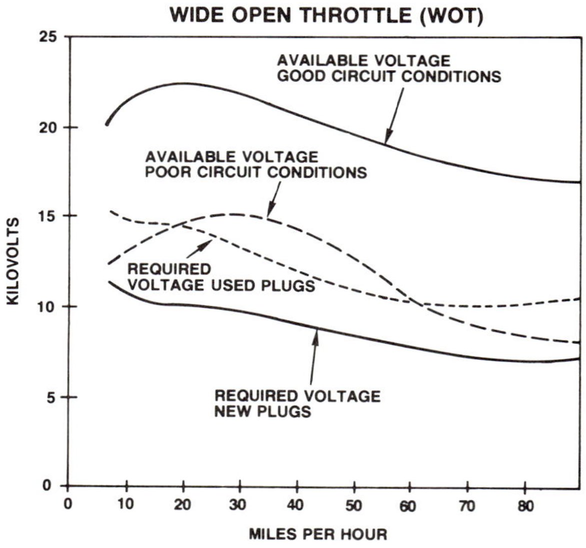

Figure 10-10. Available and required voltage levels under different system conditions.

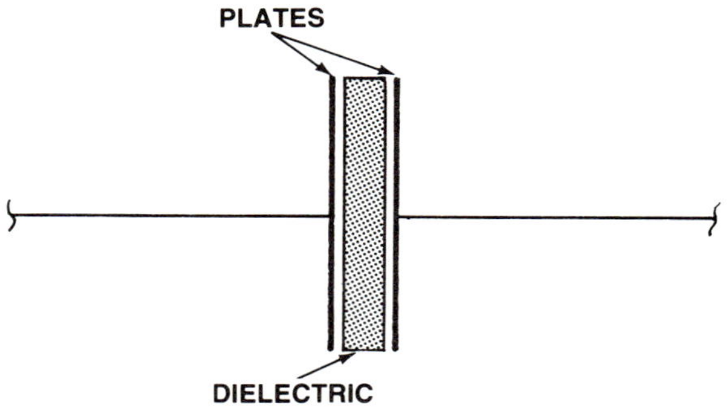

Figure 10-11. A simple condenser.

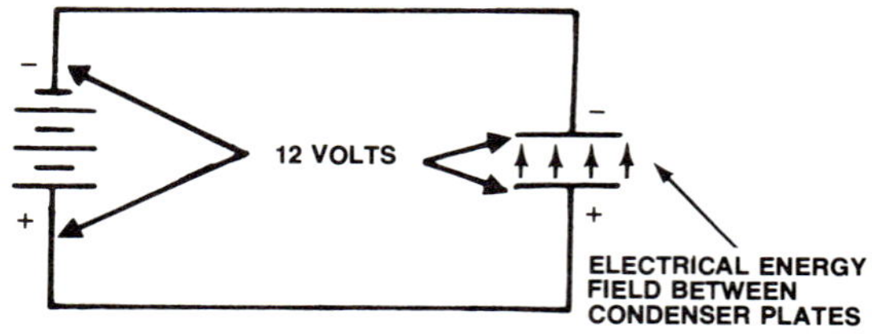

Figure 10-12. When a condenser is connected to a battery, voltage is stored in the field between the condenser plates.

Figure 10-10 shows available and required voltage levels in a particular ignition system under various operating conditions. **Voltage reserve** is the amount of coil voltage available in excess of the voltage required.

Under certain poor circuit conditions there may be no voltage reserve. At those times, the spark plugs will not fire, and the engine will run poorly or not at all. Ignition systems must be properly maintained to ensure that there is always some voltage reserve. A well-tuned ignition system should have a voltage reserve of about 60 percent of available voltage, under most operating conditions.

CONDENSERS

We said in Chapter 8 that a condenser, or capacitor, is a device that is able to store voltage. It consists of two conductive plates with an insulating material between them, figure 10-11. The insulation is called the **dielectric**. When a condenser is placed in a circuit, figure 10-12, the battery's voltage will force current to flow around the circuit. Because the current cannot flow through the dielectric insulator, one of the condenser's plates becomes positively charged and the other plate becomes negatively charged. An electric charge (voltage) exists in the field between the two plates. The voltage stored in the condenser is equal to the battery's voltage.

If the condenser is removed from the circuit right now, the energy field between the plates would still exist. The voltage stored inside the condenser is not changed or used in any way. If the charged condenser is installed in another circuit, it will act like a battery. The voltage in the condenser will force current to flow through the new circuit until the condenser's plates become discharged.

Condenser Ratings

Condensers are measured in terms of how much energy they can store. Condensers are rated in units called **farads**. The abbreviation for farad is F. One farad is a very big unit, and a condenser rated at one farad might be the size of a hot-water heater. Condensers used in automotive systems are rated in microfarads, abbreviated μF. A microfarad is one-millionth (0.000001) of a farad. Typical ignition system condensers have capacities from 0.18 to 0.32 μF.

Condenser Construction And Installation

Most automotive condensers are formed from two thin foil strips, separated by several layers of insulating paper, figure 10-13. These layers, each more than eight feet long, are tightly rolled into a cylinder. The foil strips are offset, so that the top edge of one strip protrudes past the paper on one end of the cylinder and the bottom edge of the other strip protrudes from the other end of the cylinder. These edges provide an electrical contact with the foil strips. The edges are flattened, and the cylinder is installed in a

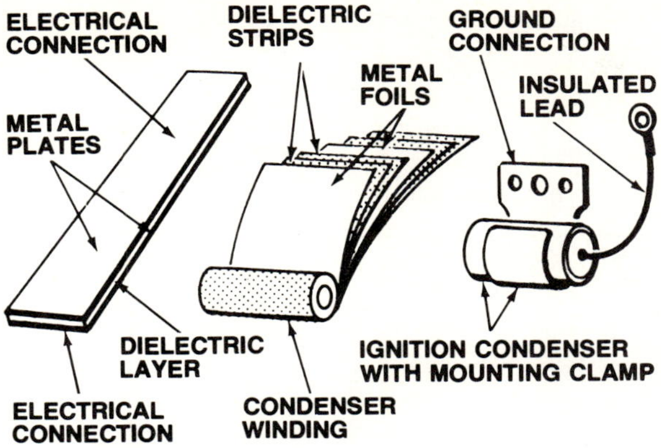

Figure 10-13. The construction of an ignition condenser.

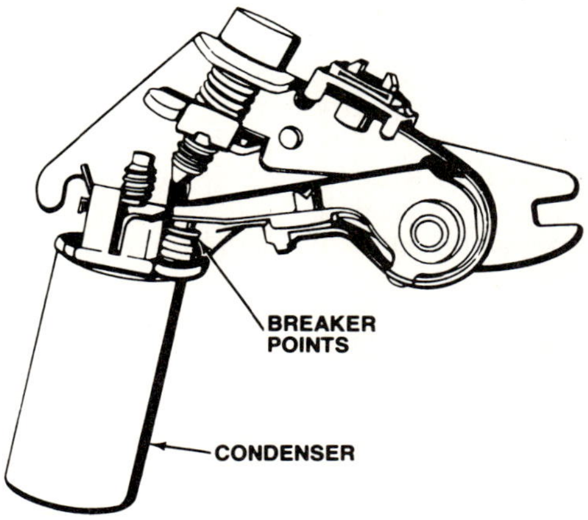

Figure 10-15. Delco-Remy's Uni-Set point and condenser assembly.

metal canister. The bottom of the canister contacts one foil edge, and grounds it. The other foil edge is connected to an insulated lead at the top of the canister.

To help insulate the foil strips, the canister is placed in a vacuum. Wax or oil is drawn into the canister, and the entire unit is sealed. Condensers cannot be adjusted or repaired, but must be replaced if they are bad.

A condenser is usually installed inside the distributor housing on the breaker plate, figure 10-14. The bracket is held to the breaker plate by a screw. This is the ground connection. The insulated lead from the top of the condenser is attached to the spring of the movable point arm, giving it an electrical connection to the movable breaker point.

The condenser also can be attached to the outside of the distributor housing. The wiring connections look different, but provide the same electrical paths.

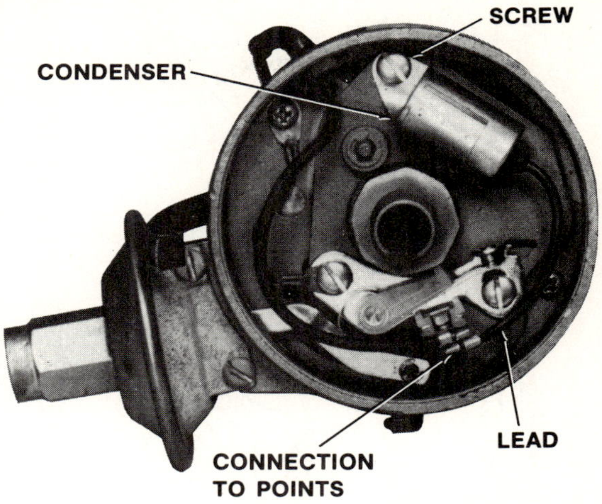

Figure 10-14. A condenser installed inside the distributor housing.

Some Delco-Remy distributors use a combined points and condenser unit called a Uni-Set, figure 10-15. This is a conventional condenser already attached to the point assembly. The assembly is installed and replaced as one unit. A similar product, the Prestolite Capaci-Point, attaches a ceramic capacitor to the point assembly. Both of these units eliminate the condenser wire lead, which is a major cause of radio interference.

Condenser Purpose

Self-induction in the coil primary winding can increase the primary voltage to as much as 400 volts. When the breaker points open, figure 10-16, position A, the high voltage would cause current to arc across the air gap.

The unwanted spark would:

- Use energy at the expense of the secondary circuit energy
- Burn and pit the contacts, causing rapid point failure
- Leave an oxidized coating on the points, increasing resistance in the primary circuit.

Voltage Reserve: The amount of coil voltage available in excess of the voltage required to fire the spark plugs.

Dielectric: The insulating material between the two plates of a condenser, or capacitor.

Farad: The unit of measurement of a condenser's ability to store electrical energy.

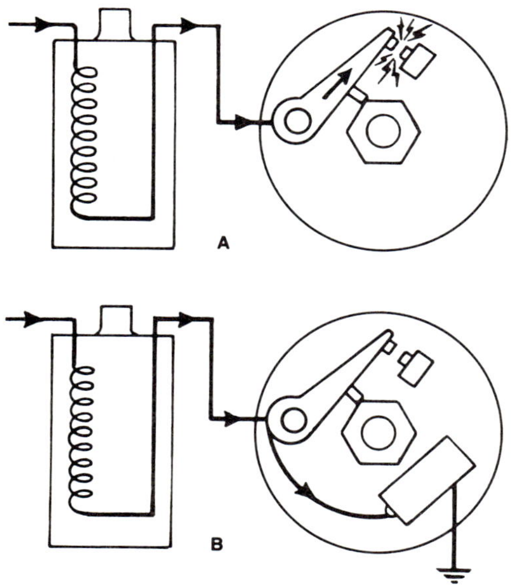

Figure 10-16. In position A, primary current arcs at the breaker points as they open. In position B, primary current flows to the condenser when the breaker points open, eliminating the arc.

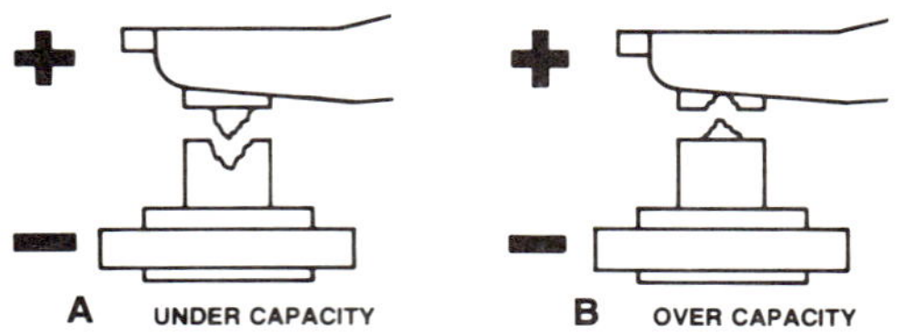

Figure 10-17. Incorrect condenser capacity can cause point pitting.

To prevent these problems, the ignition condenser is installed with the breaker points, figure 10-16, position B. When the points open, the condenser is charged by the inductive current from the primary winding. It requires a certain amount of time, about 0.1 millisecond, to charge the condenser to the peak voltage of the primary winding. By the time the condenser is fully charged, the contact points have been opened far enough so that the current cannot arc across the air gap.

The energy in the condenser is then discharged, **oscillating** between the condenser and the coil primary winding and dissipating as heat.

Because the condenser allows the primary circuit to be broken quickly and completely, the coil's magnetic field collapses rapidly. The field collapses about 20 times faster than if there were no condenser in the circuit. The faster the collapse, the greater the induced voltage in the secondary winding. The field has fully collapsed by the time the condenser begins its first discharge into the primary circuit.

At low engine speeds, the breaker points open relatively slowly. At engine speeds causing fewer than 3,000 sparks per minute, the primary windings's induced voltage is great enough to cause a slight arc across the slowly opening point air gap, despite the condenser. Automobiles driven mostly at low engine speeds will show more rapid point failure than those used for high-speed travel.

It is important to follow manufacturers' recommendations when installing ignition condensers. If a condenser with too little capacity is used, primary current will charge the condenser and still be able to arc across the point gap. This causes pitting at the points, with metal transfer from the grounded (−) point to the movable (+) point, figure 10-17, position A. If the condenser has too large a capacity, pitting and metal transfer can occur in the opposite direction, position B. The **"Minus Rule"** may help you to remember these relationships:

Minus metal on the minus point means a minus-capacity condenser.

BREAKER-POINT DISTRIBUTORS

Great demands are placed on the ignition breaker points by the rest of the ignition system. Points must be correctly installed and accurately adjusted.

Breaker Point Installation

Ignition breaker points are normally supplied as a complete unit, figure 10-18. This assembly is fastened onto the distributor breaker plate. The fasteners are either two screws or one screw and a small pin, as illustrated.

One of the mounting holes of the point assembly is elongated, so that the position of the points on the breaker plate can be adjusted. This adjustment has a great effect upon breaker point operation, as we will soon see.

The wiring connections between the points, the condenser, and the primary circuit can be made at one of two places. When the condenser is mounted inside the distributor, the connection is usually made at the bracket where the movable arm's spring is braced, figure 10-18. In some distributors, a nut and bolt hold the two wiring terminals to the bracket. Other distributors have push-on terminals. When the condenser is mounted outside of the housing, the primary lead and the condenser lead may be attached to a single slide terminal. A lead from

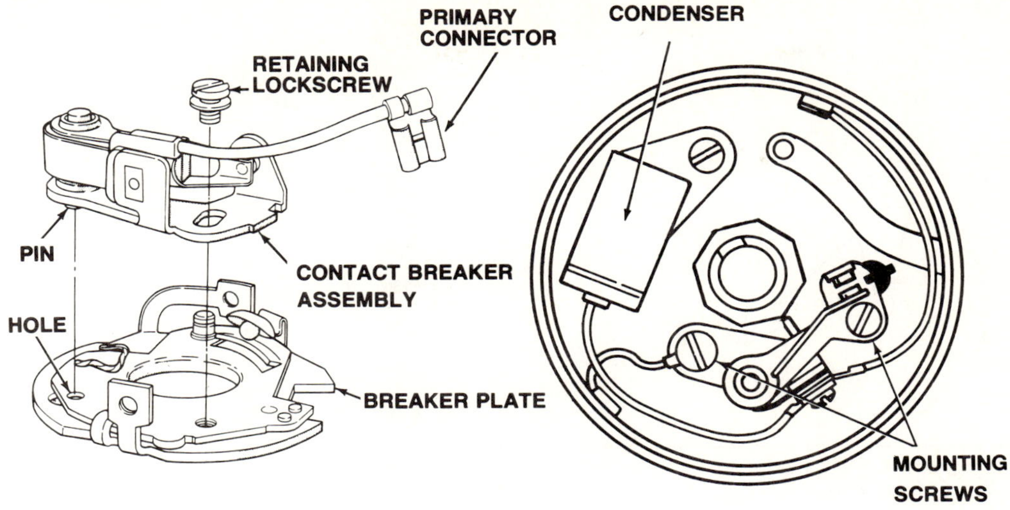

Figure 10-18. Point assemblies are usually mounted with either one screw and a holding pin (left) or with two screws (right). (Ford)

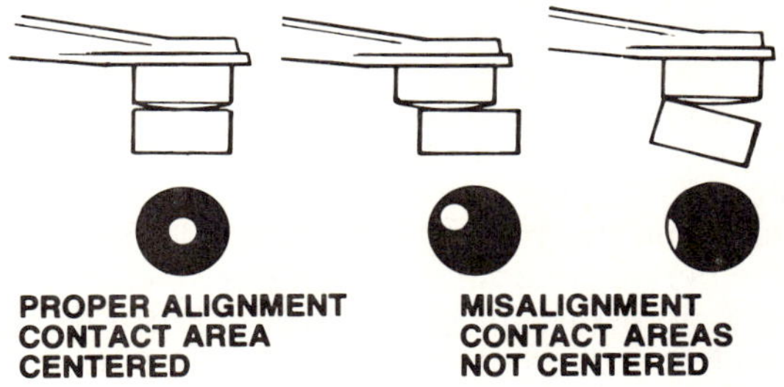

Figure 10-19. Breaker points must be properly aligned for maximum ignition system performance.

the movable breaker point attaches to this terminal.

The stationary breaker point and the condenser canister are grounded through the distributor housing and the engine.

Correct ignition point alignment and spring tension are essential for proper ignition operation and long service life. Points are correctly aligned, figure 10-19, when the mating surfaces are in the center of both contacts, the faces are parallel, and the diameters are **concentric**. This ensures maximum contact area and precise switching action by the points.

Correct spring tension also ensures precise point action. Excessive spring tension causes rapid cam and rubbing block wear. In some cases, too much spring tension can cause distributor shaft bushing wear or broken points. If spring tension is too light, the points will bounce as they open and close at high speed. This generally results in a loss of engine power. Spring tension is normally between 15 and 25 ounces.

Point Dwell Angle

Point **dwell angle**, or cam angle, measures how far the distributor cam rotates while the points are closed. In figure 10-20, the points closed when line A was at the rubbing block. The

Oscillating: Moving back and forth with a steady rhythm.

Minus Rule: Minus metal on the minus side of the distributor breaker points means a minus-capacity condenser.

Concentric: Having the same center.

Dwell Angle: Also called cam angle, or dwell. The measurement in degrees of how far the distributor cam rotates while the breaker points are closed.

■ Dual Breaker Points

Some high-performance engines have a special breaker point installation to counteract the reduction of dwell at high speed, typically above 4,500 or 5,000 rpm. Two sets of ignition points are connected in parallel. The point sets open and close at different times. One set opens the primary circuit: the other closes it. This provides an overlapping action that extends the dwell angle. The increased dwell allows higher coil voltage output during high-speed operation.

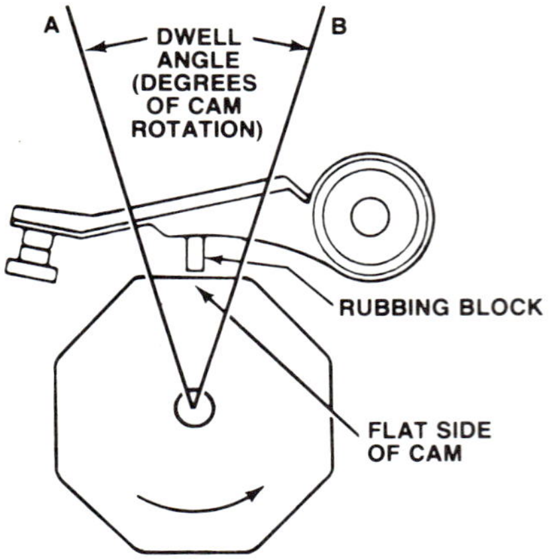

Figure 10-20. Dwell angle is the period during which the ignition points are closed. (Ford)

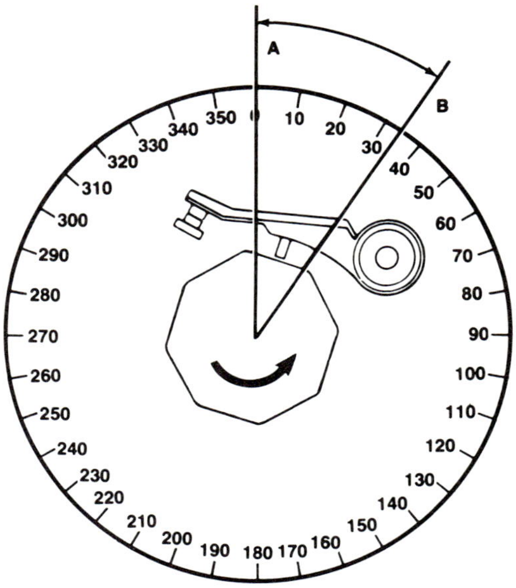

Figure 10-21. Dwell angle is measured in degrees of cam rotation. In this case, dwell is about 33 degrees.

points will open when line B reaches the rubbing block. Between line A and line B, the points stay closed. The number of degrees that the cam rotates between the point closing (line A) and the point opening (line B) is called the dwell angle.

Distributor rotation, like crankshaft rotation, is measured in degrees. If we superimpose a degree scale onto the point assembly, figure 10-21, we see that this particular dwell measures about 33 degrees.

The cam illustrated has eight lobes, so it is used with an 8-cylinder engine. Four-cylinder or 6-cylinder engines require 4-lobe or 6-lobe cams. This means that the points open four or six times during each cam revolution. The more times the points must open, the less time they can remain closed.

The ignition dwell angle is directly related to an engine's firing intervals. There are 90 degrees of crankshaft rotation between the firing intervals in an 8-cylinder engine. Because the distributor rotates at one-half the crankshaft speed, the distributor cam rotates half as far as the crankshaft between firing intervals, or 45 degrees. The theoretical dwell angle of an 8-cylinder engine would be 45 degrees, but the points are open during part of that time, and dwell equals the number of degrees during which the points are *fully closed*. About 12 to 17 degrees are required for the points to open and close on an 8-cylinder engine. Therefore, a typical dwell angle for an 8-cylinder engine is about 28 to 32 degrees.

An even-firing 6-cylinder engine fires every 120 degrees of crankshaft rotation (60 degrees of distributor rotation). A 4-cylinder engine fires every 180 degrees of crankshaft rotation (90 degrees of distributor rotation). Dwell on a 6-cylinder engine could be in the range of 45 degrees; on a 4-cylinder engine, it could be about 75 degrees. However, large dwell angles mean that the primary current would flow for long periods of time at low engine speeds. This is not necessary for full coil saturation and could lead to coil overheating. Therefore, 6- and 4-cylinder distributors are designed for less dwell than the maximum amount they might have. Dwell on a typical 6-cylinder engine is about 38 degrees; on a 4-cylinder engine it is usually about 50 degrees.

Dwell and point gap

Point gap is the maximum distance between the breaker points when they are open. Figure 10-22 shows the relationship between point gap and dwell. A small point gap means a large dwell angle. A large point gap means a small dwell angle. Point gap is usually measured in thousandths of an inch or tenths of a millimeter. Typical point gaps range from 0.015 to 0.025″ or 0.4 to 0.6 millimeter.

Point gap is adjusted by shifting the position of the breaker point assembly. This changes the position of the points in relation to the cam, altering the point gap and the dwell. Point assemblies have elongated mounting holes so that the point gap can be adjusted.

Effect of dwell on the coil

As long as the points are closed, current flows in the coil primary winding. This creates a magnetic field within the coil and makes induction possible.

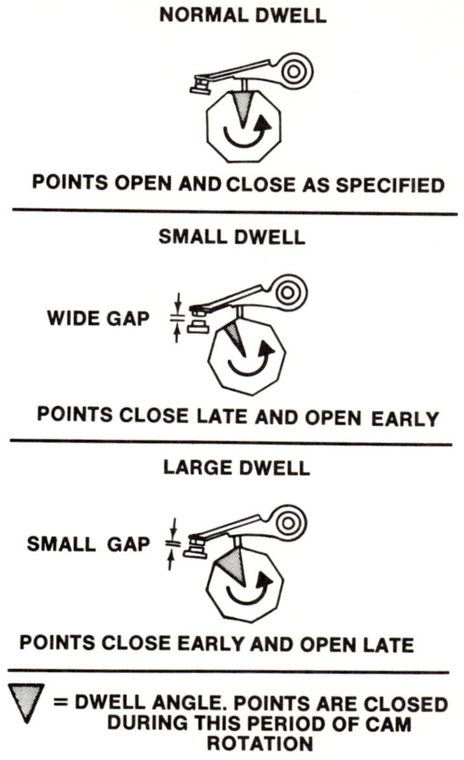

Figure 10-22. Point gap and dwell angle are related; when one increases, the other decreases.

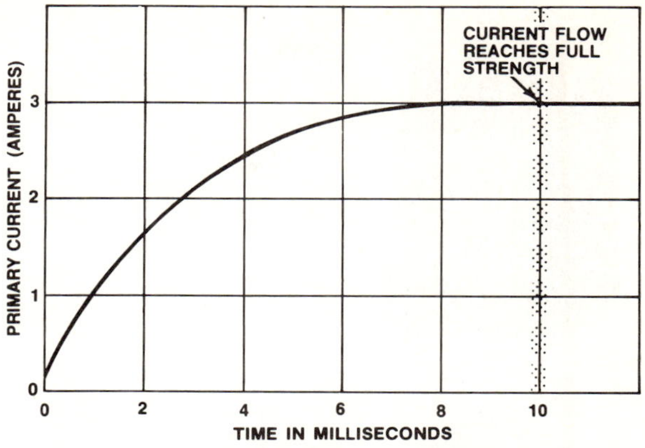

Figure 10-23. Primary winding current flow does not immediately reach full strength. (Bosch)

However, the primary winding magnetic field does not appear instantly. When current first flows through the primary winding, self-induction causes a countervoltage within the winding. This countervoltage opposes primary current flow. The magnetic field of the primary winding does not immediately reach full strength, because the primary current flow does not immediately reach full strength, figure 10-23. It generally takes from 10 to 15 milliseconds (0.010 to 0.015 second) for the primary current flow to reach full strength.

The breaker points must remain closed long enough for the primary winding's magnetic field to reach nearly full strength. If the field is not at full strength, secondary voltage is reduced. The point dwell must be great enough to allow nearly maximum coil strength.

However, if the point dwell is too great, the point gap will be very small. Primary current will be able to arc across the air gap, and the magnetic field will not collapse quickly and completely. Point dwell and gap must be carefully adjusted to the manufacturer's specifications if the ignition system is to perform at its best.

Dwell and engine speed

As engine speed increases, the distributor cam rotates faster. Dwell angle is unchanged, but the *time* it takes the cam to rotate through this angle is decreased. That is, the amount of time that the points are closed is reduced as engine speed increases.

We have seen that it takes a certain amount of time — about 0.010 second — for primary current flow to reach full strength, figure 10-23. At a 625-rpm idle speed, the 33-degree dwell angle of an 8-cylinder engine lasts about 0.0165 second. Primary current flow can reach its full strength before the points open to interrupt it.

As engine speed increases to 1,000 rpm, the 33-degree dwell period decreases to 0.0099 second. Primary current flow is interrupted just

■ Plagued By Pitted Points?

Do you have a problem with a car that burns its ignition points prematurely? The problem could be in the condenser, but if you have checked the condenser and know that it has the right capacitance, what then? Do you check the charging system voltage as part of your regular tune-up procedure? You should.

If the charging system voltage is too high, the increased voltage across the points (even after it is dropped by the ballast resistor) can cause the points to arc excessively when they open. This, of course, causes early burning and pitting. Excessive voltage can also shorten lamp bulb life, so check the charging system as part of every tune-up.

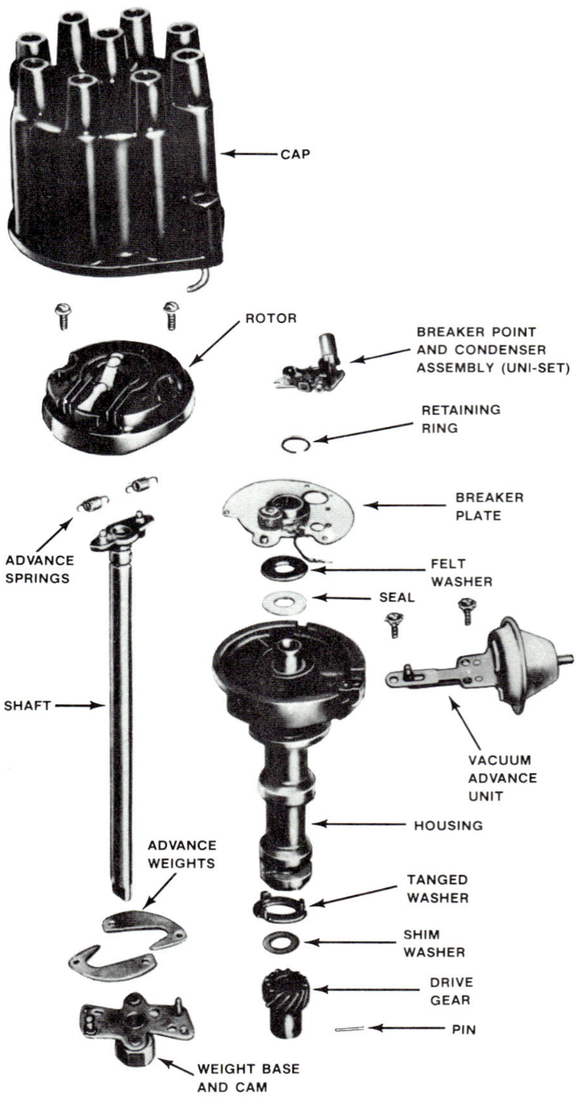

Figure 10-24. An exploded view of the Delco-Remy external-adjustment distributor. (Delco-Remy)

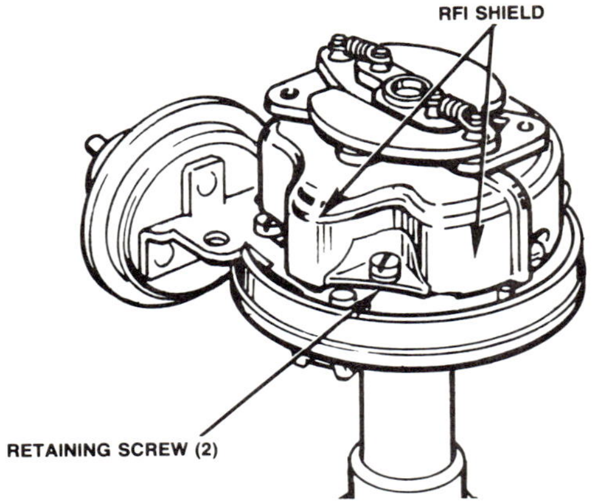

Figure 10-25. Some Delco-Remy distributors have a shield to reduce radio interference from the ignition system.

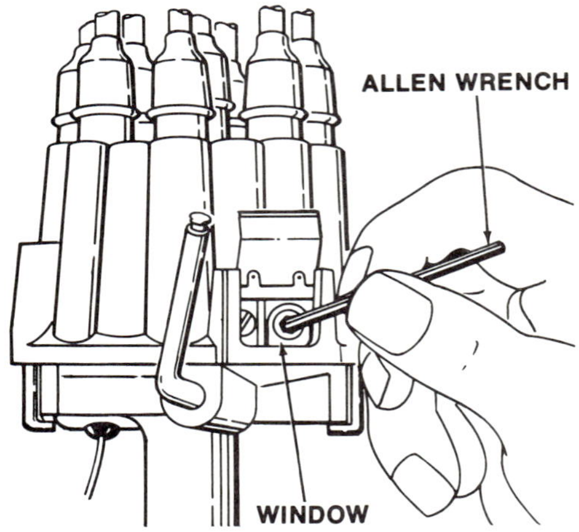

Figure 10-26. The point gap and dwell adjustment can be made from outside the distributor.

before it reaches its maximum strength. Available secondary voltage will be decreased slightly.

At 2,000 rpm, the dwell period is reduced to 0.0045 second. Primary current flow is interrupted well before it reaches its maximum strength. Available secondary voltage will be reduced considerably. If the engine and ignition system are not in excellent condition, the required voltage level may be more than the available voltage level. If this occurs, the engine will misfire.

SPECIFIC DISTRIBUTORS

The following paragraphs describe the various distributors used by major manufacturers. We will learn more about servicing these units in the *Shop Manual*.

Delco-Remy

All domestically built General Motors products used Delco-Remy breaker-point distributors until 1975, when the Delco-Remy High Energy Ignition (HEI) system became standard equipment. Many AMC products built before 1975 also used Delco-Remy breaker-point distributors.

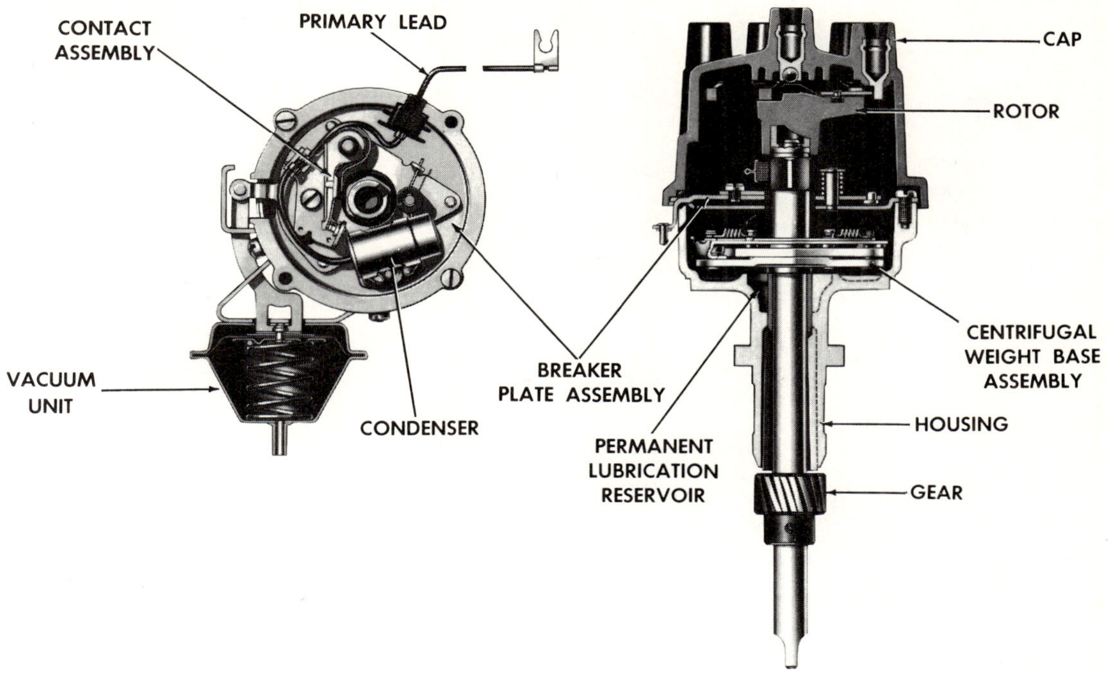

Figure 10-27. A cross-section view of the Delco-Remy internal-adjustment distributor. (Delco-Remy)

External-adjustment distributor
The distributor used with V-6 and V-8 engines, figure 10-24, has some unusual design features, as we have seen. The cap is held to the housing by two spring-loaded latches. The large, circular rotor is keyed to the weight base and cam assembly by round and square locators and is held by two screws. The centrifugal advance weights are installed above the cam and breaker plate.

On late-model cars with windshield-mounted radio antennas, a shield may be placed around the breaker-point assembly, figure 10-25. This is a radiofrequency interference (RFI) shield held to the housing by two screws. If a Uni-Set point-and-condenser assembly is used, the RFI shield is not needed.

Both Uni-Set and conventional point assemblies are held to the breaker plate by two screws. If the condenser is separate, it is held by an additional screw.

Point gap and dwell adjustment is made by turning a female hex-headed screw, figure 10-26, through a sliding window in the distributor cap. This adjustment can be made from the outside while the engine is running. These distributors are called the external-adjustment type.

Internal-adjustment distributor
Delco-Remy's internal-adjustment distributor, figure 10-27, is used with inline 4- and 6-cylinder engines. It does not have the unusual design features of the external-adjustment type.

Two clips hold the cap to the housing. A conventional rotor is pressed onto the distributor shaft. Directly below the rotor is the breaker plate, with cam, points, and condenser. The point assembly is held to the breaker plate by one screw. Point gap is adjusted by loosening the screw, shifting the point assembly, and tightening the screw. The centrifugal advance unit lies below the breaker plate.

Vacuum advance units on all Delco-Remy distributors are not adjustable. Centrifugal advance units can be adjusted by replacing the calibrated springs.

■ Useful Distributor Tool

When you're timing an engine and want to rotate the distributor to advance or retard the spark, you may find that the distributor does not want to budge. This can occur even after the holddown bolt has been loosened. If it does, try using an oil filter wrench. It may give you the grip and leverage you need.

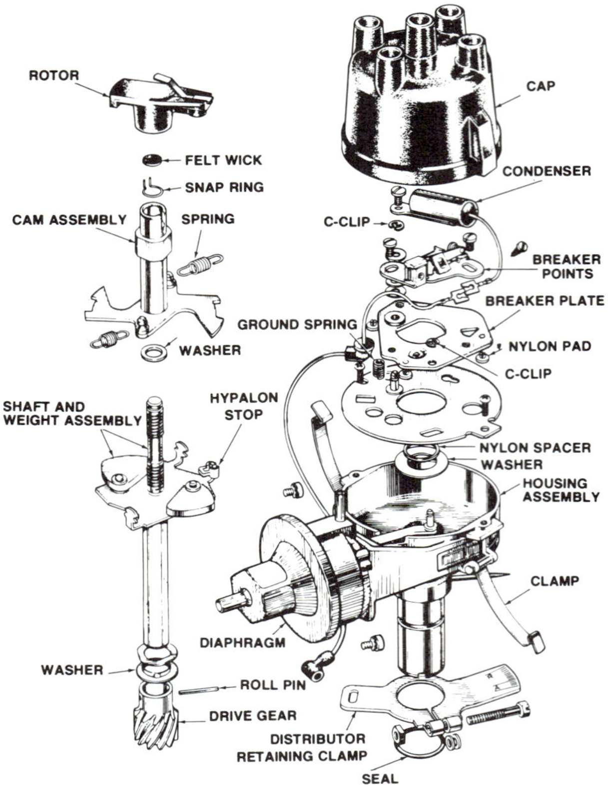

Figure 10-28. The Motorcraft distributor used with Ford's 1600-cc engine. (Ford)

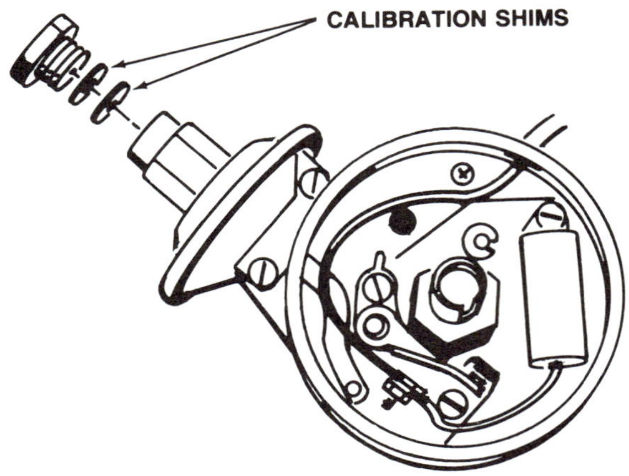

Figure 10-30. The vacuum advance units on early Ford distributors are adjusted by inserting or removing shims.

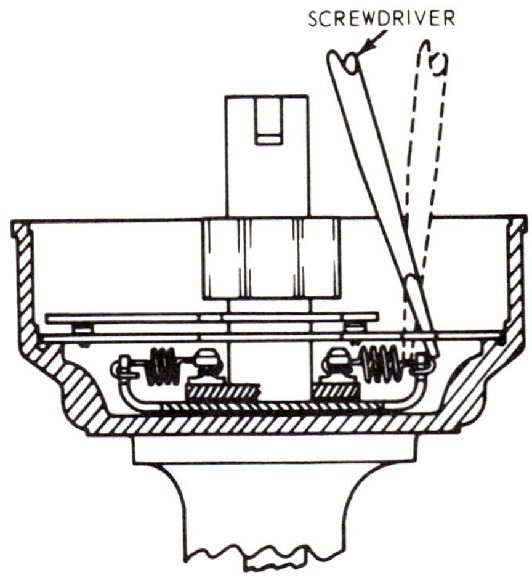

Figure 10-29. Adjusting the centrifugal advance on a Motorcraft distributor. (Ford)

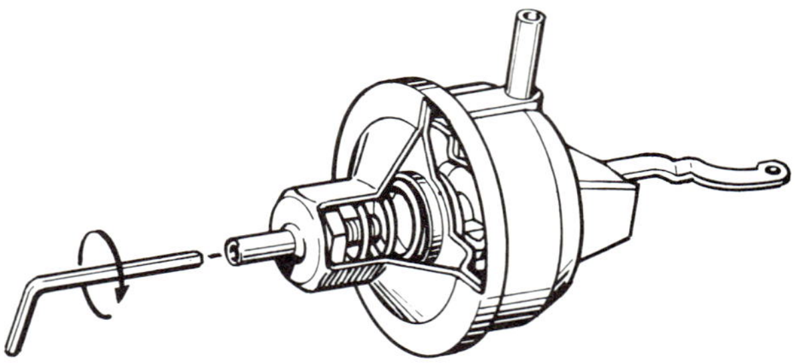

Figure 10-31. The vacuum advance units on later Ford distributors are adjusted with an allen wrench inserted through the vacuum nipple.

Motorcraft

Ford products used Motorcraft (and earlier Autolite) breaker-point distributors until 1975, when the Motorcraft solid-state ignition system became standard. The only common exception to this is the German-built Capri, which used a Bosch breaker-point distributor, until 1976, on all but the 1,600-cc engine.

The breaker-point distributor used on Ford's 4-cylinder, 1,600-cc engine, figure 10-28, has a dual-diaphragm vacuum advance unit like that studied in Chapter 9. The distributor cap is held to the housing by two spring-loaded clips. The rotor is pressed onto the shaft. Two screws hold the point assembly to the breaker plate, and another screw attaches the condenser.

Point gap is adjusted by loosening both mounting screws, shifting the point assembly, and tightening the screws. The centrifugal advance mechanism, which lies below the breaker plate, is adjusted by bending the spring brackets, figure 10-29. On early models, the

Figure 10-32. The Motorcraft 8-cylinder distributor.

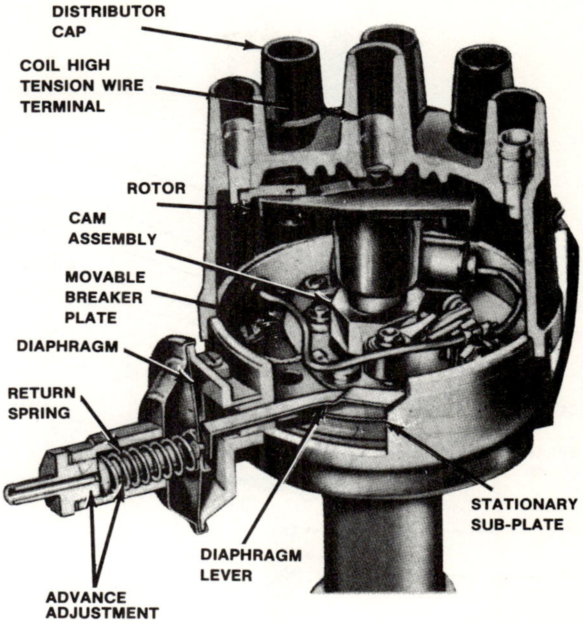

Figure 10-33. Some of the 6- and 8-cylinder Ford engines use a single-diaphragm vacuum advance unit. (Ford)

vacuum advance diaphragm is adjusted by inserting or removing calibrated shims behind the spring, figure 10-30. Vacuum advance on later models is adjusted by inserting an allen wrench through the vacuum nipple, figure 10-31, and turning a screw that controls spring tension on the diaphragm. Vacuum retard diaphragms are not adjustable.

The distributors used on 6- and 8-cylinder Ford engines, figure 10-32, are very similar to the one described above. Construction, point gap adjustment, and centrifugal advance adjustment are identical for most units.

Some 6- and 8-cylinder units have a single-diaphragm vacuum advance unit, figure 10-33, while others have the dual-diaphragm unit. Vacuum advance diaphragms on early distributors are adjusted by adding or removing calibration shims, figure 10-30. Later models are adjusted with an allen wrench, figure 10-31. Retard diaphragms on dual-diaphragm vacuum units are not adjustable.

Some distributors have a radio interference suppression cover mounted below the rotor similar to the Delco-Remy unit.

High-performance V-8 engines may have dual breaker points. The two sets of points are adjusted separately, in the same way that the single-point units are adjusted.

Chrysler

Chrysler products use breaker-point distributors made by Chrysler and by Prestolite. Chrysler units are used on all 6-cylinder engines, figure 10-34, and on many V-8 engines, figure 10-35, through 1972. Some of the Chrysler V-8 distributors have a retard solenoid, as we saw in Chapter 9.

Prestolite units are used on some V-8 engines through 1971. High-performance V-8 engines use Prestolite dual-point distributors.

The Chrysler and Prestolite distributors are very similar, except for the dual-point units. The only major difference is that, on Prestolite units, the centrifugal advance mechanism can be adjusted easily by bending the spring brackets. The Chrysler centrifugal advance unit is not designed to be adjustable. Also, the breaker plates in Chrysler-built distributors pivot eccentrically so that dwell changes as engine speed increases.

On all units, the distributor cap is held to the housing by two spring-loaded clips. The rotor is pressed onto the shaft. One screw holds the point assembly to the breaker plate, and the condenser is held by another screw.

The point gap is adjusted by loosening the screw, shifting the assembly, and tightening the screw. On Prestolite dual-point units, each set of points is adjusted separately, just as a single set is adjusted. The single-diaphragm vacuum advance unit is not adjustable.

American Motors

Most AMC products used Delco-Remy breaker-point distributors until the 1975 adoption of Prestolite's solid-state ignition system. The Delco-Remy external-adjustment distributor is used on V-8 engines, and the internal-adjustment unit is used on 6-cylinder engines. These distributors are described under the "Delco-Remy" heading.

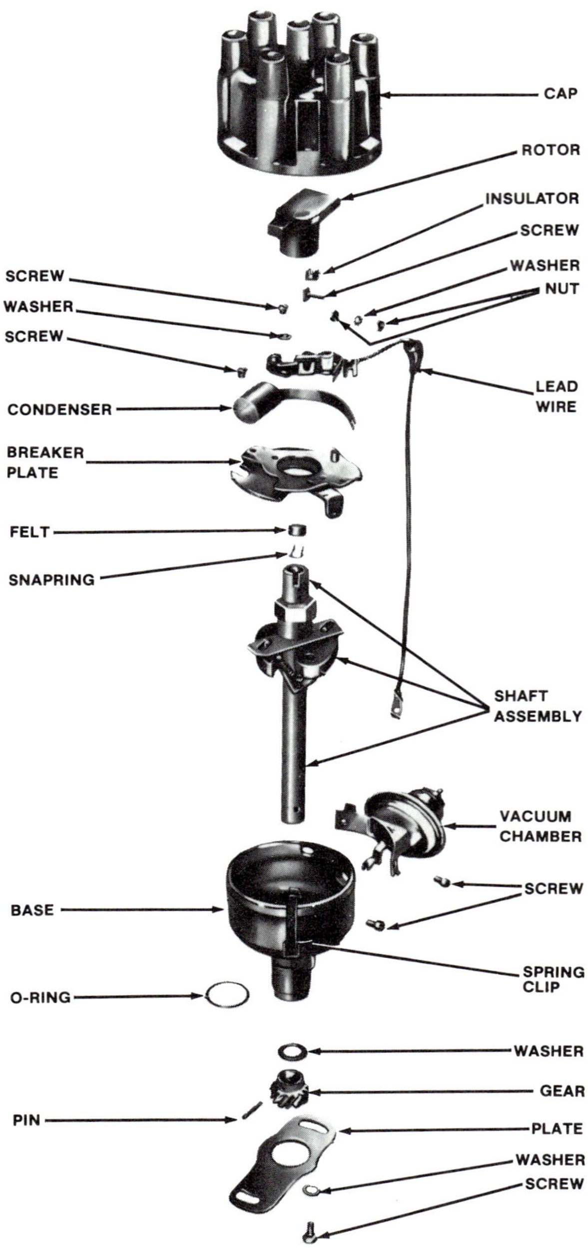

Figure 10-34. The Chrysler breaker-point distributor used with 6-cylinder engines. (Chrysler)

In 1970, some AMC V-8 engines used a Motorcraft dual-diaphragm distributor, described under the "Ford" heading.

Before 1969, some AMC 6-cylinder engines used Prestolite distributors. This unit is similar to the Prestolite unit described under the "Chrysler" heading.

Figure 10-35. The Chrysler breaker-point distributor used with 8-cylinder engines.

Bosch

A dual-diaphragm Bosch distributor is used in Ford's Capri, through 1975, on all but the 1,600-cc engine. Volkswagen models have used a similar unit with either a single- or dual-diaphragm unit.

A typical Bosch breaker-point distributor, figure 10-36, has the condenser mounted on the outside of the housing. Some rotors have a high-resistance carbon insert. The centrifugal advance can be adjusted by bending the spring brackets, but neither type of vacuum unit is adjustable. Point gap is adjusted by loosening the single holddown screw, shifting the point assembly, and tightening the screw.

Because of the unusual engine design, units on air-cooled Volkswagen engines are driven by a gear on the crankshaft, rather than by the camshaft.

SUMMARY

The ignition coil produces the high voltage necessary to ionize the spark plug gap through electromagnetic induction. Low-voltage current flow in the primary winding induces high voltage in the secondary winding. A coil must be installed with the same primary polarity as the battery to maintain proper secondary polarity at the spark plugs.

Available voltage is the amount of voltage the coil *can* produce. Required voltage is the voltage necessary to ionize and fire the spark plugs under any given operating condition. Voltage reserve is the difference between available voltage and required voltage. A well-tuned ignition system should have a 60-percent voltage reserve.

The ignition condenser is a capacitor that absorbs primary voltage when the points open. This prevents arcing across the points and premature burning.

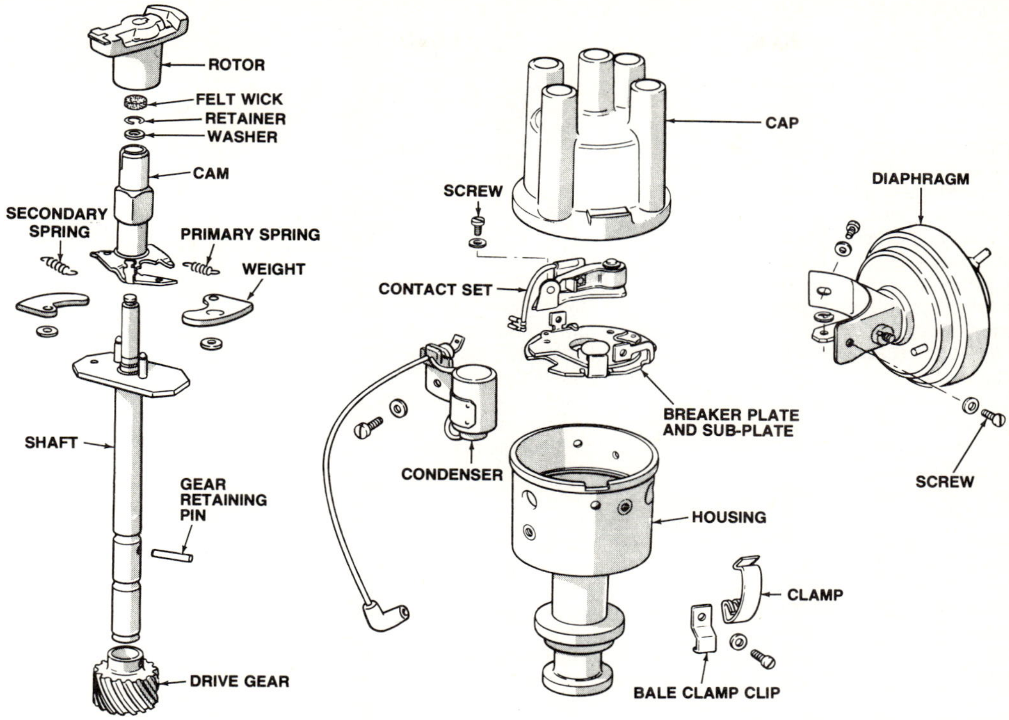

Figure 10-36. Typical Bosch 4-cylinder distributor. (Bosch)

The breaker points are a mechanical switch that opens and closes the primary circuit. The period during which the points are closed is called the dwell angle. The dwell angle varies inversely with the gap between the points when they are open. As the gap increases, the dwell decreases.

Domestic carmakers use breaker-point distributors made by Delco-Remy, Motorcraft, Chrysler, Prestolite, and Bosch. Although slightly different in physical construction, all operate on the same principles.

Review Questions

Select the single most correct answer.
Compare your answers with the correct answers on page 280.

1. Which of the following is used as insulation to protect the windings of coils?
   a. Oil
   b. Wax
   c. Paper
   d. All of the above

2. Which of the following statements is true about ignition coils?
   a. Easily repaired
   b. Adjustments are made by setscrews
   c. Cannot be replaced
   d. None of the above

3. Most of Delco-Remy's solid-state ignition systems have a coil mounted:
   a. On the engine
   b. On a fender panel
   c. On the distributor cap
   d. None of the above

4. When the coil terminals are properly connected to the battery, the grounded end of the secondary circuit is electrically:
   a. Positive
   b. Negative
   c. Neutral
   d. None of the above

5. A loss of one volt in the primary circuit can decrease available secondary voltage by______ volts.
   a. 10
   b. 100
   c. 1,000
   d. 10,000

6. The accompanying illustration shows:

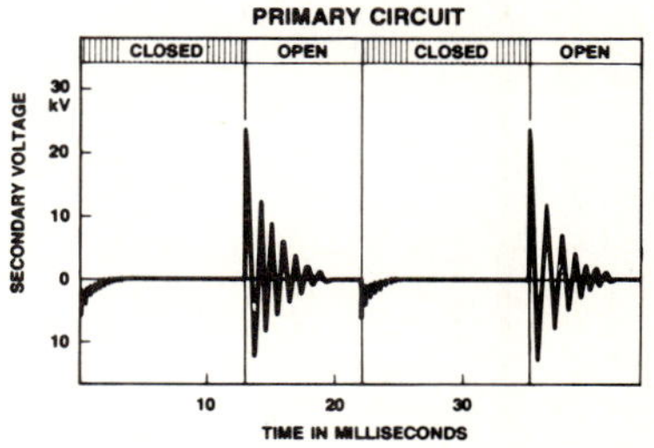

   a. Secondary circuit firing pulse trace
   b. Secondary circuit ignition pulse trace
   c. Secondary circuit no-load voltage trace
   d. Secondary circuit spark voltage trace

Review Questions

Select the single most correct answer.
Compare your answers with the correct answers on page 280.

7. Firing voltage is usually about ______ as high as spark voltage.
   a. One-fourth
   b. One-half
   c. Four times
   d. Two times

8. The voltage delivered by the coil is:
   a. Its full voltage capacity under all operating conditions
   b. Approximately half of its full voltage capacity at all times
   c. Only the voltage necessary to fire the plugs under any given operating condition
   d. Its full voltage capacity only while starting

9. The voltage reserve is the:
   a. Voltage required from the coil to fire a plug
   b. Maximum secondary voltage capacity of the coil
   c. Primary circuit voltage at the battery side of the ballast resistor
   d. Difference between the required voltage and the available voltage of the secondary circuit

10. A well-tuned ignition system should have a voltage reserve of about ______ of available voltage, under most operating conditions.
    a. 30 percent
    b. 60 percent
    c. 100 percent
    d. 150 percent

11. Which of the following is true about condensers?
    a. Easily repaired
    b. Difficult to adjust
    c. Virtually never need to be replaced
    d. None of the above

12. The accompanying illustration shows:

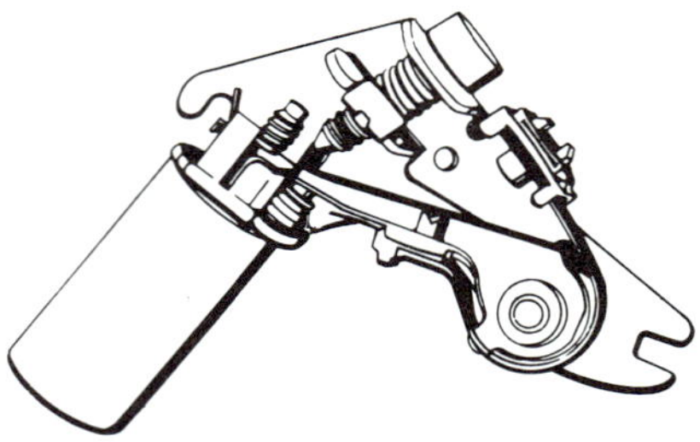

    a. Delco-Remy's Uni-Set point and condenser assembly
    b. Prestolite's Capaci-Point assembly
    c. Ignition condenser assembly, with mounting bracket
    d. None of the above

13. Condensers are rated in:
    a. Ohms
    b. Milliohms
    c. Volts
    d. Microfarads

14. The accompanying illustration shows which of the following situations with respect to the breaker points?

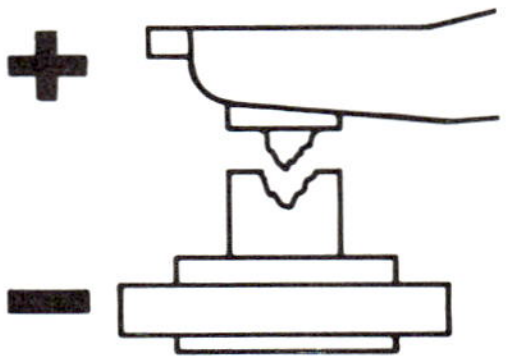

    a. Condenser with too little capacity used
    b. Condenser with too much capacity used
    c. Normal wear of points with correct condenser used
    d. None of the above

15. The accompanying illustration shows:

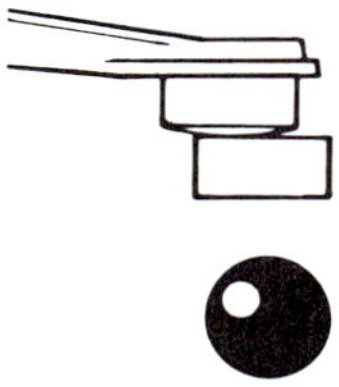

    a. Properly aligned breaker points
    b. Properly offset breaker points
    c. Misaligned breaker points
    d. Breaker points with excessive spring tension

16. Dwell angle is the period during which the ignition points:
    a. Are fully open
    b. Are fully closed
    c. Are opening and closing
    d. None of the above

17. The point gap is ______ related to the dwell angle.
    a. Directly
    b. Inversely
    c. Proportionately
    c. Reciprocally

18. The distributor rotates at ______ the speed of the crankshaft.
    a. One-half
    b. The same speed as
    c. Twice
    d. None of the above

Chapter

11

Solid-State Ignition Systems

During the 1960's, solid-state ignition systems normally were used only on high-performance engines. Solid-state systems did not appear on many domestic automobiles until the early 1970's. Now, they are the rule rather than the exception. Emission control standards were a major force behind this rapid change.

Although solid-state systems cost more than breaker-point equipment, the advantages they offer outweigh the drawback of increased cost:

- Greater available voltage, especially at high engine speeds
- Reliable system performance at all engine speeds
- Potential for more responsive and variable advance curves
- Decreased maintenance.

Engine operation and exhaust emissions are more accurately controlled by using solid-state circuitry.

BASIC SOLID-STATE SYSTEMS

Solid-state systems differ from breaker-point systems in the devices that control primary circuit current flow. Breaker-point systems use the mechanical breaker points to open and close the primary circuit. The points are operated by a cam on the distributor shaft. Most solid-state systems use an electronic switch to control primary current flow. We will learn about these solid-state devices in this chapter. The solid-state circuitry is enclosed in a sealed **module**. The module's operation is controlled by the triggering device, which is usually in the distributor.

As we learned in Chapter 8, this triggering device is usually a magnetic pulse generator. Solid-state ignitions also can be triggered by a set of breaker points, as will be explained later in this chapter.

Solid-state ignitions can be classified in terms of their primary circuit operation. There are two general types:

1. Inductive discharge
2. Capacitive discharge.

The difference between these types lies in how they use primary current flow to produce a high-voltage secondary current.

Inductive Discharge

The inductive discharge system uses battery voltage to force current flow through the coil primary winding. When a signal is sent by the triggering device, primary current flow is interrupted. This sudden *decrease* in primary current flow collapses the coil's magnetic field, inducing a high-voltage surge in the secondary circuit. This is the same method used in breaker-point

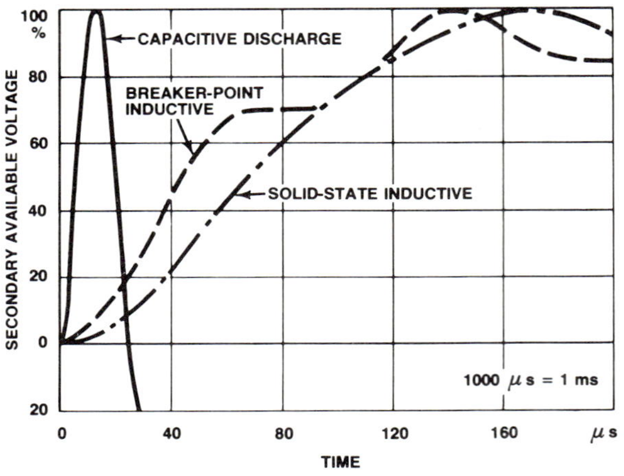

Figure 11-1. A major difference in ignition system performance is the *time* at which maximum secondary voltage is developed. (Bosch)

systems, but the device used to open the circuit is different.

Most original equipment solid-state systems use inductive discharge. It provides about 30,000 volts of available voltage and sustains a spark for about 1,800 microseconds. (A microsecond is one-millionth of a second, or 0.000001 second.)

Capacitive Discharge

The capacitive discharge (c.d.) system uses battery voltage to charge a capacitor in the control module. This capacitor can be charged to as high as 300 volts. When the triggering device signals, the capacitor is allowed to discharge through the coil primary winding. The sudden *increase* in primary current flow expands the coil's magnetic field, inducing a high-voltage surge in the secondary circuit.

Capacitive discharge systems are available as add-on units, but no major domestic manufacturer installs them as original equipment. A few imported cars, such as Audi and Mercedes-Benz, have used c.d. ignition systems. They provide a greater available voltage than inductive discharge systems but will sustain a spark for only about 200 microseconds. The spark is much more intense than that produced by an inductive discharge system and will fire plugs that are in very poor condition. Under certain engine operating conditions, however, a longer sparking time is required or the air-fuel mixture will not burn completely.

The time and voltage characteristics of a breaker-point ignition system, a capacitive discharge system, and a solid-state inductive system are shown in figure 11-1.

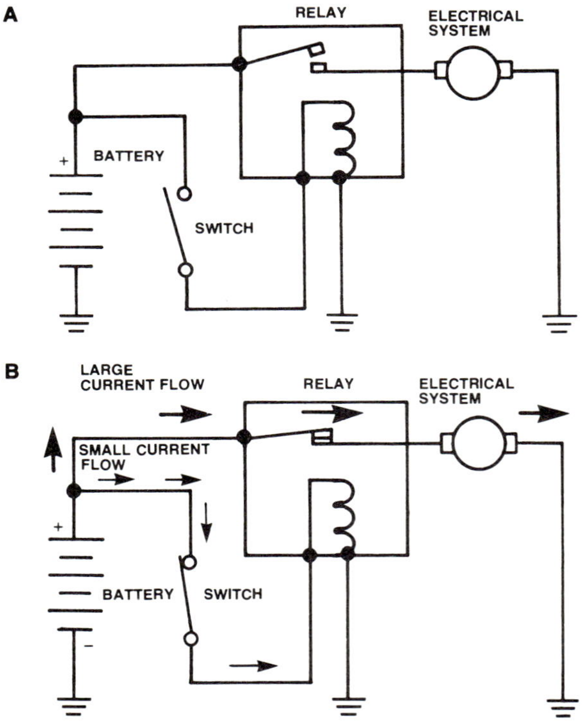

Figure 11-2. A relay in a circuit allows a small amount of current to control the flow of a large amount of current.

CONTROL MODULES AND PRIMARY CIRCUITRY

Although manufacturers use different circuitry in their ignition system control modules, the modules all do the same job. The following paragraphs explain the control module operation for both types of electronic ignition systems. Because the electronic components within the module are delicate and complex, the modules are sealed during manufacturing. If any part fails, the entire module is replaced rather than being repaired. Our explanation of module circuitry and operation will be brief, because you will never need to service the module.

Solid-State Devices

We learned about one solid-state device, the diode, in Chapter 7. Two other parts are commonly used in ignition system control modules:

- Transistors
- Silicon-controlled rectifiers.

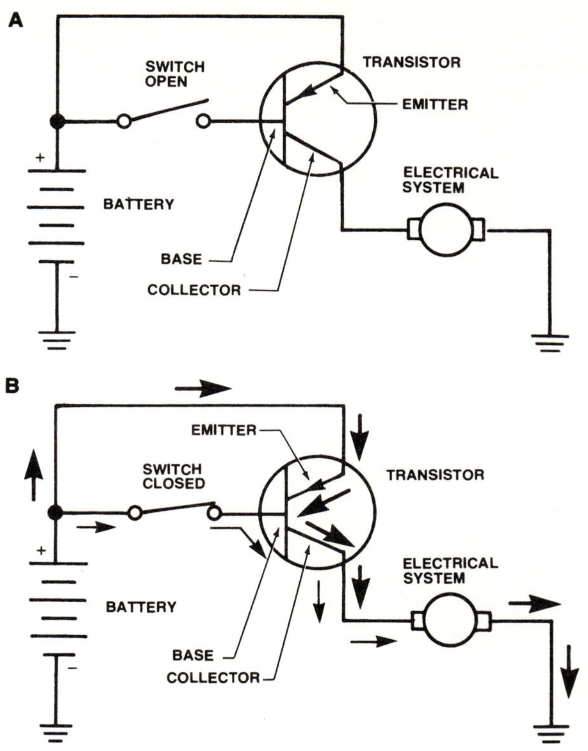

Figure 11-3. This transistor is made conductive by base-to collector current flow.

Transistor operation

Transistors act like relays because they allow a small amount of current flow to control the flow of a large amount of current. Figure 11-2 shows a relay installed in a circuit. When the contact points are open, position A, no current can flow through the circuit. When the switch is closed, position B, a small amount of current flows through the relay's electromagnetic coil. This closes the contact points, and a large amount of current can flow through the relay armature. If the switch is opened again current flow through the coil will stop. The contact points will be opened by a spring, and the large amount of current flow will stop.

Figure 11-3 shows a transistor installed in the same circuit, in place of the relay. The transistor symbol shows that it has three parts:

- The **base**
- The **emitter**
- The **collector**.

In its normal state, position A, the transistor will not allow any current flow through the emitter and collector. When the switch is closed, position B, there is a complete circuit through the base and collector. A small amount of current will flow through the base-to-collector circuit. This current flow makes the emitter-to-collector circuit complete, because when a small amount of current flows through the base, the emitter and collector become conductive. A larger amount of current can flow through the emitter-to-collector circuit. When the switch is opened and base-to-collector current flow stops, the emitter-collector circuit current flow also stops.

The switch in figure 11-3 controls base-collector current flow by opening and closing the circuit. The flow can also be controlled by varying the voltage present in the circuit. A certain amount of voltage is needed to force current to flow through the base and collector. If not enough voltage is present, or if too much voltage is present, no current will flow through any of the transistor's circuits.

Module: A self-contained, sealed unit that houses the solid-state circuits needed to control certain electrical or mechanical functions.

Transistor: A 3-terminal semiconductor used for current switching, detection, and amplification. Low current flow between one pair of terminals can control high current flow between another pair of terminals, with one common terminal.

Base: The center layer of semiconductor material in a transistor.

Emitter: The outside layer of semiconductor material, in a transistor, that conducts current to the base.

Collector: The outside layer of semiconductor material, in a transistor, that conducts current away from the base.

■ Tran(sfer) + (Re)sistor

The word, "transistor," was originally a business trademark. The word was used as a name for an electrical part that transferred electric signals across a resistor.

The point-contact transistor was invented by two scientists, John Bardeen and Walter H. Brattain, at Bell Telephone Laboratories in 1948. In 1951, the junction transistor was invented by their colleague, William Shockley. As a result of their work, these three men received the 1956 Nobel Prize for physics.

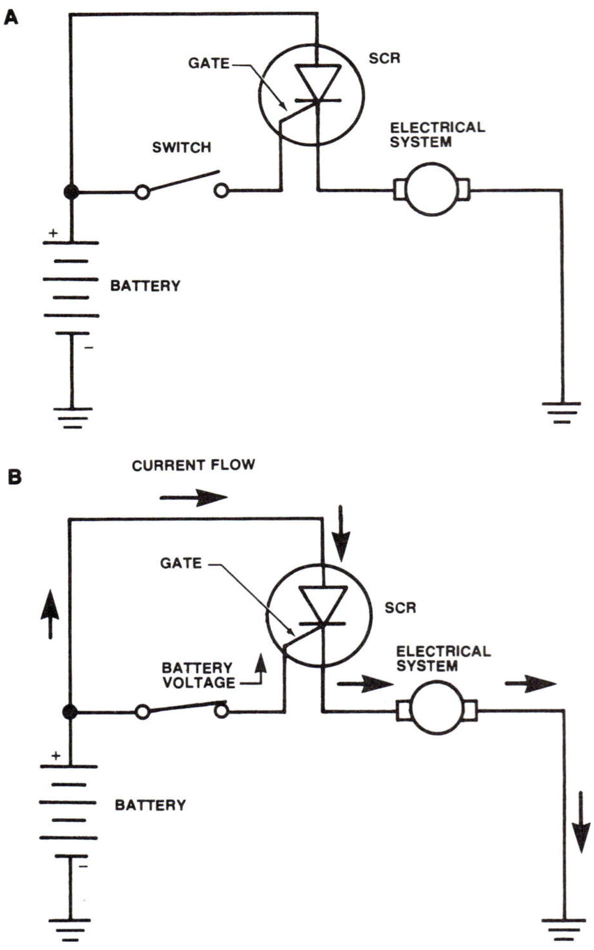

Figure 11-4. An SCR in a circuit can be changed from a total nonconductor to a diode by applying a small amount of voltage to the SCR gate.

Silicon-controlled rectifier operation

The **silicon-controlled rectifier** (SCR) is also called a thyristor. Figure 11-4 shows an SCR installed in a circuit. In its normal state, position A, the SCR blocks current flow in both directions. When a small amount of battery voltage is applied to the SCR's gate, position B, the SCR acts like a diode. That is, it allows current to flow in one direction while still blocking current flow in the opposite direction. When voltage is taken away from the gate, the SCR will again block all current flow in either direction.

Now we can see how transistors and SCRs are used in ignition system control module circuitry.

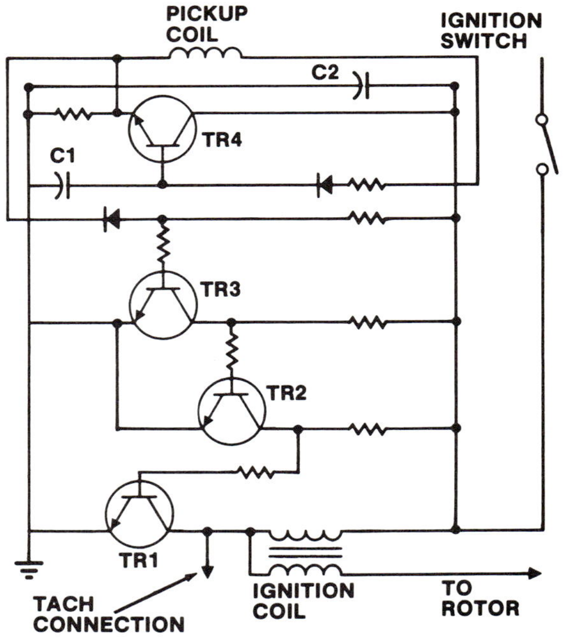

Figure 11-5. A simplified inductive discharge control module. (Chevrolet)

Inductive Discharge System

A simple inductive discharge ignition control module circuit is shown in figure 11-5. The various triggering devices that can be used in this system all produce the same type of signal: a pulsating voltage. In the illustration, the signal comes from the pickup coil of a magnetic pulse generator. This pulsating voltage is applied to the control module's transistor, and signals the module to turn primary current flow off. Turning the current off induces the high-voltage secondary current surge that fires a spark plug.

In figure 11-5, the module can either be in the dwell mode, when primary current flows, or in the firing mode, when primary current stops.

During the dwell mode, transistor number 1 (TR1) and TR3 are conducting. TR2 is nonconducting. A signal from the pickup coil turns TR3 off, charges capacitor number 1 (C1) and turns TR4 on. This results in the firing mode, as reduced primary current flow causes a high-voltage secondary surge.

TR4 stays on until C1 is discharged. When C1 is discharged, TR3 turns on and the module returns to the dwell mode. At higher engine speeds, C1 will be charged less and less. This results in reduced firing times and longer dwell periods. C2 is a capacitor in the distributor for radio noise suppression.

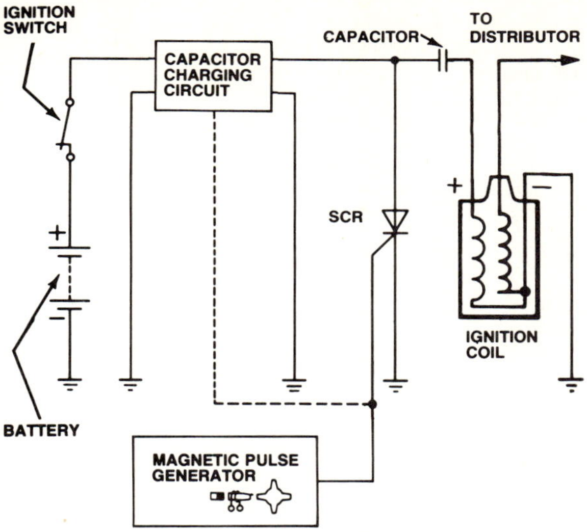

Figure 11-6. A simple capacitive discharge ignition primary circuit. (Bosch)

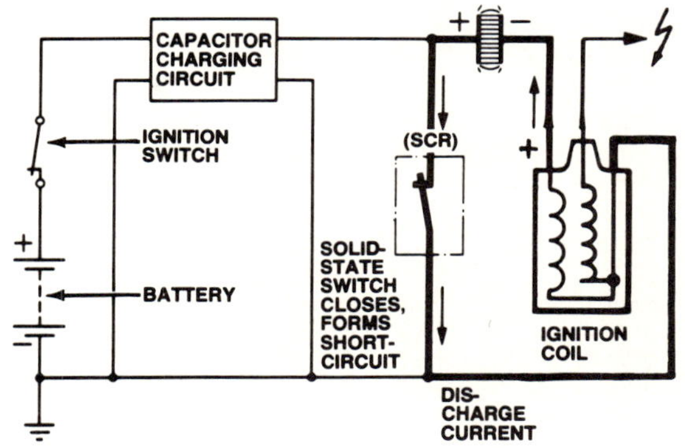

Figure 11-8. When a signal voltage makes the SCR conduct current flow, the capacitor discharges through the coil primary winding. (Bosch)

Actual modules contain other components for circuit protection and other functions.

Capacitive Discharge System

A simple capacitive discharge ignition primary circuit is shown in figure 11-6. Pulsating voltage from the triggering device is applied to a silicon-controlled rectifier.

The SCR controls the module's capacitor. When the SCR blocks all current flow, figure 11-7, the capacitor is charged by battery voltage. A charging circuit within the module amplifies battery voltage until the capacitor contains about 300 volts. When the triggering device sends a voltage pulse to the gate of the SCR, figure 11-8, discharge current flow is allowed.

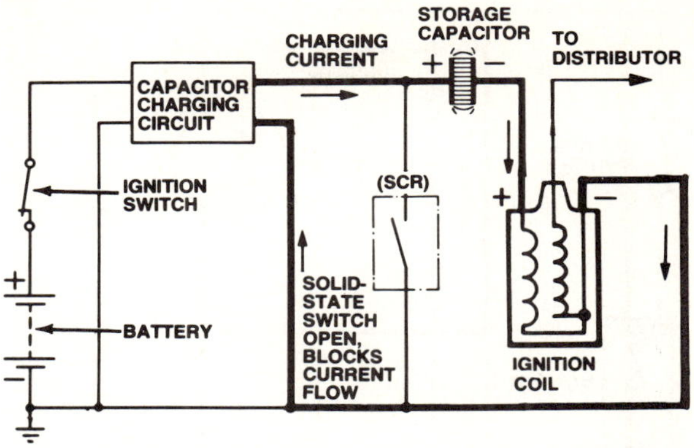

Figure 11-7. When the SCR blocks current flow, the capacitor is charged. (Bosch)

The capacitor discharges through the coil primary winding. The sudden increase in primary current flow induces a high-voltage secondary current surge, which fires a spark plug.

The main difference, then, between these two ignition systems is that in the inductive discharge system, primary current flow is turned *off* to produce the secondary voltage surge to fire a plug, while in the capacitive discharge system, primary current flow is *increased* to produce the high voltage surge needed for ignition.

TRIGGERING DEVICES AND IGNITION TIMING

We have seen that all triggering devices produce a pulsating voltage that determines when to start the ignition spark. Five types of triggering devices are commonly used in automotive ignition systems:

1. Breaker points
2. Magnetic pulse generator
3. Metal detection
4. Hall-effect
5. Optical (light detection).

Breaker Points

When solid-state systems were first developed, manufacturers adapted control modules to the existing trigger device — the ignition breaker

Silicon-Controlled Rectifier: Also called a thyristor. A solid-state device that normally blocks current flow in both directions, but which can be triggered to allow current flow in one direction while still blocking it in the opposite direction.

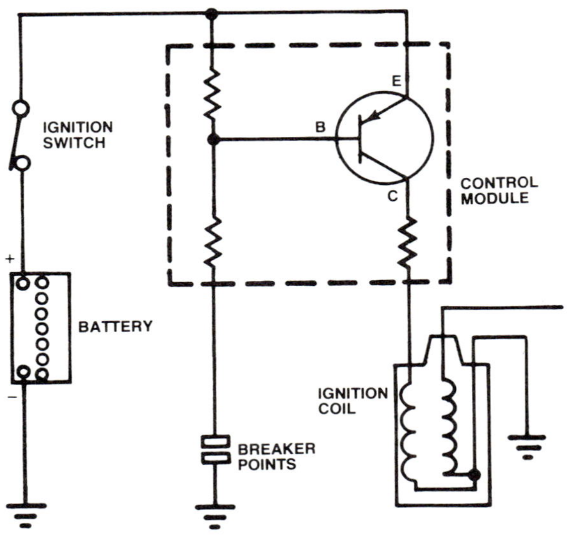

Figure 11-9. A typical breaker-point-triggered electronic ignition system.

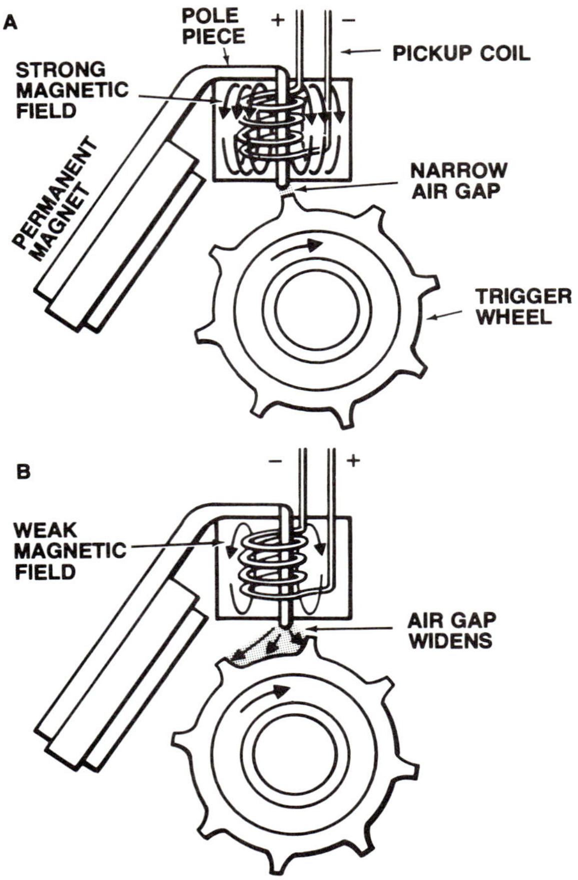

Figure 11-10. The magnetic field expands and collapses as the trigger wheel's teeth move past. This induces a varying-strength voltage in the pickup coil.

points, figure 11-9. When the points are closed, battery voltage causes current flow through the emitter-to-base circuit of the transistor. This makes the transistor conduct current through the emitter-to-collector circuit and so to the ignition coil primary winding. When the breaker points open, the emitter-to-base circuit is opened. There will be no emitter-to-base current flow, which means that the emitter-to-collector current flow must also stop. Current flow through the coil primary winding stops and a high-voltage surge is induced in the ignition secondary circuit.

Because the emitter-to-base current flow is very small, there is little current flow through the breaker points. This lengthens their life and improves their performance. However, the points are still subject to mechanical wear, and their efficiency decreases at high engine speeds.

■ Mind Those Magnets

When you are servicing a distributor from an electronic ignition system, be sure that no metal particles or iron filings get inside. The magnetic pickup coils and pole pieces in the breakerless distributors used by many carmakers will attract metallic debris which can really foul up ignition performance. Use a clean, soft-bristled brush or low-pressure compressed air to clean the inside of a distributor and keep scrap metal off the pickup coil.

Breaker-point triggering devices are not commonly installed as original equipment, but many add-on units still use this method. It requires little or no modifications of an original breaker-point distributor, and some systems retain the original ignition circuitry for emergency use.

Magnetic Pulse Generator

Magnetic pulse generators are the most common type of original equipment triggering device. Most manufacturers use the rotation of the distributor shaft to time voltage pulses. As the nonmagnetic trigger wheel turns with the shaft, figure 11-10, teeth on the wheel approach, align with, and move away from the pole piece of a permanent magnet. The low reluctance of the trigger wheel teeth causes the magnet's field to expand as a tooth approaches, position A. The field collapses as the tooth moves away, position B. This motion of the magnetic field induces a signal voltage in the pickup coil wound around the pole piece.

Figure 11-11. This Delco-Remy magnetic pulse generator has a slightly different design, but operates in the same manner as single pole piece units. (Delco-Remy)

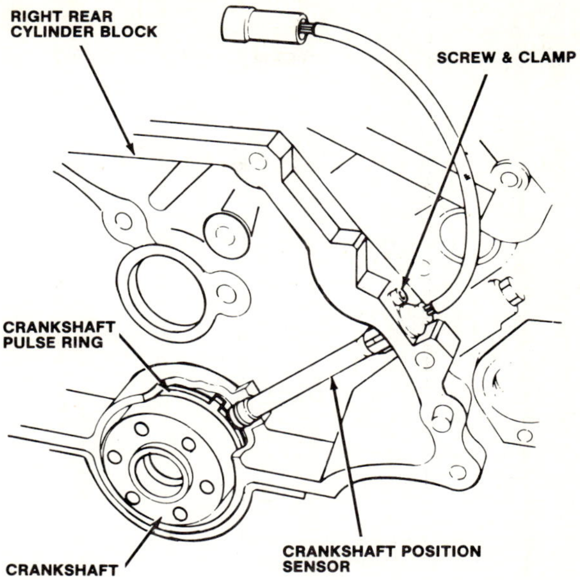

Figure 11-12. The crankshaft position sensor in Ford's EEC-I system. (Ford)

The Delco-Remy and Bosch distributor-mounted magnetic pulse generator designs have as many pole pieces as the trigger wheel has teeth, figure 11-11. The permanent magnet is a cylindrical core around the distributor shaft. The pickup coil is wound between the magnet and the pole pieces. This design operates basically as described above.

As mentioned in Chapter 9, GM and Ford have introduced magnetic pulse generators that use the rotation of the engine crankshaft to produce a signal voltage. In the 1977 GM MISAR system, the pickup coil, pole piece, and permanent magnet are mounted near a disc connected to the pulley end of the crankshaft. Teeth on the disc act like the teeth on a distributor-mounted trigger wheel to induce a signal voltage in the pickup coil. In the 1978 MISAR system, the pickup coil and trigger wheel were returned to the distributor.

The Ford EEC-I system takes a crankshaft rotation signal from the flywheel end of the crankshaft, figure 11-12.

Metal Detection

This triggering device is a variation of the magnetic pulse generator. Instead of a permanent magnet affecting the pole piece, an **electromagnet** supplies the magnetic field, figure 11-13. The control module applies a small amount of battery voltage to the electromagnet coil. The metal teeth on the trigger wheel affect the electromagnetic field and the voltage within the electromagnet coil. These voltage changes are sensed by the control module. This device produces a more reliable signal voltage at lower engine speeds than will a magnetic pulse generator. A metal-detection triggering device is used in the Prestolite electronic ignition used by American Motors from 1975 to 1977 and by International Harvester.

Hall Effect

The Hall-effect triggering assembly uses a small chip of **semiconductor** material, a permanent

Electromagnet: A soft iron core wrapped in a coil of a current-carrying conductor. Current flow in the conductor induces a magnetic force around the core.

Semiconductor: A material that is neither a good conductor nor a good insulator. Semiconductors are the raw material used to make solid-state devices such as diodes and transistors. Silicon and germanium are common semiconductors.

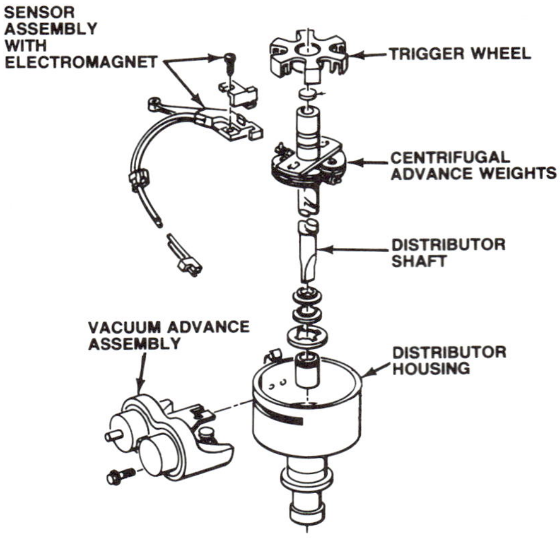

Figure 11-13. A metal-detection triggering device.

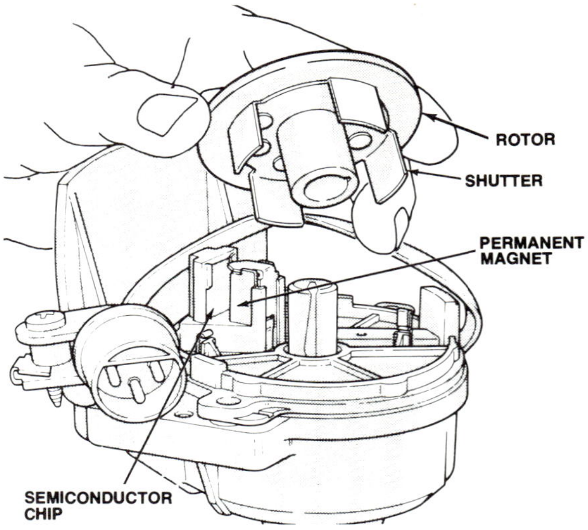

Figure 11-14. A Hall-effect triggering device. (Chrysler)

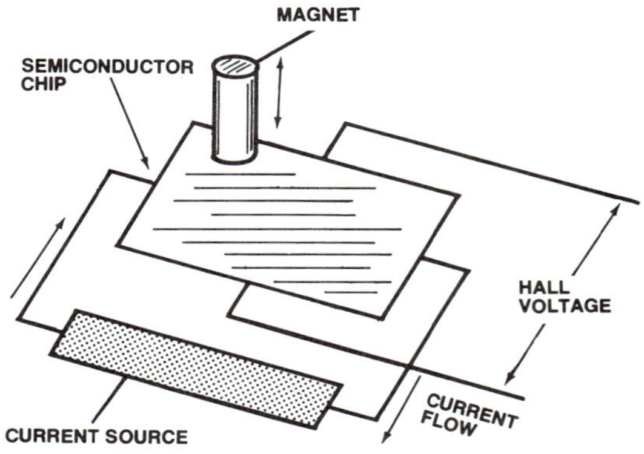

Figure 11-15. The Hall-effect principle.

magnet, and a ring of low-reluctance shutters, figure 11-14. When a small amount of current flows through the semiconductor and a magnetic field crosses the semiconductor at right angles, a voltage is developed at the edges of the semiconductor chip, figure 11-15. When the magnetic field is taken away, the voltage disappears. This Hall-effect voltage can be amplified and used to trigger an ignition system control module.

The only original-equipment system using the Hall effect is Chrysler's Lean-Burn system for a 4-cylinder engine, figure 11-14. The low-reluctance shutter blades hang down from the bottom of the distributor rotor. As they turn, they pass between a permanent magnet and the Hall-effect chip. The magnetic field will concentrate in the shutter as it passes close by, and not affect the semiconductor chip. When no shutter is close, the magnetic field will be spread out and will affect the chip. The Hall-effect voltage appears and disappears as the position of the shutters changes.

The Hall-effect voltage is not changed by changing engine speed. Magnetic pulse generators and metal-detection units depend on induction to create the signal voltage. The strength of an induced voltage varies if the magnetic lines move more quickly or slowly. The Hall effect is not induction, and the speed of the magnetic lines has no effect on the signal voltage. This constant-strength signal voltage offers more reliable ignition system performance through a wide range of engine speeds.

Optical (Light Detection)

Optical triggering devices use a **light-emitting diode** (LED) and a light-sensitive **phototransistor** (photocell) to produce signal voltage pulses. When the LED light beam strikes the phototransistor, voltage is generated. A slotted disc rotating on the distributor shaft, figure 11-16, interrupts the light beam, sending an on-off voltage signal to the control module.

This device is used on some add-on ignition systems, but no major vehicle manufacturer uses it as original equipment. The unit provides a more reliable signal voltage at much lower engine speeds than either a magnetic pulse or a metal-detection device. However, periodic LED and photocell cleaning may be required.

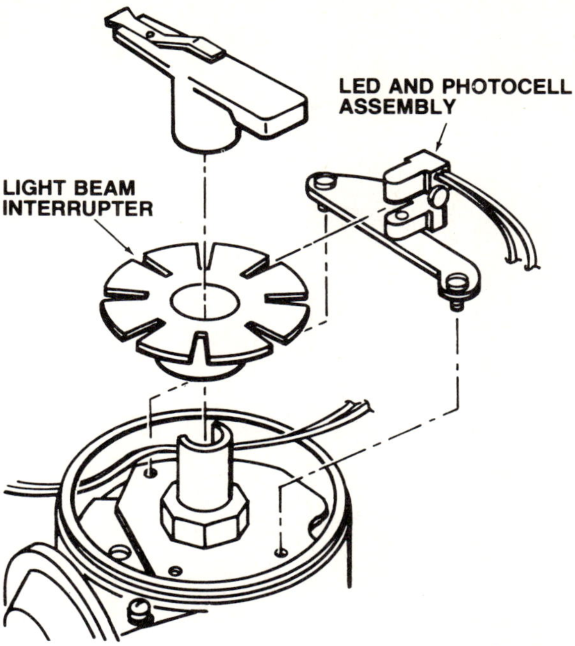

Figure 11-16. A light-detection triggering device.

Ignition Timing And Advance

With most solid-state ignition systems, basic timing is set by the methods we studied in Chapter 1: a timing mark or a magnetic probe. The systems that take signals from the crankshaft are slightly different. GM's 1977 MISAR can be timed by adjusting the stationary sensor unit containing the pickup coil. Ford's EEC-I system has no provision for adjusting the initial timing.

To advance timing, most of the distributor-mounted triggering devices use the same centrifugal and vacuum advance mechanisms that we studied in Chapter 9. Chrysler Lean-Burn distributors have no vacuum advance units, and only 1976 models have a centrifugal advance unit. GM's MISAR and Ford's EEC-I also have completely computer-controlled timing, as we saw in Chapter 9.

Dwell

In a breaker-point ignition system, dwell refers to the degree of cam rotation while the points remain closed. Although this does not apply to a solid-state system, the concept of dwell can refer to the length of time during which primary current flows.

The dwell period of a solid-state system can be measured with the use of an oscilloscope, as you will see in your *Shop Manual*. However, there is usually no provision for adjusting dwell. Often, dwell is determined by a special timing circuit in the control module.

In systems manufactured by Prestolite, using the metal-detection triggering method, the relationship of the trigger wheel to the pole piece has an effect upon dwell. This can be adjusted on the system used by International Harvester, but not on the system used by AMC.

ORIGINAL EQUIPMENT SOLID-STATE IGNITIONS

Solid-state ignition systems appeared in large numbers on domestic automobiles in the early 1970's, although some electronic ignitions were used in limited applications during the 1960's. They are now standard equipment used by all major domestic carmakers. The following paragraphs describe specific manufacturers' systems. Because solid-state technology is constantly being improved, solid-state ignition systems can change from year to year. We begin with the manufacturer's earliest standard (or commonly installed optional) system, explain its changes, and offer some glimpses into future technology. All descriptions apply to domestically built products.

Light-Emitting Diode: A diode that gives off visible light when current flows through it.

Phototransistor: Also called a photocell. A type of solid-state device that will generate a voltage when exposed to light.

■ Mr. Hall's Discovery...100 Years Ago

While the Hall-effect distributor pioneered as original equipment by Chrysler Corporation in 1978 is new, the basic principles are not at all new.

In 1879, long before the first automobile was plodding the roads, a man named Edward H. Hall of Johns Hopkins University discovered the principle that was given his name. He found that when a thin rectangular conductor made of gold and carrying an electric current is crossed at right angles by a magnetic field, a difference in potential can be detected at the edges of the conductor.

For several generations, this idea was left largely unused. Only when mass-produced semiconductors became readily available was the Hall effect put into practice, when it was discovered that the semiconductors can take the place of the gold plate, and at a fraction of the cost.

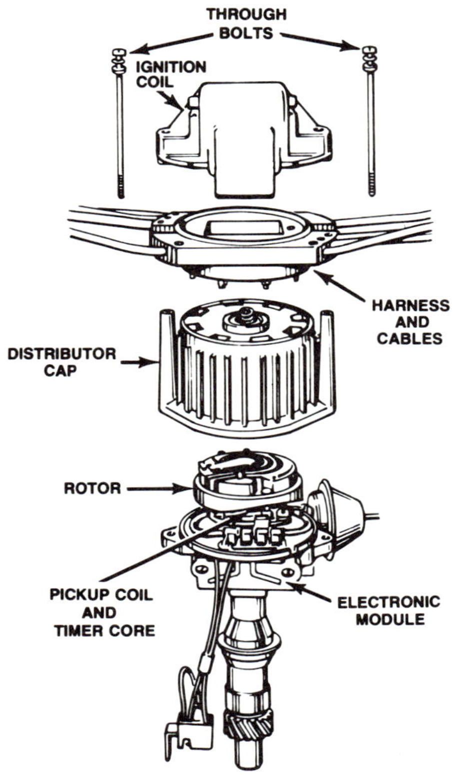

Figure 11-17. Delco-Remy's Unitized electronic ignition system.

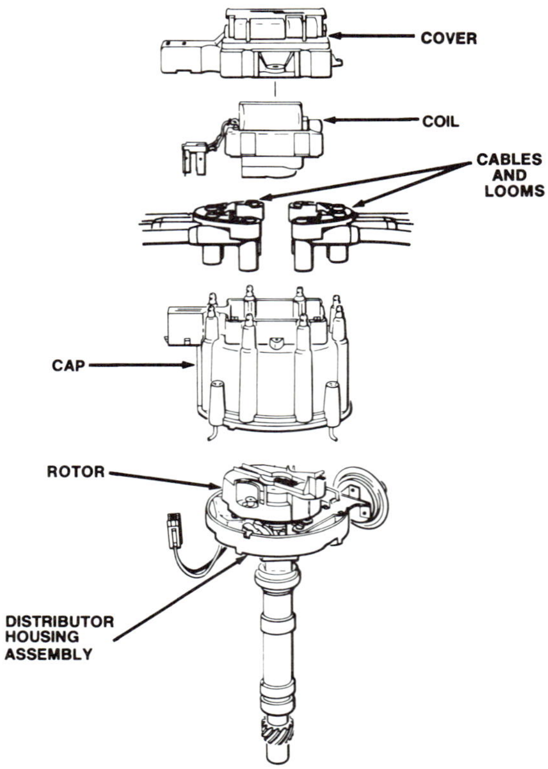

Figure 11-18. The HEI distributor for V-6, V-8, and some inline engines.

We have seen that manufacturers have different names for the components of a triggering device. To avoid confusion, review this list of terms:

Manufacturer	*Stationary Pickup Coil*	*Rotating Trigger Wheel*
AMC	Sensor	Trigger Wheel
Chrysler	Pickup Coil	Reluctor
Ford	Stator	Armature
GM	Pole Piece and Magnetic Pickup	Timer Core

Delco-Remy

Unitized ignition

Delco-Remy's Unitized ignition is a solid-state electronic system for 8-cylinder engines. It was introduced in 1972 on several General Motors cars and continued through 1974 as an option on some Pontiac models. This ignition system contains all components in one assembly, figure 11-17. The assembly consists of:

- A breakerless, magnetic pulse distributor
- A distributor cap
- The ignition coil
- A secondary wiring harness containing all spark plug cables
- An electronic control module.

The assembly is held together by two through-bolts.

The Unitized distributor uses conventional centrifugal and vacuum advance mechanisms to control spark timing. The electronic module is mounted on the side of the distributor housing, rather than in a remote location.

When the engine is cranked with the spark plugs disconnected, such as when an engine compression check is done, the ignition switch connector should be disconnected from the distributor. Open-circuit operation can damage the control module.

High-energy ignition (HEI)

The Delco-Remy HEI system was introduced in 1974 as optional equipment on GM V-8 engines. In 1975, it became standard equipment on all GM engines.

The HEI system has two different versions, but both work in the same way. V-type and late model inline engines have the distributor, the

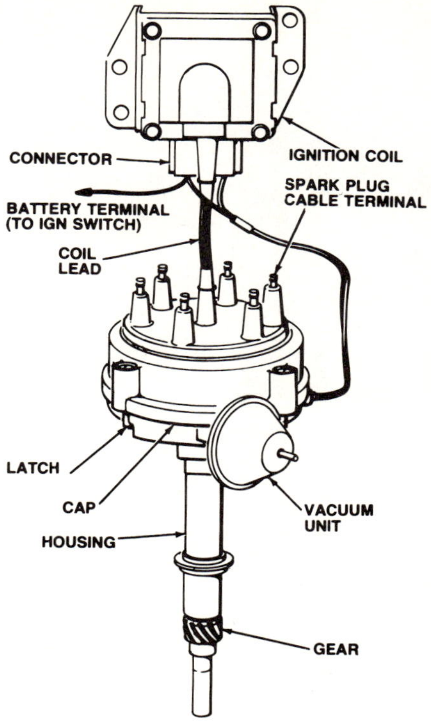

Figure 11-19. The HEI distributor for some inline engines.

coil, and the control module assembled in one unit, figure 11-18. Early-production inline 4- and 6-cylinder engines have the coil mounted separately from the distributor and control module, figure 11-19.

Distributors for all engines except GM uneven-firing V-6 engines have as many equally spaced teeth on the pole piece and timer core as there are engine cylinders. Because of the unevenly spaced cylinder firing intervals, uneven-firing V-6 distributors have three evenly spaced teeth on the timer core and six teeth alternately spaced 45 degrees and 75 degrees apart on the pole piece.

The electronic control module of the HEI system delivers full battery voltage to the ignition coil. Current flow is less with the ignition on and the engine off or cranking. No primary circuit ignition resistor is used.

The electronic control module automatically controls the ignition dwell period, lengthening with increasing engine speed. Dwell is not adjustable. A capacitor is installed in the distributor to suppress radio noise.

Most engines that use this system also use wide spark plug gaps to take advantage of the high voltage output. Special 8-mm spark plug cables are used that have silicone insulation to minimize crossfiring. The silicone insulation is softer than rubber and must be handled carefully. Standard 7-mm cables *cannot* be used with the HEI system.

The system uses conventional centrifugal and vacuum advance mechanisms to control spark timing. Engine tests must be performed correctly because open-circuit operation can harm the control module.

MISAR

The Delco-Remy "microprocessed sensing and automatic regulation" (MISAR) system was introduced on the 1977 Oldsmobile Toronado. It uses a computer to control spark timing advance, as we learned in Chapter 9. A modified HEI distributor is used, containing the rotor, the coil, the control module, and the spark plug electrodes. The magnetic pulse generator is mounted at the front of the crankshaft on 1977 systems. In 1978 it was relocated to the distributor.

Electronic spark selection (ESS)

An electronic spark timing system called ESS was introduced on the Cadillac Seville in 1978. The system has a solid-state decoder that is connected to the HEI module in the distributor. During certain engine operating conditions, the decoder signals the HEI module to retard the ignition timing. Timing is retarded during cranking and other specific conditions such as during cold-temperature operation of California engines.

The decoder receives engine rpm signals from the HEI pickup coil and on-off signals from various engine sensors. These include the EGR solenoid, which signals coolant temperature; the ignition switch, which signals cranking; and the fuel economy switch, which signals engine vacuum level.

Chrysler

Electronic ignition

The Chrysler electronic ignition was introduced in 1971 and became standard on all 6- and 8-cylinder engines in 1973, figure 11-20. The system contains a magnetic pulse distributor, an electronic control module, and a dual ballast resistor as well as a conventional coil. The distributor is externally similar to a breaker-point distributor. The cap, the rotor, the housing, the advance mechanisms, and the spark plug cables are the same as for a breaker-point ignition.

The air gap between the pickup coil and the reluctor is adjustable, but has no effect upon the dwell period. Dwell is determined by the control unit.

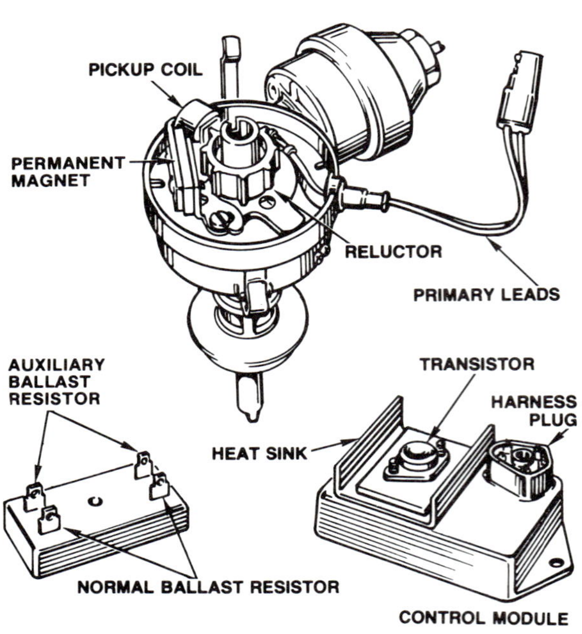

Figure 11-20. The basic Chrysler electronic ignition system.

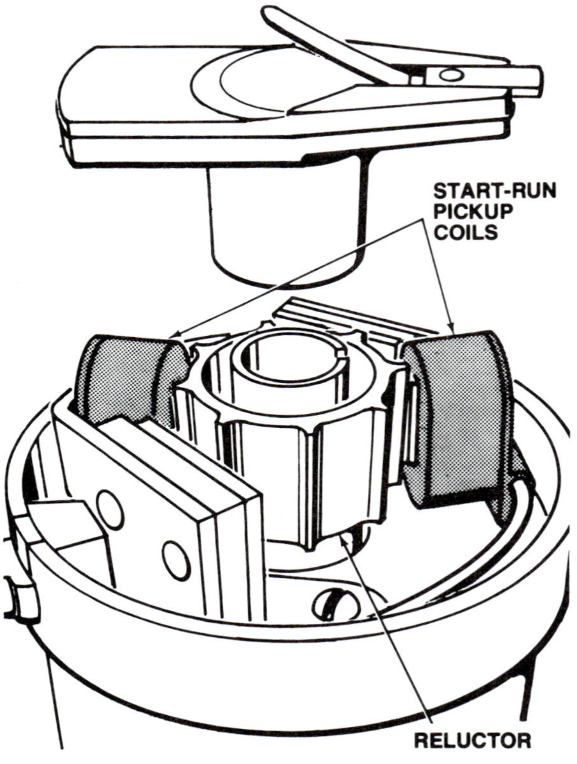

Figure 11-22. The Lean-Burn distributor for 1976 and 1977 (except 318 V-8) systems has two pickup coils.

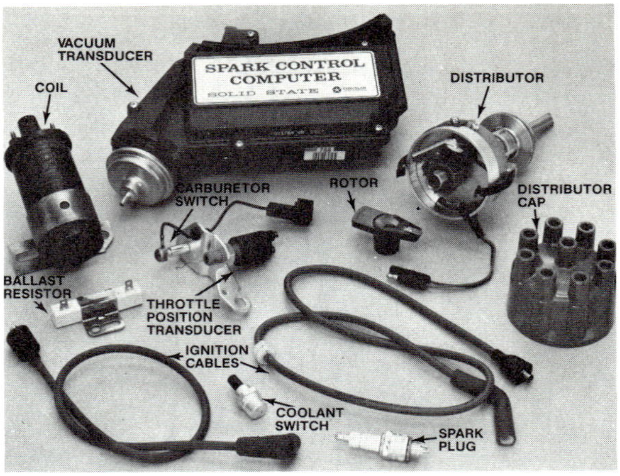

Figure 11-21. The parts of a late-model V-8 engine Lean-Burn system. (Chrysler)

The control unit is mounted on the firewall or fender panel and contains a switching transistor to control primary current flow. The switching transistor is mounted on top of the control unit and should not be touched when the ignition switch is on. Enough voltage is present to produce a shock. Some control units used on pre-1973 engines contained circuitry to limit engine speed.

The dual ballast resistor, mounted on the firewall, serves two functions. The 0.5-ohm resistor tends to maintain constant uniform primary current flow with variations in engine speed, as in a breaker-point ignition. This resistor is bypassed during cranking. The 5-ohm resistor protects the electronic part of the system by limiting current flow through the electronic circuitry. A radio noise suppression capacitor was added to the system in 1973. This capacitor is mounted on the firewall near the dual ballast resistor.

Electronic Lean-Burn

The Chrysler Lean-Burn system, which we studied in Chapter 9, controls the ignition system so that an engine can operate efficiently on air-fuel ratios as lean as 18 to 1.

The timing computer, figure 11-21, receives signals from various engine sensors, determines the optimum time for ignition, and interrupts the primary circuit accordingly.

The distributor for 1976 and most 1977 systems has two pickup coils for various timing requirements, figure 11-22. The Start pickup, which works only when the engine is cranking, provides a fixed amount of ignition advance. The Run pickup provides signals to the computer at all other times. Lean-Burn systems for

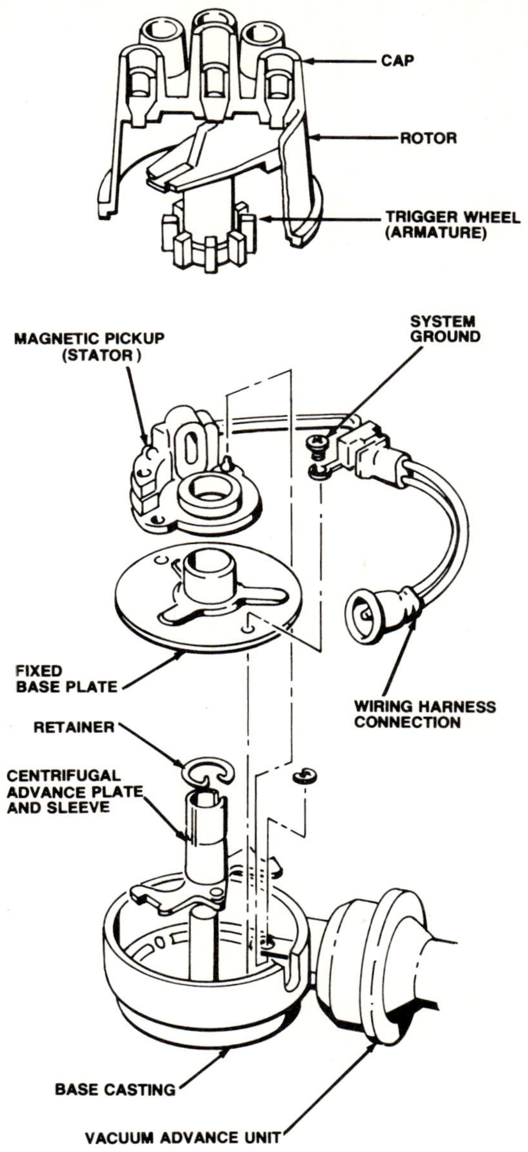

Figure 11-23. Ford's breakerless ignition system uses conventional advance mechanisms.

1977 318 V-8 models and for 1978 and later V-8 systems have only one pickup coil in the distributor for all timing signal functions. (The 4-cylinder engine system uses the Hall-effect distributor assembly.) Also, these later Lean-Burn systems have a redesigned computer module that has all ignition control and timing functions programmed into one circuit board. Wire harness connectors at the computer were redesigned, and the auxiliary 5.0-ohm ballast resistor was eliminated.

No vacuum advance mechanism is used on Lean-Burn distributors. The distributors for 1976 systems have centrifugal advance mechanisms, but these have been eliminated from 1977 and later systems, and all spark advance is provided by the computer.

In 1978, Chrysler extended the use of the Lean-Burn system to the 4-cylinder engines in the Dodge Omni and Plymouth Horizon. Before these models were introduced, the Lean-Burn system had been used only on V-8 engines.

Ford (Motorcraft)

Breakerless solid-state ignition

The Ford Motorcraft breakerless ignition was introduced in mid-1973, was available on all V-8 and inline 6-cylinder engines in 1974, and became standard on all engines in 1975.

It consists of a magnetic pulse distributor, figure 11-23, an electronic control module, and a special coil. The distributor uses a conventional cap, rotor, and body. The centrifugal and vacuum advance mechanisms also are retained.

The coil is a special oil-filled type that is identified by a blue case or tower. The coil primary terminals are labeled BAT and DEC

■ Simple Simulator

If you've done much solid-state ignition service, you know that pinpointing an intermittent problem — a random misfire — can be time consuming and frustrating. You often spend a lot of time on the road, driving and waiting for the bug to appear. However, there is an ignition pulse simulator as near as your toolbox that may reduce your troubleshooting time considerably.

An electric engraving tool, driven by a vibrating electromagnet, will produce a high-frequency signal that can trigger the pickup coil in some electronic ignitions. It works fine on Ford solid-state ignitions and may work as well on similar systems.

Simply remove the coil high-tension lead from the distributor cap and set up about a half-inch gap to ground. Then remove the distributor cap and lay the electric engraver across the top of the distributor near the pickup coil. Be sure it doesn't touch the shaft, pickup coil, or armature (reluctor). Then turn the ignition switch to ON and switch on the engraving tool. A strong, high-frequency spark should be produced at the coil lead. You can use your scope to test coil output and you can poke, wiggle, and otherwise check all connectors while the ignition is "running."

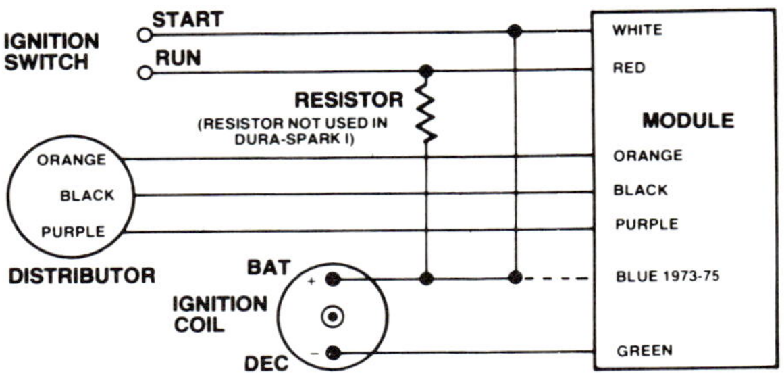

Figure 11-24. The wiring color code for Ford's electronic ignition module from 1973 to 1978.

(Distributor Electronic Control). A standard primary circuit resistor wire is used.

Since its introduction, the Motorcraft system has undergone several circuit design changes. These changes center around the wire color coding and connector identification of the seven wires that feed out of the control module. They feed out in two groups, one ending in a 2-wire or a 3-wire connector and the other in a 4-wire connector. Both connectors plug into the main vehicle wiring harness.

The wire color coding is the key to tracing the ignition circuit on cars of different years, figure 11-24. Power from the ignition switch reaches the module through the *white* wire when the engine is cranking but through the *red* wire once it is running. Distributor signals are received through the *orange* and *purple* wires. Primary current from the coil is carried by the *green* wire to the module. The *black* wire is a ground connection between the module and the distributor. On 1973-75 systems, the *blue* wire protects the system against voltage surges.

Wire color coding and connector arrangements are the same for 1973 and 1974 systems. In 1975 systems, the color coding is unchanged, but the wire arrangement at the connectors and the connector shapes are altered. In 1976, the blue wire for system protection was eliminated, and the module circuitry was redesigned to dump temporary voltage surges. The connector shapes and the terminal arrangements were also changed in 1976. The 3-wire connector became a 2-wire connector. Because of these changes, modules and wiring harnesses are not interchangeable between model years.

In 1978, American Motors adopted what is essentially the 1976 Ford Motorcraft ignition system, replacing the Prestolite system used from 1975 through 1977.

Dura-Spark

In 1977, Ford introduced a second-generation solid-state ignition. Called the Dura-Spark ignition, it is a direct evolution of the breakerless

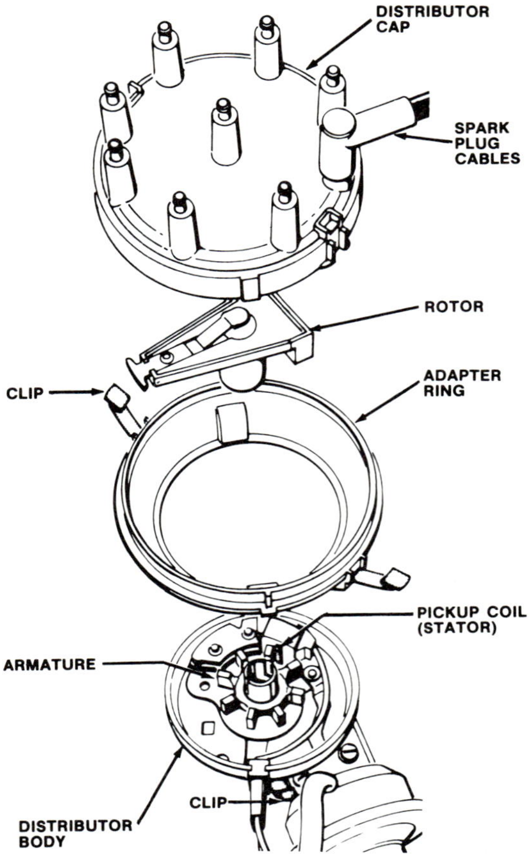

Figure 11-25. A Dura-Spark distributor.

system just described.

The Dura-Spark system has higher secondary voltage capabilities, and electrical values of some components are different from earlier Ford systems. However, the primary circuit and its test procedures are similar to those for the 1973-76 systems.

There actually are two versions of the Dura-Spark system. Dura-Spark I was first used on all California V-8 passenger cars in 1977. Dura-Spark II was first used on all other cars in 1977, including California 4- and 6-cylinder engines. A comparison of secondary voltage capabilities illustrates the difference between earlier Ford ignitions and the Dura-Spark systems:

	Available Kilovolts	
	Start	Run
1976 Solid-State Ignition	32	26
Dura-Spark I	47	42
Dura-Spark II	42	36

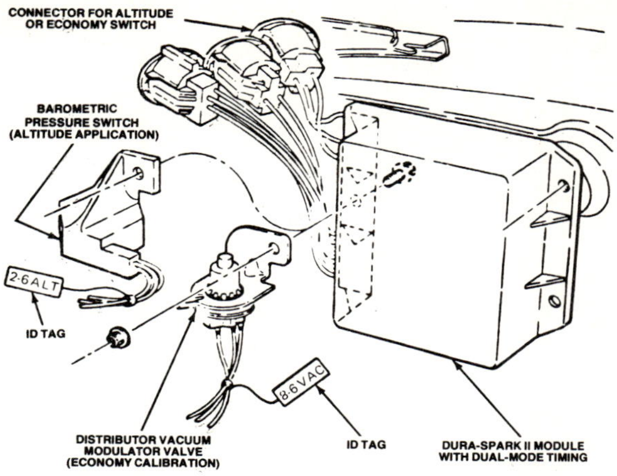

Figure 11-26. The control module used with Ford's Dual-Mode timing has an additional 3-wire connector and either a pressure or a vacuum switch mounted on the module. (Ford)

The Dura-Spark I system has a new ignition module and a new coil. Integrated circuitry in the new module adjusts the ignition dwell in relation to current value through the coil at the time of a spark. The module also has a stall-shutdown feature for circuit protection. If the engine stalls, the module shuts off the primary circuit even though the ignition switch is in the On position. The switch must be turned off and then back to Start to close the primary circuit.

Dura-Spark I has no primary resistor. The Dura-Spark I ignition coil has a gray tower and different terminals to prevent its use in other systems.

Dura-Spark II more closely resembles the 1976 Ford ignition system. It retains the primary resistor, but resistance values are changed to provide higher secondary voltage capabilities. The coil is the same as that used in 1976.

All Dura-Spark distributor caps, figure 11-25, have male terminals for the 8-mm spark plug cables. The caps on 6- and 8-cylinder distributors are much larger than previous models and have adapter rings to mate with the distributor housing. Larger rotors are also used with the big caps.

To remove the new cap, first unlatch and remove the top section. Then remove the rotor, followed by the adapter ring. Do not try to unlatch or unscrew the adapter from the housing and remove the entire assembly, or it will jam on the rotor.

The Dura-Spark systems have a 2-wire and a 4-wire connector at the control module. The connectors for the 1976 system and for Dura-Spark II are identical. The index lugs of the

Figure 11-27. The distributor used with Ford's Electronic Engine Control system.

Dura-Spark I 4-wire connector are different to prevent installation of the module in any other system. Additionally, two terminals in the connector have been transposed. Wire color coding for both Dura-Spark systems is the same as that used for 1973-76 systems.

Most spark plugs gaps for use with Dura-Spark systems are wider than those used with previous Ford solid-state ignitions.

Some 1978 Dura-Spark systems also have another timing modification system called Dual-Mode timing. There are two versions of Dual-Mode timing: one for increased fuel economy and one for operation at varying altitudes. Each type has a switch mounted on the ignition control module, figure 11-26. The switch is connected to the module by an additional 3-wire connector. The fuel economy system switch is exposed to intake manifold vacuum. The altitude compensating switch is exposed to atmospheric pressure.

When the switch signals the ignition module, ignition timing is retarded by 3 to 6 degrees from the initial setting. The fuel economy switch signals the module to retard timing when there is less than 6 inches of manifold vacuum. The altitude switch signals the module to retard timing when the vehicle is operated at or below 4,300 feet elevation.

Electronic engine control (EEC)

In 1978, Ford introduced an electronic engine control system known as EEC-I. As we learned in Chapter 9, a solid-state control module, similar to the Dura-Spark I module, receives signals from the EEC-I computer and adjusts spark timing. A blue, high-voltage Dura-Spark I igni-

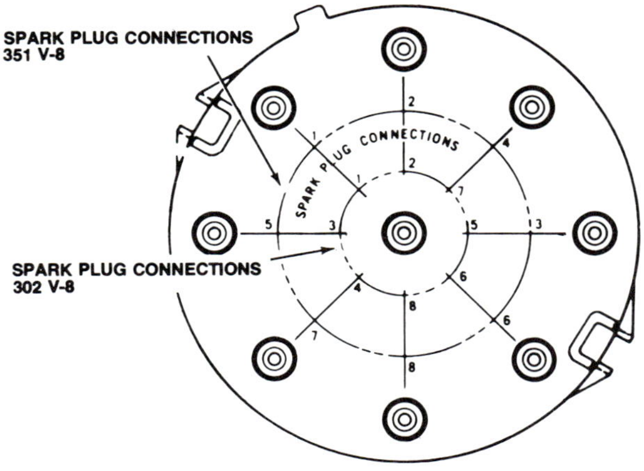

Figure 11-28. The distributor cap used with EEC-I and EEC-II has spark plug cable connections identified. These are *not* the firing orders. (Ford)

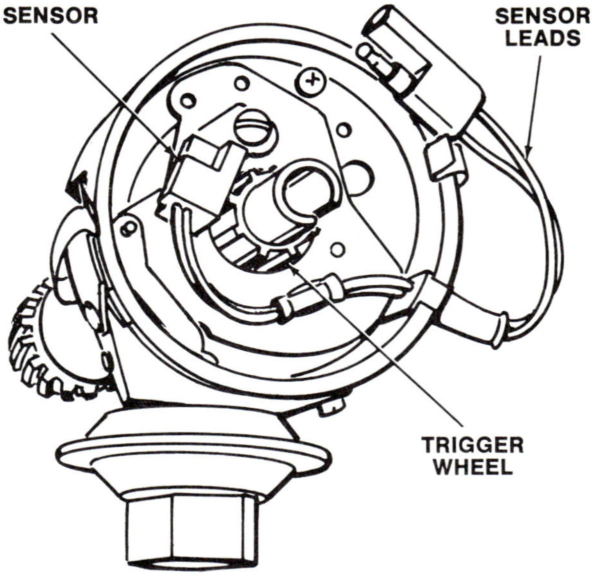

Figure 11-29. Prestolite's electronic ignition distributor as used on IHC engines.

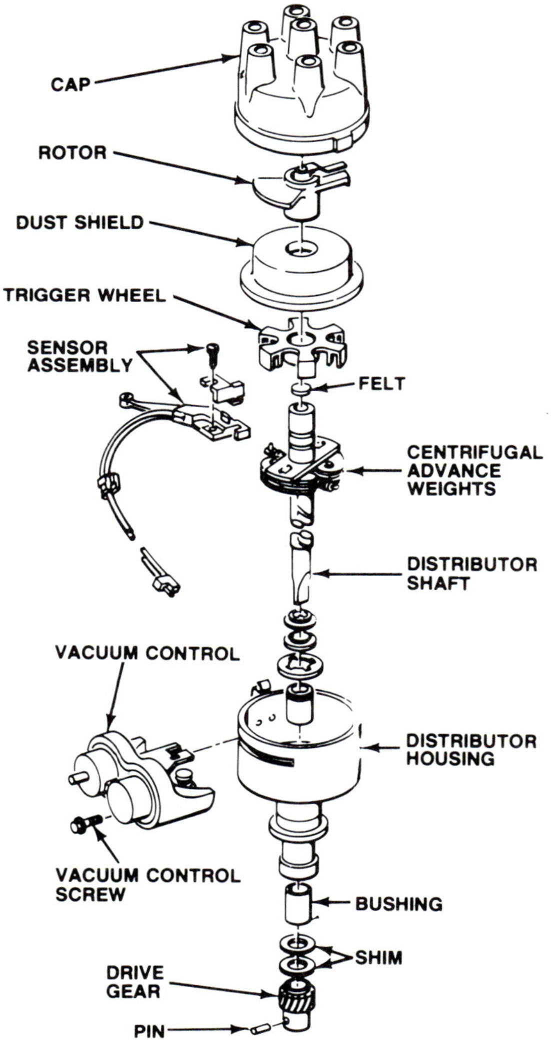

Figure 11-30. Prestolite's electronic ignition distributor as used on 1975-77 AMC engines.

tion coil is used, with no primary resistor. The EEC distributor, figure 11-27, contains only the rotor, the coil electrode, and the spark plug electrodes.

EEC-II, introduced in 1979, uses the same ignition system parts as EEC-I. Stamped onto the distributor cap is the proper spark plug cable connection order for each engine, figure 11-28. These are *not* the firing orders, because of the distributor's 2-level rotor design.

Prestolite

Breakerless inductive discharge (BID) ignition
The Prestolite BID ignition was first used on some 1974 V-8 engines from International Harvester. In 1975, this system was continued on 4- and 8-cylinder engines, and a variation was introduced as standard equipment on 6- and 8-cylinder AMC engines.

The Prestolite system consists of a conventional ignition coil, a control unit, and a breakerless distributor. Standard spark plugs and cables are used.

The control unit is a solid-state module which protects against reverse polarity and voltage surges. Because the control unit has built-in current regulation, no ballast resistor is used in the primary circuit. The circuit design allows full battery voltage to the coil whenever the ignition switch is at On or Start.

The breakerless distributor uses a conventional cap and rotor and conventional centrifugal and vacuum advance mechanisms. A sensor and trigger wheel use metal detection to provide the signal to the control unit to interrupt primary current flow.

Although the AMC and International systems are about the same in the way they work and the way they are designed, the internal parts, such as the sensor and trigger wheel, are different.

The trigger wheel in the IHC distributor, figure 11-29, is similar to Chrysler's reluctor or Ford's armature. The trigger wheel has four or eight external teeth, depending on the number of engine cylinders. The sensor is located alongside the rotating wheel. The dwell angle on the Prestolite-International ignition is affected by the adjustable air gap between the sensor and the trigger wheel. The dwell decreases as the sensor is moved closer to the wheel and increases as the sensor is moved away.

The sensor and trigger wheel in the AMC distributor, figure 11-30, are different from those in the IHC distributor. The sensor is pin-shaped, and trigger wheel teeth are shaped like an inverted U. The teeth pass over the sensor, and air gap is not adjustable. Dwell is not adjustable, and is determined by the control unit and the angle between the trigger wheel teeth.

SUMMARY

Solid-state electronic ignition systems came into widespread use on domestic vehicles in the early 1970's. Today, they are standard equipment on almost all U.S.-built cars and light trucks. Solid-state ignitions provide greater available voltage, reliable performance for longer periods and under varying operating conditions, and decreased maintenance.

Solid-state ignitions are either the inductive discharge type or the capacitive discharge type. Most original equipment systems are the inductive discharge type. The primary circuits of solid-state ignitions can be triggered by breaker points, a magnetic pulse generator, a metal detector, a Hall-effect assembly, or an optical triggering device. Most original equipment systems use a magnetic pulse generator in the distributor to trigger the primary circuit.

The primary circuit is switched on and off by transistors in the ignition control module. The solid-state control module also controls the length of the ignition dwell period.

Review Questions

Select the single most correct answer.
Compare your answers with the correct answers on page 280.

1. Which of the following is true of solid-state ignition systems?
   a. Provide greater available voltage than breaker-point systems
   b. Give reliable system performance at all engine speeds
   c. Require less maintenance than breaker point systems
   d. All of the above

2. Inductive discharge systems use a triggering device to provide a sudden _____ in primary current.
   a. Increase
   b. Decrease
   c. High-voltage surge
   d. Low-voltage pulse

3. Which of the following is *not* true of inductive discharge systems?
   a. Used as original equipment by most car manufacturers
   b. Provide about 30,000 volts of available voltage
   c. Sustains a spark for only 200 microseconds
   d. All of the above

4. Capacitive discharge systems provide _____ secondary voltage when compared to inductive discharge systems.
   a. About the same
   b. Less
   c. Greater
   d. None of the above

5. The most common type of original equipment solid-state triggering devices are:
   a. Breaker points
   b. Light detectors
   c. Metal detectors
   d. Magnetic pulse generators

6. The accompanying illustration shows a:

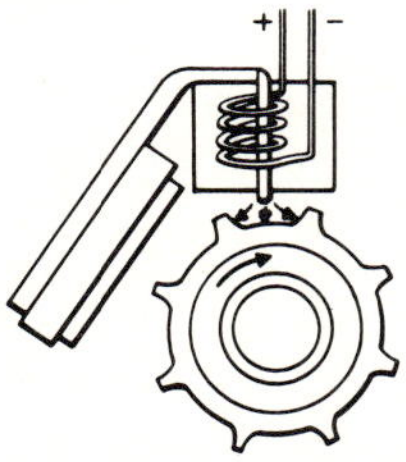

   a. Magnetic pulse generator
   b. Metal detector
   c. Light detector
   d. Breaker point assembly

7. Metal detection triggering devices have been used by:
   a. General Motors Corporation
   b. Ford Motor Company
   c. American Motors Corporation
   d. Chrysler Corporation

8. In the Delco High Energy Ignition (HEI) system, the dwell period is controlled by the:
   a. Pole piece
   b. Timer core
   c. RFI filter capacitor
   d. Electronic control module

9. The Delco HEI system uses:
   a. A single ballast resistor
   b. A dual ballast resistor
   c. No ballast resistor
   d. Calibrated resistance wire

10. The Chrysler electronic ignition system uses a:
    a. Dual ballast resistor
    b. Single ballast resistor
    c. No ballast resistor
    d. Calibrated resistance wire

11. Ford's Dura-Spark I system uses:
    a. A single ballast resistor
    b. A dual ballast resistor
    c. No ballast resistor
    d. A calibrated resistance wire

12. In the IHC-Prestolite electronic distributor, the sensor-to-trigger wheel gap is:
    a. Not adjustable
    b. Adjustable and does affect dwell
    c. Adjustable but does not affect dwell
    d. Variable with engine speed

PART THREE

The Fuel System

Chapter

12

Fuel System Operation

Fuel and air must be delivered to the carburetor before the air-fuel mixture can be created. This is the job of the fuel system. The system contains:

- The fuel storage tank
- Fuel delivery lines
- Evaporative emission controls
- The fuel pump
- The fuel filters
- Air cleaners and filters
- Thermostatic air cleaner controls.

This chapter will examine the basic operation of the fuel system and its effects on engine performance, economy, and emission control. Chapter 13 describes how the carburetor and intake manifold distribute the air-fuel mixture to the cylinders. Chapters 14 and 15 look at two variations on the traditional carburetor and manifold combination — electronic fuel injection systems and superchargers and turbochargers.

TANKS AND FILLERS

The automobile fuel tank, figure 12-1, is made of two corrosion-resistant steel halves, which are ribbed for additional strength and welded together. Exposed sections of the tank may be made of heavier steel for protection from road damage and corrosion.

Some cars and light trucks may have an optional fuel tank. Some of these auxiliary tanks — particularly on Ford products — are made of polyethylene plastic, and greater use of plastics in fuel tank construction is likely in the future.

Tank design and capacity are a compromise between available space, filler location, fuel expansion room, and fuel movement. Some late-model tanks deliberately limit tank capacity by extending the filler tube neck into the tank low enough to prevent complete filling, figure 12-1. A vertical **baffle** in this same tank limits fuel sloshing as the car moves.

Regardless of its size and shape, a fuel tank must have the following:

- An inlet or filler tube, through which fuel can enter the tank, and a filler cap
- An outlet to the fuel line leading to the fuel pump
- A vent system.

Tank Location And Mounting

Except for station wagons, most domestic cars use a horizontally mounted fuel tank. It is usually suspended below the rear of the floor pan, figure 12-2, between the frame rails, and behind the rear axle. Many station wagons use a vertical tank, located on one side of the car between the outer and inner rear fender panels.

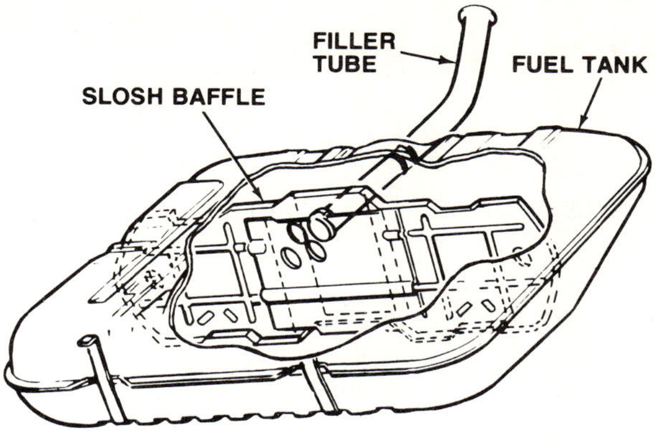

Figure 12-1. The filler tube is located in this tank so that the tank cannot be filled completely. The air space at the top of the tank allows room for fuel expansion. (Oldsmobile)

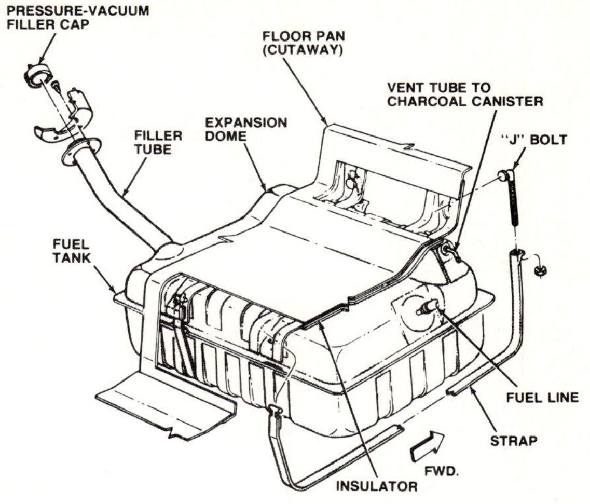

Figure 12-2. Typical fuel tank installation. (Chrysler)

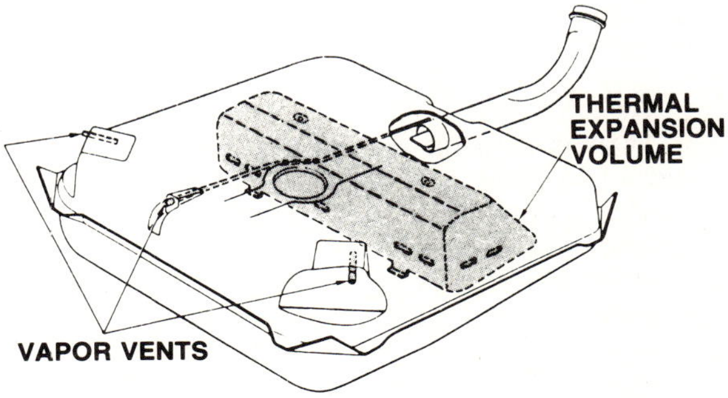

Figure 12-3. This fuel tank has an internal expansion tank to allow for changes in fuel volume due to changes in temperature.

Filler Tubes

All 1975 and later cars that require unleaded fuel (cars with catalytic converters and a few without) have a special filler tube. This filler has a restriction in it so that the only nozzles that can be inserted are the smaller-diameter nozzles of pumps dispensing unleaded fuel. A decal reading "Unleaded Fuel Only" is located beside the filler cap and on the instrument panel.

Tank Venting Requirements

Fuel tanks must be vented or a **vacuum lock** will keep the fuel from being delivered to the engine. Before 1970, a vent line to the atmosphere, or a vented filler cap, provided the venting. However, these systems added to air pollution by passing fuel vapors into the air. To reduce evaporative hydrocarbon (HC) emissions, controls have been installed on all 1971 and later cars (1970 in California).

Because fuel tanks no longer are vented directly to the atmosphere, the tank must allow for fuel expansion, contraction, and overflow that results from changes in temperature. One way is to use a separate expansion tank within the main fuel tank, figure 12-3. Another way is to provide a dome in the top of the tank, figure 12-2. As we mentioned earlier, some tanks are limited in capacity by the angle of the fuel filler tube, figure 12-1.

Rollover Leakage Protection

All 1976 and later cars have one or more devices to prevent fuel leaks in case the car turns over. These include check valves, float valves, and redesigned fuel tank filler caps.

AMC uses several different rollover leak protection devices. All of them work on the **check valve** principle, figure 12-4. An inline check valve is in the fuel return line at the carburetor. A rollover check valve is in the fuel vapor vent line between the tank and the charcoal canister. Some AMC cars also have a liquid check valve in the fuel tank outlet to the vapor vent line.

Baffle: A plate or obstruction that restricts the flow of air or liquids. The baffle in a fuel tank keeps the fuel from sloshing as the car moves.

Vacuum Lock: A stoppage of fuel flow caused by insufficient air intake to the fuel tank.

Check Valve: A valve that permits flow in only one direction.

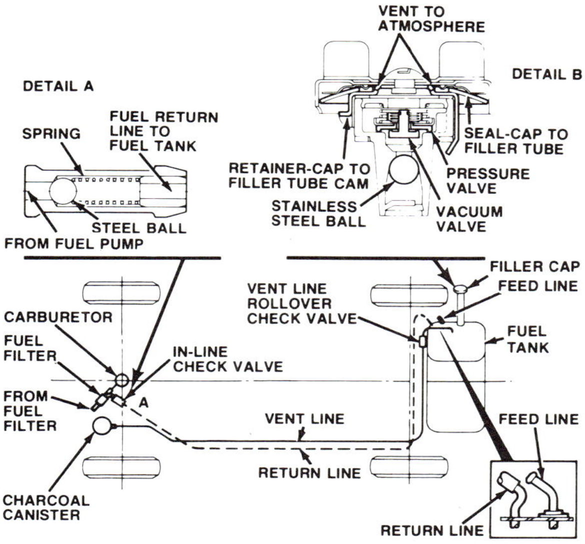

Figure 12-4. American Motors rollover leakage protection devices. (AMC)

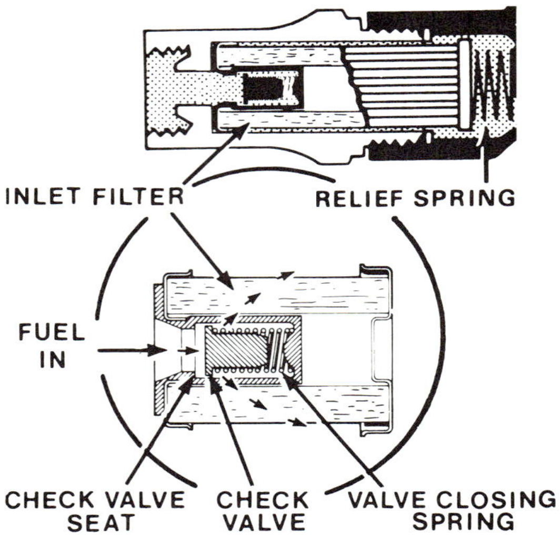

Figure 12-5. The rollover protection check valve is built into the fuel filter on late-model GM cars. (Chevrolet)

Chrysler uses a rollover valve that is similar to the basic AMC check valve. This valve replaced the overfill limiting valve in the vapor vent line between the vapor canister and the fuel tank. Cars sold in California have a return line from the fuel filter to the tank. These have a special filter with a built-in check valve.

■ **Fuel System Development**

The first automobiles used a gravity-feed system to provide fuel to the engine. The fuel tank was mounted in the cowl, higher than the engine, and gravity drew fuel out of the tank and carried it to the engine. But front-mounted fuel tanks were limited in capacity and were dangerous.

Moving the fuel tank to the rear of the car solved the problems of safety and storage capacity, but required the use of a vacuum tank. This was a small fuel tank, still positioned above the engine in the cowl, but connected to the rear tank as well. Suction created by engine vacuum provided fuel for the vacuum tank from the larger rear-mounted tank.

If the car was not driven for long periods of time, the gasoline in the vacuum tank would eventually evaporate. In this case, it was necessary to prime the engine in order to start it and create vacuum which would move fuel through the system. With the appearance of the mechanical fuel pump after World War I, the vacuum tank was retired.

GM uses a spring-loaded plastic check valve in the carburetor inlet fuel filter, figure 12-5. The valve is pushed off its seat by fuel pump pressure and returns to its seat when the engine is off. The valve is built into all replacement filters for Rochester and Holley carburetors used in 1976 and later GM cars. While the new filters are interchangeable with earlier ones, the proper filter should always be used.

Ford's rollover protection includes a spring-operated float valve in the vapor separator, figure 12-6. It closes whenever the car is tipped 90 degrees or more. A redesigned fuel pump is also used on 1976 and later cars to reduce fuel spillage during an accident.

FUEL LINES

The parts of the fuel system are connected by fuel and vapor lines and hoses. These are used to supply fuel to the carburetor, to return excess fuel to the tank, and to carry fuel vapors.

Fuel lines must be located to remain as cool as possible. If any part of the line is close to too much heat, the gasoline passing through it will vaporize more rapidly than the fuel pump can create suction. **Vapor lock** will then occur. When it does, the fuel pump will pump only vapor, which passes into the carburetor and out the bowl vent without the engine receiving any gasoline. Depending on their placement, fuel and vapor lines may be rigid or flexible.

Rigid Lines

All fuel lines that are fastened to the body, frame, or engine are made of seamless steel

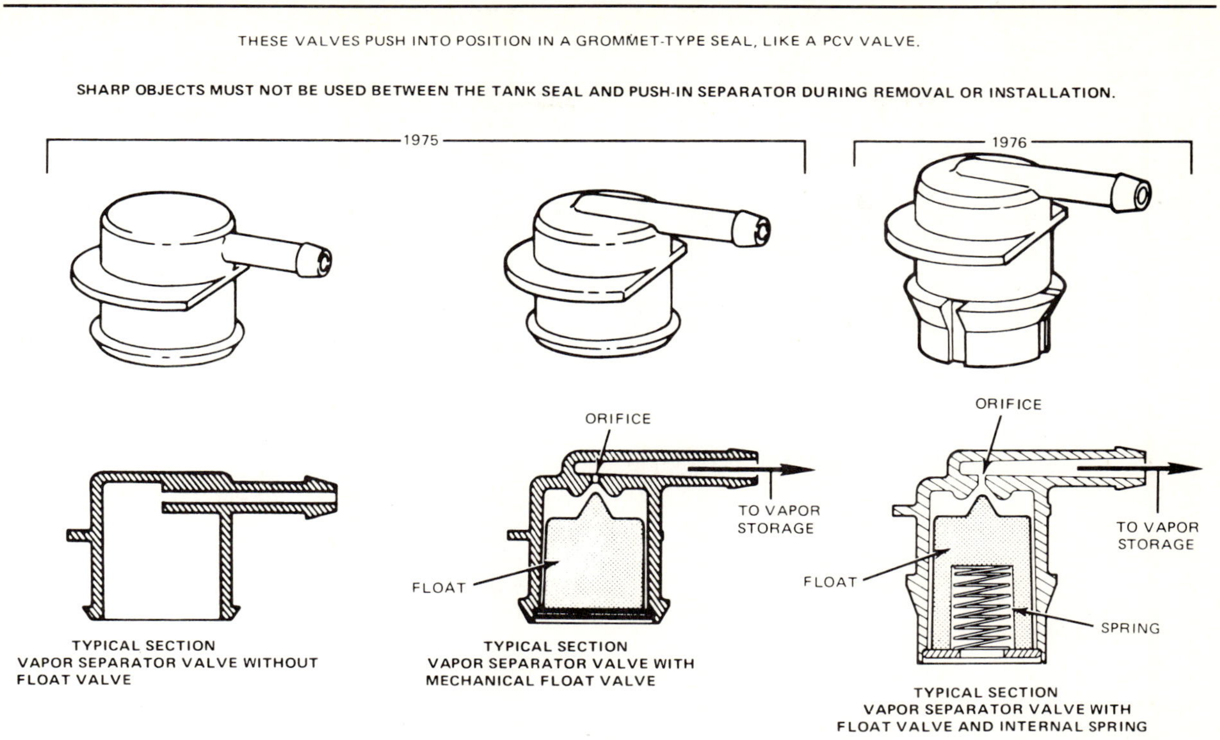

Figure 12-6. Orifice-type vapor separators on some 1974 and later Ford products have a mechanical check valve that also provides rollover leakage protection. (Ford)

tubing. Steel springs may be wound around the tubing at certain points to protect against damage.

When fuel line replacement is necessary, only steel tubing should be used. *Copper and aluminum tubing should never be substituted for steel tubing.* These materials will not withstand normal vehicle vibration, and they could combine with gasoline to form a chemical reaction.

Flexible Lines

In most fuel systems, synthetic rubber hose sections are used where flexibility is needed. Connections between steel fuel lines and other parts of the system are often made with short hose sections. The inside diameter of fuel delivery hose is generally larger (5/16 to 3/8 of an inch) than that of fuel return hose (1/4 of an inch).

Fuel system hoses must be made of special, fuel-resistant material. Ordinary rubber hose, such as that used for vacuum lines, would deteriorate when exposed to gasoline. Only hoses made for fuel systems should be used for replacement. Similarly, vapor vent lines must be made of material that will resist attack by fuel vapors. Replacement vent hoses are usually marked with the designation EVAP to indicate their correct use.

Fuel Line Mounting

Fuel supply lines from the tank to the carburetor are routed to follow the frame along the underbody of the car. Vapor and return lines may be routed with the fuel supply lines, but usually are on the frame rail opposite the supply line. All rigid lines are fastened to the frame rail or underbody with screws and clamps or clips.

EVAPORATIVE EMISSION CONTROL SYSTEMS

California's stringent emission laws brought the first **evaporative emission controls** (EEC) on 1970 cars sold in that state. Use of EEC systems was extended to all 1971 cars, regardless of where they were sold.

Vapor Lock: A condition in which bubbles are formed in a car's fuel system when the fuel gets hot enough to boil. Flow is stopped or restricted as a result.

Evaporative Emission Control (EEC): A way of controlling HC emissions by collecting fuel vapors from the fuel tank and carburetor fuel bowl vents and directing them through an engine's intake system.

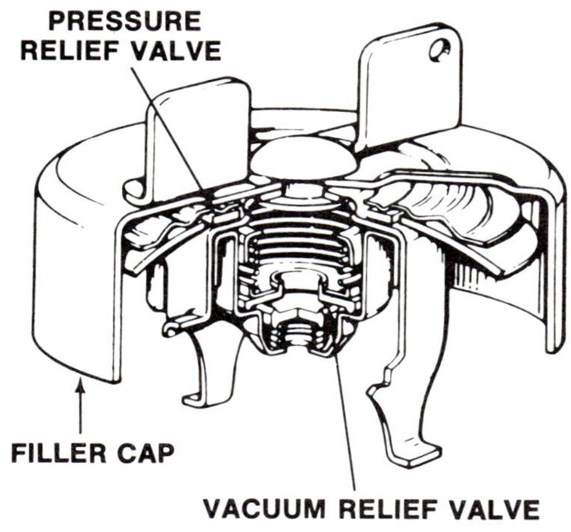

Figure 12-7. Fuel tank caps for EEC systems have vacuum and pressure relief valves.

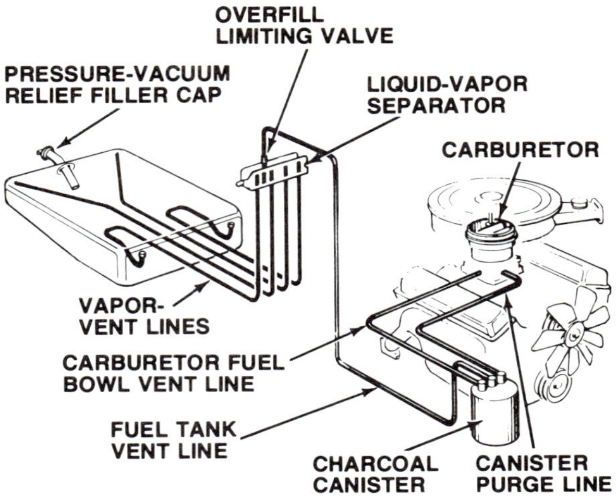

Figure 12-8. This EEC system has a liquid-vapor separator mounted separately from the tank.

Common Components

The fuel tank filler caps used on cars with EEC systems differ from those used on non-EEC cars. Most EEC caps have pressure-vacuum relief, figure 12-7, built into them. If a sealed cap is used on an EEC system that requires a pressure-vacuum relief design, a vacuum lock may develop in the fuel system, or the fuel tank may be damaged by fuel expansion or contraction. Fuel tanks are protected in various ways against fuel expansion and overflow caused by heat. An overfill limiter, or temperature expansion tank, was used on many 1970-73 EEC systems to prevent total filling of the tank. This is attached to the inside of the fuel tank and contains small holes which open it to the fuel area. When the fuel tank appears to be completely full — it will hold no more and the fuel gauge reads full — the expansion tank remains virtually empty. This provides enough space for fuel expansion and vapor collection if the vehicle is parked in the hot sun after filling the tank.

The dome design of the upper fuel tank section used in some later-model cars, or the overfill limiting valve contained within the vapor-liquid separator, eliminates the need for the overfill limiter tank used in earlier systems.

All EEC systems use some form of **liquid-vapor separator** to prevent liquid fuel from reaching the engine crankcase or vapor storage canister. Some liquid-vapor separators are built into the tank and use a single vapor vent line from the tank to the vapor canister. When the separator is not built in, figure 12-8, it is usually mounted on the outside of the tank or on the frame near it. In this case, vent lines run from the tank to the separator and are arranged to vent the tank, regardless of whether the car is level or not. Liquid fuel entering the separator will return to the tank through the shortest line.

Carburetor Venting

Carburetors must be vented to keep atmospheric pressure in the fuel bowl, which provides the pressure differential necessary for precise fuel metering. Two types of vents are used: internal and external.

Internal vents

Carburetors may be vented internally through the balance tubes that connect the fuel bowl to the airhorn, figure 12-9. The main purpose of this tube is to keep atmospheric pressure pushing down on the fuel in the bowl. This causes the fuel to flow from the bowl, through the circuits and jets, to the lower-pressure area created by the carburetor venturi. The internal vent, or balance tube, also helps to compensate for an air pressure drop caused by a dirty air cleaner. This helps prevent overly rich air-fuel mixtures.

We will learn more about this in Chapter 13. This balance tube also allows vapors from the fuel bowl to collect in the air cleaner when the engine is off. Collecting fuel vapors in this way controls evaporative emissions.

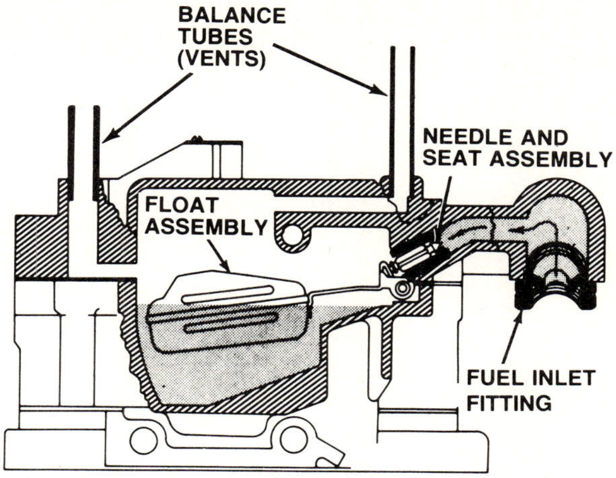

Figure 12-9. Internal carburetor vents. (Chrysler)

External vents

Many carburetors have external vents for the fuel bowl. On older cars without EEC systems, these vents opened directly to the atmosphere. They released vapors from the fuel bowl to prevent the buildup of vapor pressure, which could cause **percolation**. On 1970 and later cars with EEC systems and external carburetor vents, the vents are connected to the vapor storage canister by a rubber hose.

Vapor Storage

The EEC system traps gasoline vapors from the fuel tank and carburetor. These trapped vapors are fed into the engine intake system when it is running, or stored until the engine is started. Vapors are stored in one of two places: the engine crankcase, or a charcoal-granule-filled canister.

A few early EEC systems stored the vapors in the engine crankcase. When the engine is started, the vapors are drawn from the crankcase through the PCV system and added to the incoming air-fuel mixture.

Since 1972, most cars have used a vapor canister for storage. The canister is located under the hood, figure 12-8, and is filled with activated charcoal granules which will hold up to one-third their own weight in fuel vapors. A vent line connects the fuel tank to the canister. Carburetors with external bowl vents are also vented to the canister.

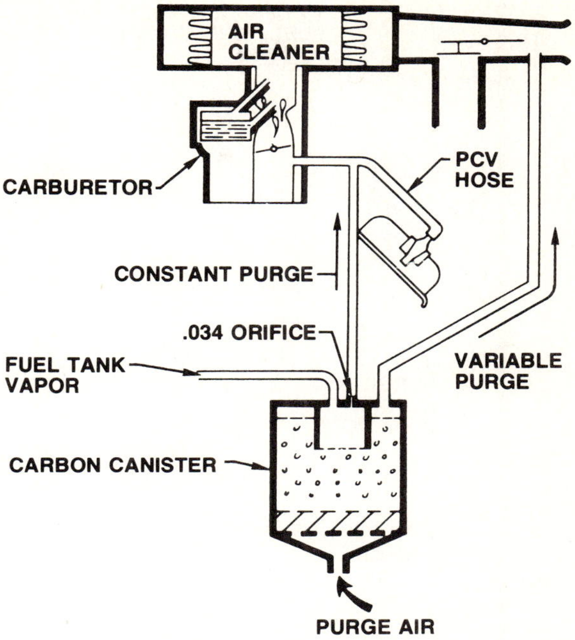

Figure 12-10. In this EEC system, a variable-purge hose runs from the canister to the air cleaner, and a constant-purge hose runs to the intake manifold. (Ford)

Vapor Purging

During engine operation, the stored vapors are drawn from the canister to the engine through a hose connected either to the carburetor base or to the air cleaner. This process is called purging the canister. Fresh air is drawn through the charcoal, and the vapors are carried with the airflow back into the engine air-fuel mixture. The fresh airflow is created by engine vacuum and can be controlled in several ways.

The constant purge method creates a fixed amount of airflow that does not vary with engine air consumption. The vapor canister is connected to intake manifold vacuum through an orifice that provides a constant flow rate, figure 12-10.

In the variable-purge method, the air flow rate varies with the amount of air the engine takes in. The vapor canister is connected to the

Liquid-Vapor Separator Valve: A valve in EEC fuel systems that separates liquid fuel from fuel vapors.

Percolation: The bubbling and expansion of a liquid. Similar to boiling.

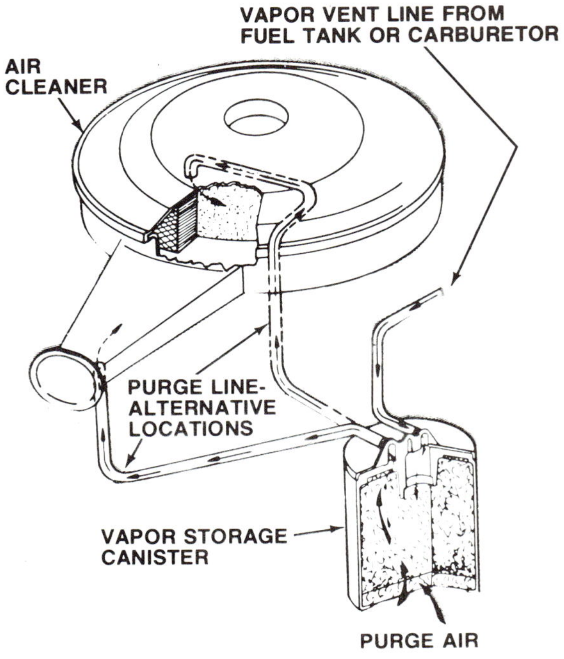

Figure 12-11. Purging the vapor storage canister through the air cleaner results in a variable purge rate. (Ford)

air cleaner, figure 12-11. The more air that passes through the air cleaner, the more air is drawn through the vapor canister.

Sometimes the amount of airflow created by the air cleaner connection is not enough. Some systems use a 2-stage purge method, with a purge valve that opens a second passage from the canister to the intake manifold. The purge valve is controlled by a vacuum signal from the carburetor, figure 12-12. Another 2-stage purge system combines a constant purge line with a line to the carburetor, above the throttle plates. As the throttle plates open and vacuum increases, this second line adds to the purging air flow.

FUEL PUMP OPERATION OVERVIEW

The fuel pump and the fuel lines, figure 12-13, deliver gasoline from the tank to the carburetor. The pump moves the fuel with a mechanical action that creates a low-pressure, or suction, area at the pump inlet. This causes the higher atmospheric pressure in the fuel tank to force fuel to the pump. The pump spring also exerts a force on the fuel within the pump and delivers it under pressure to the carburetor. All pumps, except electric turbine pumps, develop this mechanical action through a reciprocating, "push-pull," motion. Various kinds of fuel pumps are described in detail in the following paragraphs.

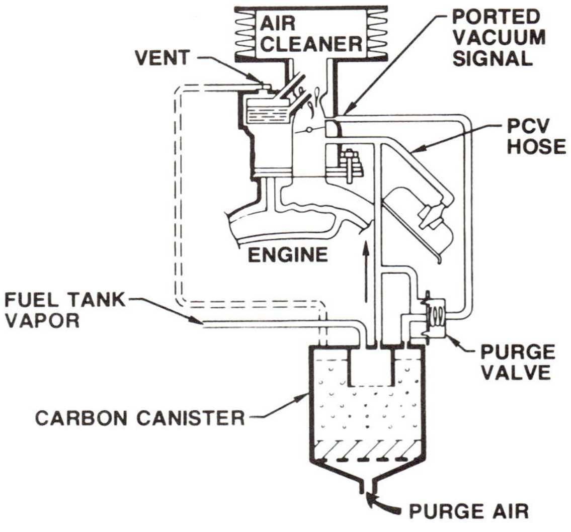

Figure 12-12. The two-stage purge arrangement in this EEC system uses a vacuum-operated valve to open a second purge line from the canister to the manifold. (Ford)

PUMP TYPES

While all pumps deliver fuel through mechanical action, they are generally divided into two groups:
1. Mechanically driven by the car engine
2. Electrically driven by an electric motor or vibrating armature.

MECHANICAL FUEL PUMPS

The most common type of fuel pump used by domestic and foreign carmakers is the single-action, diaphragm-type mechanical pump, figure 12-14. The rocker arm is driven by an eccen-

■ The Good Old Days?

Well, yes, prices have increased in the automotive industry. Just take a look at this 1926 price list for doing repairs on the hugely popular Ford Model T.

These are the flat-rate labor charges recommended by Ford to all its dealers and mechanics:

Overhaul motor and transmission	$25.00
Overhaul motor only (includes R&R, rebabbitting, reboring)	$20.00
Overhaul transmission only	$14.00
Grind valves, clean carbon	$ 3.75
Replace head gasket	$ 1.00
Overhaul carburetor	$ 1.50
Adjust transmission bands	$ 0.40

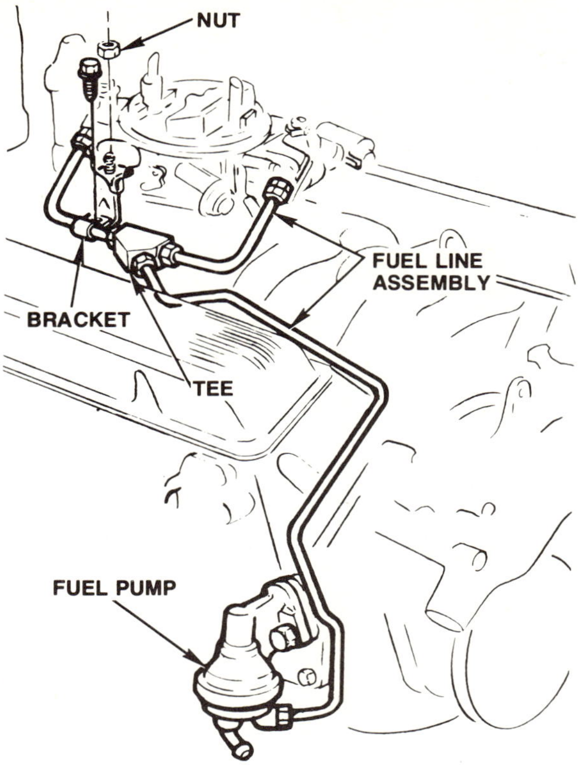

Figure 12-13. Typical fuel pump and line installation on a V-8 engine. (Chevrolet)

tric lobe on the camshaft. The pump makes one stroke with each revolution of the camshaft. The eccentric lobe (often called simply "the eccentric") may be part of the camshaft.

Sometimes the rocker arm is driven directly by the eccentric. Other engines have a pushrod between the eccentric and the rocker arm, figure 12-15. The most common example of this arrangement is the small-block Chevrolet V-8.

Mechanical Pump Operation

The fuel intake stroke begins when the rotating camshaft eccentric pushes down on one end of the pump rocker arm. This raises the other end, which pulls the diaphragm up and tightens the diaphragm spring. Pulling the diaphragm up creates a vacuum, or low pressure area, in the fuel chamber. Since there is a constant high pressure in the fuel lines, the inlet check valve in the pump is forced open and fuel enters the fuel chamber.

As the camshaft eccentric continues to turn, it allows the outside end of the rocker arm to "rock" back up. This, along with the push given by the diaphragm spring, allows the diaphragm

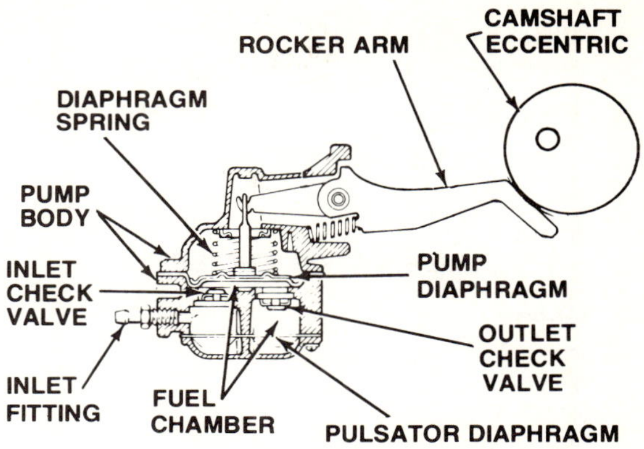

Figure 12-14. Typical diaphragm-type mechanical fuel pump.

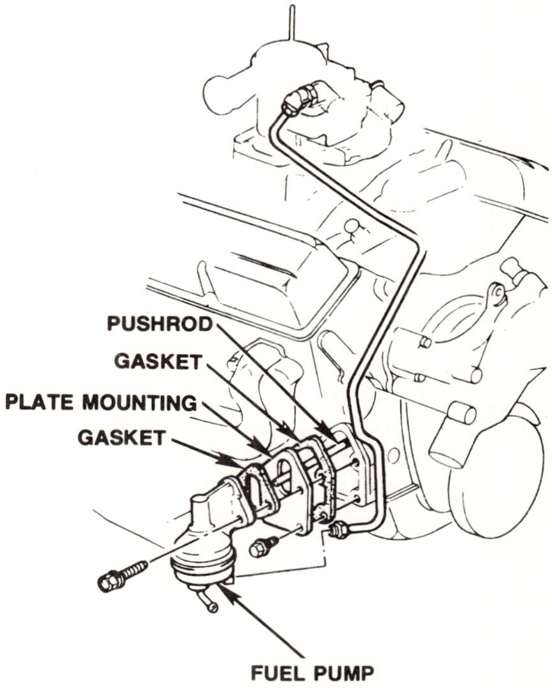

Figure 12-15. A pushrod is installed between the camshaft eccentric and the fuel pump rocker arm on this engine. (Chevrolet)

to relax back down. This is the start of the fuel output stroke. As the diaphragm relaxes, it causes a pressure buildup in the fuel chamber.

This pressure closes the inlet check valve and opens the outlet check valve. The fuel flows out of the fuel chamber and into the fuel line on the way to the carburetor. The outlet check valve keeps a constant pressure in the outlet line and prevents fuel from flowing back into the pump.

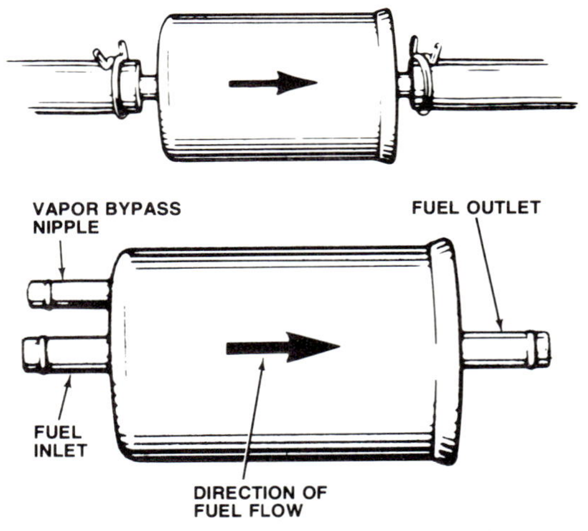

Figure 12-16. A simple inline fuel filter (top) and a vapor bypass filter (bottom).

We measure a fuel pump by the *pressure* and the *volume* of the fuel it delivers. Delivery pressure is controlled by the diaphragm spring. Delivery rate, or volume, is controlled by the carburetor.

The fuel pump delivery rate is proportional to the fuel required by the carburetor. When the carburetor inlet needle valve is open, fuel will flow from the pump, through the lines, and into the carburetor. When the needle valve is closed (due to the fuel bowl in the carburetor being full), then no fuel flows through the lines.

With the needle valve closed and the pressure in the fuel line increasing, the fuel pump diaphragm stays up, even though the rocker arm continues to go up and down in a "freewheeling" motion. No fuel is pumped until the fuel level in the carburetor bowl drops enough for the inlet needle valve to open again.

During different operating conditions, the fuel level in the carburetor bowl varies. Therefore, the position of the inlet needle valve varies between fully open and fully closed. The opening of the needle valve and the rate of fuel flow into the carburetor is always controlled by, and proportional to, the rate of fuel flow out of the carburetor.

When an engine with a mechanical fuel pump is shut off, pressure in the fuel line to the carburetor is maintained by the pump diaphragm spring. If engine compartment heat expands the gasoline in the fuel line, the fuel will push the carburetor inlet needle valve open and pass through. The result is too much fuel in the carburetor, and the engine will not restart easily. This is known as **flooding** the carburetor.

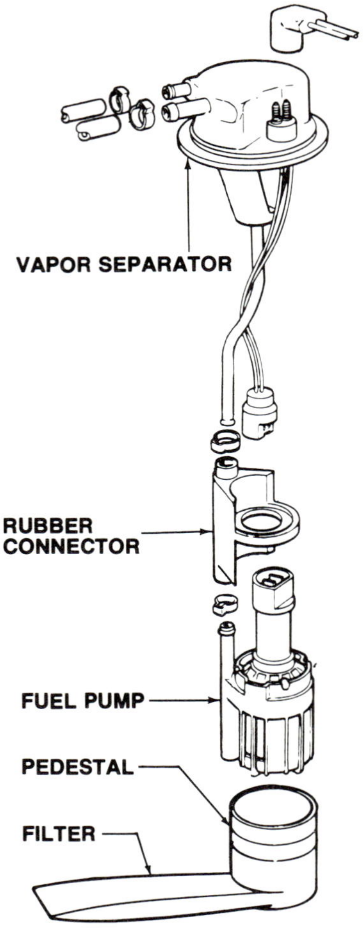

Figure 12-17. This impeller-type electric pump assembly is installed inside the fuel tank. (Ford)

Also, since fuel expands when it is hot, it may turn from a liquid into a vapor. This causes a vapor lock in the pump and lines. Four methods have been used to maintain fuel pressure and to prevent flooding and vapor lock:

- Some older pumps use an air chamber on the outlet side of the pump to separate vaporized fuel and return it to the tank through a separate line.
- A vapor separator can be installed between the pump and the carburetor. The carburetor draws liquid fuel from the bottom of the separator while the vapor at the top is returned to the tank.
- A vapor bypass filter is often used on air-conditioned cars. It combines a fuel filter and a vapor separator in a single unit, figure 12-16.
- Many pumps now have small bleed-down holes drilled through each check valve, allowing excess pressure in the outlet line to bleed back to the inlet line.

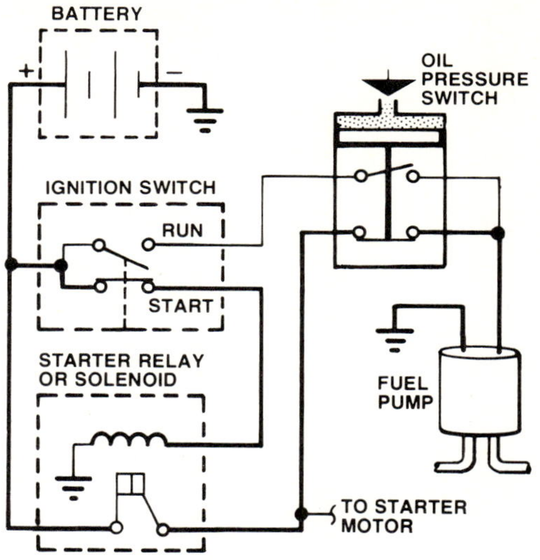

Figure 12-18. During cranking, the fuel pump receives current through the starter relay (solenoid) and the normally closed contacts of the oil pressure switch.

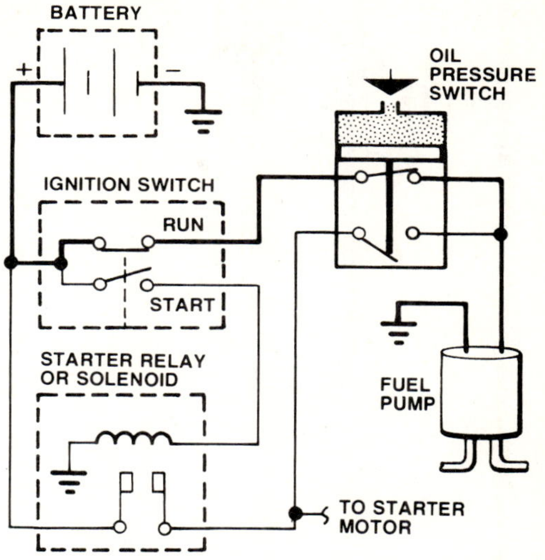

Figure 12-19. When the engine starts, the oil pressure switch opens one set of contacts and closes another. Current then flows through the ignition switch, through the oil pressure switch, and to the fuel pump as long as the oil pressure is above a minimum level.

ELECTRIC FUEL PUMPS

The impeller, or turbine, pump is driven by a small electric motor.

The impeller-type pump, figure 12-17, is sometimes called a rotary vane pump. It sucks fuel into the pump, then pushes it out through the fuel line to the carburetor. Since this type of pump uses no valves, the fuel is moved in a steady flow rather than the pulsating motion of all other electric and mechanical pumps.

Electric Fuel Pump Location

The electric fuel pump is principally a pusher unit: it pushes the fuel through the supply line. Because it does not rely on the engine camshaft for power, an electric pump can be mounted in the fuel line anywhere on the vehicle — including inside the fuel tank.

Pusher pumps are most efficient when they are mounted as near as possible to the fuel tank and at or below its level. This allows the pump to use gravity to transfer fuel from the tank to the pump. It also eliminates the problem of vapor lock under all but the most severe conditions. With the pump mounted at the tank, the entire fuel supply line to the carburetor is pressurized. Regardless of how hot the fuel line gets, it is unlikely that vapor bubbles will form to interfere with fuel flow. Having the pump close to the tank also allows the pump to stay cooler because it is away from engine heat. It is therefore less likely to overheat during hot weather.

Electric Fuel Pump Operation

Most original equipment electric fuel pumps are controlled by a pressure switch in the engine oil system. This switch opens the electric circuit to the pump motor when the engine is off, and controls the operation of the pump when the engine is started and while it is running.

The pressure switch has two sets of contact points. One set is normally closed, and allows current to flow from the battery, through the starter solenoid or relay, to the fuel pump. The other set is normally open. When closed, it allows current to flow from the battery, through the ignition switch, to the fuel pump.

Turning the ignition key to Start energizes the pump by providing current through the normally closed contact points, figure 12-18. Once the engine is running, the pump receives current through the normally open contacts (which have been closed by the oil pressure), figure 12-19.

Flooding: A condition caused by heat expanding the fuel in a fuel line. The fuel pushes the carburetor inlet needle valve open and fills up the fuel bowl even when more fuel is not needed. Also, the presence of too much fuel in the intake manifold.

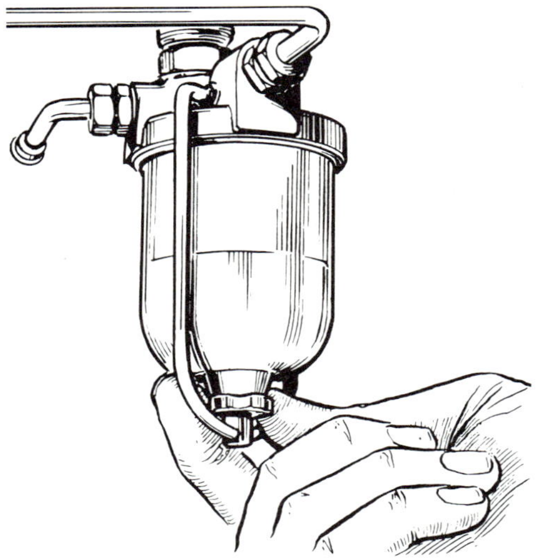

Figure 12-20. Sediment bowls may contain a paper, fiber, ceramic, or metal filter element.

Engine oil pressure opens the normally closed contacts and closes the normally open contacts to keep the pump energized. When the ignition switch is turned off, the pump circuit is deenergized. If oil pressure drops below the specified level (usually 2 psi) for any reason, contact is broken at the pressure switch and the fuel pump stops operation immediately.

In-Tank Pumps

GM compacts of the Astre, Monza, Skyhawk, Starfire, Sunbird, and Vega class use an impeller-type electric fuel pump located in the fuel tank. This is part of the fuel pickup unit and is similar to the in-tank pumps used on 1969-70 Buick Rivieras and 1972-74 Fords with 460 police engines, figure 12-17.

■ Tools For Making Fuel Lines

With few exceptions, replacement fuel lines cannot be bought preformed. Tubing is stocked in large rolls and must be shaped and formed however the mechanic wants it. Ordinary hand tools cannot be used to properly make a replacement fuel line. When tubing is cut with a hacksaw, its shape is frequently distorted, and the cut edge will be jagged instead of clean and smooth.

Four special tools are required: a tube cutter, a tube reamer, a tube bender, and a tube flaring device. The tube cutter uses sharpened metal discs to make a smooth, distortion-free cut. After cutting, a tapered reamer is necessary to remove any burrs which could make a bad seal. The tube bender shapes the tubing without kinking or bending it. Flaring tools are available to make either single or double flares. It is essential to use them to properly shape the connecting ends of any new fuel line.

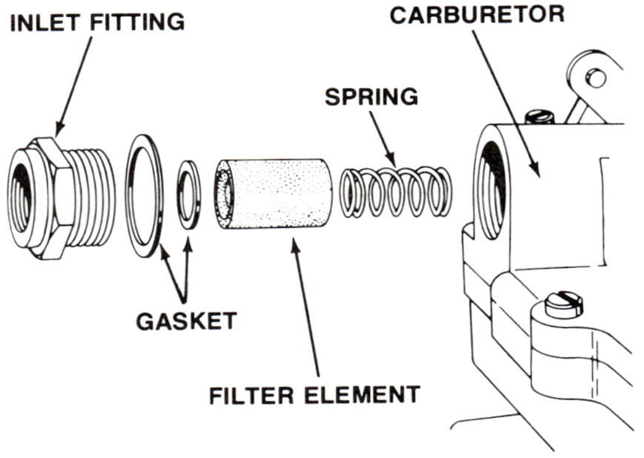

Figure 12-21. Ford and GM vehicles use carburetor inlet filters like this one.

FUEL FILTERS

Despite all the care taken in refining, storing, and delivering gasoline, some impurities get into the automotive fuel system. Fuel filters are designed to remove dirt, rust, water, and other harmful materials from the gasoline before it can reach the carburetor. The useful life of all these filters is limited. The filters will become clogged and restrict fuel flow if they are not cleaned or replaced according to the maker's recommendations.

Several different types of fuel filters are used, and some systems may contain two or more. Filters can be located in several places within the fuel system.

Fuel Tank Filters And Strainers

A sleeve-type filter of woven Saran is usually fitted to the end of the fuel pickup tube inside the fuel tank. This filter prevents sediment, which has settled at the bottom of the tank, from entering the fuel line. It also protects against water contamination by plugging itself up. If enough water somehow enters the fuel tank, it will accumulate on the outside of the filter and form a jelly-like mass. If this happens the filter must be replaced; except for this, no maintenance is required.

Inline Filters

Used primarily by AMC and Chrysler, the inline filter, figure 12-16, is in the line between the fuel

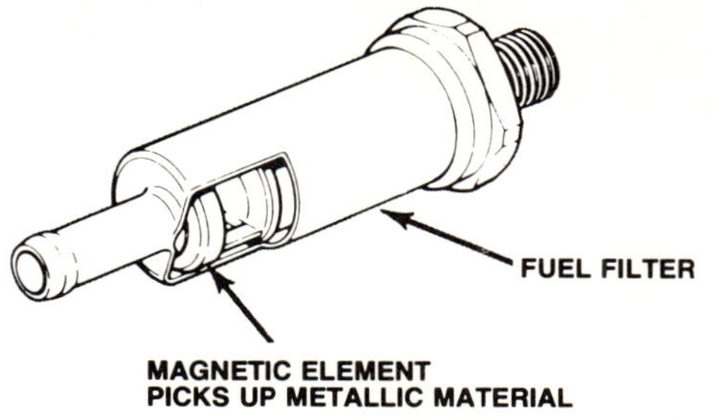

Figure 12-22. This late-model Ford filter has a magnetic element to remove metallic contamination from the fuel. (Ford)

pump and carburetor. This protects the carburetor from contamination but does not protect the fuel pump. The inline filter is usually a throw-away plastic or metal container with a pleated paper element sealed inside.

Inline filters must be installed so that gasoline flows through them in the direction shown by the arrow. If an inline filter is installed backwards, it can restrict fuel delivery to the carburetor.

Some imported cars and older domestic cars have a sediment bowl between the fuel pump and carburetor. The bowl contains a pleated paper, ceramic, fiber, or metal filter element. The filter element works much like an inline filter. The bowl cover is held in place by a wire bail and clamp screw. It can be removed for filter cleaning or replacement, figure 12-20. Ceramic and metal elements can be cleaned and reused if necessary. Paper and fiber filter elements must be replaced when they are dirty.

Carburetor Inlet Filters

Ford and General Motors equip most of their cars with carburetor inlet filters, figure 12-21. The Ford filter is a 1-piece throw-away metal unit. It has a filter screen and magnetic washer, figure 12-22, to trap dirt and metal particles. The filter screws into the carburetor fuel inlet and clamps to the inlet hose at the other end.

Rochester carburetors on GM vehicles use a throw-away pleated paper element. Older cars had a bronze filter element, but reusing was not recommended. Filters on 1976 and later Rochester carburetors must contain the new rollover check valve as described earlier. A similar paper filter is used on 1977 and later Motorcraft 5200 and 2700 VV carburetors.

Pump Outlet Filters

Some cars have fuel filters in the outlet side of the fuel pump. Those pumps used on Chrysler 6-cylinder engines during the early Seventies, figure 12-23, contain a throw-away filter ele-

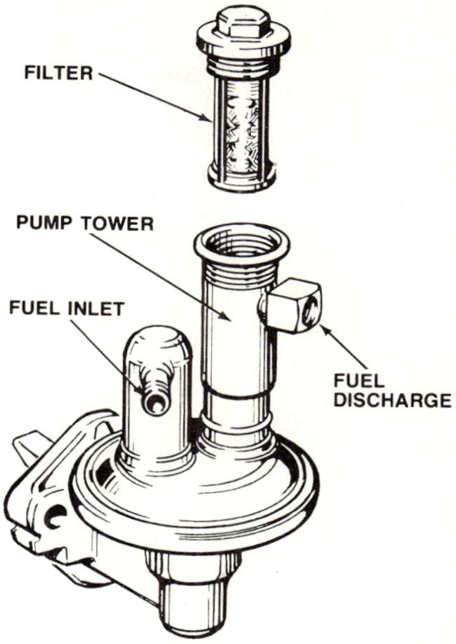

Figure 12-23. Some 6-cylinder Chrysler engines have a filter in the fuel pump outlet tower.

■ Why Vapor Lock?

When gasoline vapors form in the fuel system, vapor lock occurs. This is the partial or complete stoppage of fuel flow to the carburetor. Partial vapor lock will lean the air-fuel mixture and reduce both the top speed and the power of an engine. Complete vapor lock will cause the engine to stall, and make restarting impossible until the fuel system has cooled off.

Four factors usually cause vapor lock:

1. High gasoline temperature and pressure in the fuel system
2. Vapor-forming characteristics of a particular gasoline
3. The fuel system's inability to minimize vapors
4. Poor engine operating conditions, such as overheating.

Vapor may form anywhere in the fuel system, but the critical temperature point is the fuel pump.

Engineers have improved fuel pumps and fuel systems to make today's cars less likely to have vapor lock. Oil companies have succeeded in reducing the vapor-locking tendencies of gasoline by adjusting its volatility according to weather requirements. But vapor lock may still occur during long periods of idle (such as bumper-to-bumper driving), or when the car's fuel system is not properly maintained. Periodically inspect the fuel system and correct all air leaks and defects to prevent vapor lock with today's cars.

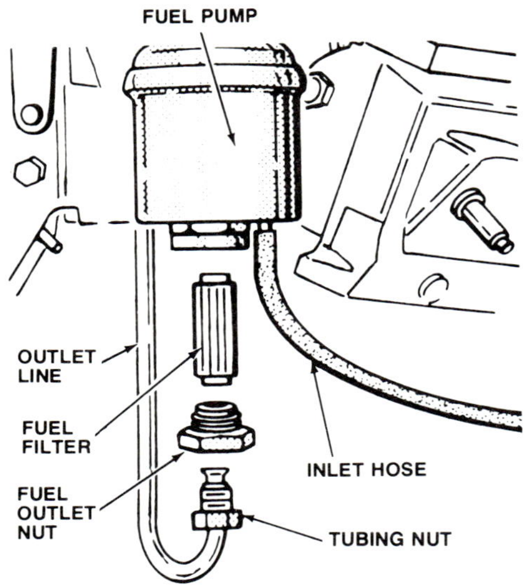

Figure 12-24. Some Cadillacs have a filter in the outlet side of the fuel pump.

ment in the fuel outlet tower. Cadillacs through 1974 use a fuel pump outlet filter, located on the bottom of the pump, figure 12-24.

Disposable Element Filters

Screw-on, throw-away element filters, figure 12-25, look much like replaceable oil filters. Ford has used this type on the fuel pump of some V-8 engines. The Cadillac Seville uses a disposable filter mounted to the frame near the left rear wheel. Other fuel-injected Cadillacs have the filter mounted to a bracket on the lower left front of the engine.

ENGINE AIR FILTERING REQUIREMENTS

The air cleaner and filter have three primary functions:
1. They clean the air before it is mixed with fuel.
2. They silence intake noise.
3. They act as a flame arrester in case of a backfire.

In 1957, engineers at the Lincoln-Mercury Division of Ford Motor Company were searching for ways to improve driveability under different weather conditions. They discovered that by using the air cleaner to provide warm air to the carburetor at low temperatures, driveability is improved. It also allows more efficient carburetor adjustments which:
- Reduce exhaust emissions without reducing engine performance
- Permit leaner air-fuel ratios
- Give better fuel economy
- Reduce **carburetor icing** in cold weather.

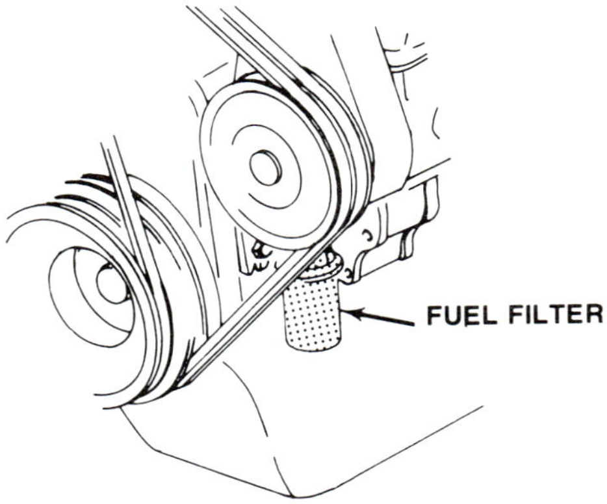

Figure 12-25. Disposable element filters may be mounted on the engine or near the fuel tank.

Since then, the air cleaner has become a separate emission control system for intake air temperature control. It also has become a part of other emission controls, such as the PCV system.

■ No One Misses The Good Old Fuel Pump

Today's fuel pump may seem to be a simple device, but pump manufacturers have worked hard to make it so. Pump designs, capacities, pressures, and performance requirements make the modern pump a rather sophisticated device. This is especially true when you consider that a fuel pump is expected to transport large amounts of gasoline for thousands of miles without failure.

Back in the thirties, fuel pump breakdown and replacement was a common occurrence every few thousand miles. Remember the '36 Ford V-8? Its fuel pump operated from a pushrod As the pushrod wore, the pump stroke lessened. Most mechanics and a lot of owners kept the fuel pump operating with a wad of chewing gum or tinfoil stuffed into the pushrod cup to compensate for wear. Rather simple, but it worked.

The vacuum booster fuel pump was the first big change in pump design. Since no one could keep windshield wipers running at a constant speed, pump designers provided additional vacuum with a dual pump design. But super highways, higher horsepower, and emission controls brought new approaches to pump design. Windshield wipers went electric, and the vacuum booster fuel pump disappeared. Intake electric pumps are gradually supplementing and even replacing the traditional mechanical pump design. Automakers now build modern fuel pumps to supply at least 30,000 trouble-free miles.

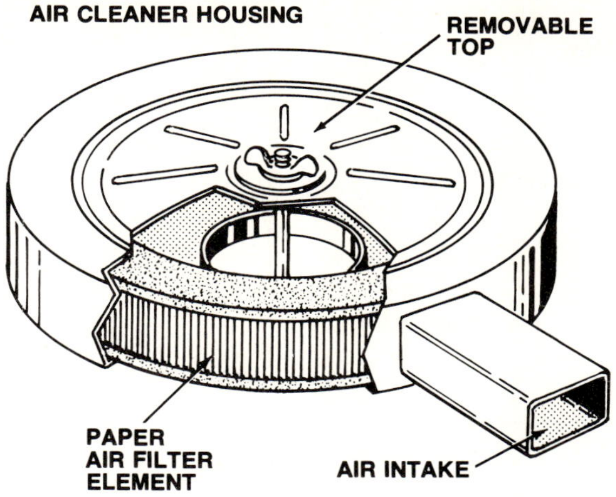

Figure 12-26. A simple air cleaner and filter.

A modern car will take in about 200,000 gallons of air to mix with each 20-gallon tank of fuel. This air is filled with dust and dirt that must be removed before reaching the carburetor. If the air is not properly filtered, the particles can upset the carburetor's operation. They can even damage engine parts and shorten engine life.

The amount of dirt, dust, and carbon particles in the air will vary from region to region. In constant bumper-to-bumper traffic, an engine will take in many abrasive carbon particles from other cars' exhaust. Dust and dirt particles increase in agricultural or construction areas.

THE AIR CLEANER AND FILTRATION

Intake air is cleaned by a filter in the air cleaner, figure 12-26. The 2-piece, circular air cleaner housing of stamped steel is usually located on top of the carburetor and generally has a snorkel or air intake tube. The snorkel draws air from the engine compartment or from outside the car into the air cleaner; from there it passes to the carburetor. The air cleaner housing has a removable top section.

Most air filter elements are the throw-away type, but some can be cleaned and reused. In areas with high particle concentrations, air filters should be cleaned or replaced at more frequent intervals. A dirty or clogged filter may restrict airflow into the engine, affecting engine performance and life.

AIR FILTER ELEMENTS

Three general types of air filter elements are used on passenger cars and light trucks:

1. Paper filters
2. Polyurethane filters
3. Oil bath filters.

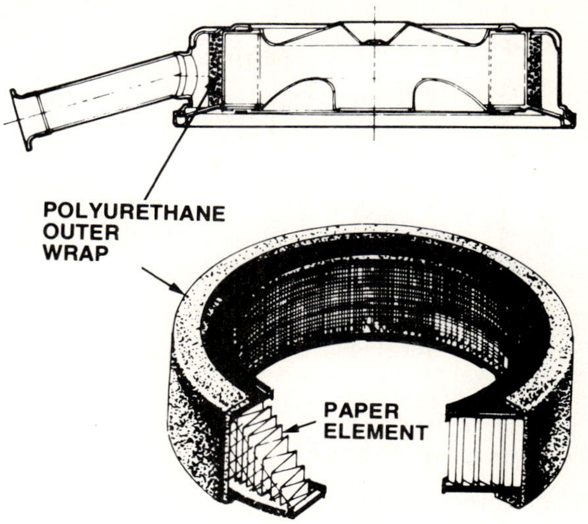

Figure 12-27. Paper air cleaner element with polyurethane wrapper. (AC Delco)

Paper Filter Element

The paper air filter element, figure 12-26, is the most common type used on late-model cars and trucks. It is made of a chemically treated paper stock that contains tiny passages in the fibers. These passages form an indirect path for the airflow to follow. This causes the airflow to pass through several fiber surfaces, each of which traps microscopic particles of dust, dirt, and carbon.

The paper is pleated and formed into a circle, with the top and bottom edges sealed with heat-resistant plastic. A fine wire mesh screen may be used on the inside of the filter ring to reduce the possibility of the element catching fire from an engine backfire. A similar, but coarser, wire mesh screen may be used on the outside of the filter ring for extra strength.

While these filter elements are generally made of dry paper, an oil-wetted paper stock may also be used. The light oil coating helps prevent contaminants from working their way through the paper. It also increases the dirt-holding capacity over the same area of dry paper stock. An outer wrapper of polyurethane sometimes is used to make the filter work better, figure 12-27. Paper element filters are disposable and should be replaced at the recommended intervals.

Carburetor Icing: A condition that is the result of the rapid vaporization of fuel entering a carburetor; the temperature drops enough to freeze the water particles in the airflow.

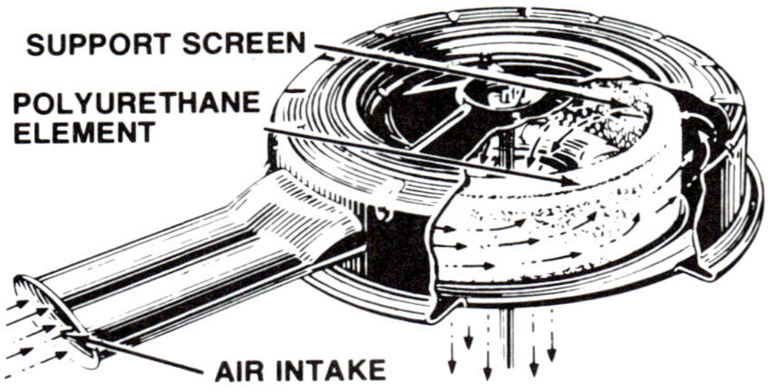

Figure 12-28. Air cleaner with polyurethane filter element over a metal support screen.

Polyurethane Filter Element

Sometimes called a foam filter, the polyurethane element is similar to the paper element type. It consists of a polyurethane wrapper stretched over a metal support screen, figure 12-28. Polyurethane contains thousands of pores and interconnecting strands that create a maze-like dirt trap, while allowing air to flow through. Properly maintained, the polyurethane element has a capacity and efficiency equal to that of the paper type. It may be used dry, or it can be lightly oiled. It can be cleaned and reused or replaced if necessary. Polyurethane filter elements are not common as original equipment on late-model vehicles, but some are sold as aftermarket equipment.

Oil Bath Filter

This kind of filter is used mainly on older cars, some trucks, and some off-road vehicles. It is most efficient at high airflow rates. The oil bath cleaner, figure 12-29, contains a wire mesh filter which rests in an oil reservoir in the air cleaner housing. Air entering the housing is deflected down, where it strikes the oil in the reservoir and deposits heavier particles of dirt. Picking up an oil mist from the reservoir, the air flows back up and across the surface of the filter, where it leaves the mist with finer particles entrapped. The oil then drains back to the reservoir from the filter, carrying the entrapped dirt with it in a self-cleaning action.

AIR INTAKE DUCTS AND FRESH AIR INTAKES

The main source of air intake to the carburetor is the air cleaner snorkel. Some air cleaners use a second snorkel to provide additional air intake at full throttle. The snorkel passes air to the filter and then to the carburetor from the engine compartment. The snorkel also increases the velocity of the air entering the air cleaner. Temperatures in the engine compartment will often

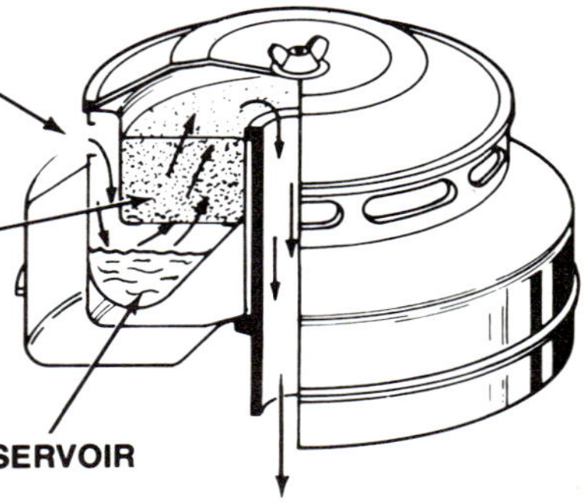

Figure 12-29. Oil bath air filter.

exceed 200° F (93° C) on a hot day, and hot air can thin out the air-fuel mixture enough to cause detonation and possible engine damage.

Allowing the engine to breathe cooler air from outside the engine compartment will prevent such problems. Cooler air is provided by a cold air duct, or induction (zip) tube. The tube runs from the snorkel to a fresh air intake at the front of the car, figure 12-30. The fresh air intake is normally open at all times, but may have a screen to prevent insects and other foreign matter from being drawn into the air cleaner.

Some fresh air intakes are in the cowl or in the rear area of the hood.

THERMOSTATICALLY CONTROLLED AIR CLEANERS

Late-model cars all have **thermostatic** controls on their air cleaners to regulate intake air temperature. This is important to maintain the precise air-fuel ratios required for exhaust emission control.

The thermostatically controlled air cleaner has the usual sheet metal housing mounted on the carburetor airhorn. Another sheet metal duct, called a heat stove or shroud, is fastened around the exhaust manifold. The heat stove is connected to the air cleaner intake snorkel by a flexible hose or metal tube, figure 12-31. Heat radiating from the exhaust manifold is kept in by the heat stove, and sent to the air cleaner snorkel to provide heated air to the carburetor.

An air control valve, or damper, in the snorkel permits the intake of:

1. Heated air from the heat stove
2. Cooler air from the snorkel or cold-air duct
3. A combination of both.

The damper maintains intake air at a specified temperature, usually between 90° and 100° F. The damper may be operated by a vacuum motor or diaphragm or by a thermostatic bulb.

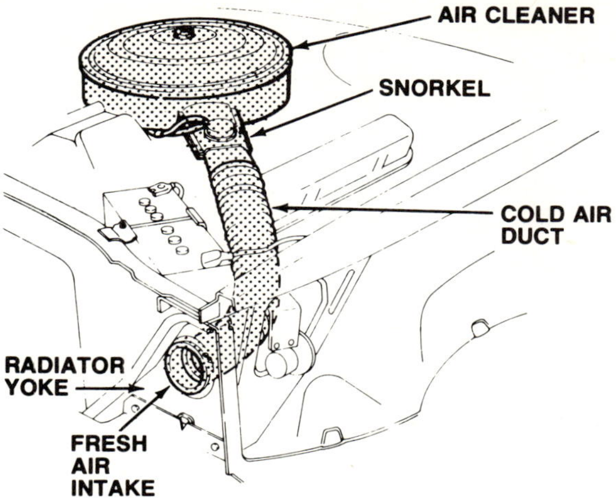

Figure 12-30. Air cleaner with fresh air intake mounted on the radiator yoke.

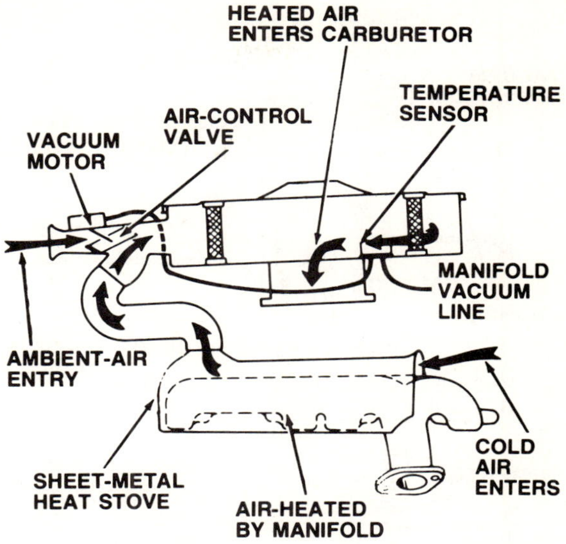

Figure 12-31. A thermostatically controlled air cleaner with a heat stove.

In an air cleaner with a vacuum motor, a bimetal temperature sensor and a vacuum bleed in the air cleaner housing regulate vacuum supply to the vacuum motor. Vacuum is supplied from the intake manifold. When intake air temperature is below about 100° F (38° C) the temperature sensor holds the vacuum bleed closed and full manifold vacuum is applied to the vacuum motor. The motor holds the air control valve in the full-hot-air position, figure 12-32, position A.

As intake air warms up, the sensor begins to open the vacuum bleed. This decreases the vacuum going to the motor. A spring in the motor then starts to move the air-control valve from the hot-air to the cold-air position, figure 12-32, position B.

As air temperature continues to rise, the vacuum bleed continues to open. Vacuum to the motor is further reduced. At high air temperatures, vacuum to the air cleaner motor is completely shut off, and the air control valve is in its full-cold-air position, figure 12-32, position C. The air control valve will also open to the full-cold-air position during heavy acceleration, regardless of air temperature. This provides maximum airflow through the air cleaner to the carburetor when it is needed the most.

The thermostatic bulb, used to control some air cleaners, is inside the air cleaner snorkel and connected by linkage to a spring-loaded air control valve, or damper. The air control valve is normally held in its closed position by the spring, allowing heated intake air to enter the snorkel, figure 12-33, position A. As the temperature rises, the thermostatic bulb begins to expand. This expansion is more than air valve spring tension, and the valve gradually opens to its cold-air position, figure 12-33, position B.

AMC uses a thermostatic bulb control on its 6-cylinder and early V-8 engine air cleaners. Late-model V-8 engines have a vacuum motor. All Chrysler engines use a vacuum motor. Thermostatic bulbs are used on all Ford vehicles

Thermostatic: Referring to a device that automatically responds to temperature changes.

■ Air Cleaner Filter Maintenance

What causes engine wear? Contrary to popular belief, it is not how many miles the engine has been driven, or how old the engine is. Providing that the engine is kept properly lubricated, wear is mainly caused by the dust and dirt which enters the engine. A teaspoonful of gritty dirt will ruin the piston rings; a cupful will virtually destroy the entire engine. A properly maintained air cleaner filter is the primary line of defense against the gallons of dirty air sucked into the engine with each tiny sip of fuel.

To prevent dirt from entering the engine, some mechanics say paper filter elements should be discarded when dirty. Others say the element can be reused if properly cleaned with low pressure compressed air (under 100-psi nozzle pressure).

Avoid rough handling or damage to filters. Never try to clean a paper filter by striking or rapping the filter assembly. Washing the filter element is also out, since even drying a wet filter will not restore the air passages to normal.

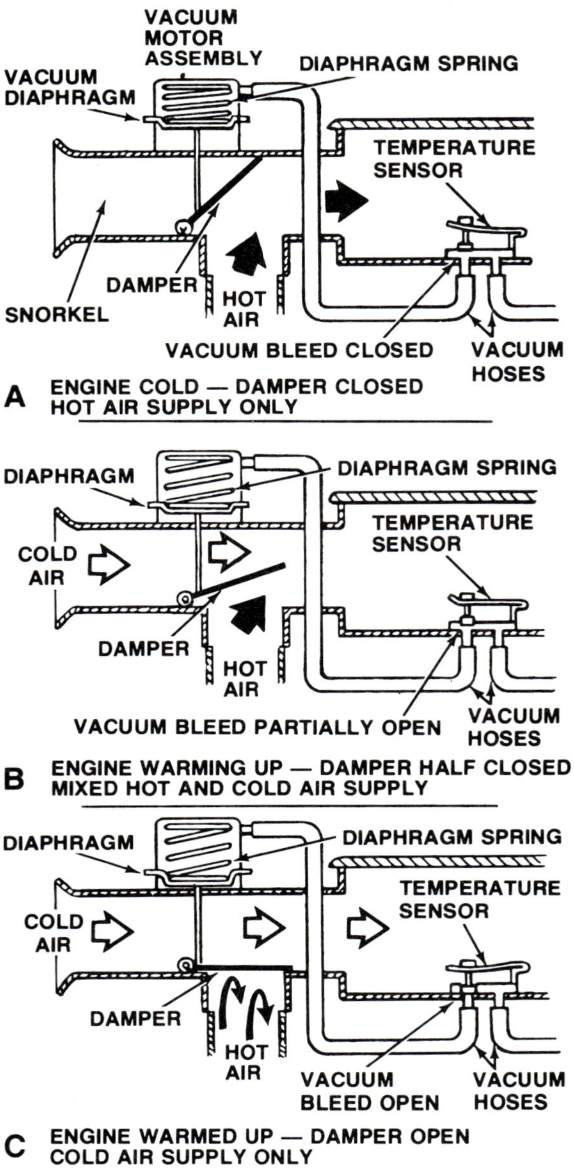

Figure 12-32. A thermostatically controlled air cleaner with vacuum motor control.

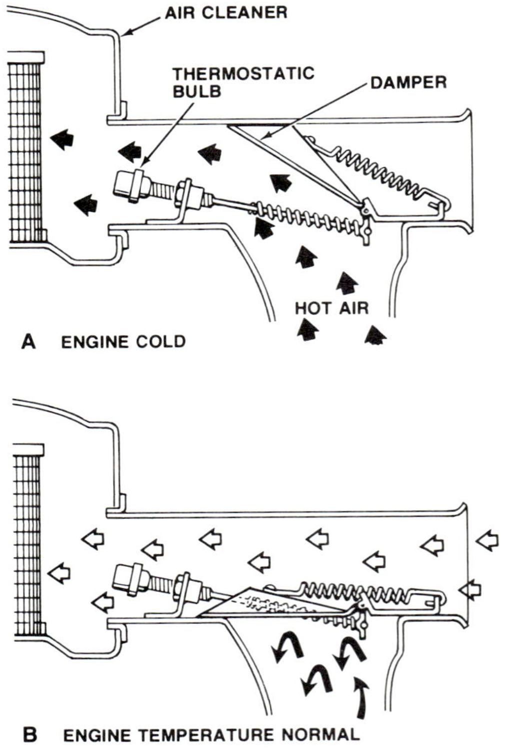

Figure 12-33. A thermostatically controlled air cleaner with thermostatic-bulb control.

through 1972, except those with 351C, 390GT, and 400-cid engines. Bulbs are also used on pre-1977 1,600-cc and 2,000-cc Pinto engines. Most GM engines have a vacuum motor, but a thermostatic coil that works similarly to a bulb is used on pre-1977 4-cylinder engines.

SUMMARY

Automobile fuel tanks can be mounted either vertically or horizontally, depending on how much room there is. Filler tubes, besides allowing the tanks to be filled, also are used as fill limiters, leaving room for fuel to expand. Cars requiring unleaded gasoline will not accept a leaded gas nozzle in the tube. Tanks must be vented, but evaporative emission control (EEC) requirements state that the fuel vapors must not be vented to the atmosphere. Carmakers have devised many ways to make sure all the vapors stay within the fuel system. They have provided rollover leak protection, pressure-vacuum relief valves, liquid-vapor separators, and positive crankcase ventilation. The EEC systems all use vapor storage canisters. Vapors stored in the canisters are purged into the engine; each carmaker has a slightly different way to do this.

Fuel pumps move the fuel from the tank to the carburetor. All pumps do this through a mechanical action that creates a low-pressure area into which the fuel will flow. With check valves and high pressure, the fuel is then forced out of the pump and into the carburetor. There are two types of fuel pumps: mechanical and electrical. Mechanical pumps use the engine camshaft eccentric for power. Electric pumps push the fuel, rather than pull it, so they are frequently installed in the fuel tank. Many types of filters are used in the fuel system, and they must be replaced or cleaned as directed by the manufacturer. Filters are used as inline filters, or at the fuel tank or the carburetor.

Like fuel, the air used by an engine has tiny particles of dirt and other materials that damage an engine if they are allowed to be breathed in. Air cleaners and their filters screen out this material. Air cleaners are also part of the emission control system, since they help reduce emissions and increase performance and fuel economy. Since 1968, most domestic passenger cars and light trucks have used thermostatically controlled air cleaners, which provide warm air to the carburetor at low temperatures. Each carmaker has a slightly different design, but all these devices work about the same.

Review Questions

Select the single most correct answer.
Compare your answers with the correct answers on page 280.

1. When the fuel filter shown below is installed, the arrow must point toward the:

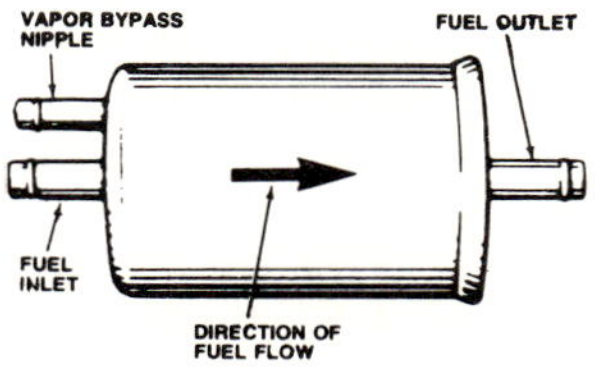

   a. Carburetor
   b. Fuel tank
   c. Canister
   d. Fuel pump

2. Fuel tank filters eliminate:
   a. SO_2
   b. NO_x
   c. Sediment and water
   d. HC

3. Inline filters:
   a. Are used mostly by AMC and Chrysler
   b. Are located in the fuel line between the pump and carburetor
   c. Protect the carburetor but not the pump
   d. All of the above

4. Oil bath filters:
   a. Are efficient at all speeds
   b. Cause large particles to deposit in an oil reservoir
   c. Have an auxiliary paper element
   d. Pass oil through the carburetor

5. The primary source of air intake to the carburetor is the:
   a. Air cleaner snorkel
   b. Venturi
   c. Fuel bowl
   d. Heat stove

6. A heat stove may be located on the:
   a. Carburetor
   b. Snorkel
   c. Exhaust manifold
   d. Intake manifold

7. When intake air temperature is low in a thermostatic air cleaner:
   a. The vacuum bleed is closed
   b. The vacuum bleed is open
   c. The air control valve is in the cold-air position
   d. Little vacuum is applied to the vacuum motor

8. To prevent the paper air filter element from catching fire:
   a. It is chemically treated
   b. It is soaked in oil
   c. A fine wire mesh screen is used on the inside of the filter ring
   d. A coarse wire mesh screen is used on the outside

9. Airborne contaminants can:
   a. Change air-fuel ratios
   b. Damage piston rings and cylinder walls
   c. Enter the engine oil
   d. All of the above

10. Which of the following is *not* true about a dirty air filter element:
    a. It will pass contaminants
    b. It does not affect fuel consumption
    c. It will change air-fuel ratio
    d. It may keep the engine from running

11. All fuel tanks must:
    a. Be vertically mounted
    b. Be horizontally mounted
    c. Have a vent system
    d. Contain a vertical baffle

12. Filler necks with restricted openings:
    a. Provide better venting
    b. Prevent entry of leaded fuel dispensers
    c. Reduce emissions
    d. Act as rollover check valves

13. Fuel lines fastened to the frame, body, or engine are made of:
    a. Steel
    b. Aluminum
    c. Copper
    d. Rubber

14. Ordinary rubber hose can be used for:
    a. Fuel supply lines
    b. Vapor vent lines
    c. Vacuum lines
    d. Fuel return lines

15. Electric fuel pumps are most efficient when located near:
    a. The engine
    b. The fuel tank
    c. The carburetor
    d. The camshaft

16. Electric fuel pumps are controlled by:
    a. An oil pressure switch
    b. The starter relay
    c. The ignition switch
    d. Fuel line pressure

17. All EEC systems eliminate vapors containing:
    a. NO_x
    b. CO
    c. SO_2
    d. HC

18. The most common type of fuel pump is:
    a. The double-action diaphragm type
    b. The single rocker arm type
    c. The single-action diaphragm type
    d. The pushrod rocker arm type

19. The intake stroke in the fuel pump:
    a. Exerts pressure in the fuel tank line
    b. Creates a vacuum in the fuel chamber
    c. Opens the outlet valve
    d. All of the above

20. The output stroke of the fuel pump:
    a. Increases pressure on the diaphragm spring
    b. Opens the inlet valve
    c. Increases pressure in the pump chamber and opens the outlet valve
    d. Draws fuel into the fuel tank line

21. Fuel pump pressure is controlled by:
    a. The carburetor inlet needle valve
    b. The strength of the diaphragm spring
    c. The carburetor float
    d. None of the above

13 Basic Carburetion And Manifolding

So far in our study of a car's fuel system, we have examined enough hardware to be able to dump raw fuel into the combustion chambers. But adding a spark to that raw fuel will not create the controlled combustion we need to power the car.

We need something more: a method to mix the fuel with air, and then to feed it in equal amounts to the combustion chambers. The air-fuel mixing is done by a carburetor, figure 13-1, and the resulting mixture is delivered to the cylinders by the intake manifold. We will study these two items in this chapter.

There are tremendous differences between carburetors. The linkage, assist devices, and controls on carburetors can vary widely even between two cars of the same make and model but with different engines.

Yet all carburetors work on the same basic principles, and all manifolds have the same design factors. We will cover the similarities in designs for carburetors and intake manifolds, briefly identify some of the differences between manufacturers, and discuss how carburetion and manifolding work together to deliver the air-fuel charge to the cylinders where it can be used for power.

Once you know how these systems work, you will better be able to understand the relationship between a good tune-up and good performance, or good mileage. Carburetor adjustments are an especially important part of any thorough tune-up, but you'll need to know how the systems work in order to adjust them properly.

PRESSURE DIFFERENTIAL

Since air is a substance with weight, the air outside an engine has a specific weight, and so does the air inside the engine. The *weight* of air exerts *pressure* on whatever it touches: the greater this weight, the greater the pressure. When the weight of air outside the engine is greater than the weight of air inside the engine, we say that there is a pressure differential, or difference, between the two.

Atmospheric Pressure

The weight of air is not always the same, but changes with temperature and height above sea level. Air pressure is measured in pounds per square inch (psi) at sea level at an average temperature of 68° F. In metric units, atmospheric pressure at sea level is measured in millimeters of mercury (mm Hg) at 20° C

At sea level and at an average temperature, one cubic foot of air weighs about 1¼ ounce. This seems light enough, but remember that the

Figure 13-1. The modern carburetor is a complex device, but it works on simple principles of airflow and pressure differential.

blanket of air surrounding the earth extends many miles into the earth's atmosphere, figure 13-2. These many cubic feet of air piled one on top of another all have weight. This increases the weight of air pressing down on an object at sea level to about 14.7 psi or 760 mm Hg.

Effects of temperature
Air expands and becomes lighter as its temperature rises. This reduces the pressure it exerts. As its temperature falls, air contracts. This makes it heavier and increases its pressure. Variations in air temperature account for changing weather conditions. Direct heat from the sun and reflected heat from the earth's surface warm the air. As its temperature increases, air becomes lighter and rises. Cooler air sinks and takes its place, resulting in a constant motion. This motion creates wind and weather patterns.

Effect of height
As you climb above sea level, the amount of air pressing down on you becomes smaller. Since this air weighs less, it exerts less pressure. Air pressure gradually decreases with increased distance above sea level. At 30,000 feet above sea level, air pressure is reduced to about 5 psi. A few hundred miles above the earth, the atmosphere ends and is replaced by a vacuum, or a complete lack of pressure.

Manifold Pressure — Vacuum

With each intake stroke of an engine piston in its cylinder, a partial vacuum is produced. As the piston moves down, it creates a larger space for the air to fill. The air expands to fill the space, which reduces its pressure, or weight. This reduced pressure is a partial vacuum.

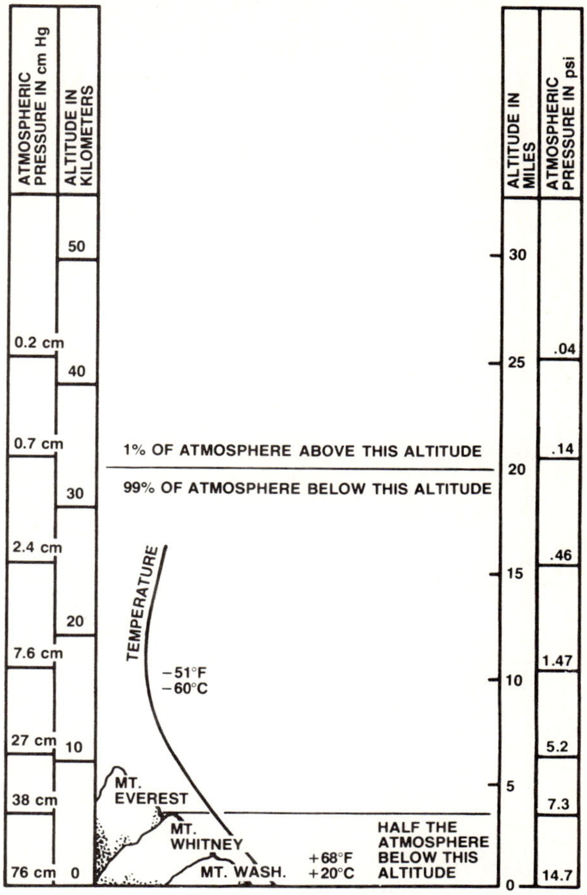

Figure 13-2. The blanket of air surrounding the earth extends many miles into the atmosphere. Atmospheric pressure decreases at higher altitudes.

As the piston moves further down, it increases the vacuum and lowers the air pressure in the cylinder and intake manifold above it. This causes a pressure differential between the air inside and the air outside the engine. To equalize the pressure, outside air rushes into the engine. As it passes through the carburetor, it is mixed with gasoline to form an air-fuel mixture. This combustible vapor is then drawn by vacuum through the intake manifold and open intake valve into the cylinder, figure 13-3. Here it is compressed, burned, and removed.

AIRFLOW AND THE VENTURI PRINCIPLE

Opening the carburetor throttle valve causes air to move from the higher pressure area outside the engine, through the carburetor, to the lower

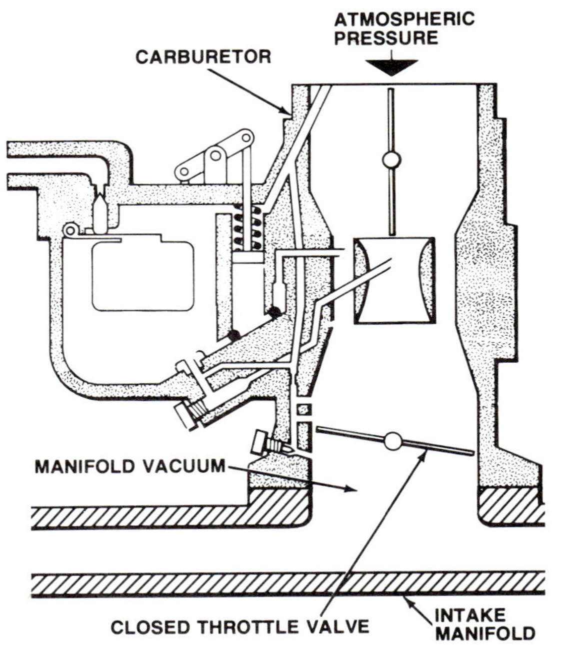

Figure 13-3. Manifold vacuum is the low pressure created in the intake manifold by the downward movement of the engine's pistons.

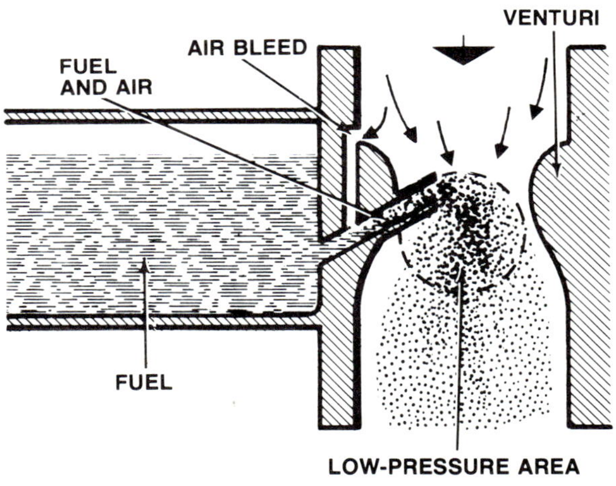

Figure 13-5. Air flowing through the venturi increases in speed. This lowers the pressure within the venturi to draw in more fuel.

pressure area of the manifold. How much and how fast the air will travel is determined by the opening of the throttle valve.

The pressure of air passing rapidly through a carburetor barrel is lower along the sides of the barrel than it is in the center of the airflow. By putting a small hole in the side of the barrel,

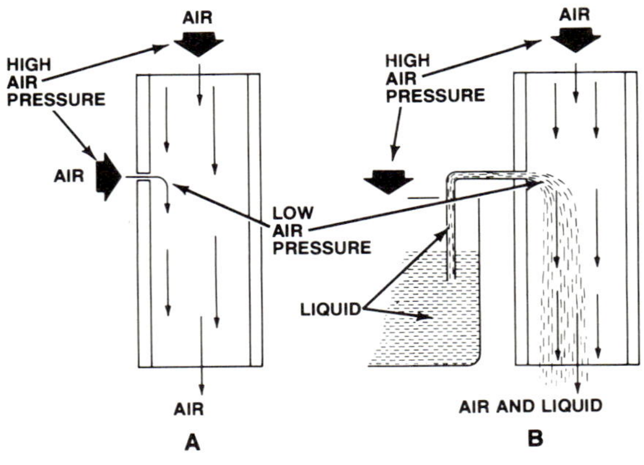

Figure 13-4. Airflow through a tube creates low pressure along the sides of the tube that can draw in more air (A). The low pressure can also draw liquid into the tube (B).

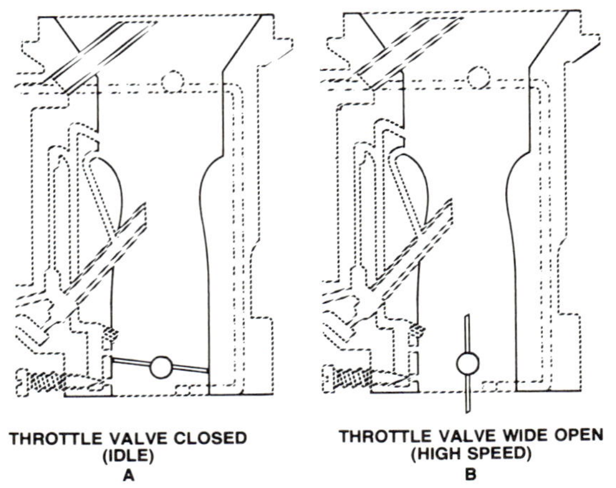

Figure 13-6. The carburetor throttle valve controls engine speed and power by regulating the amount of air and fuel entering the engine.

more air can be drawn into the stream of air rushing through the barrel, figure 13-4, position A. If a hose is used to connect the hole to a liquid-filled container or bowl, then the liquid will be forced through the hose and into the stream of air rushing by, figure 13-4, position B.

This is caused by the higher air pressure on the liquid, which forces it to the lower air pressure area inside the barrel. How much liquid passes through the hose will depend on the airflow velocity, or how fast the air is flowing through the inside of the barrel. The higher the airflow velocity, the lower the pressure will be at the inlet hole, and the more liquid will flow.

To make the carburetor work better, we must increase the air velocity through the barrel. This can be done by placing a restriction called a

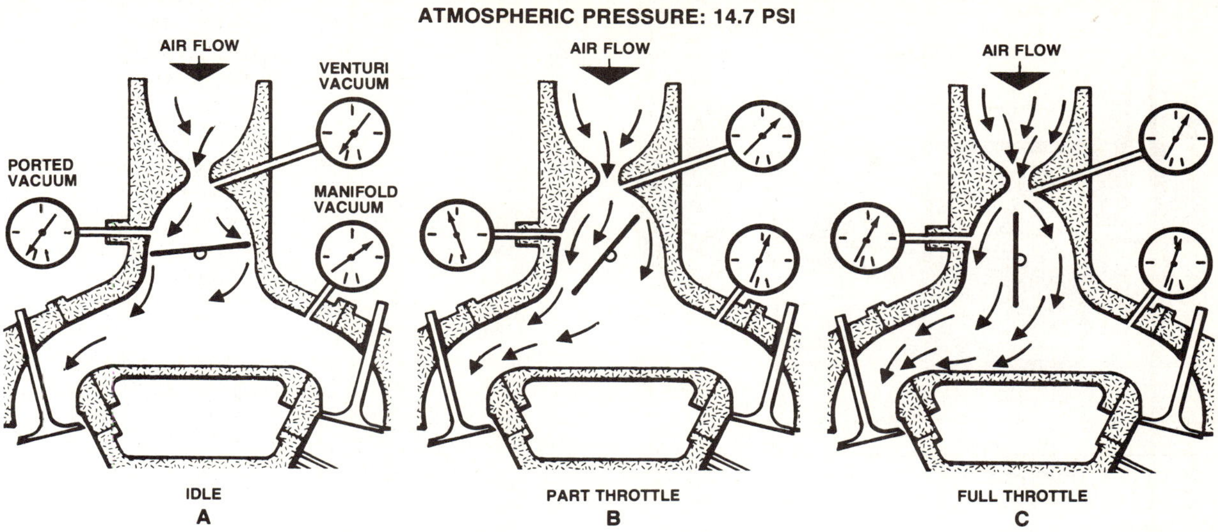

Figure 13-7. At idle (position A), manifold vacuum is high, but vacuum in the carburetor barrel is low because the throttle is closed. At part throttle (position B), venturi vacuum is high; ported vacuum is slightly lower, almost equal to manifold vacuum. At full throttle (position C), venturi vacuum is still slightly higher, but all three are almost equal.

venturi inside the barrel, figure 13-5. When air flows through the venturi restriction, it speeds up. This lowers the pressure inside the carburetor barrel and permits more liquid fuel to be drawn into the airflow.

In addition to mixing liquid fuel with air, the carburetor must also vaporize the liquid as much as possible. To help break up the liquid fuel for better vaporization, a small opening called an air bleed is put in the fuel inlet passage, figure 13-5. This mixes air with the liquid fuel as it is drawn into the main airflow.

The carburetor must also change the air-fuel mixture automatically. It must deliver a rich mixture for starting, idle, and acceleration; and a lean mixture for part-throttle operation. Engine speed and power are regulated by the position of the carburetor throttle valve, which controls the flow of the air-fuel mixture, figure 13-6.

Carburetor Vacuum

There are four measurements of air pressure, or vacuum, that are important here:

1. Atmospheric pressure — the pressure of the air outside the carburetor
2. Manifold vacuum — the low pressure beneath the carburetor throttle valve
3. Venturi vacuum — the low pressure area created by airflow through the venturi restriction in the carburetor barrel
4. Ported vacuum — the low pressure area in the carburetor just above the throttle valve.

Pressure and vacuum measurements at these four areas are shown in figure 13-7.

Atmospheric pressure is always present outside the carburetor and engine. It remains constant at any given altitude and temperature. Manifold vacuum is produced by the engine and is always present when the engine is running. **Venturi vacuum** is created by airflow through the venturi and increases with the speed of the airflow. Venturi vacuum is present whenever the throttle valve is open to allow air to flow through the carburetor. Because the throttle valve regulates the speed of the airflow, venturi vacuum increases as the throttle is opened. Ported vacuum is present whenever the throttle is opened to expose the lower portion of the carburetor barrel to manifold vacuum.

Ported vacuum exists just above the throttle valve. Vacuum taken from this point often is used to operate distributor vacuum advance units and other vacuum-operated devices. Small ports, or holes, in the sides of the carburetor are connected to hoses or tubes which are connected to the vacuum devices.

Venturi Vacuum: Low pressure in the venturi of a carburetor, caused by fast airflow through the venturi.

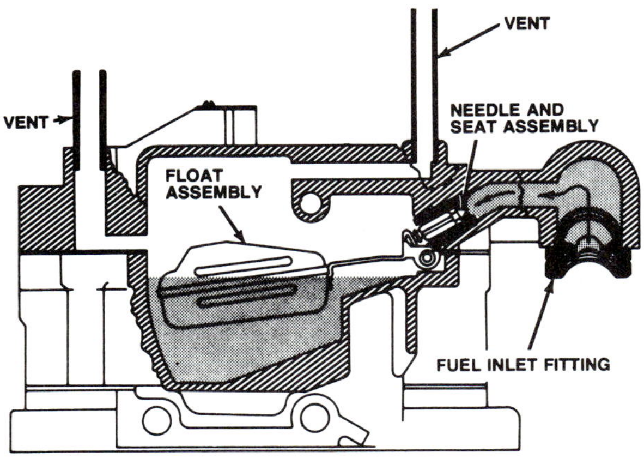

Figure 13-8. Fuel level in the fuel bowl is controlled by the float and needle valve acting against fuel pump pressure.

BASIC CARBURETOR SYSTEMS

To mix the fuel and regulate engine speed, the carburetor has a series of fixed and variable passages, jets, ports, and pumps which make up the fuel-metering systems or circuits. There are seven basic systems common to all carburetors:

1. Float system
2. Idle system
3. Low-speed system
4. High-speed, or main metering, system
5. Power system
6. Accelerator pump system
7. Choke system.

The Float System

Gasoline from the fuel tank is delivered by the fuel pump to the carburetor fuel bowl, where it is stored for use. Once in the fuel bowl, the gasoline must be kept at a precise, nearly constant level. This level is critical, since it determines the fuel level in all the other passages and circuits within the carburetor. A fuel level that is too high in the bowl will produce an air-fuel mixture that is too rich. One that is too low will produce an overly lean mixture. For this reason, float level is one of the most critical adjustments needed on a carburetor.

The main fuel discharge nozzle for the high-speed system is connected directly to the bottom of the fuel bowl. Because liquids seek their own level in any container, the fuel level in the bowl and in the nozzle is the same. If the level is too high, venturi vacuum at the tip of the nozzle will draw too much fuel into the high-speed system. If the fuel level is too low, too little fuel will be drawn in.

Fuel level is controlled by the float and the inlet needle valve, figure 13-8. As gasoline is drawn from the bowl, the float lowers in the remaining fuel. Fuel pump pressure then opens the needle valve and allows more fuel to enter the bowl. As the fuel level rises, so does the float, until it forces the inlet needle back against its seat. This closes the inlet valve and shuts off both fuel pump flow and pressure to the carburetor bowl. During many operating conditions, fuel flow into and out of the fuel bowl is about equal. The needle stays in a partly open position to maintain the required flow rate.

The float and needle valve regulate fuel flow, as well as fuel level. Since the needle valve is like a door between the carburetor fuel bowl and the fuel pump, it maintains an air space above the fuel in the bowl. The needle valve and float allow an air space to remain above the fuel in the bowl. Air pressure in this space is equal to atmospheric pressure, since the bowl has at least one vent tube running to the carburetor airhorn. Atmospheric pressure pushing down on the fuel in the bowl provides the pressure differential needed for precise fuel metering into the venturi vacuum area of the carburetor barrel.

Float and needle valve design and location in the fuel bowl vary with different carburetor designs, figure 13-9. Some floats have small springs to keep them from bobbing up and down when the car travels over rough roads. Many fuel bowls have baffles, which keep the fuel from sloshing on rough roads and sharp turns. The needle valves and their seats in older carburetors were usually made of stainless steel. The steel often attracted metallic particles in the fuel. The particles would collect between the needle and seat and allow the valve to leak. The needles and seats in most modern carburetors are made of brass, and the needles often have plastic tips that conform to any rough spots on the seat and still provide a good seal when the valve is closed. If the float and the needle valve do not keep the correct fuel level in the bowl and too much fuel enters, the carburetor will flood.

When the engine is shut off, engine heat causes the fuel in the bowl to evaporate. This was no problem in pre-emission control days, but with the installation of vapor canister systems, the amount of evaporation from a large fuel bowl can easily overload the canister. Thus, late-model carburetors use a somewhat smaller float bowl. Some carburetors use a molded plastic float bowl to reduce heat evaporation because plastic does not conduct heat as much as metal does. Others use an insulator, figure 13-10, between the intake manifold and the carburetor to reduce heat.

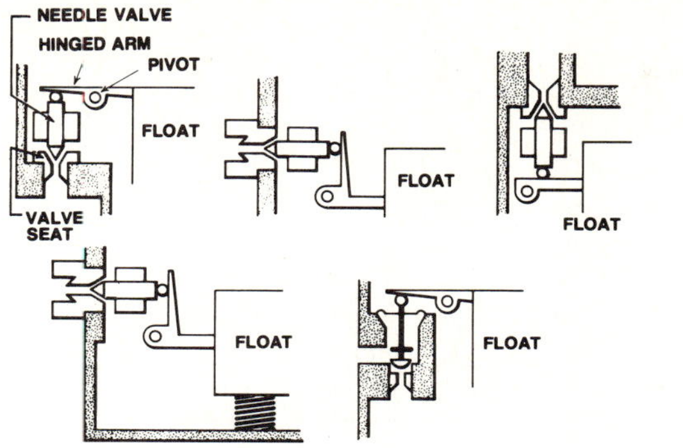

Figure 13-9. Float and needle valve designs vary with different carburetors.

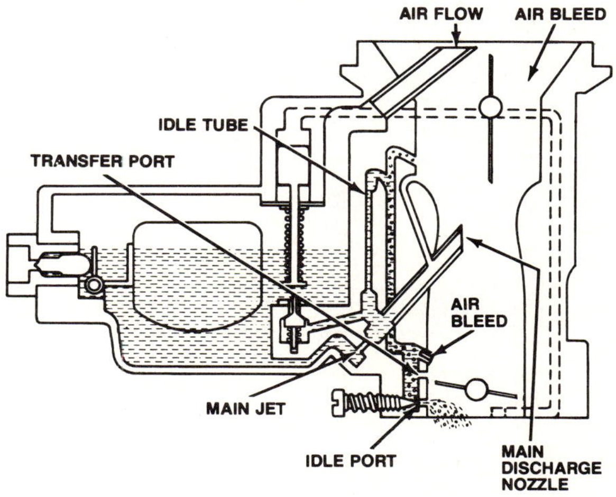

Figure 13-11. Air and fuel for the idle system are mixed inside the carburetor passages and delivered to the idle port below the throttle.

The Idle System

When an engine is idling, the throttle is just cracked open and airflow through the carburetor barrel and venturi is reduced. Since there is little or no venturi effect, no fuel flows from the main discharge nozzle. The idle system, figure 13-11, supplies enough air and fuel to keep the engine running under these conditions.

Intake manifold vacuum is high at idle, so idle ports are located just below the closed throttle. The pressure differential between the fuel bowl and the vacuum at the idle ports forces fuel through the ports. Gasoline flows from the bowl, through the main jet, and to the idle tube. Because the fuel must be well mixed with air for proper distribution, air bleeds in the idle tube let in air for the idle mixture. The air bleeds also prevent fuel **siphoning** at high speeds or when the engine is stopped.

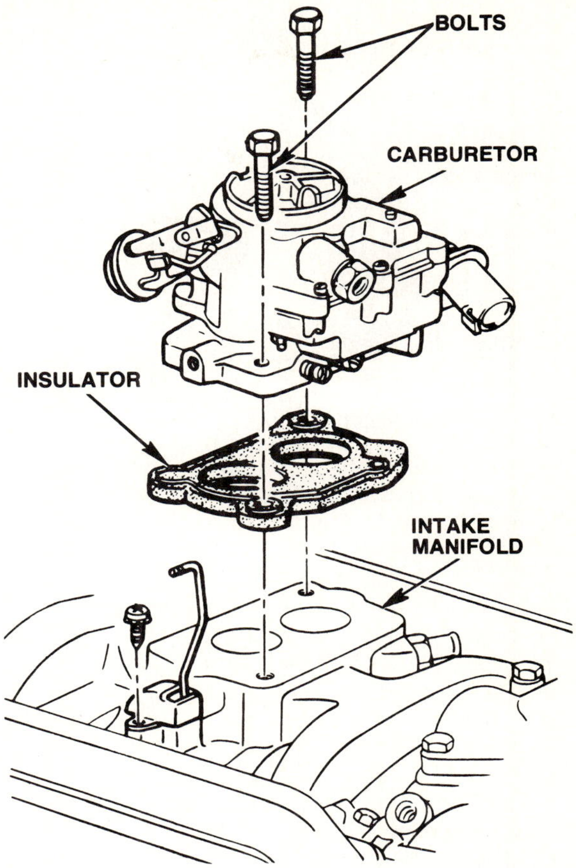

Figure 13-10. This insulator between the carburetor and the intake manifold reduces heat that causes fuel evaporation in the fuel bowl.

Siphoning: The flowing of a liquid as a result of a pressure differential, without the aid of a mechanical pump.

■ A Tube Tip

Sometimes, when you are working on a carburetor still in the engine, you'll bust your knuckles trying to reach the idle mixture screws. Most mechanics know that one way to get at those screws is to use a length of rubber tubing. Slip it tightly over the idle screws to make the adjustments. Similar tubing, but a bit larger on the inside diameter, can be used to loosen or tighten hard-to-get-at spark plugs.

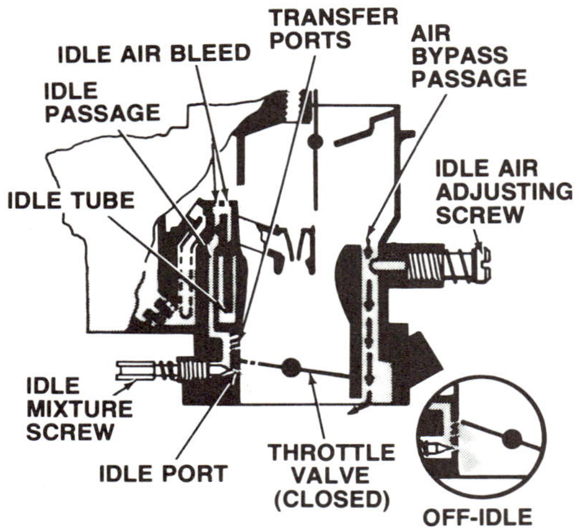

Figure 13-12. Air for the idle circuit in this carburetor passes through a bypass passage and is controlled by an idle air adjusting screw.

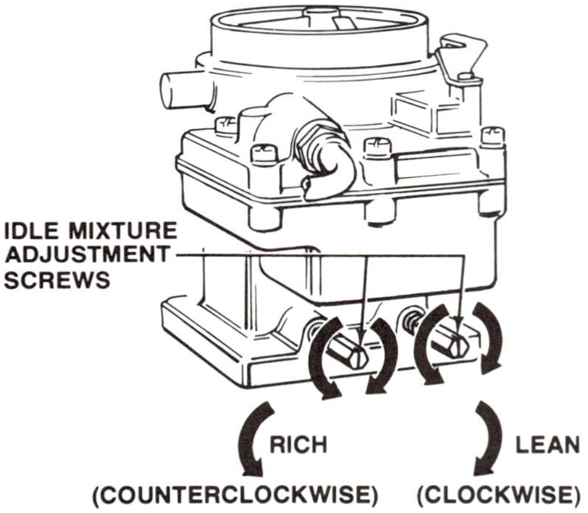

Figure 13-14. Idle mixture adjustment screws control the amount of gasoline in the air-fuel mixture.

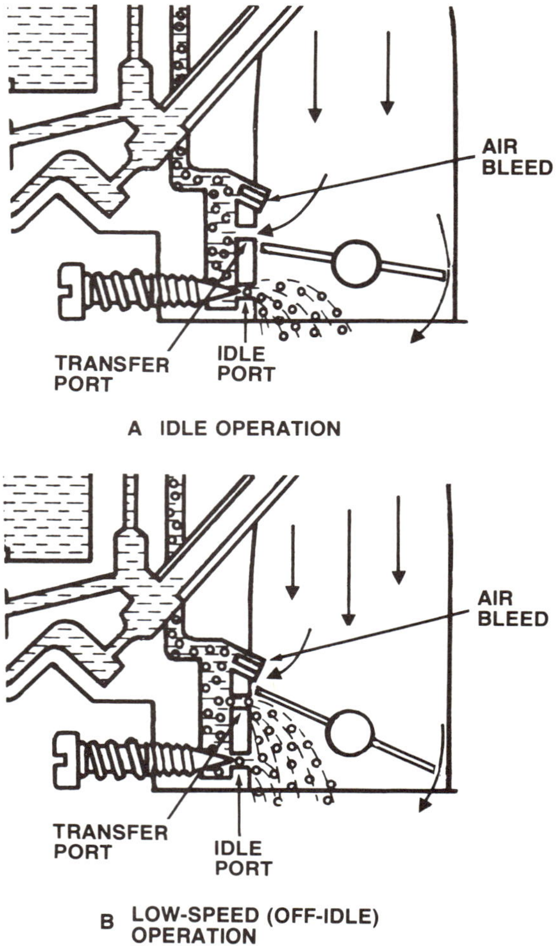

Figure 13-13. At idle, air flows in through the transfer port to mix with the idle air-fuel mixture. As the throttle opens, flow reverses through the transfer port. Fuel and air now flow out for low-speed operation.

Extra air for the idle air-fuel mixture is provided in several different ways, depending on the carburetor design. In many carburetors, the throttle valve does not close completely, but remains slightly open to let in a small amount of air, figure 13-11. A few designs draw air for the idle circuit through a separate air passage in the carburetor body called the idle air bypass, figure 13-12.

Additional small openings called transfer ports, figure 13-13, are located just above the closed throttle valve in the carburetor barrel. At idle, the transfer ports suck air from the barrel into the fuel flow in the idle system. A small amount of air and fuel is released just below the throttle valve. When the engine is under slight acceleration, the throttle valve opens a little and exposes the transfer port to manifold vacuum. This draws the fuel out into the barrel to mix with the air. We will discuss this more under The Low-Speed System.

Adjustable needle valves, called idle mixture screws, figure 13-14, control the amount of gasoline used in the idle air-fuel mixture. One adjustment screw generally is used for each primary barrel. The screw tips stick out into the idle system passages and are turned inward (clockwise) to create a lean mixture, or outward (counterclockwise) to richen the mixture.

Some carburetor mixture screws have plastic limiter caps, figure 13-15. These caps restrict the amount of adjustment to prevent the idle mixture from being too rich. They are required on

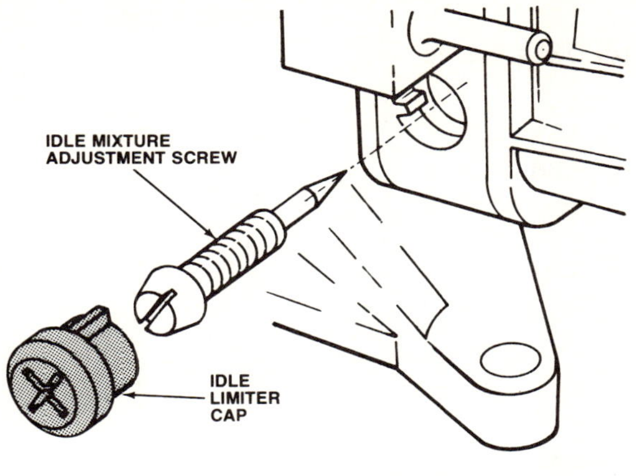

Figure 13-15. Idle limiter caps restrict the amount of adjustment allowed for the idle mixture.

some engines in order to meet emission standards and must be put back on whenever a carburetor is overhauled.

Engine idle speed is adjusted by changing the amount of air going to the idle system. This idle speed adjustment is usually done by a screw that changes the position of the throttle valve in the carburetor. Carburetors which use idle air bypass passages are adjusted by a large screw that varies the opening in the air passage to change the airflow, figure 13-12.

The Low-Speed System

Once the throttle valve begins to open for low-speed operation, the engine needs more fuel than the idle port alone can provide. The airflow passing through the venturi is still not strong enough to develop fuel flow through the main discharge nozzle. To provide more fuel, the transfer port comes into operation as the low-speed system, figure 13-13.

The transfer port is located above the throttle at idle, and the air pressure there is about equal to atmospheric pressure. Air from the barrel flows *into* the transfer port to mix with the fuel going to the idle port. As the throttle opens, the transfer port is exposed to intake vacuum, and the flow reverses. Extra fuel flows *out* of the transfer port to meet the engine's needs during the switch from idle to low-speed operation. Fuel continues to flow from the idle port, but at a reduced rate. This permits an almost constant air-fuel mixture during this transition period.

The High-Speed (Main Metering) System

When the throttle valve opens wider, airflow increases through the carburetor. At the same time, the partial vacuum area of the intake manifold moves up in the carburetor barrel. This

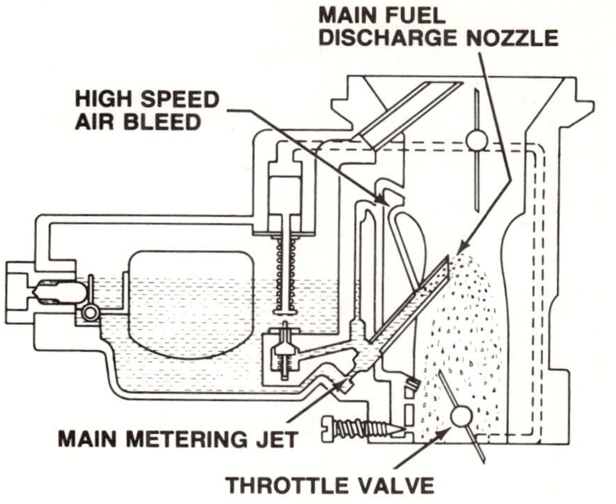

Figure 13-16. Fuel for the high-speed or main metering system flows through the main jet and out the fuel discharge nozzle in the venturi.

airflow and pressure change strengthens the venturi effect, causing gasoline to flow from the main discharge nozzle in the venturi. This is the high-speed or main metering system, figure 13-16.

For better mixing of the fuel and air, most carburetors have multiple, or boost, venturis placed one inside another, figure 13-17. The main discharge nozzle is located in the smallest venturi to increase the partial vacuum effect on the nozzle. Fuel flows from the bowl, through the main jet and main passage, and into the discharge nozzle. A high-speed air bleed, figure 13-16, mixes air into the fuel before it is discharged from the nozzle.

The primary or upper venturi produces vacuum, which causes the main discharge nozzle to spray fuel. The secondary venturi creates an airstream which holds the fuel away from the barrel walls where it would slow down and condense. The result is air turbulence, which causes better mixing and finer atomization of the fuel.

As the throttle continues to open wider, fuel flow from the low-speed system tapers off, while flow from the high-speed system increases. The engine's fuel needs are now supplied entirely by the main discharge nozzle during high-speed, light-load cruising.

The Power System

The high-speed system delivers the leanest air-fuel mixture of all the carburetor systems. When engine load increases during high-speed operation, this mixture is too lean to deliver the

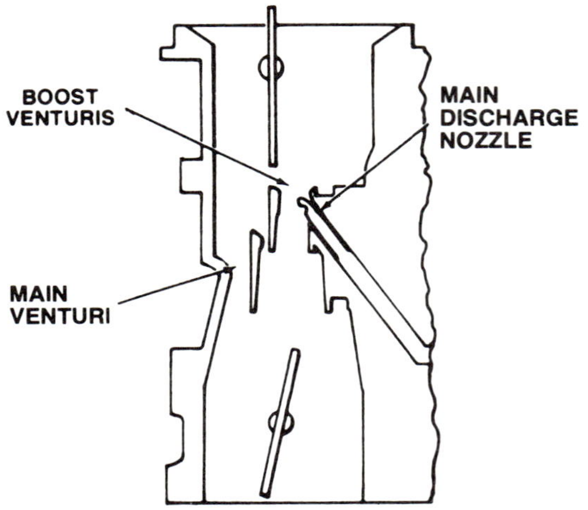

Figure 13-17. Most carburetors have multiple (boost) venturis for better air and fuel mixing.

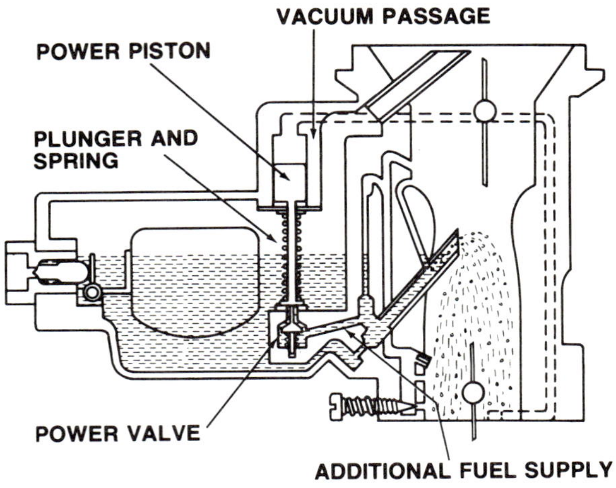

Figure 13-18. This power valve is operated by a vacuum-controlled piston and plunger. When vacuum decreases, the spring moves the plunger to open the power valve.

power required by the engine. The extra fuel needed is provided instead by another system called the power system, or power valve. It supplements main metering fuel delivery. The power system or valve can be operated by vacuum or mechanical linkage. The exact type differs according to carburetor design, but all provide a richer air-fuel mixture.

One type of power valve, figure 13-18, is located in the bottom of the fuel bowl with an opening to the main discharge tube. A spring holds a small poppet valve closed, while a vacuum piston holds a plunger above the valve. Since manifold vacuum decreases as the engine load increases, a large spring moves the plunger downward to open the valve.

Another type of vacuum-operated power valve uses a diaphragm, figure 13-19. Manifold vacuum against the diaphragm holds the valve closed. As vacuum decreases under an increased load, a spring opens the valve.

Metering rods also can be used as a power system, figure 13-20. These may be controlled by vacuum pistons and springs, or by mechanical linkage connected to the throttle. The ends of the rods are tapered, or stepped, and installed in the main jet opening. The rods restrict the area of the main jets and reduce the amount of fuel that flows through them during light load operation of the main metering system. Extra fuel for full-throttle power is provided by moving the rods out of the jets to increase the flow.

■ What's Next In Emission Controls?

Emission control devices and systems now in use have been refined about as much as possible. Further control of emissions will have to be made by new approaches to the problem. One device that looks promising for use in the near future is the exhaust manifold thermal reactor.

Thermal reactors have been under study for several years. In its basic form, a thermal reactor is an oversized exhaust manifold that has sheet metal tubing and a series of port liners. The tubing is centrally located in the cast iron reactor housing, but is insulated from the housing. The port liners are located in the cylinder head's exhaust ports. They are aligned with the reactor inlet ports, but do not touch them.

The thermal reactor is basically an oxidation unit. Air is injected into it to maintain the high (1,600-1,800° F) temperature needed to completely oxidize the HC and CO in exhaust gas. The thermal reactor is a "holding area" to get rid of the unwanted emissions by thermally oxidizing them to form water and carbon dioxide.

The thermal reactor will not replace the catalytic converter, but will work with it to reduce emission levels below those now considered acceptable. So far, the main drawback to the thermal reactor is the higher underhood temperatures it causes. But engineers are studying how to move the battery and other temperature-sensitive parts around so that the heat will not affect them.

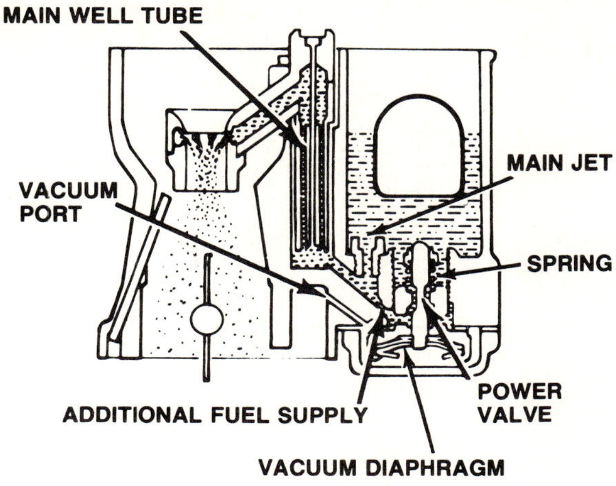

Figure 13-19. The vacuum diaphragm holds the power valve closed. When vacuum decreases, a spring opens the valve to allow more fuel into the main passage.

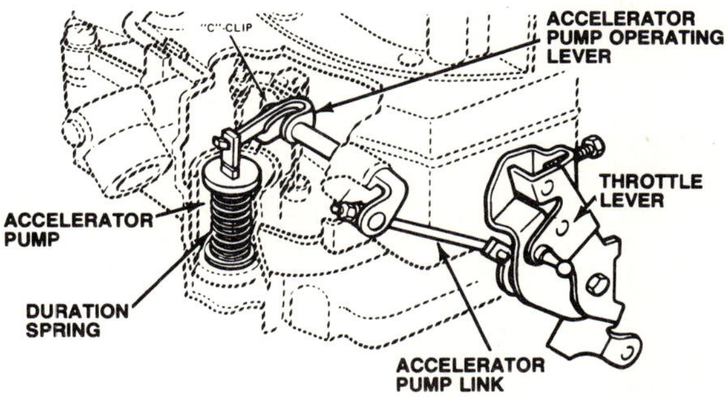

Figure 13-21. Typical plunger-type accelerator pump. (Ford)

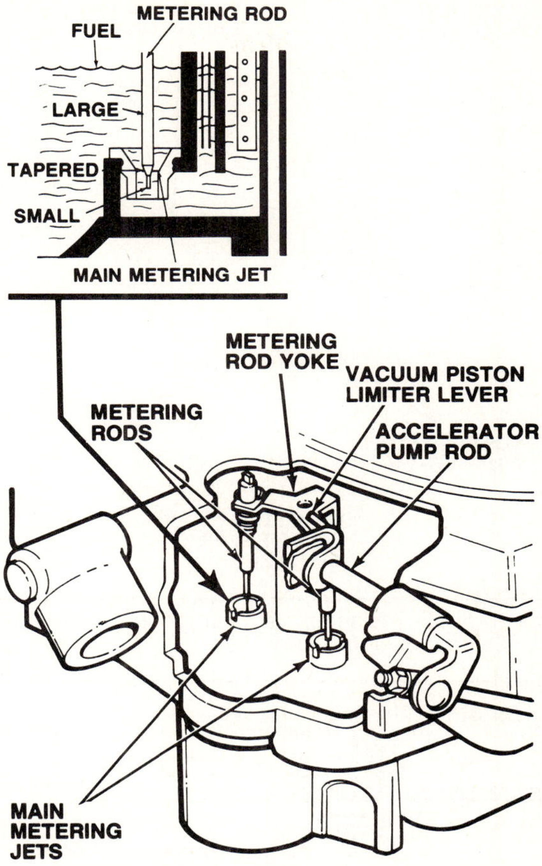

Figure 13-20. Some power systems consist of metering rods placed in the main jets. Mechanical or vacuum linkage moves the rods upward to allow more fuel to flow through the jets when required.

Vacuum-controlled metering rods, also called step-up rods, are held in the jets by manifold vacuum applied to pistons attached to the rods. When vacuum drops under heavy load, springs, working against the pistons, move the rods out of the jets.

Mechanically operated metering rods are controlled directly by mechanical linkage connected to the throttle linkage. Most metering rods are tapered, or stepped, to increase the extra fuel flow gradually.

The Accelerator Pump System

This system provides fuel for some engine operating conditions. If the throttle is opened suddenly from a closed, or nearly closed, position, airflow increases more rapidly than fuel flow from the main discharge nozzle. This "dumping" of air into the intake manifold reduces manifold vacuum suddenly and causes a lean air-fuel mixture. This very lean mixture results in a brief hesitation, or stumble, sometimes called a flat spot. To keep the mixture rich enough, extra fuel must be provided by the accelerator pump.

The accelerator pump, figure 13-21, is a plunger or diaphragm in a separate chamber in the carburetor body. It is operated by a linkage connected to the carburetor throttle linkage, figure 13-22. When the throttle closes, the pump draws fuel into the chamber. An inlet check valve opens to allow fuel into the chamber, figure 13-23, position A, and an outlet check valve closes so that air will not be drawn through the pump nozzle. When the throttle is opened quickly, the pump moves down or inward to deliver fuel to the nozzle in the barrel, figure 13-23, position B. The pump outlet check opens, and the inlet check closes. The inlet check ball is usually (but not always) larger than the outlet check ball.

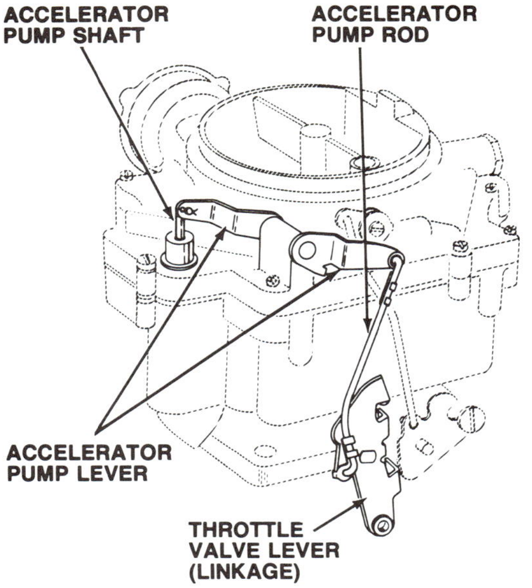

Figure 13-22. The accelerator pump linkage is connected to the throttle linkage.

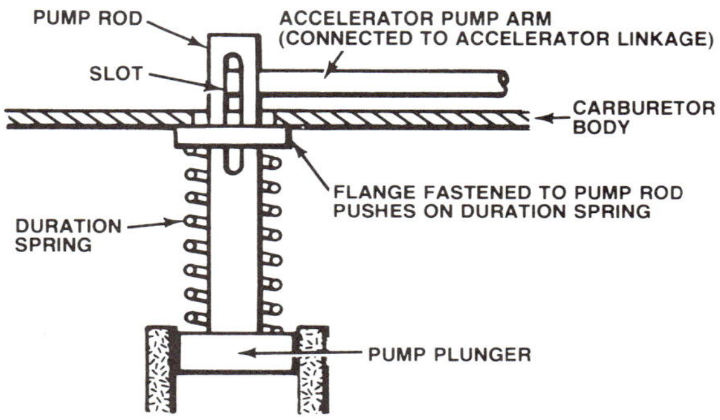

Figure 13-24. The duration spring provides uniform pump delivery regardless of the speed at which the throttle linkage moves.

The pump outlet check may be a steel ball or plunger. The inlet check may be a steel ball, a rubber diaphragm, or part of the pump plunger. Not all pumps have inlet checks. Some rely on an inlet slot in the pump well, or chamber, that is closed by the plunger on the downward stroke.

Most pump plungers or diaphragms are operated by a duration spring, figure 13-24. The throttle linkage holds the pump in the returned position. When the throttle opens, the linkage releases the pump, and the spring moves the plunger for a steady and uniform fuel delivery. The accelerator pump operates during the first half of the throttle travel from the closed to the wide-open position.

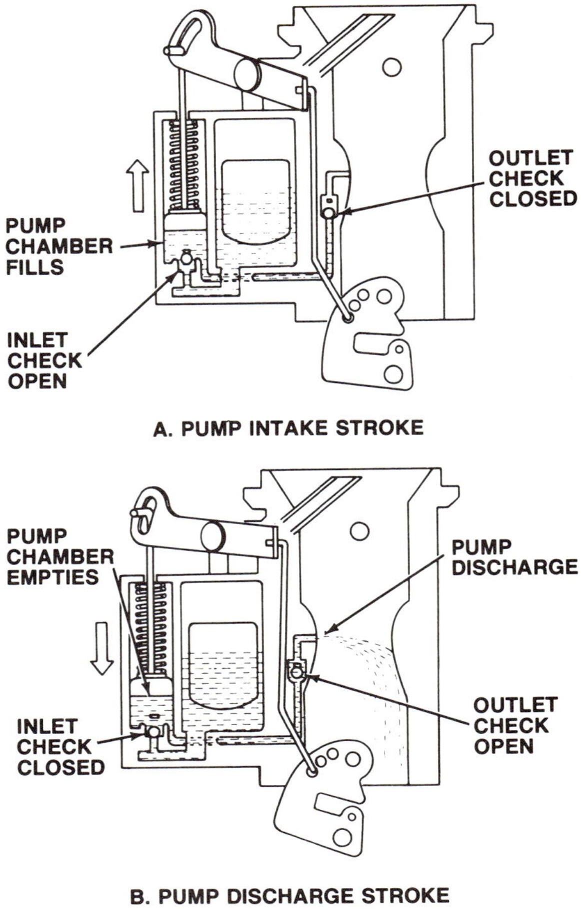

Figure 13-23. Accelerator pump operation.

During high-speed operation, the vacuum at the pump nozzle in the carburetor barrel may be strong enough to unseat the outlet check and siphon fuel from the pump. This is called pump pullover, or siphoning. In a few carburetors, this extra fuel is included in the high-speed system adjustment. In most carburetors, air bleeds are placed in the pump discharge passages to prevent the siphoning. In still other carburetors, an extra weight is added to the outlet check to resist the siphoning. The pump plungers in some carburetors have antisiphon check valves.

The Choke System

The choke provides a very rich mixture for starting a cold engine. This extra-rich mixture is needed because:

- Engine cranking speed is slow.
- Airflow speed is slow.
- Cold manifold walls cause gasoline to condense from the air-fuel mixture, and less vaporized fuel reaches the combustion chambers.

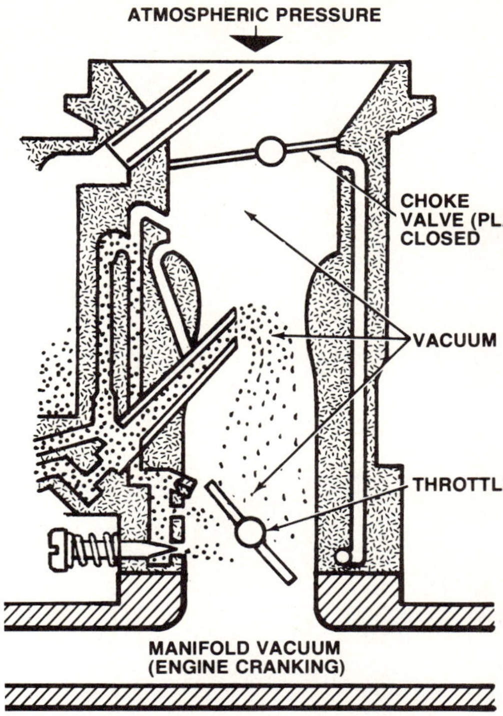

Figure 13-25. Vacuum is present throughout the carburetor barrel below the closed choke. This draws fuel from the idle, low-speed, and high-speed circuits for starting the engine.

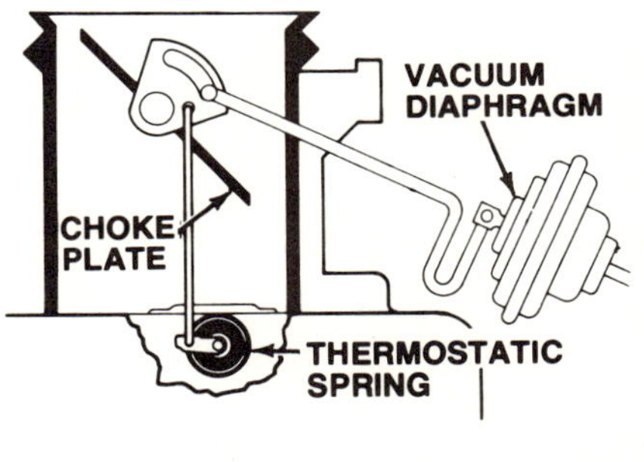

Figure 13-27. In a remote (well-type) choke, the thermostatic bimetal spring is in a heated well on the intake manifold.

To make the mixture richer, a choke plate, or butterfly valve, is placed above the venturi in the carburetor barrel. This choke plate can be tilted at various angles to restrict airflow, figure 13-25. Cranking the engine with the choke plate closed creates a partial vacuum throughout the carburetor barrel below the plate. This airflow reduction and partial vacuum area work together to allow more fuel to be drawn into the mixture.

The choke plate can be controlled by hand with a cable running to the driver's compartment or automatically by a thermostatic

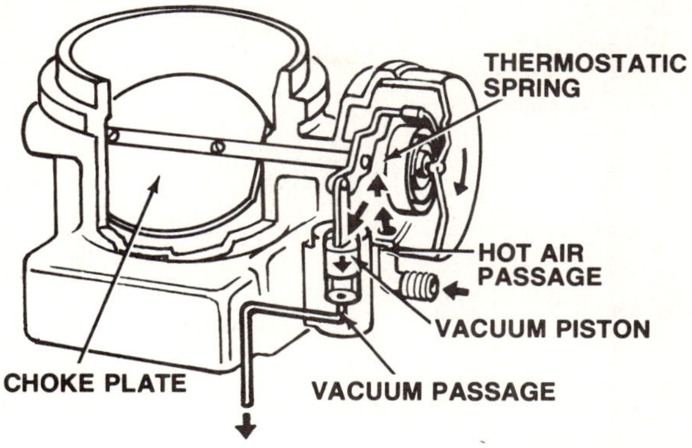

Figure 13-26. In an integral (piston-type) choke, the thermostatic bimetal spring is in a housing on the carburetor airhorn.

spring. Chokes on most domestic carburetors for the past 25 or 30 years have been operated by a bimetal thermostatic spring. The choke plate shaft is connected to the spring by linkage. The bimetal spring is normally located in one of two places:

1. In a round housing on the carburetor airhorn, figure 13-26. This is called an integral, or piston-type, choke.
2. Off the carburetor in a well on the intake manifold, figure 13-27. This is called a remote, a well-type, or a vacuum-break, choke.

Regardless of type and location, the thermostatic spring forces the choke closed when the engine is cold. Running the engine heats up the spring, which then opens the choke. With an integral choke, hot air from a source near the exhaust manifold, or hot coolant from the cooling system, is routed to the choke housing to heat the spring. Remote, well-type chokes are normally heated by exhaust routed through a crossover passage in the intake manifold. Both integral and remote chokes also may have electric heating elements to heat the spring faster and speed the choke opening.

When a cold engine is cranked, the choke must be completely closed. As soon as the engine starts, the choke must open slightly for sufficient airflow. This is done in two ways. First, the choke plate shaft is offset in the carburetor so that airflow will tend to open the plate. Second, manifold vacuum is applied to a vacuum piston or a vacuum-break diaphragm that pulls the choke open a few degrees. Some integral chokes have a vacuum piston built into the choke housing, figure 13-26, which gives them their name of piston-type chokes. A remote choke has the vacuum-break diaphragm

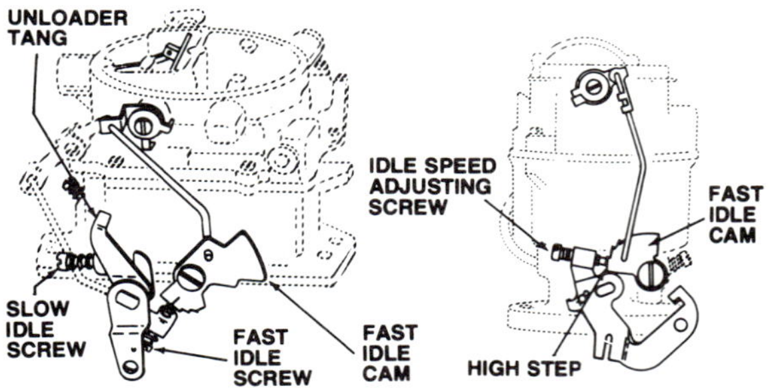

Figure 13-28. The fast-idle cam opens the throttle wider for faster engine speed when the choke is operating. It may work on the slow-idle speed adjusting screw or on a separate fast-idle screw.

mounted on the side of the carburetor, figure 13-27. Most integral chokes have a vacuum-break diaphragm rather than a piston to open the choke.

A cold engine must idle faster than a warm engine, or the lack of air and fuel flow will cause it to stall. A fast-idle cam and screw, figure 13-28, provide enough air and fuel to prevent engine stalling. The cam is linked to the choke plate, and the screw is located on the throttle valve shaft. Pushing down on the accelerator to start a cold engine allows the choke to close. This moves the fast-idle cam to allow the screw to rest against a high step of the cam. The cam may contact the normal slow-idle adjusting screw or a separate fast-idle screw. In either case, the throttle is held open slightly more than for a normal slow idle, and idle speed increases between 400 and 800 rpm.

A mechanical link or choke unloader, figure 13-28, opens the choke about halfway when the throttle is fully open. If the engine is accidentally flooded during starting, this provides the extra airflow necessary to help clear out the fuel. Because the choke system provides a very rich mixture, external devices are used to modify its

■ Check For A Cold Carburetor

Cold outside air can cause carburetors to freeze up, especially when low temperature combines with high humidity. This carburetor icing can, in turn, cause the choke valve to stick or bind in the carburetor. If you are servicing a car whose owner has complained about poor engine performance during cold weather, always be sure to check for carburetor icing. If it's your car, you can throw a blanket over the engine and carburetor when it is left to sit for a long time in cold weather.

Figure 13-29. A 2-barrel carburetor has one airhorn but two venturis and throttle plates.

operation in various ways to improve driveability and reduce emissions. These will be discussed later.

CARBURETOR TYPES

The operation of the basic carburetor systems has been explained in terms of a carburetor which uses a single barrel and throttle valve. But carburetors also are made with two or more barrels. Various carburetor types are used to match fuel flow to engine requirements. Domestic engines all use downdraft carburetors, while some imports use sidedraft carburetor designs. Carburetors are usually classified by the number of barrels or venturis used.

One-Barrel

The 1-barrel carburetor has a single outlet through which all systems feed to the intake manifold. This type of carburetor may also be known as a single-venturi design. These carburetors are generally used on 4- and 6-cylinder engines.

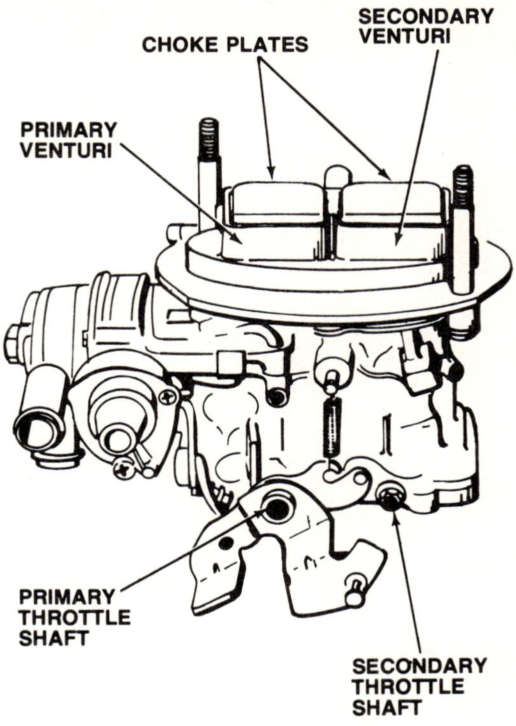

Figure 13-30. The 2-stage, 2-barrel carburetor was developed in response to emission control requirements.

Single-Stage Two-Barrel

This carburetor contains two barrels and two throttles which operate together, figure 13-29. Since the various fuel discharge passages in each barrel operate at the same time, it can be considered as two 1-barrel carburetors sharing the same body. The two throttle plates are mounted on the same shaft, and operate together. The two barrels share a common float, choke, power system, and accelerator pump. Single-stage 2-barrel carburetors are used on many 6- and 8-cylinder engines.

Two-Stage Two-Barrel

This carburetor, figure 13-30, is a relatively recent development, brought about by emission control requirements. It differs from the single-stage, 2-barrel design in that its two throttles operate independently. The primary barrel is generally smaller than the secondary, and handles engine needs at low-to-moderate speeds and loads. The larger secondary opens as necessary to handle heavier load requirements.

The primary stage usually includes the idle, the accelerator pump, the low-speed, the main metering, and the power systems. The secondary stage usually has a transfer, a main metering, and a power system. Both stages draw fuel

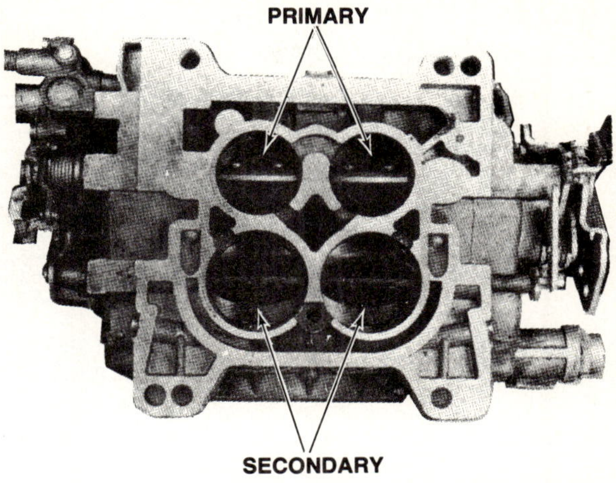

Figure 13-31. A 4-barrel carburetor is made of two 2-barrel carburetors.

from the same fuel bowl. Some designs use a common choke for both barrels. In others, only the primary stage is choked.

A 2-stage 2-barrel carburetor is used primarily on 4- and 6-cylinder engines and a few V-8's.

Four-Barrel

Used on V-8 engines, the 4-barrel, or quad, carburetor uses two primary barrels and two secondary barrels in a single body, figure 13-31. The two primaries operate like a 1-stage 2-barrel at low-to-moderate engine speeds and loads. The secondary barrels open at about half to three-quarter throttle to provide the increased fuel and airflow required for high-speed operation. The primary barrels contain the choke, the idle, the low-speed, and the high-speed systems, as well as an accelerator pump and a power system. The secondary barrels have their own high-speed and power systems and may use their own accelerator system.

Two methods provide airflow through the secondary barrels: venturi action or air velocity valves. Air velocity valves look like large choke plates located in the secondary barrels. They are opened by the low pressure created in the secondary barrels when the throttles are opened. Venturis and air valves also may be combined in one carburetor to modulate the airflow through the barrels.

The primary barrels supply all eight cylinders during low-to-moderate speeds and loads. The secondary barrels provide additional fuel and airflow for high speeds and heavy loads.

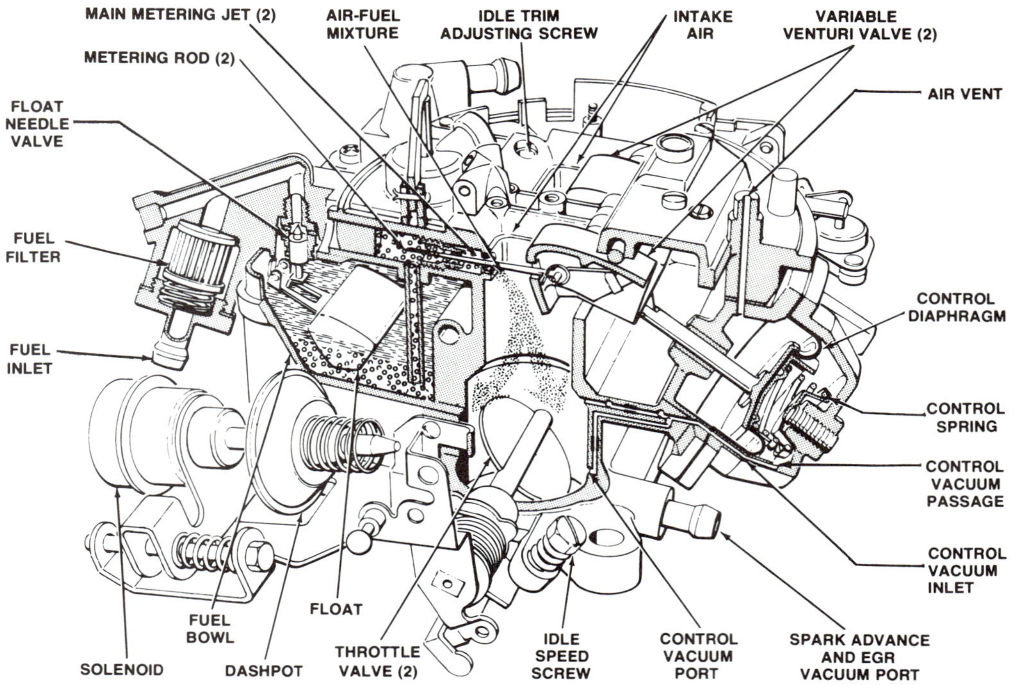

Figure 13-32. The Motorcraft 2700 VV variable-venturi carburetor uses vacuum to control the movement of the venturi valves.

■ Ford's First Variable Venturi Carburetor

Ford's first variable venturi carburetor was a one-barrel model made by Detroit Lubricator. It was used on the original flathead Model 18 V-8's in 1932 and early '33. The variable venturi was formed by two air vanes in the barrel that responded directly to airflow, rather than to vacuum linkage control as in the current Motorcraft 2700 VV carburetor.

The air vanes were linked to a movable main jet (metering valve) that slid up and down on a fixed metering rod (pin) in the center of the venturi. The metering rod was adjustable to obtain the proper idle mixture and main system fuel flow.

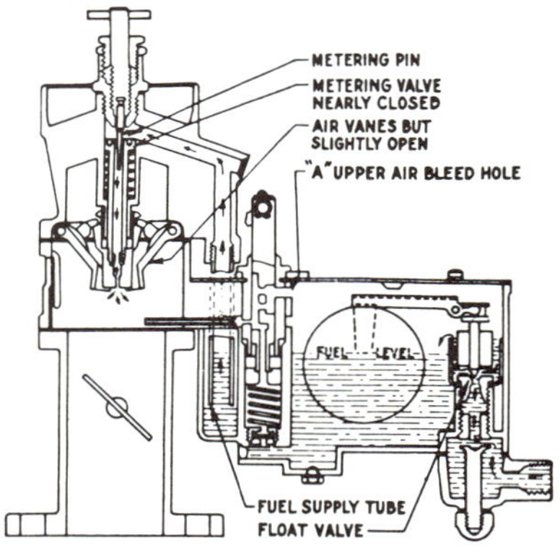

Normal Running

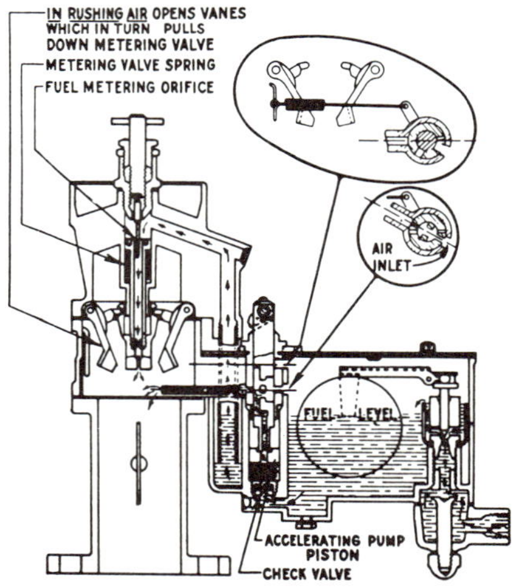

Maximum Power

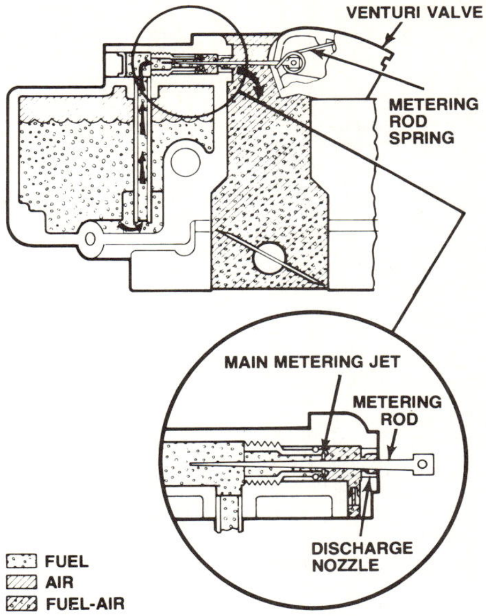

Figure 13-33. Motorcraft 2700 VV main metering system. (Ford)

VARIABLE-VENTURI (CONSTANT DEPRESSION) CARBURETORS

As we have seen, carburetors meter fuel by using a venturi to create a partial vacuum in the barrel. Airflow through the venturi increases velocity which decreases pressure. This pressure drop causes fuel to flow through the discharge nozzle into the barrel. Since both the carburetor barrel and venturi are fixed in size, the volume and velocity of air passing through will be correct for some operating conditions but not correct for others. To produce better performance under *all* operating conditions, auxiliary circuits such as the choke, power, and idle systems must be added to the main metering systems.

A carburetor with a venturi whose size changes according to the demands of engine speed and load does not need these extra systems. At the same time, air-fuel mixtures can more closely be controlled for better fuel economy and emission control. Changing the size of the venturi in relation to engine speed and load results in an even pressure drop across the venturi under all operating conditions. This gives a variable venturi carburetor its other name, a "constant depression" carburetor. Variable-venturi carburetors, such as the SU, Hitachi, and Stromberg CD models, have been used on some imported cars for many years. In 1977, Ford Motor Company introduced the first variable-venturi carburetor used on domestic cars in 45 years. (Ford's first V-8 engines in 1932 had variable-venturi, 1-barrel carburetors.)

Called the Motorcraft 2700 VV, figure 13-32, Ford's variable-venturi, 2-barrel carburetor was first used on V-6 and 302-cid V-8 engines. It has a fuel inlet system with a replaceable filter in the inlet housing. Throttle plates and an accelerator pump are also used. However, the variable venturis and the different fuel metering systems make this carburetor unique.

The variable venturis are formed by two rectangular valve plates (actually one casting) that slide back and forth across the tops of the two barrels. Movement is controlled by a spring-loaded vacuum diaphragm, regulated by a vacuum signal taken below the venturis, but above the throttle plates, in the carburetor barrels. As the throttle opens, the vacuum increases, opening the venturis and allowing more air to enter.

The front edge of each venturi valve has a tapered metering rod. Each rod moves in and out of a fixed main jet on the other side of the barrel, figure 13-33. This arrangement meters fuel in proportion to the airflow through the venturis. Because of the variable venturis, there are fewer fuel metering systems than in a fixed-venturi carburetor. However, the main metering system of the 2700 VV carburetor cannot handle all operating conditions without help from secondary systems. For example, at full throttle under heavy load, vacuum may not be strong enough to override the diaphragm spring. In this case, a limiter lever on the throttle shaft pushes the venturi valves fully open. Other auxiliary systems in this carburetor are the accelerator pump, the idle trim system, the cranking enrichment system, and the cold enrichment system.

Cold Enrichment System

The 2700 VV carburetor does not have a conventional choke. Instead, a bimetal thermostatic choke control, with Ford's electric-assist heater (explained later in this chapter), controls the cold enrichment system, figure 13-34. This system contains a fast-idle cam, a cold enrichment auxiliary fuel passage and metering rod, and a control vacuum regulator rod. A unique feature of the fast-idle cam is a vacuum-operated "high cam speed positioner" (HCSP). When a cold engine is started, a lever slides between the fast-idle cam and the fast-idle lever to provide more throttle opening. When the engine starts, vacuum is applied to the HCSP diaphragm and the lever is pulled back.

When the engine is started below 95° F (35° C), the choke spring pushes the control vacuum regulator rod to block the ported control vacuum and send manifold vacuum to the venturi valve diaphragm. This opens the venturis wider than normal. The fast-idle cam is touched by the HCSP for a wider throttle opening.

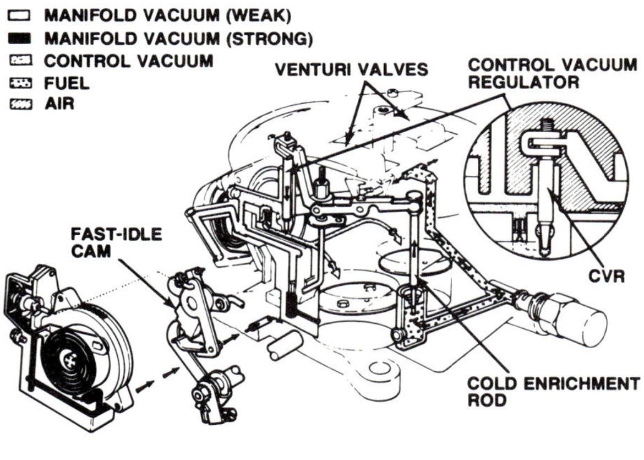

Figure 13-34. Motorcraft 2700 VV cold enrichment system operation with a cold engine. (Ford)

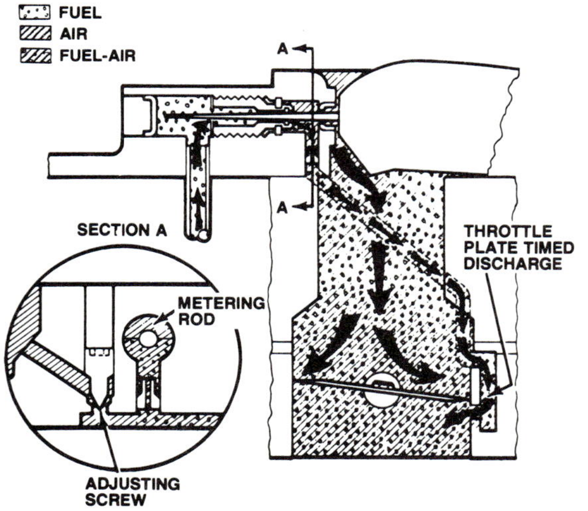

Figure 13-36. Motorcraft 2700 VV idle trim system. (Ford)

Cranking Enrichment System

The cranking enrichment system, figure 13-35, also provides extra fuel only for starting a cold engine. It uses an electric solenoid, energized by the ignition switch, to open an auxiliary fuel passage. When the engine starts, the solenoid is deenergized and closes the cranking fuel passage. The main metering system and the cold enrichment system then maintain the fuel flow.

Idle Trim System

The idle fuel flow is controlled by the main jets, but an additional small amount of fuel is drawn by manifold vacuum through internal passages to discharge ports below the throttles. This is called the idle trim system, figure 13-36.

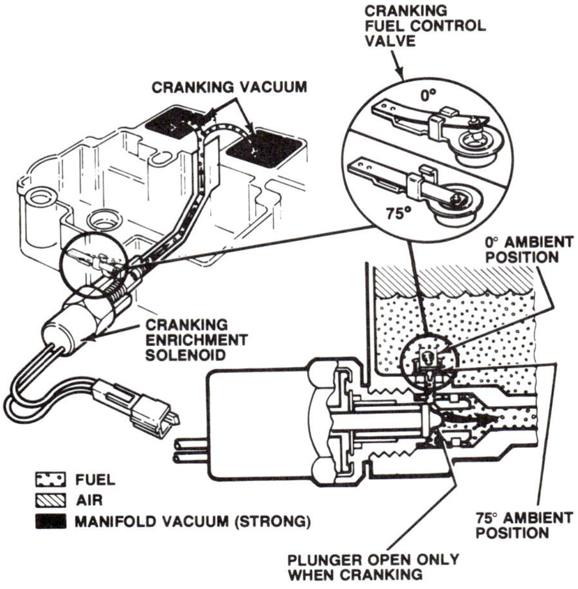

Figure 13-35. Motorcraft 2700 VV cranking enrichment system. (Ford)

CARBURETOR LINKAGE

Car speed is controlled by the accelerator pedal, which moves rods, cables, levers, and springs to operate the carburetor throttle valve. The throttle linkage on some cars is a combination of solid rods and levers, figure 13-37. Other cars use a cable to link the accelerator to the carburetor throttle, figure 13-38.

Other carburetor linkage pieces operate the accelerator pump, the automatic choke, the fast-idle cam, secondary throttle valves, and the automatic transmission downshift points on some cars. Because of the wide variety of linkages between car models, or even between carburetors on the same car model, we will not attempt to describe each one. Instead, we will look briefly at the major types.

Accelerator Pump Linkage

The accelerator pump piston or plunger is connected to the throttle by a small rod, or rods and levers. On some carburetors, the linkage holds the pump in the retracted position, against a compressed duration spring. When the throttle opens all the way, the linkage releases the pump, and the duration spring moves the pump piston through its stroke. On some carburetors, the accelerator pump linkage also opens a vent for the fuel bowl when the throttle is closed.

Figure 13-37. This throttle linkage is a combination of solid rods and levers.

Figure 13-38. A cable from the accelerator to the carburetor is used for this throttle linkage.

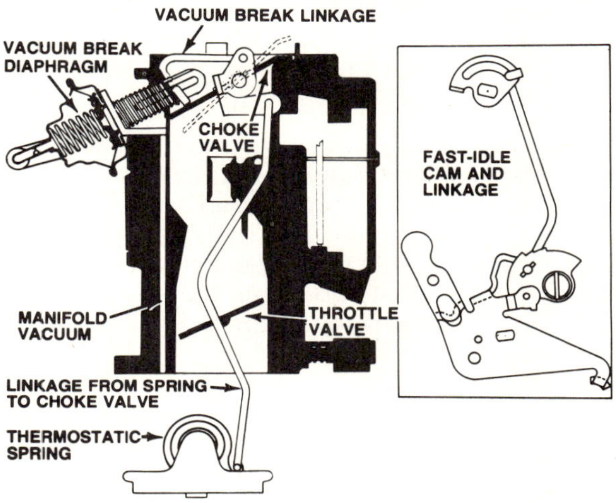

Figure 13-39. Several pieces of linkage are used on this choke system. (Chevrolet)

On other carburetors, the accelerator pump linkage operates a diaphragm to deliver fuel to the pump system. The accelerator pumps on other carburetors are operated directly by mechanical linkage. Pump operation may be balanced by a duration spring and a return spring.

Accelerator pump and linkage designs vary from carburetor to carburetor, but all work on these same principles. On most carburetors, the accelerator pump ends its stroke at the half-throttle position. From this point to the full-throttle position, the high-speed and the power systems can supply enough fuel. The pump linkage on some carburetors can be installed in two or three different positions on the carburetor throttle linkage to provide different pump strokes.

Automatic Choke Linkage

Automatic chokes vary in design, but all require connecting linkage between the thermostatic spring that closes the valve and the vacuum piston or diaphragm that opens it. A remote, well-type choke with a vacuum-break diaphragm is shown in figure 13-39. The rod linkage from the spring to the choke valve closes the choke. The vacuum-break diaphragm opens the choke through its linkage as soon as the engine starts. With most integral, piston-type chokes, the thermostatic spring acts directly on a lever on the end of the choke shaft, inside the choke housing, figure 13-26. The vacuum piston also acts directly on this lever to open the choke when the engine starts. Some carburetors are built with an integral-type choke housing mounted away from the choke valve. This arrangement requires an external linkage rod from the thermostatic spring to the choke valve, as well as a separate vacuum-break diaphragm.

Fast-Idle Cam Linkage

The fast-idle cam is linked to the choke valve, figure 13-39, to provide a faster than normal idle and prevent stalling when the engine is cold. As the choke closes, the linkage turns the fast-idle cam to adjust idle speed. When the choke opens, the fast-idle cam reduces idle speed step by step. To begin the choking operation with a cold engine, the accelerator must be pressed down to release the fast-idle cam and linkage and allow the choke to close.

Secondary Throttle Linkage

The secondary throttles of 4-barrel and 2-stage 2-barrel carburetors may be operated by mechanical linkage or by vacuum. Secondary throttle valves operated by mechanical linkage have an operating rod to connect the primary throttle shaft to the secondary throttle shaft,

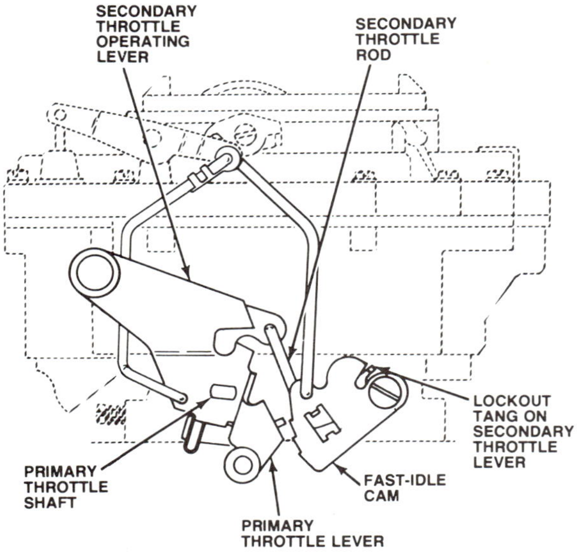

Figure 13-40. Mechanical secondary throttle linkage on a 4-barrel carburetor.

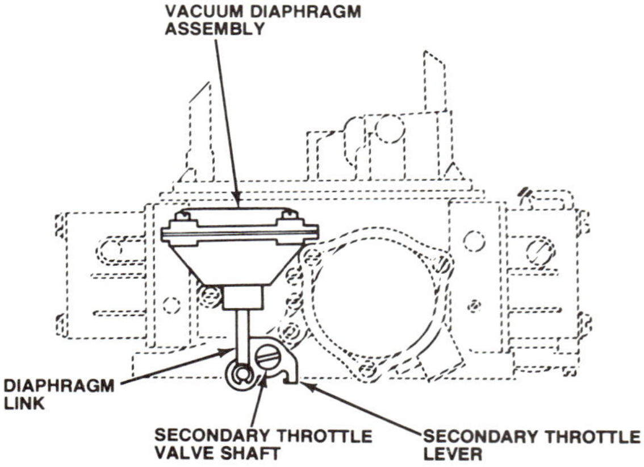

Figure 13-42. Vacuum-operated secondary throttles are controlled by a diaphragm like this.

figure 13-40. Secondary throttles begin to open when the primary throttles are about one-half open. The secondaries continue to open along with the primaries, but at a faster rate. The primary and secondary throttles then reach the fully open position at the same time.

On some carburetors, the primary and secondary throttle shafts are mechanically linked, but the secondary throttles are not visible through the carburetor airhorn. This type of carburetor has secondary air velocity valves, called auxiliary throttle valves, figure 13-41.

Figure 13-41. Air velocity valves in a 4-barrel carburetor.

These valves have offset shafts and counterweights so that they remain closed until air velocity through the carburetor barrels is strong enough to open them. These auxiliary throttle valves operate only when the mechanical secondary throttles are open, but they do not rely on the mechanical movement of the secondary throttles.

Vacuum-operated secondary throttle valves are controlled by a vacuum diaphragm that is mounted on the side of the carburetor, figure 13-42. At low cruising speeds, the secondary throttle valves are closed, and the engine's air-fuel requirements are met by the primary half of the carburetor. At higher speeds, when more fuel and air are needed, the secondary throttles are opened by linkage connected to the vacuum diaphragm.

The vacuum diaphragm responds to increasing vacuum within the primary venturi as engine airflow increases, figure 13-43. Linkage from the diaphragm opens the secondary throttles. Diaphragm action is changed by another vacuum port, or air bleed, in the secondary barrels. The amount and rate at which the secondary throttles open are determined by the vacuum signal at the diaphragm.

All secondary throttle linkage, whether mechanical or vacuum operated, includes a secondary throttle or air valve lockout device to keep the secondary throttles from opening when the choke is closed.

Transmission Linkage

Chrysler Corporation cars with automatic transmissions use an adjustable throttle rod connection, figure 13-44, between the transmission and the carburetor to control shift points and shift quality. Other automatic transmissions accomplish this with vacuum control.

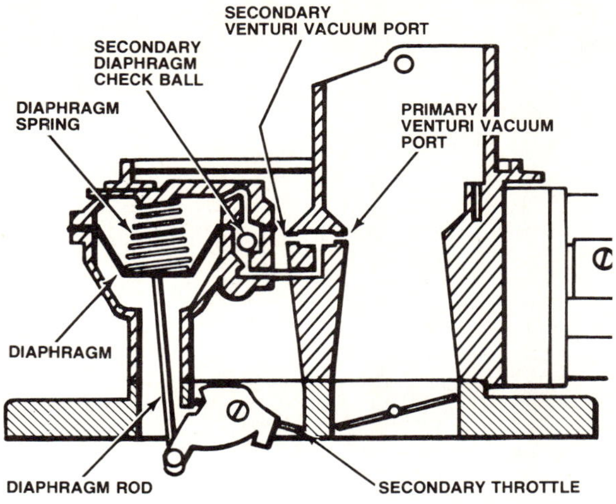

Figure 13-43. The secondary diaphragm responds to vacuum from ports within the carburetor barrels.

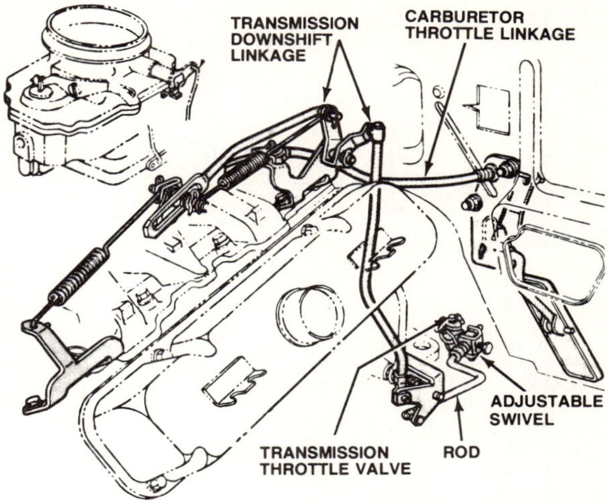

Figure 13-44. Chrysler products with automatic transmissions have mechanical linkage from the carburetor to the transmission throttle valve, which controls shift points. (Chrysler)

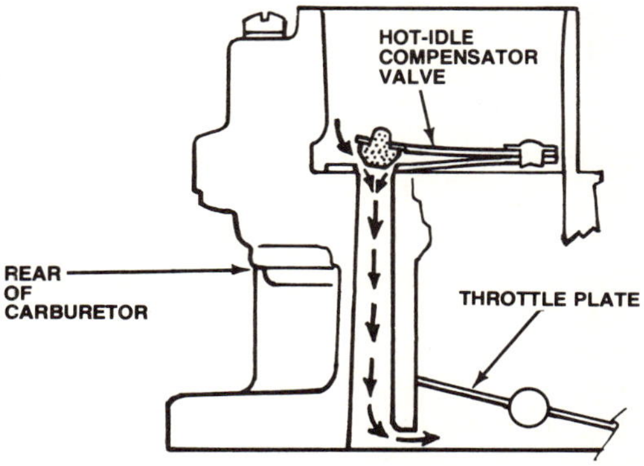

Figure 13-45. The hot-idle compensator is a thermostatic valve that opens at high temperature to admit more air to the idle circuit.

CARBURETOR CIRCUIT VARIATIONS AND ASSIST DEVICES

Variations in the basic carburetor systems we just talked about are used by all carmakers. One or more add-on devices also may be used to improve economy, driveability, and emission control. We will discuss the most common of these.

Hot-Idle Compensator Valve

High carburetor inlet air temperature causes gasoline to vaporize rapidly, which can cause an overly rich idle mixture. To prevent this, many carburetors use a hot-idle compensator valve, figure 13-45. The compensator is a thermostatic valve consisting of a bimetal spring, a bracket, and a small poppet. The compensator valve usually is located either in the carburetor barrel or in a chamber on the rear of the carburetor bowl. A dust cover is placed over the chamber. A third location, used primarily in Autolite 2-barrel carburetors on older air-conditioned cars, is an external mounting in the PCV valve hose near the carburetor.

The hot-idle compensator valve is normally closed by spring tension and engine vacuum. As temperature rises, the bimetal strip bends. This uncovers an auxiliary air passage, or air bleed, through which air enters the carburetor below the throttle plates. As this extra air mixes with excess fuel to lean out the idle mixture, it prevents stalling and rough idling. Once the carburetor temperature returns to normal, the compensator valve closes to shut off the extra air supply.

Idle Enrichment Valve

To reduce emissions, late-model carburetors are adjusted for leaner mixtures. For good cold-engine operation, the idle mixture must be enriched in some cases. All 1975 and later Chrysler cars with automatic transmissions use an idle enrichment system. This works opposite to a hot idle compensator valve.

A small vacuum diaphragm mounted near the carburetor top, figure 13-46, controls idle circuit air. When control vacuum is applied, the diaphragm reduces idle system air. This increases fuel and reduces the air in the air-fuel mixture. Diaphragm vacuum is controlled by a

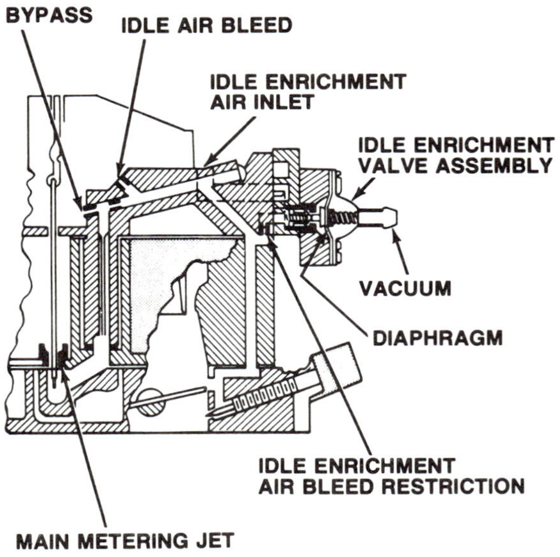

Figure 13-46. Chrysler idle enrichment valve.

temperature switch in the radiator. As the engine warms, this switch stops the vacuum signal, returning the airfuel mixture to its normal lean level.

Fast-Idle Pulloff (Choke Pulloff)

Long periods of choke and fast idle can damage catalytic converters with rich air-fuel mixtures. Some converter-equipped GM cars use a fast-idle pulloff to avoid converter overheating. In one system, manifold vacuum acts on a vacuum-break diaphragm at the rear of the carburetor. Another system uses the front vacuum diaphragm to pull the throttle down on the fast-idle cam. A third method uses an electric solenoid instead of a vacuum diaphragm to open the choke.

Regardless of the system used, fast-idle pulloff has no effect on engine warmup during ordinary operation because normal throttle movement will disengage the fast-idle cam. The fast-idle pulloff system only operates when the car is warming up while parked.

Throttle Stop Solenoid

Engine **dieseling** or after-run results when combustion chamber temperatures remain hot enough to ignite an idle air-fuel mixture after the ignition is turned off. Dieseling is caused by several aspects of late-model engines: higher operating temperatures, faster idle speeds, retarded ignition timing at idle, and lean air-fuel mixtures. Closing the throttle more than it would close for the engine's normal slow idle speed will prevent dieseling. By shutting off

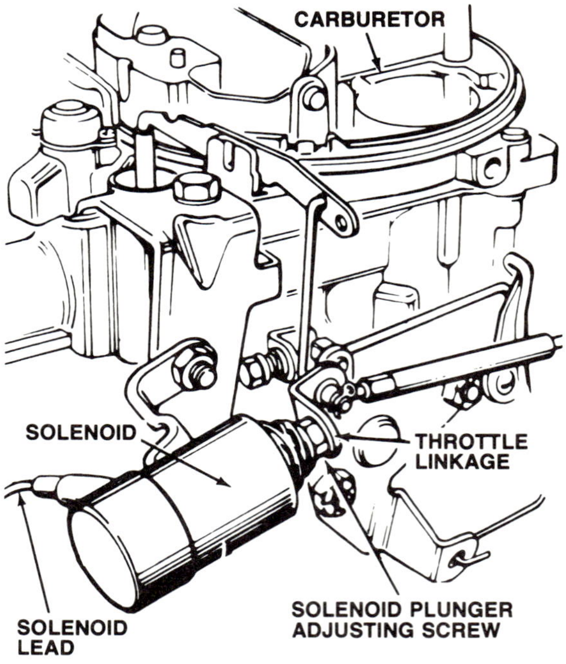

Figure 13-47. The throttle stop solenoid holds the throttle open for normal slow idle and allows the throttle to close farther when the engine is shut off.

airflow past the throttle valve, the idle circuit is closed.

A throttle stop solenoid, figure 13-47, provides the new stop position for the throttle during normal slow idle. Turning on the ignition energizes the solenoid, and its plunger moves out to contact the idle speed adjusting screw or a bracket on the throttle shaft. This holds the throttle open slightly for a normal slow idle until the ignition is shut off. The solenoid is then deenergized, its plunger retracts, and the throttle closes to block airflow.

Air Conditioning Throttle Solenoid

Late-model Ford, GM, and Chrysler engines with air conditioning may use a solenoid which looks exactly like a throttle solenoid, figure 13-48, and even may carry the same factory part number. But, it should not be confused with the throttle stop solenoid we just discussed, since it is energized *only* through the air conditioner switch. Its plunger moves forward to contact a bracket on the throttle shaft *only* when the air conditioning is turned on. This raises engine idle speed to prevent the engine from stalling due to the increased load.

Choke Hot Air Modulator

Some GM engines use a choke hot air modulator check valve (CHAM-CV) in the air cleaner, figure 13-49. At air cleaner temperatures below

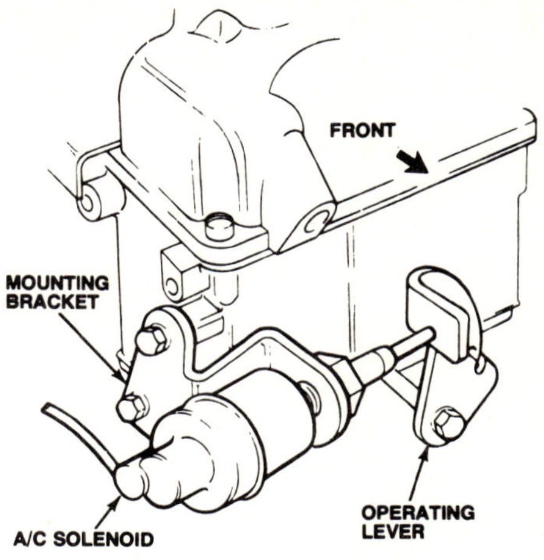

Figure 13-48. The air conditioning throttle solenoid opens the throttle slightly when the air conditioner is on. This maintains a uniform idle speed, even with the increased engine load.

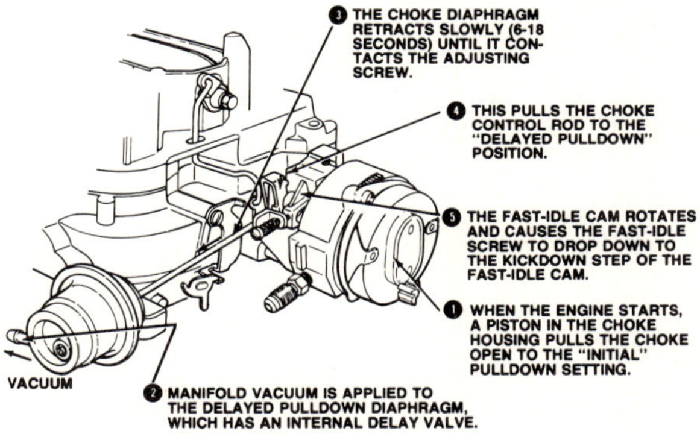

Figure 13-50. The delayed choke pulldown diaphragm provides rich initial choking, rapid choke release, and fast-idle modulation.

68° F (20° C), this valve is closed. Air that is to be heated by the heater coil passes through a tiny hole in the modulator. This restricts hot airflow over the bimetal thermostatic coil and results in a slower choke warmup.

Staged Choke Pulldown

Found only on 1972 Ford engines, this system has a small double-chamber housing connected to the choke linkage at one end and to manifold vacuum at the other. Inside the housing, a divider with an orifice separates the two chambers. The chamber facing the carburetor contains a spring and silicone fluid; the other chamber holds a diaphragm and a bimetal valve. The valve shuts off manifold vacuum to the diaphragm at temperatures below 60° F (16° C), allowing the choke to close.

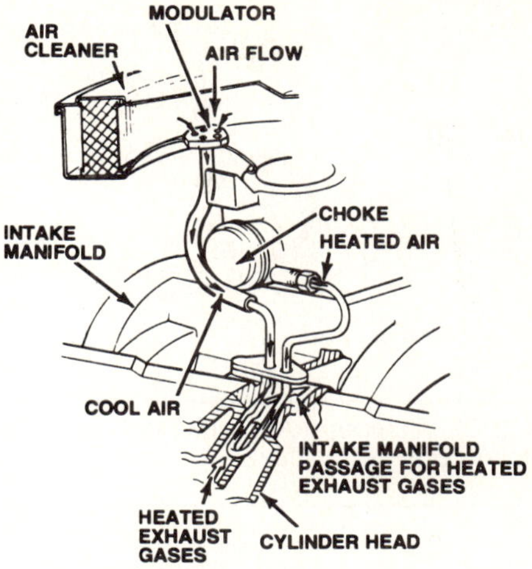

Figure 13-49. General Motors' choke hot air modulator.

Delayed Choke Pulldown

Some 1975 and later Motorcraft carburetors have a delayed choke pulldown operated by a vacuum diaphragm. This opens the choke to a wider setting 6 to 18 seconds after the engine starts. As the pulldown diaphragm operates, the fast-idle screw is pulled from the top to the second step of the fast-idle cam. This reduces cold engine idle speed. Figure 13-50 shows the exact sequence of operation.

Temperature-Controlled Vacuum Break

Since 1975, many Rochester carburetors use two vacuum-break diaphragms to provide better mixture control by choking the engine more when it is cold and less when warm. Slightly different systems are used on GM V-8 engines, figure 13-51, and 6-cylinder engines, figure 13-52.

On V-8 engines, the primary (front) diaphragm opens the choke to keep the engine from stalling when first started.

On 6-cylinder engines, the primary diaphragm (choke coil side) opens the choke to keep the engine from stalling when first started. The auxiliary diaphragm (throttle lever side) opens the choke wider when engine coolant is warm. It operates on manifold vacuum provided by a two-nozzle temperature vacuum

Dieseling: A condition in which extreme heat in an engine's combustion chamber continues to ignite excess fuel after the ignition has been turned off.

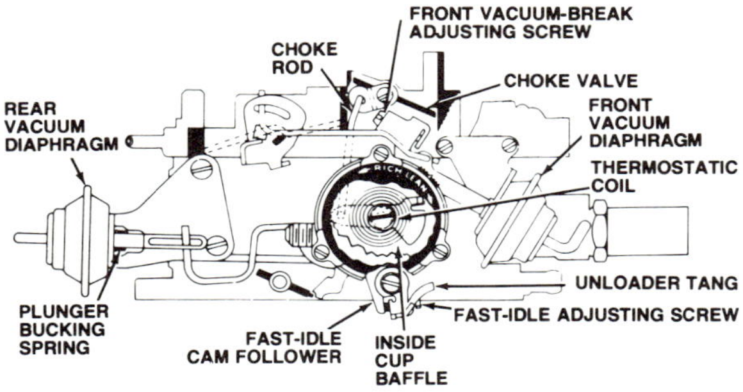

Figure 13-51. Temperature-controlled vacuum break on a GM V-8 engine. (Pontiac)

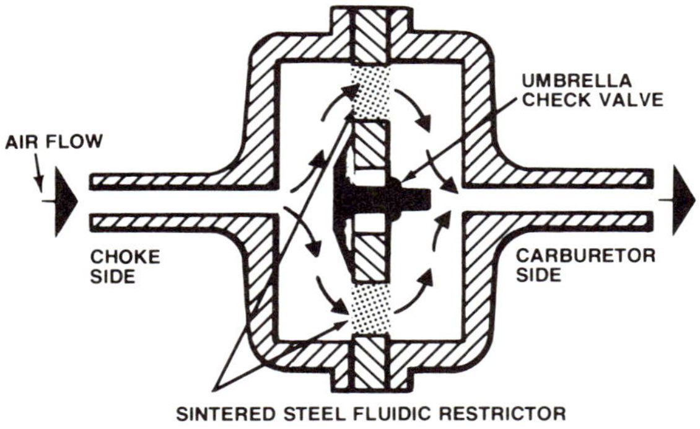

Figure 13-53. Choke delay valve. (Ford)

switch on the cylinder head. The dual vacuum-break diaphragm system was discontinued on several 1977 carburetors in favor of a single vacuum break.

Choke Delay Valves

Choke delay valves have various designs, but all delay the opening of the choke for a period of time to improve driveability and cold engine warmup.

The Ford choke delay valve, in a hose between intake manifold vacuum and the choke vacuum piston or diaphragm, figure 13-53, is the same valve used for spark delay in the ignition system. The black side must face the vacuum source, and the colored side the choke. The amount of delay is indicated by the valve color and is the same interval as when used as a spark delay valve.

GM uses an internal bleed check valve in the rear vacuum break diaphragm unit.

A slightly different system is used on the 151-cid, 4-cylinder GM engine. This engine has a choke vacuum-break and vacuum-delay valve. The valve delays manifold vacuum

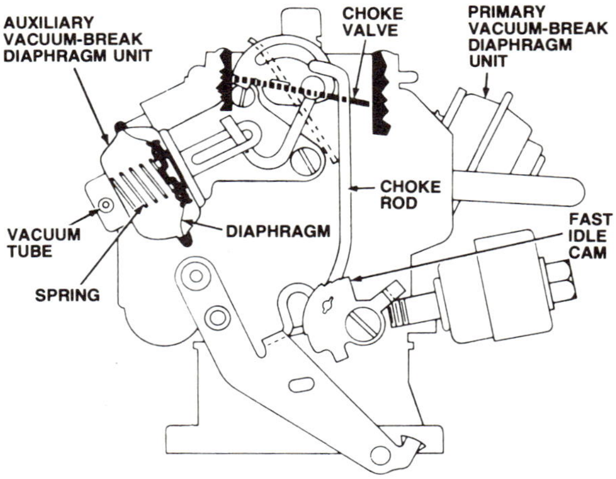

Figure 13-52. Temperature-controlled vacuum break on a GM 6-cylinder engine.

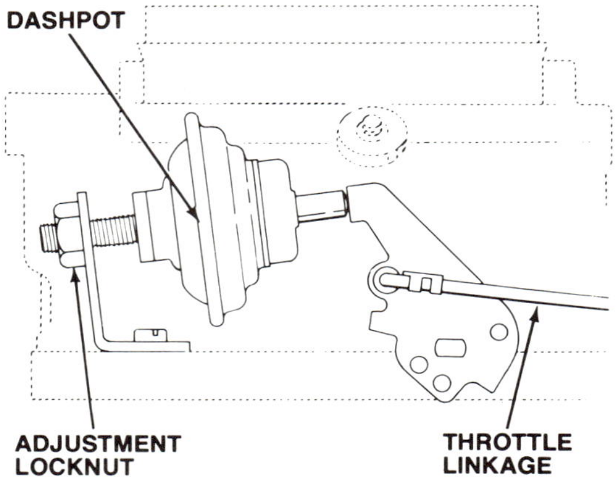

Figure 13-54. The dashpot slows throttle closing.

against the choke vacuum-break unit for approximately 40 seconds when starting the engine. If the engine should stall after being started, a relief feature in the valve releases vacuum to close the choke quickly.

Dashpots

These small chambers with a spring-loaded diaphragm and plunger have been used for over 20 years on some cars with automatic transmissions and on a few with manual transmissions. Hydraulic and magnetic dashpots also have been used on older cars.

A link from the throttle touches the dashpot plunger as the throttle closes, figure 13-54. As force is applied to the plunger, air slowly bleeds out of the diaphragm chamber through a small hole.

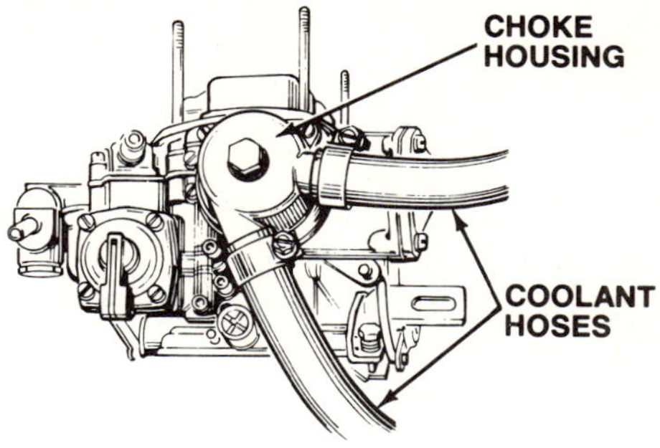

Figure 13-55. Hot-water-heated choke.

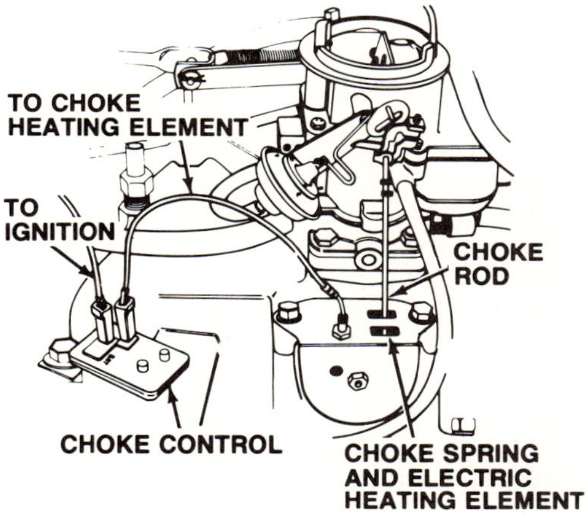

Figure 13-57. Chrysler electric-assist choke.

Dashpots were originally used to prevent an excessively rich mixture on deceleration, which can cause stalling. They now work as emission control devices on late-model cars, reducing HC emissions on deceleration.

Choke Heaters

Choking the engine can only be done for a brief time to keep emissions low. This can be done in several ways, all of which apply heat to the thermostatic spring to warm it up quickly and cause choke release.

Hot water choke

One way to apply heat to the thermostatic spring is by routing part of the engine coolant through the choke housing, figure 13-55. Once the engine reaches normal operating temperature, coolant heat helps release the choke. Older Ford models used an external coolant bypass hose held against the choke housing cap by a spring clip. This supplemented the exhaust manifold air passed through the choke housing.

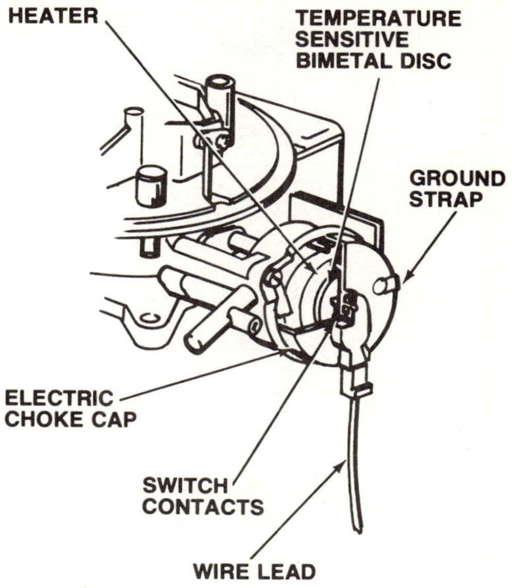

Figure 13-56. Ford and AMC electric heater for the automatic choke.

Electric-assist chokes

Electrically assisted heater elements are used on many late-model cars. These speed up the choke opening when underhood temperature is above about 60° F (16° C) by heating the choke bimetal thermostatic spring.

AMC and Ford choke heater elements are located in the thermostatic spring housing on the side of the carburetor, figure 13-56. Electric current is constantly supplied from the alternator directly to the choke cover temperature sensing switch. At underhood temperatures above 60° to 65° F (16° to 18° C), the switch closes to pass current to the choke heater. The circuit is grounded through a strap connected to the carburetor.

The Chrysler electric-assist choke heater is located in the intake manifold choke well and is regulated by a control switch which receives power from the ignition switch, figure 13-57. The control switch energizes the heater at temperatures above 63° to 68° F (17° to 20° C), and deenergizes it when the switch warms to between 110° and 130° F (43° to 55° C).

A 2-stage heater control, introduced on some cars in 1975, provides three levels of heat, depending on outside temperature. The low heat level is provided by a resistor on the control switch.

GM cars from 1975 on may use an electric choke cover heater. This 2-stage heater receives

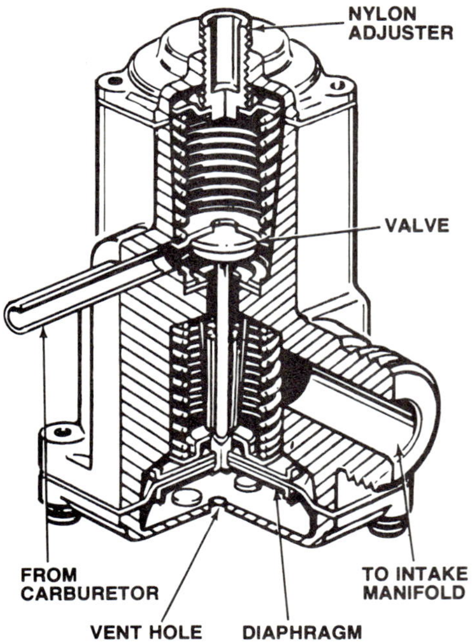

Figure 13-58. Fuel deceleration (decel) valve.

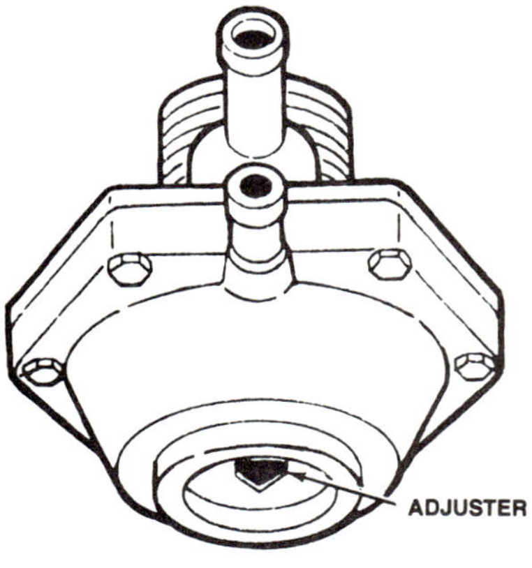

Figure 13-60. In 1974, the decel valve was redesigned. Operation remains unchanged. (Ford)

current from the engine oil pressure switch. Below specified air temperatures (usually 50° to 70° F) (10° to 20° C), a bimetal sensor in the cover turns off current to the larger heater stage. Both stages operate at higher temperatures. This choke heater receives current as long as the engine is running.

Fuel Deceleration (Decel) Valve

A decel valve, figure 13-58, is used with some Ford 4-cylinder and V-6 engines to momentarily

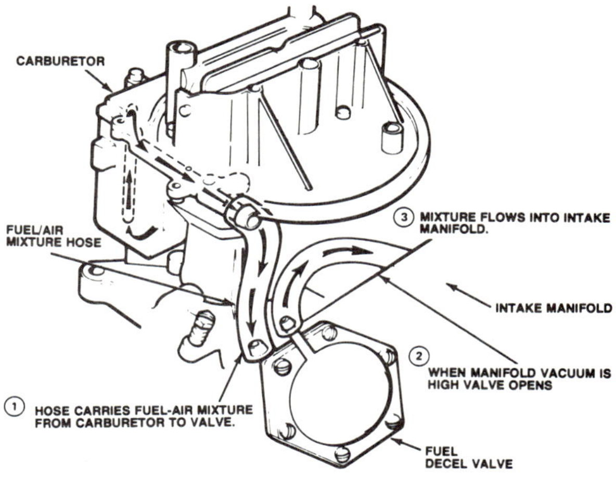

Figure 13-59. Decel valve installation. (Ford)

provide extra fuel and air during deceleration. This prevents cylinders from misfiring during deceleration and sending an unburned charge of hydrocarbons through the exhaust. The extra air and fuel provided by the valve ensure total combustion. The valve is located on or near the intake manifold. One end of the valve is connected to the manifold, either directly or by a hose. The other end connects to a deceleration section inside the carburetor. This contains a fuel pickup tube and an air bleed.

The decel valve operates only during deceleration, when manifold vacuum is high enough to operate the spring-loaded diaphragm and open the valve, figure 13-59.

Some decel valves are adjustable. A plastic screw in the valve top can be adjusted to control the valve's opening time and duration. A round valve design, figure 13-60, was first used on 1974 2,300-cc engines. For that model year, it was not adjustable, but it was modified to permit adjustment on 1975 and later engines.

Decel valves on 1975 and later V-6 engines use a speed lockout to shut off valve operation at speeds below 11 mph. The lockout system consists of a vacuum solenoid, an electronic control module, and a speed sensor in the speedometer cable. Decel valves on 1977 Mustangs with V-6 engines are interconnected with the air conditioner, operating only when the air conditioning is on.

Deceleration Throttle Opener

This is a solenoid which holds the throttle open during deceleration to prevent too much fuel from being pulled through the idle system by high manifold vacuum. It is used on catalytic-converter-equipped cars to prevent converter overheating. GM cars with the combination emission control (CEC) system use a throttle-positioning solenoid that is energized on deceleration. Some Chrysler Corporation cars with converters use a solenoid to keep the throttle from closing completely at engine speeds above 2,000 rpm.

ALTITUDE-COMPENSATING CARBURETORS

Earlier in this chapter, you learned that atmospheric pressure is greatest at sea level. Suppose that you drive a car from a low elevation into an area where the elevation is 7,000 feet. As the altitude increases, atmospheric pressure decreases and less air enters the carburetor. This means that the air-fuel mixture passing into the engine becomes richer as altitude increases. The result is poor driveability and high CO emissions at the higher elevation. For the driver who is only passing through high-altitude areas, the poor driveability is mostly a temporary inconvenience. But for driving at that elevation for an extended time, the car will run better if the engine is tuned for high elevation by leaning the air-fuel mixture.

However, if the retuned engine is driven back to lower elevations without once again adjusting the mixture, performance will suffer. As the car descends from the higher elevation to sea level, the air-fuel mixture receives increasing amounts of air, leaning the mixture too much. If the engine is to operate properly at the lower elevation, it will have to be retuned to restore the proper air-fuel mixture. Unfortunately, it is not always possible or even desirable to tune the engine for such driving conditions. As a result, performance is bad, driveability is also bad, and emissions are high under such circumstances.

To keep the right air-fuel mixture while the car is driven in an altitude other than that for which it is adjusted, GM and Chrysler introduced the altitude-compensating carburetor on some 1975 models. In the 1977 model year, the U.S. Federal EPA designated 167 counties in 10 western states (not including California) as high-altitude emission control areas. These counties are entirely above 4,000 feet in elevation. These high-altitude emission control requirements were suspended in 1978 but will be reintroduced in a modified form in 1980.

Because the major problem at high altitude is CO emission due to richer mixtures, the heart of the special emission controls is the carburetor. High-altitude carburetors have auxiliary systems or devices to provide more air or less fuel when operating at higher elevation than when

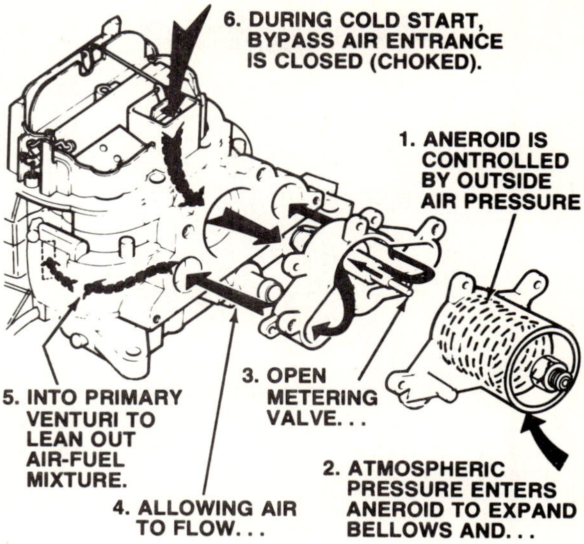

Figure 13-61. Motorcraft 2150 and 4350 automatic altitude compensator.

operating at sea level. Many of the altitude-compensating systems are automatic, responding to changes in atmospheric pressure. Other devices require manual adjustment or operation.

Automatic Compensation

The most widely used altitude-compensating device is the aneroid bellows. An aneroid bellows is an accordion-shaped bellows that responds to changes in atmospheric pressure by expanding and contracting. As pressure decreases at high altitude, the bellows expands.

Aneroid bellows are used on the following carburetors to open auxiliary air tubes into the main metering circuit at high altitude:

- Motorcraft 2150, 2-barrels, used on some AMC and Ford V-8's
- Motorcraft 4350, 4-barrels, used on other Ford V-8's
- Some Carter Thermo-Quads, used on Chrysler V-8's

The air bypass passage on the Motorcraft carburetors has its own air intake and choke valve, figure 13-61. The aneroid bellows on these carburetors have adjusting screws and locknuts that appear to be for adjusting the tension on the bellows. However, these screws are for original factory adjustments only, and should not be changed while overhauling the carburetor. The bellows and air valve can be removed for carburetor cleaning without upsetting the adjustment.

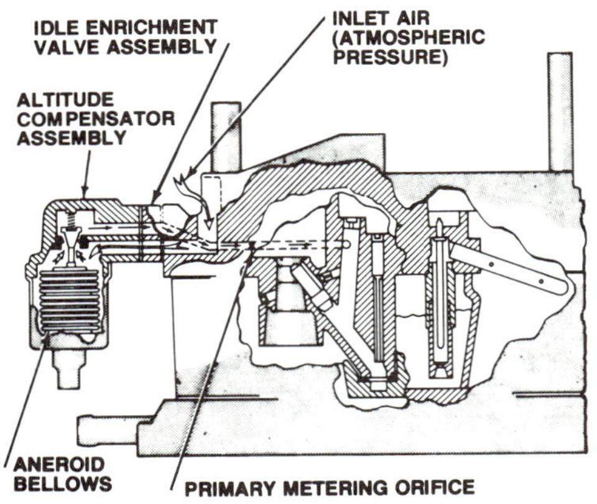

Figure 13-62. Carter Thermo-Quad automatic altitude compensator. (Chrysler)

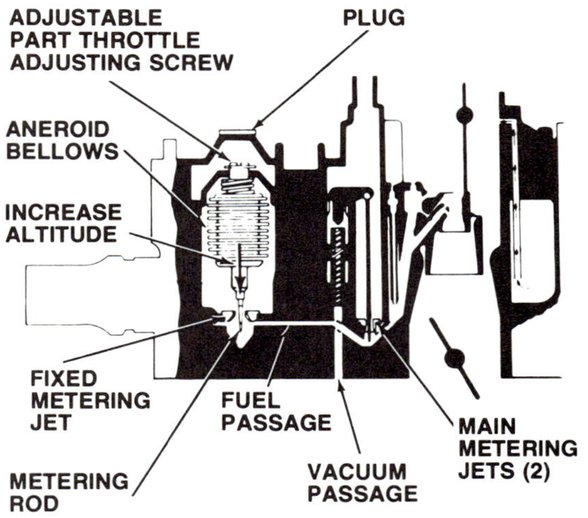

Figure 13-63. Rochester Quadrajet automatic altitude compensator.

The altitude compensator on the Carter Thermo-Quad, figure 13-62, is also automatic and requires no service. It was introduced on some 1975 cars, and its use has increased since then.

The aneroid bellows is also part of the Rochester M4MEA Quadrajet used on 1976 and later high-altitude Cadillacs. On this carburetor, the aneroid bellows controls the position of the metering rods in the primary main jets, figure 13-63. At high altitude, the bellows expands and moves the rods into the jets to reduce fuel flow and keep the mixture from getting overly rich.

Manual Compensation

American Motors' high-altitude 6-cylinder engines use a Carter YF 1-barrel carburetor with a manually adjustable auxiliary air bleed. The adjustment plug is located on the side of the airhorn, near the fuel inlet, figure 13-64.

For operation above 4,000 feet, use a screwdriver to turn the plug fully *counterclockwise*. This opens an extra air passage for the idle and main metering circuits. Below 4,000 feet, the plug must be turned fully *clockwise* to close the air passage. The ignition timing must be readjusted whenever this carburetor compensator is reset.

Carter BBD 2-barrel carburetors used on 318-cid Dodge and Plymouth high-altitude truck engines in 1977 also have manually adjustable air bleeds. The adjustment screw for this air bleed is inside the airhorn, above the venturi clusters, figure 13-65. For high-altitude operation turn the screw counterclockwise from the lightly seated position. For operation below 4,000 feet, lightly seat the screw to close the air bleed.

Holley 5200 2-stage, 2-barrel carburetors have a driver-controlled fuel valve to compensate for changes in altitude, figure 13-66. These carburetors are used on Ford's 2.3-liter engine in Pintos, Bobcats, Mustangs, and Capris sold at high altitude in 1977.

The carburetor valve is controlled by a lever mounted below the instrument panel. The lever has two positions. When it is set at SEA LEVEL,

■ Funny-Looking Gaskets Aren't Funny

It's easy to overlook gaskets. They are small, not too expensive, and they don't seem very important.

Don't you believe it. Gaskets that don't fit right will not work, and a bad gasket can rob your engine of a lot of power. Here is a tip from the professionals: be careful of how you store gaskets. Store them flat, never standing on edge. Don't hang them on a nail; that would most certainly pull them into funny shapes. Don't store gaskets near heat, since that also could cause them to warp into some interesting shapes. All non-metal gaskets have some water content, so they can't be allowed to dry up. If you do have a dry one, soak it in warm water for just a few minutes. Too much water will, again, cause it to warp.

The best thing to do is to store gaskets flat in a protective wrapper. Leave them in a drawer or cabinet where they will not be disturbed, and won't have other parts set on top of them.

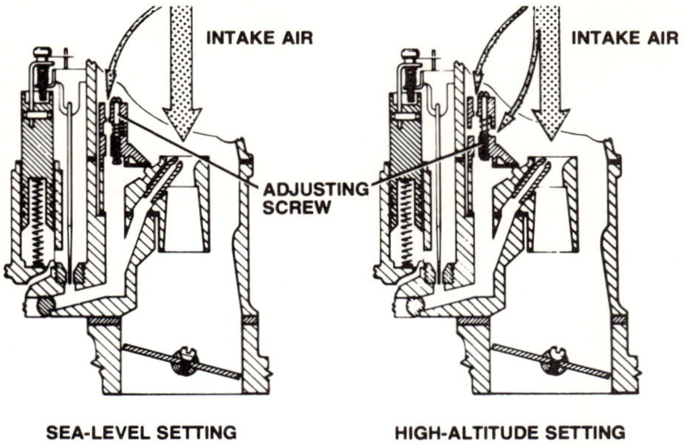

Figure 13-64. Carter YF altitude compensator. (AMC)

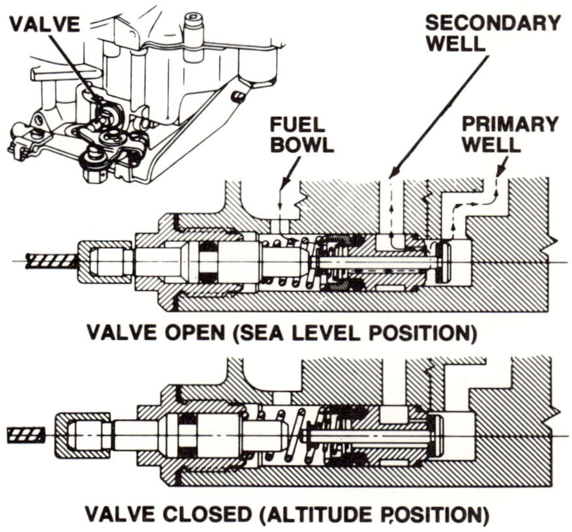

Figure 13-66. Motorcraft 5200 altitude compensator.

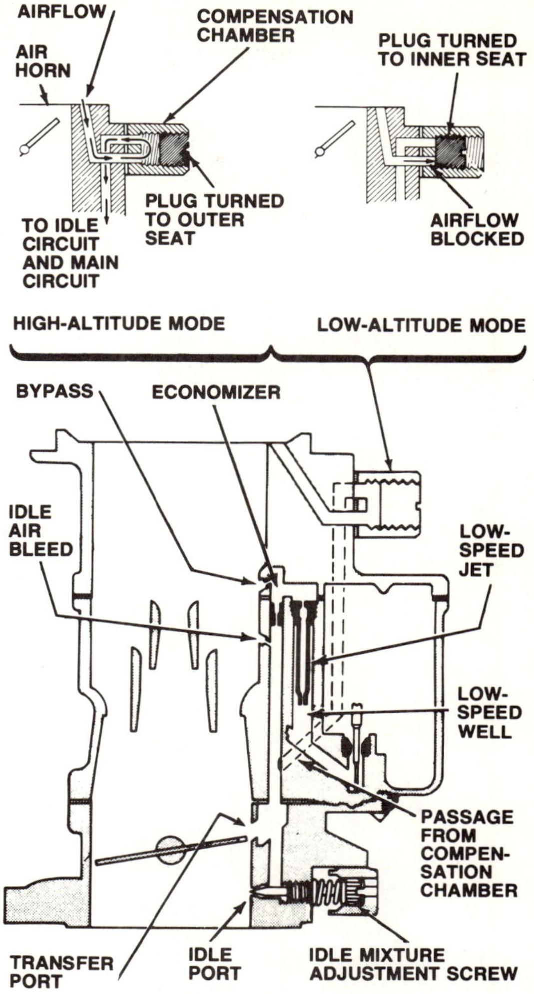

Figure 13-65. Carter BBD altitude compensator. (Chrysler)

the fuel valve in the base of the carburetor is opened to meter extra gasoline from the fuel bowl to the primary and secondary metering wells. When the lever is set at ALTITUDE, the fuel valve is closed to maintain a leaner air-fuel mixture.

INTAKE MANIFOLDS AND CARBURETOR-TO-MANIFOLD RELATIONSHIP

The vaporized air-fuel mixture flowing through the carburetor must be evenly distributed to each cylinder. The intake manifold does this with a series of carefully designed passages to connect the carburetor with the engine's intake valve ports, figure 13-67. To do its job, the intake manifold must be designed for efficient vaporization and air-fuel delivery.

Liquid gasoline is composed of various hydrocarbon compounds, which vaporize at different temperatures. If all of the hydrocarbons in gasoline vaporized at the same rate, the job of the intake manifold would be simple. But they do not, and so the intake manifold must be heated to keep the air-fuel mixture properly vaporized. This is done with exhaust manifold heat, engine coolant heat, or both, depending on engine type.

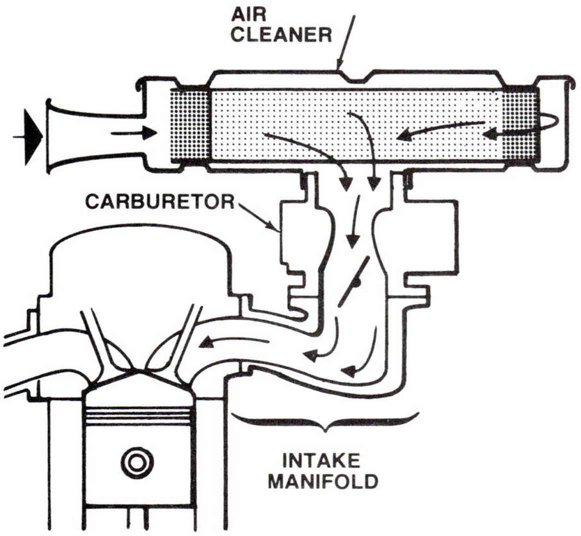

Figure 13-67. The intake manifold has passages that route the air-fuel mixture from the carburetor to the intake valve ports.

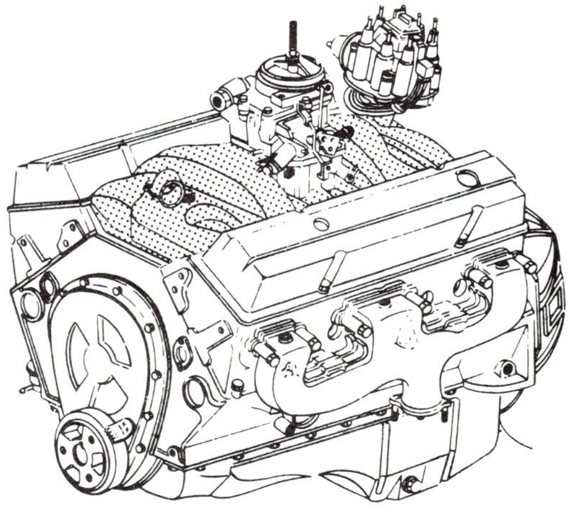

Figure 13-68. V-type engines normally have the intake manifold between the two banks of cylinders (shaded area in the illustration).(Chevrolet)

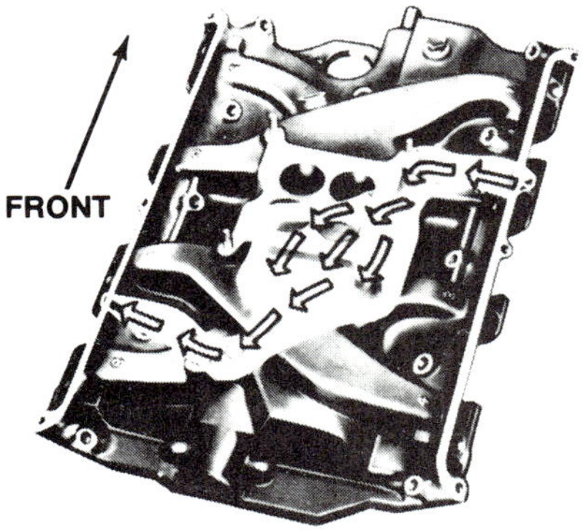

Figure 13-69. Exhaust gases are routed from ports in the cylinder heads through separate passages in the intake manifold to form the manifold hot spot. (Ford)

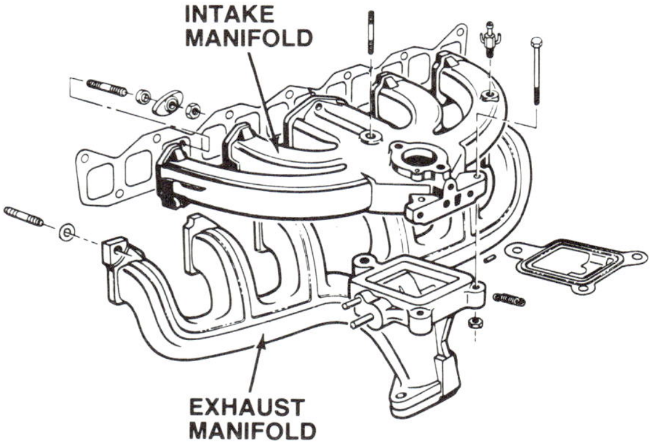

Figure 13-70. Most inline engines have both manifolds on one side of the engine. (Chrysler)

V-type engines have the intake manifold in the valley between the cylinder banks, figure 13-68. An exhaust crossover passage inside the manifold carries hot exhaust gases near the base of the carburetor, figure 13-69. In a few manifolds, engine coolant is routed near the carburetor to heat the air-fuel mixture.

Most inline engines have both manifolds on the same side of the engine, with the intake manifold on top of the exhaust manifold, figure 13-70. A chamber between the two fills with exhaust gases and creates a hot spot to improve fuel vaporization in the intake manifold passages. Some inline engines have the two manifolds mounted on opposite sides of the cylinder head. This is called a crossflow design. Coolant passages and a heat jacket on the intake manifold supply the heat.

Even with the help of heat from the exhaust manifold and engine coolant, the air-fuel mixture usually does not completely vaporize in the intake manifold. This results in an unequal mixture distribution among the cylinders, with some cylinders receiving more fuel and developing more power than others. This problem is greater during engine warmup, when less than normal heat is available to vaporize the fuel.

Causes Of Unequal Distribution

The air-fuel mixture reaching the engine cylinders may vary in amount and ratio for several reasons:

- The mixture flow is directed against one side of the manifold by the carburetor throttle valve, figure 13-71. To some extent, the choke plate influences flow in a similar manner.

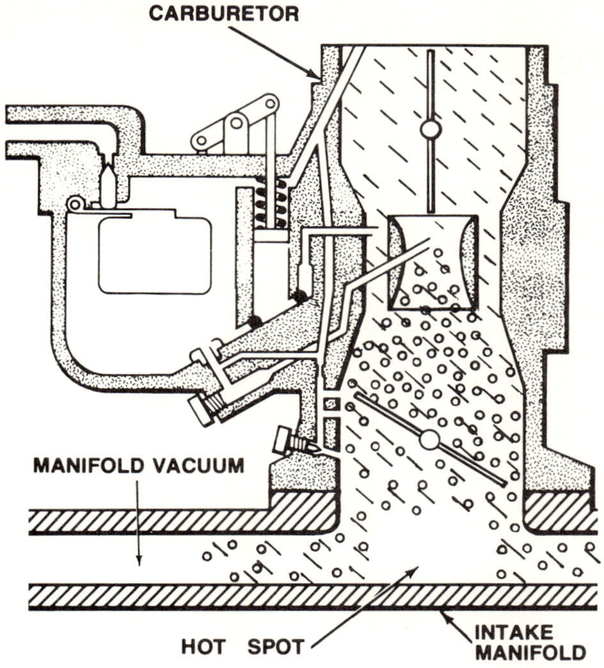

Figure 13-71. The angle of the carburetor throttle plate can affect the flow of the air-fuel mixture above the hot spot.

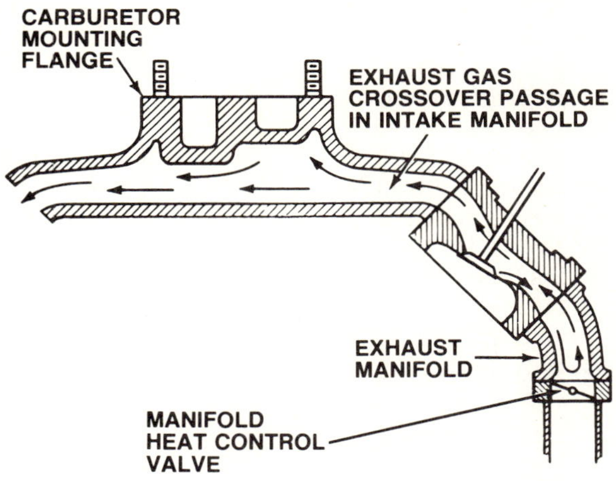

Figure 13-73. A thermostatic heat control valve or heat riser can be installed between the manifold and the exhaust pipe.

• Lighter particles of the air-fuel mixture will turn corners in the manifold easily, while heavier particles tend to continue in one direction, figure 13-72.
• Cylinders closer to the carburetor will receive a leaner mixture than will those farther from the carburetor. Carburetor and manifold design can minimize this, though.

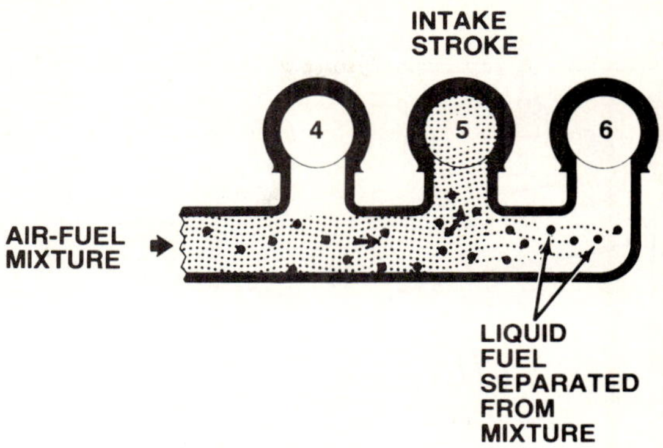

Figure13-72. An intake manifold with large passages and sharp angles will cause liquid fuel to separate out of the air-fuel mixture. (Chevrolet)

Depending upon manifold design, the use of a 2- or 4-barrel carburetor can improve mixture distribution, because each carburetor barrel supplies fewer cylinders.

The efficiency of a manifold is determined by the shape, surface, and size of its passages. They should be as short as possible, and with no sharp corners, bends, or turns to interfere with mixture flow. Smooth passage surfaces speed mixture flow. Rough surfaces aid vaporization by slowing down and breaking up mixture flow. Passages must be large enough in diameter to supply all cylinders with equal amounts of mixture. Passages that are too large or too small will slow down flow.

INTAKE MANIFOLD HEAT CONTROL

When the engine is cold, the incoming air-fuel mixture must be heated for complete vaporization. Hot exhaust gases are the most efficient heat source. A heat control valve, or heat riser, in the exhaust system routes a small part of the exhaust through passages in the intake manifold, in order to provide the heat source.

Heat control valves can be operated by:

• A thermostatic spring
• A vacuum diaphragm.

Each of these reacts to engine heat to control the valve operation, as we will see.

Thermostatic Heat Control Valve

Used by automakers for decades, this type of valve is located between the exhaust manifold and the exhaust pipe, figure 13-73, or in the manifold itself, figure 13-74. The valve is held closed by a thermostatic spring when the engine is cold, figure 13-75. This directs the hot exhaust gases around the intake manifold to preheat and help vaporize the air-fuel mixture, figure

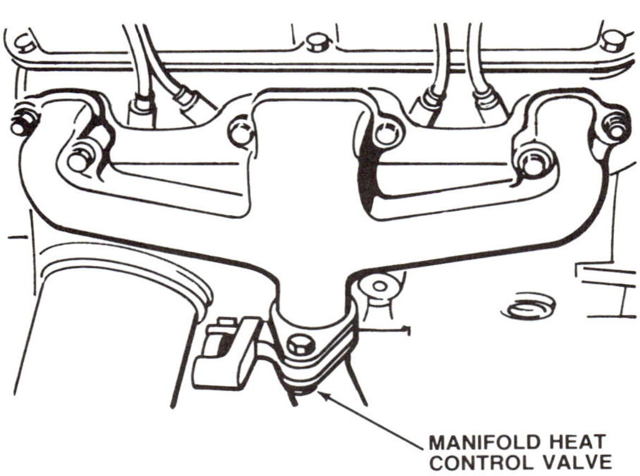

Figure 13-74. The heat control valve can also be within the manifold itself.

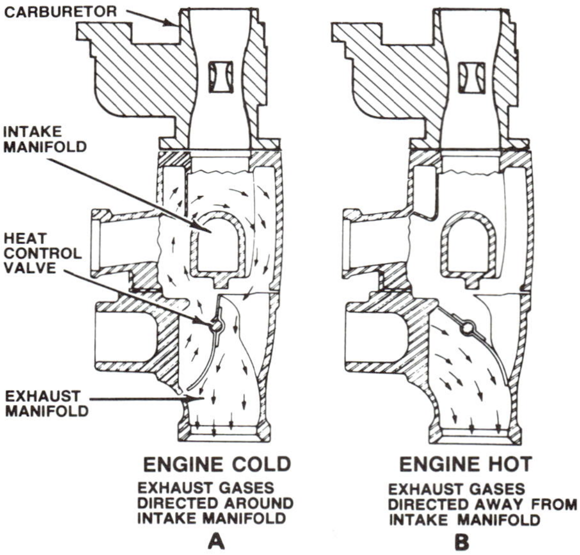

Figure 13-76. In this cross section of an inline engine manifold, the heat control valve forces the exhaust gases to flow either around the intake manifold (position A), or directly out the exhaust system (position B).

13-76, position A. As the engine warms up, the thermostatic spring unwinds. The counterweight and the pressure of exhaust gas open the valve. Exhaust gases now pass directly out through the exhaust system, figure 13-76, position B.

The thermostatic heat control valve has a tendency to stick, because the exhaust gases rust and corrode it. If the valve sticks open, it can cause increased fuel consumption, poor performance during warmup, and too much emissions because the choke stays on. If the valve

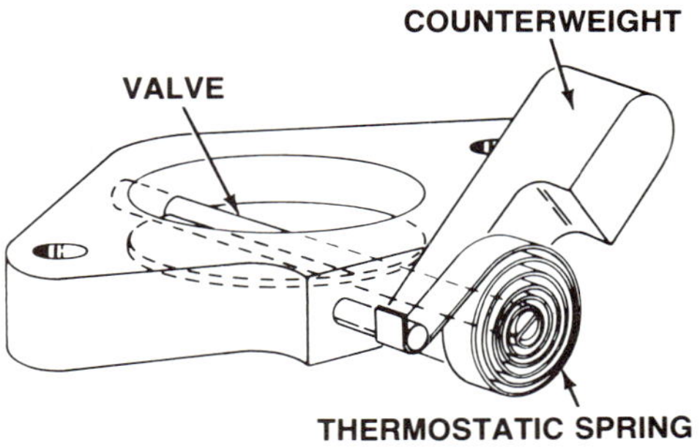

Figure 13-75. The manifold heat control valve can be operated by a thermostatic spring and a counterweight.

sticks closed, it can cause poor acceleration, a lack of power, and poor high-speed performance.

Vacuum-Operated Heat Control Valve

GM and Ford introduced vacuum-operated manifold heat control valves, figure 13-77, on some 1975 engines. Chrysler followed suit in 1977 with vacuum-operated valves on its California V-8 engines. AMC continues to use the thermostatic type on all engines. GM calls its device an early fuel evaporation (EFE) valve; Ford's is called a vacuum-operated heat control valve (HCV); Chrysler's is a power heat control valve. All work in the same way to provide more precise control of the manifold heat and to reduce emissions while improving driveability.

A rotating flapper valve is contained in a cast-iron body, figure 13-77. This valve body is installed between the manifold and exhaust pipe. The valve shaft extends through the valve body and is linked to a diaphragm in a vacuum motor. Intake manifold vacuum operates the vacuum diaphragm. The manifold vacuum source is controlled by a switch that reacts to either coolant temperature or oil temperature. Ford, Chrysler, and some GM engines have a thermal vacuum switch (TVS) installed in the cooling system; GM's inline 6-cylinder engines use an oil temperature switch.

When an engine is cold, vacuum is applied to the heat control valve diaphragm, closing the valve. This directs exhaust gas through the intake manifold passage. As the engine reaches normal operating temperature, vacuum is shut off from the valve. In systems using a thermostatic vacuum switch, the TVS closes when coolant temperature reaches a specified level, figure 13-78. This blocks vacuum from the valve. Where an oil temperature switch is used, the rising oil temperature opens a normally closed

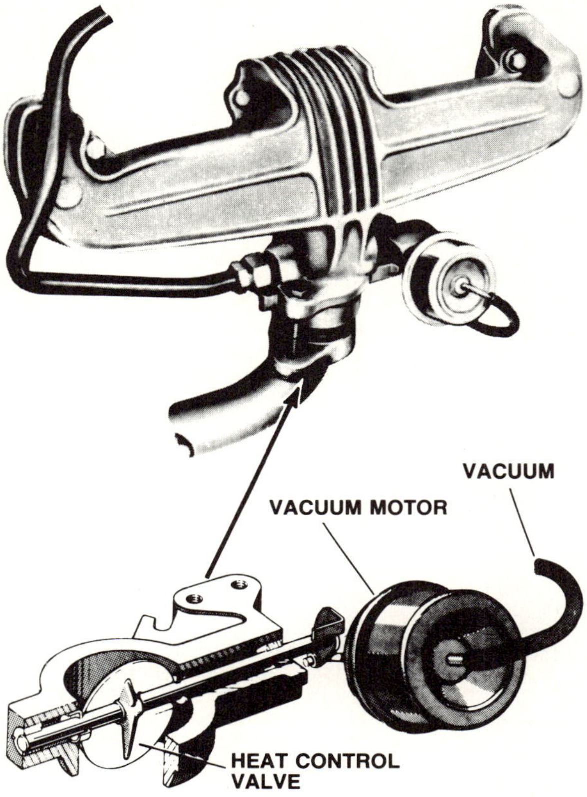

Figure 13-77. A vacuum-operated heat control valve installed in the exhaust manifold. (Chrysler)

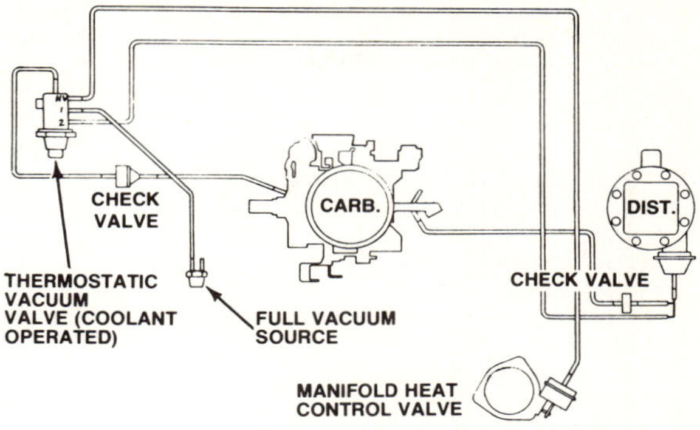

Figure 13-78. A vacuum diagram of the heat control valve installation. This valve responds to engine coolant temperature. (Buick)

thermostatic switch. This deenergizes the vacuum solenoid, and a spring in the diaphragm opens the valve. Exhaust now flows through the exhaust system instead of bypassing to the intake manifold.

SUMMARY

The carburetor is the all-important device that mixes the air with the fuel so the intake manifold can deliver the mixture to the combustion chambers. Carburetors must operate under all types of driving conditions, and in all temperatures. For this reason, they have a wide variety of assist devices, linkages, and control elements to regulate the air-fuel flow.

Yet all carburetors operate on the one basic principle of pressure differential: the air pressure inside the engine is less than it is outside the engine. This is because the downstroke of the piston creates more space for the air to fill in the cylinder. This sucks more air in through the carburetor and the intake manifold.

In spite of the differences between carburetor designs, most carburetors have the same seven basic systems of passages, ports, jets, and pumps. Venturis, or restrictions, help speed the airflow through the throat, and throttle valves help control the rate of flow.

Once the air-fuel charge leaves the carburetor, it enters the intake manifold on its way to the cylinders. Good manifold design is critical for smooth performance and even air-fuel mixture delivery. To promote better vaporization, a heat control valve is used to route exhaust gases through the intake manifold while a cold engine warms up.

Review Questions

Select the single most correct answer.
Compare your answers with the correct answers on page 280.

1. Which of the following is *not* true of air pressure:
   a. It is measured in millimeters of mercury (mm Hg)
   b. It is measured in pounds per square inch (psi)
   c. It is always constant
   d. It results from the weight of air pressing on a surface

2. Air pressure:
   a. Increases with height
   b. Decreases with height
   c. Remains the same regardless of height
   d. Increases with warming temperature

3. A pressure differential is created between outside air and engine air by:
   a. An increase in cylinder volume
   b. A decrease in cylinder volume
   c. Air rushing through the carburetor
   d. The intake manifold

4. The restriction in an airflow tube or barrel is called:
   a. A throttle
   b. An air bleed
   c. A vaporizer
   d. A venturi

5. The air pressure along the sides of a barrel in which there is an airflow:
   a. Is higher than at the center of the flow
   b. Is higher than atmospheric or outside pressure
   c. Is lower than at the center of the flow
   d. Increases with airflow velocity

6. Ported vacuum:
   a. Is the low-pressure area in the carburetor just above the throttle valve
   b. Is the low pressure beneath the throttle valve
   c. Is the air pressure in the venturi
   d. Is equal to atmospheric pressure

7. At idle:
   a. Venturi vacuum is high
   b. Manifold vacuum is high
   c. Ported vacuum is high
   d. All of the above

8. Which is *not* a fuel-metering system:
   a. Float system
   b. Idle system
   c. Power system
   d. EEC system

9. The carburetor float controls fuel level:
   a. By closing the needle valve when the level is low
   b. By opening the needle valve when the level is high
   c. By closing the needle valve when the level is high
   d. All of the above

10. At engine idle:
    a. Venturi vacuum is high
    b. Fuel enters the carburetor barrel above the throttle
    c. Air is provided by an air bleed in the idle tube
    d. The throttle is one-third open

11. Which is *not* used to provide extra air at idle:
    a. The throttle valve
    b. The choke
    c. Transfer ports
    d. Idle air bypass

12. At low off-idle speeds, extra fuel is provided by:
    a. The transfer port
    b. The main nozzle
    c. The idle air bleed
    d. The idle air adjust screw

13. In a high-speed system, better fuel and air mixtures are obtained with:
    a. Transfer ports
    b. A single venturi
    c. Multiple venturis
    d. None of the above

14. Power circuits are operated by:
    a. Vacuum diaphragms
    b. Vacuum pistons
    c. Mechanical metering rods
    d. All of the above

15. Which is *not* part of the accelerator pump circuit:
    a. The metering rod
    b. The inlet check
    c. The outlet check
    d. The duration spring

16. Which carburetor circuit makes the fuel mixture richer when starting an engine:
    a. Power circuit
    b. Choke circuit
    c. High-speed circuit
    d. Accelerator pump circuit

17. The illustration below shows:

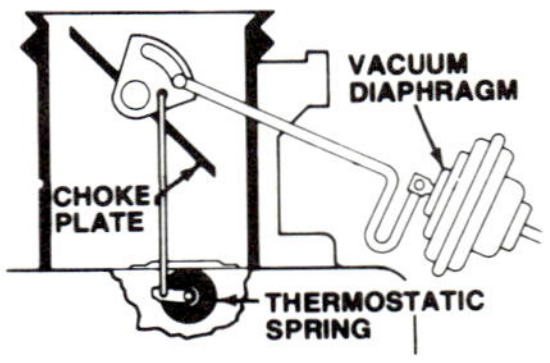

    a. A power valve diaphgram
    b. An integral choke
    c. A remote choke
    d. None of the above

18. The intake manifold is heated by:
    a. Exhaust manifold heat
    b. Engine coolant heat
    c. Both a and b
    d. Neither a nor b

19. Unequal distribution of the air-fuel mixture occurs because of:
    a. Fuel flow from the throttle
    b. Sharp turns in the manifold
    c. Distance of travel to the cylinders
    d. All of the above

20. Manifold heat control valves:
    a. Can be located between the carburetor and the intake manifold
    b. Can be located between the exhaust manifold and exhaust pipe
    c. Are open when the engine is cold
    d. None of the above

21. GM's thermostatic vacuum switch reacts to:
    a. Oil temperature
    b. Coolant temperature
    c. Exhaust gases
    d. The vacuum solenoid

Chapter

14

Electronic Fuel Injection

Automotive engineers have always known that a carburetor is neither really very accurate nor particularly fast at adapting to changing engine needs. But until stricter emission requirements forced greater efficiency, the carburetor was the accepted means of delivering the air-fuel charge to the combustion chambers.

Several times since the internal combustion engine was invented, other means of delivering the fuel have been used. Chief among these is fuel injection. However, early attempts to provide fuel injection were largely unsuccessful. Early injection systems used mechanical pumps to pressurize the fuel to as much as 750 psi (5, 171 kPa), but the pumps and designs were not capable of trouble-free operation.

In Germany in the 1930's and during the years of World War II, fuel injection for airplane use made rapid and spectacular advances. Yet it was not until the mid-1950's that fuel-injected automobile engines began to be mass produced. In 1957, some General Motors and Chrysler cars were sold with injected engines, but the high cost and poor reliability still did not make them popular with car buyers. Again in the early 1970's some "experts" were predicting that by 1975, every car made in America would be fuel injected. However, that year only one car — the Cadillac Seville — had fuel injection as a standard feature.

Cadillac still uses fuel injection, as do a number of foreign carmakers. Injected engines again seem to be a wave of the future, as the emission standards become stricter and the engineers refine their designs. In this chapter, we will study the systems designed by the Bendix Corp. and the Robert Bosch Corp.

FUEL INJECTION ADVANTAGES

Fuel injection systems offer several major advantages over carburetors:

- Injection can precisely match fuel delivery to engine requirements under all load and speed conditions. This reduces fuel consumption with no loss of engine performance.
- Because intake air and fuel are mixed at the engine port or in the combustion chamber, keeping an even mixture temperature is not as much of a problem with fuel injection systems. There is no need for manifold heat valves.
- The manifold in an injection system carries only air, so there is no problem of the air and fuel separating.
- By maintaining a precise air-fuel ratio according to engine requirements, exhaust emissions are lowered. The improved air-fuel flow in an injection system also helps to reduce emission levels.

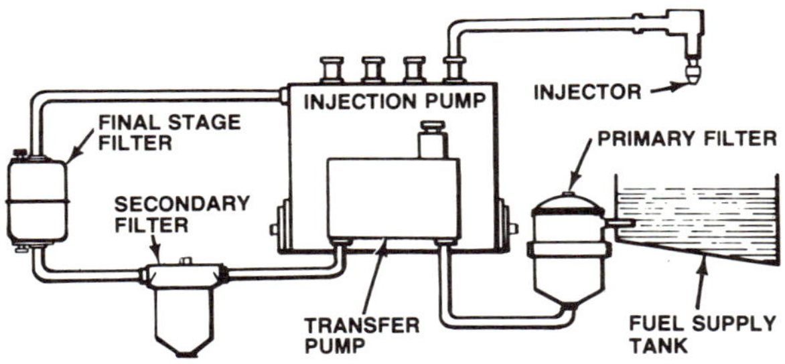

Figure 14-1. A typical diesel fuel injection system. (Fram)

MECHANICAL FUEL INJECTION SYSTEMS

Mechanical fuel injection is used with all diesel engines. Before solid-state electronic research made electronic fuel injection possible, mechanical systems were used with some gasoline engines. There are two general types of mechanical fuel injection systems:

- Combustion chamber injection
- Continuous port injection.

Combustion Chamber Injection

This type of fuel injection is used with diesel engines, figure 14-1. When air is compressed, it gets very hot very fast. With the high compression ratios used by diesel engines, air temperature in the cylinder can range from 1,000° to 1,650° F (380° to 740° C). This is hot enough to ignite a fine spray of fuel injected into the combustion chamber.

Fuel is drawn from the tank by a transfer pump and then sent to the injection pump. The injection pump boosts system pressure before delivering the fuel through separate lines to the injectors.

During the compression stroke, air temperature increases greatly. At the start of the power stroke, fuel is injected under pressure directly into the hot, highly compressed air in the cylinder, and combustion takes place, figure 14-2. Combustion chamber injection can also be used with gasoline engines, but today's emission and fuel economy requirements make it impractical.

Continuous Port Injection

Most mechanical injection systems for gasoline engines use continuous port injection. A mechanical pump delivers fuel to a manifold near the intake ports. The air and fuel are drawn into each cylinder with each intake stroke. Because

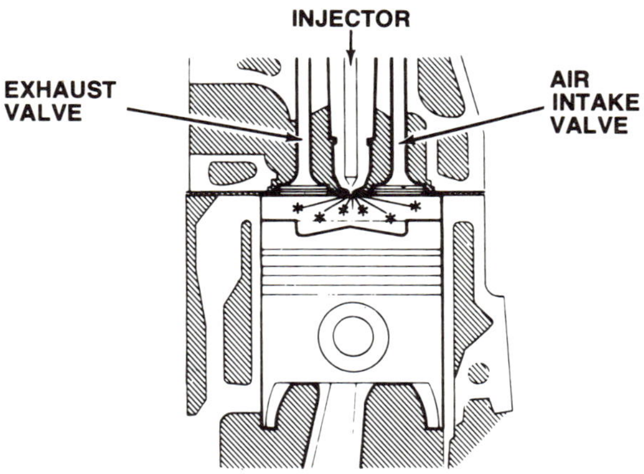

Figure 14-2. Diesel combustion occurs when fuel is injected into the hot, highly compressed air in the cylinder. (Cummins)

the system depends upon engine vacuum, it is difficult to control the fuel flow at idle and low-speed operation.

Two common continuous port mechanical injection systems are the Rochester Ramjet and the Bosch Continuous Injection System (CIS).

ELECTRONIC FUEL INJECTION

Mechanical injection systems are suitable for diesel engines, but not for most gasoline engines. Until recently, the injection systems were not accurate or reliable enough, and were too expensive, to be used widely.

In the early 1950's, the Bendix Corporation began work on electronic fuel injection systems. They used vacuum-tube electronics, and again proved unsuitable for mass production.

With the invention of solid-state semiconductors in the early 1960's, EFI became practical. Bendix holds many patents covering basic EFI technology, and all EFI systems made today are produced under license from Bendix. Three commonly used systems are Bosch's EFI-D and EFI-L, and the Bendix system used by Cadillac.

Each of these uses continuous port injection, and each has the same four basic subsystems: the fuel delivery, the air induction, the sensor, and the electronic control unit subsystems.

Instead of a mechanical metering valve controlling the amount of fuel entering the manifold, the electronic systems have solenoid-operated injectors controlled by computer. Engine sensors feed information on operating conditions to the computer, which processes the information to find the exact amount of fuel needed.

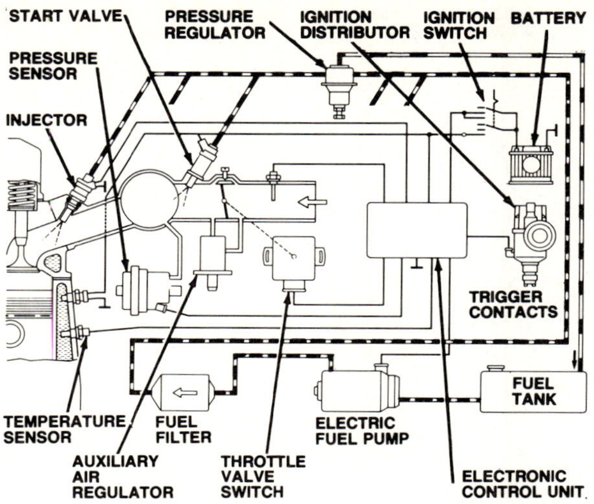

Figure 14-3. A schematic diagram of Bosch's EFI-D system. (Bosch)

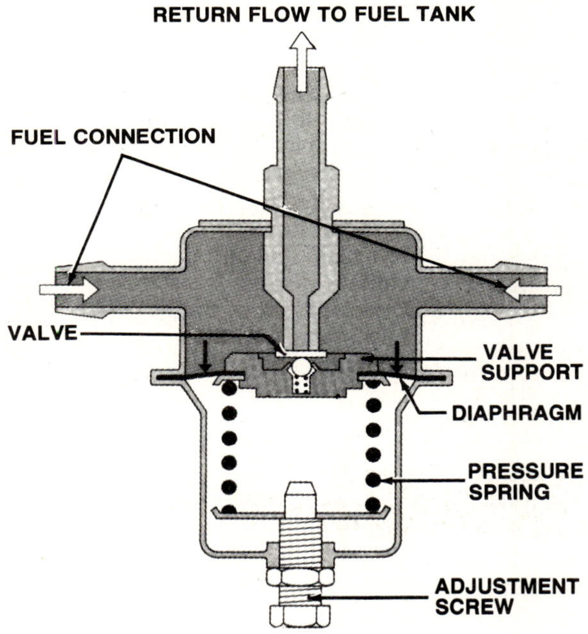

Figure 14-5. This adjustable pressure regulator assures the same fuel pressure at each EFI-D injector. (Bosch)

Bosch EFI-D System

This is an intermittent (start-and-stop) low-pressure injection system, figure 14-3. It is used on 1968-73 Volkswagens, as well as some Saab, Mercedes, and Volvo models. It is controlled by intake manifold pressure and engine speed.

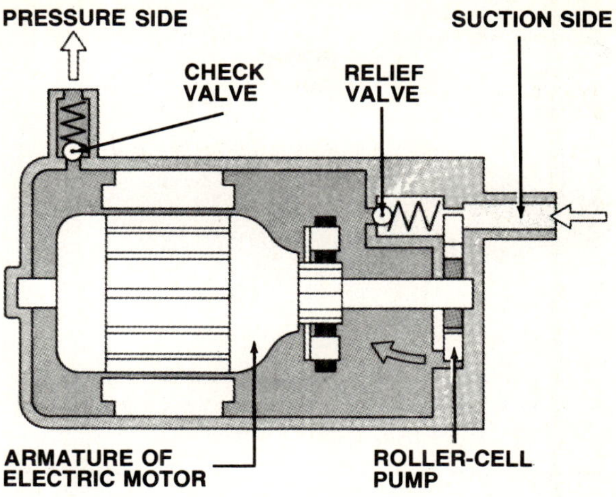

Figure 14-4. The Bosch EFI-D system uses an electric fuel pump, shown in cross-section. (Bosch)

Fuel delivery subsystem

An electric fuel pump, figure 14-4, delivers fuel from the tank to the solenoid-operated injection valves under pressure. Fuel pressure at the injectors is maintained by an adjustable pressure regulator, figure 14-5. Excess fuel is returned to the tank at atmospheric pressure. This recirculation results in the delivery of cool fuel, and prevents vapor bubbles in the fuel system. A filter between the pump and the injectors removes any dirt or contamination which might plug the injector nozzle valves.

The injectors are located in the intake manifold and spray fuel in front of each intake valve.

■ Bosch's Mechanical Injection

The Bosch Continuous Injection System (CIS) is a hydraulically controlled, mechanical injection system. In the CIS system, air enters the engine through the air venturi and is measured by the airflow sensor. The fuel distributor receives fuel from an electric fuel pump. It meters fuel to each injector according to the measured airflow. In this way, the correct amount of fuel is delivered as required.

A cold-start valve injects extra fuel into the intake manifold for cold engine starts. A pressure regulator stabilizes fuel flow at engine operating temperature. A fuel warm-up regulator makes extra fuel available during engine warmup. The additional air required by engine warmup is supplied by the auxiliary air regulator.

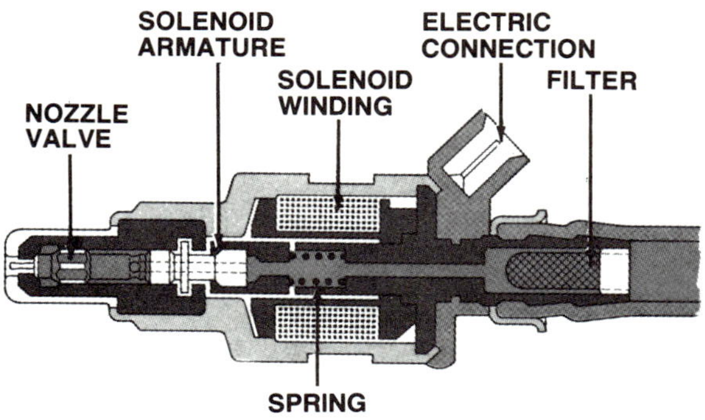

Figure 14-6. A cross-sectional view of the injector used by the Bosch EFI-D system. (Bosch)

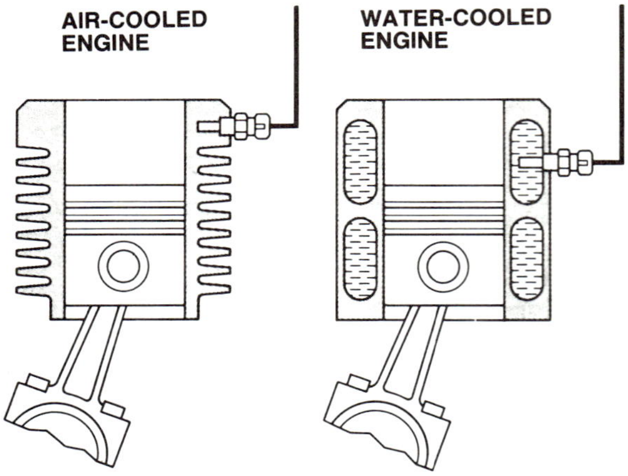

Figure 14-8. Temperature sensors installed in an air-cooled and in a water-cooled engine. (Bosch)

Each solenoid-operated injector, figure 14-6, consists of a valve body, a nozzle valve, a solenoid armature, and a filter. Electric signals from the control unit generate a magnetic field in the solenoid winding. The movable solenoid armature is drawn back, lifting the nozzle valve from its seat. Fuel under pressure passes through the nozzle valve channel until the solenoid valve is deenergized by the control unit. The armature spring then pushes the nozzle valve tightly against the valve seat to shut off fuel flow.

Injectors used in the EFI-D system are divided into two groups and each group is connected electrically in parallel and open at the same time. Only one group operates at a time.

Air induction subsystem

Combustion air is supplied to an air distributor through an air filter and throttle valve. This air distributor is a common intake manifold, figure 14-7. Individual equal-length intake manifold

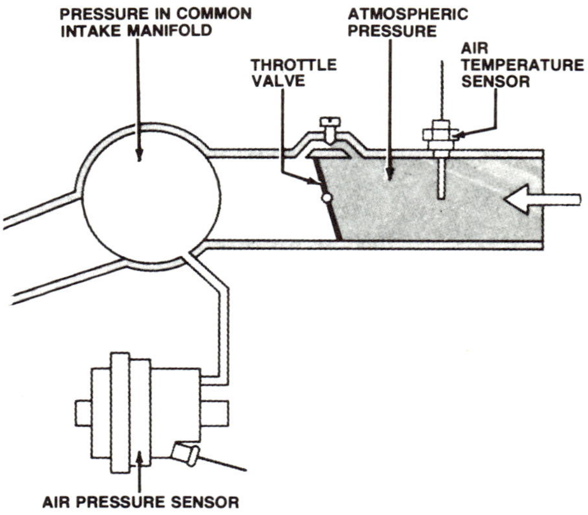

Figure 14-7. In the EFI-D system, the air distributor serves as a common intake manifold for each cylinder. (Bosch)

branches connect the air distributor with each cylinder. The throttle valve switch signals idle, acceleration, and full-load conditions to the control unit. A start valve sprays extra fuel into the intake manifold when the engine is started at low temperatures.

A second air regulator meters extra air for mixing with the additional fuel provided by the start valve. As engine temperature increases, the airflow through this second air regulator decreases. When engine coolant reaches normal operating temperature, the regulator is completely closed.

Sensor subsystem

A temperature sensor controls fuel enrichment for cold starting and engine warm-up. When the EFI-D system is used with an air-cooled engine, an engine temperature sensor is installed in the cylinder head, figure 14-8, and an air temperature sensor is installed in the intake manifold, figure 14-7. Water-cooled engines have the engine temperature sensor in the thermostat housing or water jacket to sense coolant temperature, figure 14-8.

The air pressure sensor is located in a sealed housing connected to the air distributor, figure 14-7. As engine load changes the manifold pressure changes, and the pressure sensor sends an electrical signal to the control unit. The length of this signal determines how long the injectors will remain open.

Exactly when the injectors will open is determined by a pair of trigger contacts located in the ignition distributor, figure 14-9, under the centrifugal advance mechanism. These contacts are

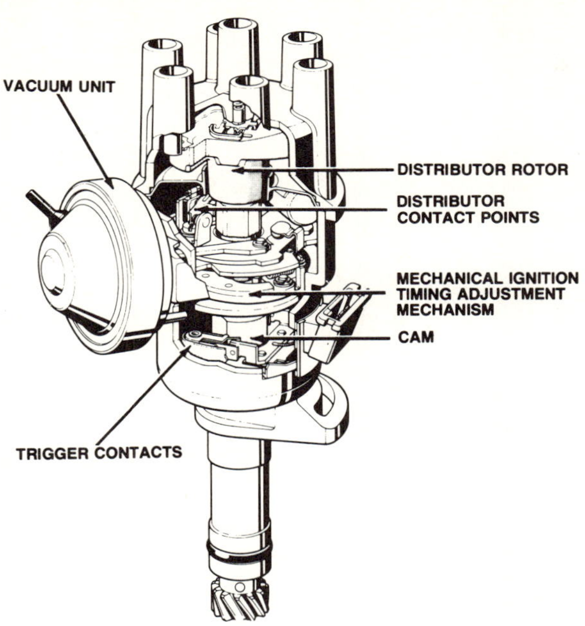

Figure 14-9. In Bosch's EFI-D system, fuel injection is triggered by a special set of contact points in the ignition distributor. (Bosch)

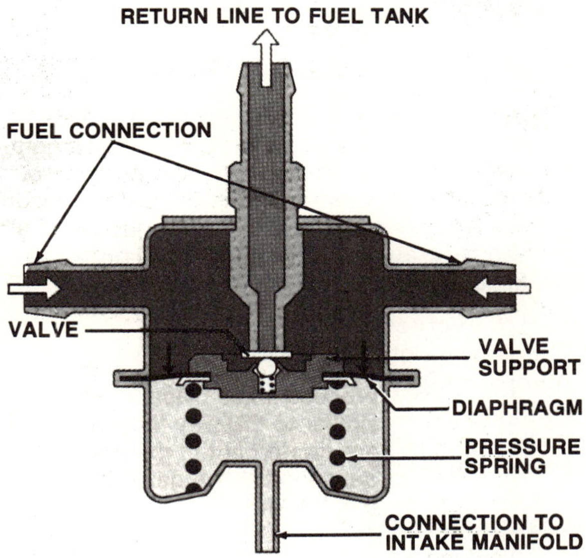

Figure 14-11. The EFI-L fuel pressure regulator is not adjustable. (Bosch)

operated by a cam on the distributor shaft, and send an electrical signal to the control unit to begin injection.

Electronic control unit
This programmed analog computer receives data about intake manifold pressure, intake air temperature, coolant and cylinder head temperature, movement and position of the throttle

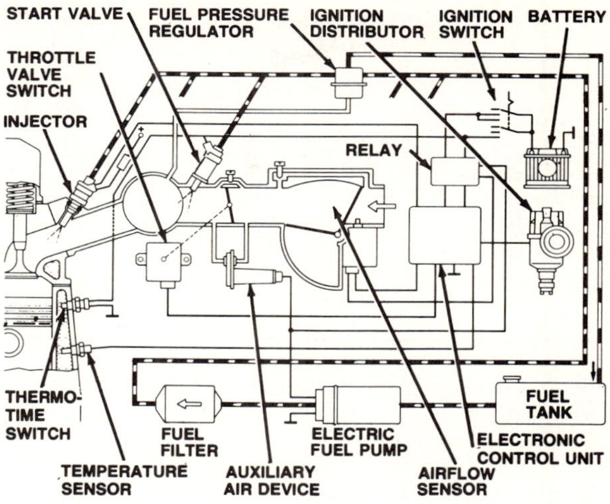

Figure 14-10. Bosch's EFI-L system responds to airflow instead of manifold pressure, as the EFI-D does. (Bosch)

valve, the starting cycle, engine speed, and the start of injection. The ECU processes this information to switch the fuel pump on and to control the length of injection.

Bosch EFI-L System

This is also an intermittently (start-and-stop) operating, low-pressure injection system, figure 14-10. EFI-L is a refinement of the EFI-D system and was developed by Bosch to reduce exhaust emissions. It first was used in 1974, when it replaced EFI-D on many foreign cars. With the EFI-L system, injection is controlled by the amount of air drawn into the engine, instead of engine manifold pressure as in EFI-D. This reduces emissions, and results in a simpler system with fewer parts. The EFI-L system is also known as the Jetronic-L or airflow-controlled system.

EFI-L uses the same basic subsystems as EFI-D, with some operating changes described in the following paragraphs.

Fuel delivery subsystem
The fuel pressure regulator, figure 14-11, is not adjustable. Its spring chamber is connected to the intake manifold. This keeps a constant difference between the intake manifold pressure and the fuel pressure. The pressure drop across the injectors is equal under all load conditions.

EFI-L injectors are about the same as those used with EFI-D but they have a smaller nozzle valve opening. This smaller opening is necessary since the injectors are combined into a

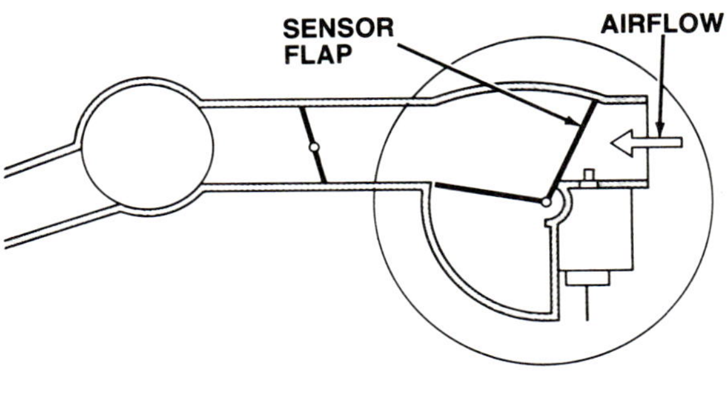

Figure 14-12. This airflow sensor generates an electric signal in proportion to the air drawn into the engine. (Bosch)

single group. All injectors operate at the same time. This means that one-half of the required fuel for an operating cycle is injected twice during each camshaft rotation. With this system, half the cylinders receive fuel for combustion on or near their intake cycle. The other half briefly stores the injected fuel in the intake manifold until their intake stroke occurs.

Air induction subsystem
The major difference in the EFI-L air intake system is the use of an airflow sensor, figure 14-12. This sensor generates an electric current in proportion to the air drawn into the engine. The air movement pushes against a spring-loaded sensor flap. A damping flap attached to the sensor flap reduces backpressure vibrations. The position of the sensor flap is measured by a **potentiometer**, figure 14-13. Since the airflow sensor measures only the fresh air drawn into the system, exhaust gas can be recirculated in the EFI-L system to lower the combustion chamber temperatures.

Sensor subsystem
Since all injectors operate at the same time, the trigger contacts in the ignition distributor are not necessary. Injector opening is triggered instead by every other opening of the distributor breaker points. The breaker points also send signals to the control unit, in order for it to count engine rpm. The temperature sensor in the EFI-L system is a temperature-time switch that controls the length of time the cold-enrichment is activated for starting.

Electronic control unit
The control unit receives battery voltage from a relay when the ignition is switched on. The same relay turns on the electric fuel pump, the start valve, the temperature-time switch, and the auxiliary air device. In the EFI-D system,

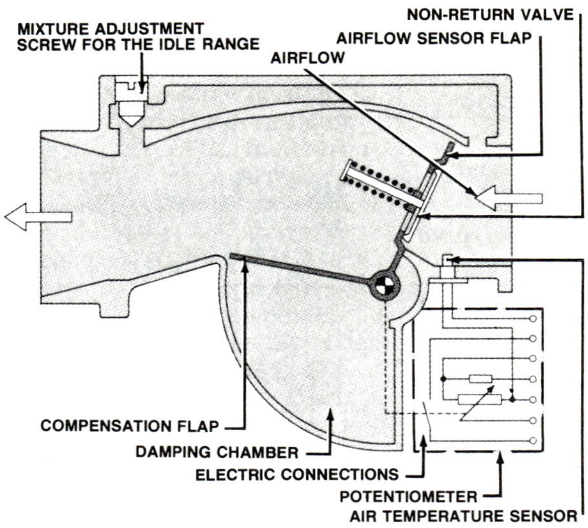

Figure 14-13. A schematic diagram of the airflow sensor used with Bosch's EFI-L system. (Bosch)

these relay jobs are done by the electronic control unit. In most other ways, the control units of the EFI-D system and the EFI-L system operate the same. Because the EFI-L system uses **integrated circuits**, fewer parts are required.

Cadillac EFI System

The EFI system made by Bendix and used by Cadillac since 1975 is similar to the Bosch EFI-D system. Intake manifold pressure and engine speed control its operation. The Cadillac EFI version also uses the four major subsystems, but the parts are different from those found in the Bosch system.

Fuel delivery subsystem
This consists of the fuel tank, two electric fuel pumps, a filter, the fuel rails (or tubes) with eight injectors, a fuel pressure regulator, supply lines, and return lines, figure 14-14. The fuel tank contains a special "bathtub" reservoir under the in-tank fuel booster pump. This ensures enough fuel pickup at all times.

The in-tank pump supplies fuel to the chassis-mounted main pump and prevents vapor lock on the suction side of the system. The chassis-mounted pump is a constant displacement type. A check valve in this pump prevents backflow and maintains fuel pressure when the pump is off. An internal relief valve protects against too much pressure. The pump is located under the car, ahead of the rear wheels. Both pumps are controlled by, and wired in parallel with, the electronic control unit (ECU). They are protected by a 10-ampere inline

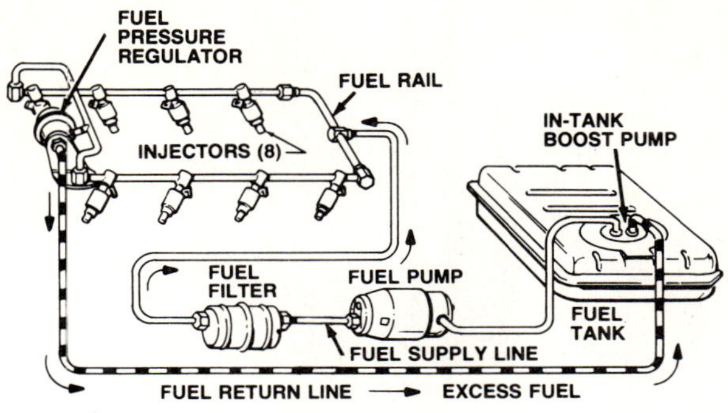

Figure 14-14. The Cadillac EFI fuel delivery subsystem. (Cadillac)

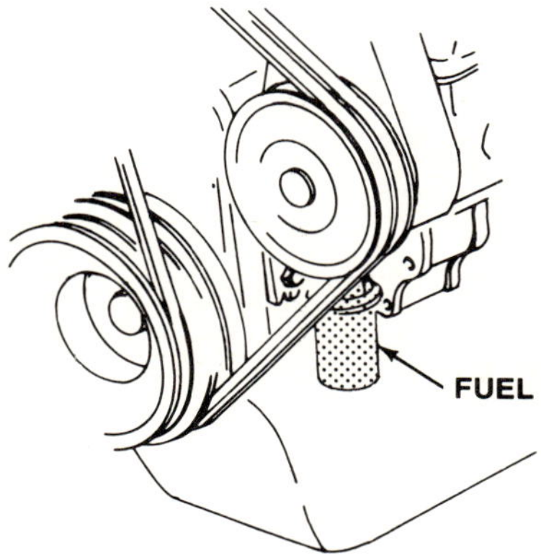

Figure 14-16. On Cadillac models other than the Seville, the fuel filter is mounted in place of the mechanical fuel pump.

fuse, located in the wiring harness near the ECU.

The fuel filter has a casing which contains a throw-away paper filter element. The filter is mounted near the fuel pump on the Seville, figure 14-15, and in place of the mechanical fuel pump, figure 14-16, on other Cadillac models.

The pressure regulator is mounted on the fuel rail at the front of the engine. It has air and fuel chambers separated by a spring-loaded diaphragm. The air chamber is connected to the throttle body by a hose. Pressure in the regulator is the same as that in the intake manifold, similar to the pressure regulator used in the more recent Bosch EFI-L system.

Manifold pressure changes cause the diaphragm valve in the pressure regulator to open or close the fuel chamber orifice. This keeps a constant 39 psi (271 kPa) fuel pressure differential across the injectors. Excess fuel is routed back to the bathtub reservoir in the fuel tank by a return line.

Figure 14-15. On the Cadillac Seville, the fuel filter is mounted near the electric fuel pump under the rear of the car.

The injectors are solenoid-operated valves controlled by the ECU. Like the EFI-D system, the eight injectors are divided into two groups. All four injectors in a group open at the same time. The two groups operate alternately. Each of the injector groups operates once for every camshaft revolution.

Air induction subsystem

This has a throttle body, a fast-idle valve, and an intake manifold, figure 14-17. Air enters the throttle body and is distributed to each cylinder through the intake manifold. The throttle body contains two bores with a throttle valve in each one. The valves are connected to the accelerator pedal linkage and are used to control primary airflow rate. A screw on the front of the throttle body adjusts warm engine idle speed.

An electrically operated fast-idle valve on top of the throttle body is connected to the fuel pump circuit of the control unit. It allows extra air to bypass the throttle when a cold engine is started. A heater in the valve warms a thermostatic element, which expands as it is heated. This closes a plunger in the air orifice. This reduces the flow of extra air, and engine rpm

Potentiometer: A variable resistor.

Integrated Circuits: Electronic circuits in which the parts have been miniaturized and assembled together to reduce the number of parts in the circuit.

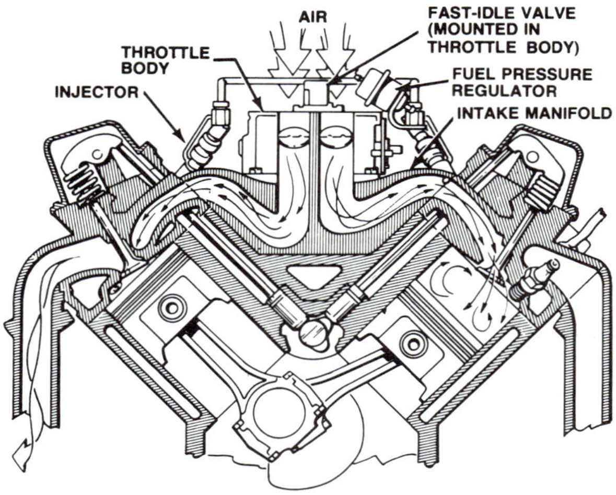

Figure 14-17. Cadillac's EFI intake manifold resembles the 2-place manifolds used with carbureted engines. (Cadillac)

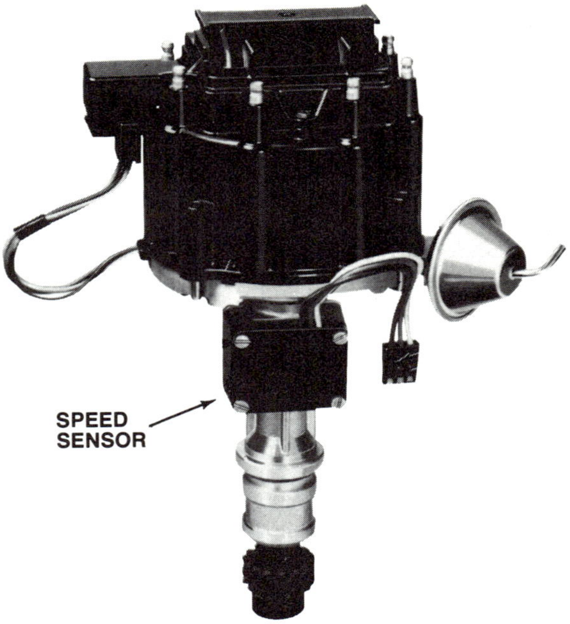

Figure 14-18. In the Cadillac EFI system, injection timing is controlled by a speed sensor mounted near the ignition distributor shaft. (Delco-Remy)

drops to the normal idle speed. The warmer the air temperature, the faster the valve will close to reduce engine speed.

The intake manifold looks like the 2-plane manifold used with a carbureted engine. Injectors are located in a port above each cylinder. This allows the fuel to be injected directly toward the top of the cylinder intake valve.

Sensor subsystem

Just as in the Bosch systems, sensors monitor various engine operating conditions, and transmit electrical signals to the control unit. The manifold absolute pressure (MAP) sensor is located inside the control unit and is connected to the throttle body by a vacuum line. It monitors manifold pressure changes which result from changes in engine speed, load, and atmospheric pressure. As manifold pressure increases, the sensor signals the control unit to increase the length of injection. A decrease in pressure causes a shorter injection time.

The throttle position switch on the throttle body is connected to the throttle shaft. This switch signals the control unit about the throttle shaft position. Two sensors provide the ECU with air and coolant temperatures. The air temperature sensor is located on the intake manifold. The coolant temperature sensor either is in the intake manifold or below the thermostat. The sensors are identical and interchangeable.

A speed sensor is built into the distributor, figure 14-18. This provides the same information as the cam-operated trigger contacts in the EFI-D system. The speed sensor is a combination external-internal unit. The external part is a plastic housing that contains two reed switches. This is attached to the distributor shaft housing. The internal unit is a rotor with two magnets attached to the distributor shaft. As the shaft rotates, the rotor with its magnets causes the reed switches to open and close, creating signals that time the correct injector group with intake valve timing. It also provides engine rpm information for fuel delivery.

Electronic control unit

The ECU is a programmed analog computer housed in a metal case and located inside the passenger compartment. Control units are adjusted differently according to car model, and cannot be interchanged. The ECU receives its power from the battery and its signals from the engine sensors. It processes these signals to determine the air-fuel requirements for all driving conditions.

When the ignition is turned on, the ECU turns on the fuel pumps, all the engine sensors, and the fast-idle valve. If the engine is not cranked or started, the pumps and fast-idle valve will shut down after about one second. As the engine is cranked, the ECU receives signals from the sensors and begins to activate the right systems. The EGR solenoid is activated to block the vacuum signal to the EGR valve until coolant temperature reaches a certain level. On some California engines, a vacuum retard solenoid is activated by the same signal used to control the EGR valve operation. This sends vacuum to the retard side of the distributor diaphragm. Once the engine is started, the ECU continues to receive and process sensor signals to control injector duration.

SUMMARY

Fuel injection offers several advantages over carburetion, such as more precise mixture control, better mixture distribution, and lower emissions. Mechanical injection systems have seldom been used on mass-production gasoline engines because of cost and complexity. However, solid-state electronics have made electronic fuel injection practical.

EFI systems are controlled by either manifold pressure or airflow. All contain the same basic subsystems, but may use different components. Engine sensors feed basic information on engine speed, load, and manifold pressure or airflow to an electronic control unit. The ECU determines air-fuel requirements for the engine's operating conditions. It controls the time during which the injectors remain open. The instantaneous response of electronics in an EFI system makes it far more responsive to the engine's needs than a carbureted system can be.

Review Questions

Select the single most correct answer.
Compare your answers with the correct answers on page 280.

1. Diesel engines use:
   a. Either combustion chamber injection or continuous port injection
   b. Only combustion chamber injection
   c. Only continuous port injection
   d. Only electronic fuel injection

2. A major drawback to continuous port injection is that:
   a. It is hard to control fuel flow at idle and low speeds
   b. It is hard to control fuel flow at high speeds
   c. It is hard to control fuel flow at low ambient temperatures
   d. It is hard to control fuel flow at high ambient temperatures

3. Which of the following is *not* a part of an EFI system:
   a. The fuel delivery subsystem
   b. The transfer pump subsystem
   c. The air induction subsystem
   d. The sensor subsystem

4. 1968-73 Volkswagens, as well as some Saabs, Volvos, and Mercedes, used:
   a. The Bosch EFI-D system
   b. The Bosch EFI-L system
   c. The Bendix system later used by Cadillac
   d. Both a and b

5. Which of the following is *not* true:
   a. Both the EFI-D and the EFI-L are intermittently operating
   b. Both the EFI-D and the EFI-L are low pressure systems
   c. Both the EFI-D and the EFI-L control injection through manifold pressure
   d. Both the EFI-D and the EFI-L use solid-state electronics

6. Which of the following is *true*:
   a. The EFI-D uses two electric fuel pumps
   b. The EFI-L uses two electric fuel pumps
   c. The Cadillac EFI uses one mechanical fuel pump and one electric fuel pump
   d. The Cadillac EFI uses two electric fuel pumps

7. Shown below is:

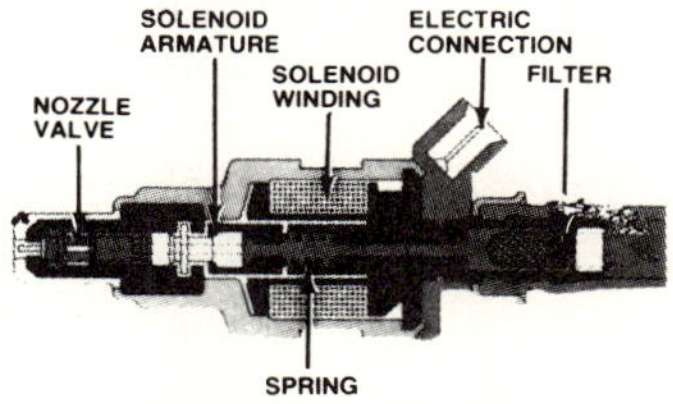

   a. A vacuum breaker switch
   b. A fuel injector
   c. A pressure sensor
   d. An adjustable pressure regulator

8. Which of the following is *true* for EFI injectors:
   a. EFI-D and EFI-L injectors are interchangeable
   b. EFI-L injectors are located in the intake manifold
   c. EFI-D injectors are solenoid-operated
   d. The Cadillac injectors are divided into four groups

9. Cadillac has used EFI since:
   a. 1969
   b. 1971
   c. 1973
   d. 1975

10. A major difference between Bosch EFI-D and Cadillac EFI systems is that:
    a. Only the Cadillac system uses an ECU
    b. The subsystems are similar but the parts are different
    c. The EFI-D system uses intake manifold pressure and engine speed to control its operation; the Cadillac system does not
    d. Only the EFI-D system uses the MAP sensor

15 Supercharging and Turbochargers

Engines with carburetors rely upon atmospheric pressure to push an air-fuel mixture into the combustion chamber vacuum created by a piston downstroke. The mixture is then compressed before ignition to increase the force of the burning, expanding gases.

The greater the mixture compression, the greater the power resulting from combustion. One way to increase mixture compression is with a high engine compression ratio. In the late 1960's, compression ratios on high-performance auto engines reached 11:1. High compression ratios have two major benefits. First, volumetric efficiency is improved because the piston displaces a larger percentage of the total cylinder volume on each intake stroke. Second, because temperature increases as pressure increases, thermal efficiency is higher with a high-compression engine.

Since 1971, emission control requirements have brought compression ratios down to the range of 8:1 or 8.5:1, because higher-compression engines tend to emit too much NO_x. Lower compression ratios have also been required because of the reduced lead content in gasoline.

SUPERCHARGING

Another way to increase mixture compression is to deliver the air-fuel charge to the cylinders at greater than atmospheric pressure. This is called **supercharging** the engine. The result is much the same as a high compression ratio, but the effect can be controlled during idle and deceleration to avoid high emissions. These control methods will be explained later in the chapter.

Boost

In a supercharger, air is pressurized to greater than atmospheric pressure. This amount of pressurization above atmospheric pressure is called **boost**. Boost can be measured in the same way as atmospheric pressure. Normal sea-level atmospheric pressure is 14.7 psi, or 760 mm Hg. Some superchargers almost double that pressure, providing about 12 psi or 600 mm Hg boost *above* atmospheric pressure. Superchargers used on racing engines can double or even triple the horsepower of a **normally aspirated** engine. On passenger cars, where reliability and driveability are of more concern, the horsepower increases by about 50 percent. For example, the 1978 Buick V-6 engine with a 2-barrel carburetor has a 40 percent increase in horsepower when turbocharged (another form of supercharging). The same engine, with a 4-barrel carburetor, gains about 60 percent horsepower when turbocharged.

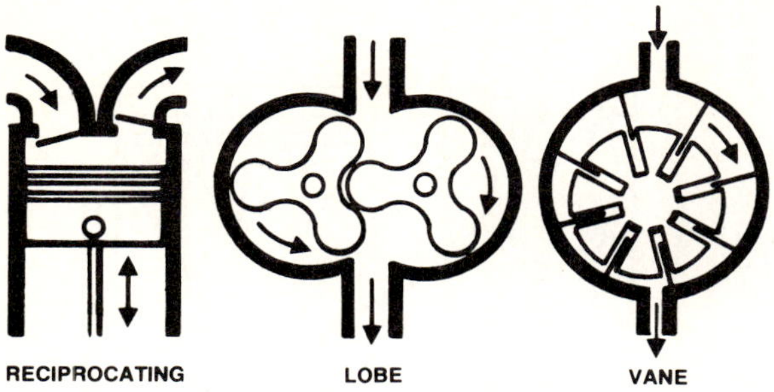

Figure 15-1. These are three common positive-displacement pump designs.

Superchargers

The term supercharger usually applies to an air pressurizing pump that is mechanically driven by the engine itself. Gears, shafts, chains, or belts from the engine crankshaft turn the pump. These mechanical linkages rob a lot of power from the engine. Some supercharger designs must often be driven at speeds of 50,000 to 90,000 rpm. Engine-driven superchargers were never extensively used for passenger car engines.

There are two general types of supercharger pumps:

- Positive displacement
- Centrifugal.

Positive displacement pumps include the reciprocating, the lobe (roots), and the vane designs, figure 15-1. These are called positive-displacement pumps because every revolution pumps the same volume of air. They compress air by taking in a large volume and forcing it into a smaller area. Some of these designs are very inefficient, and the efficient designs are usually too expensive for passenger car use. They more often can be found on diesel truck or racing engines.

The centrifugal pump, figure 15-2, contains an impeller that looks like a many-bladed fan. As the impeller turns at high speed, air pulled into the center of the impeller is speeded up, and then thrown outward from the blades by centrifugal force. This fast-moving air enters a diffuser passage. As the volume of the passage increases, the air diffuses, or slows down. The air's deceleration causes an increase in its pressure. The diffuser passage routes the air to the engine intake manifold.

Engine-driven centrifugal pumps have been tried, but the mechanical drive is always a weak link. Also, the mechanical drive can require up to 20 percent of the engine's horsepower output. However, the development of an exhaust-gas turbine-driven supercharger (usually called a turbocharger) made supercharging practical.

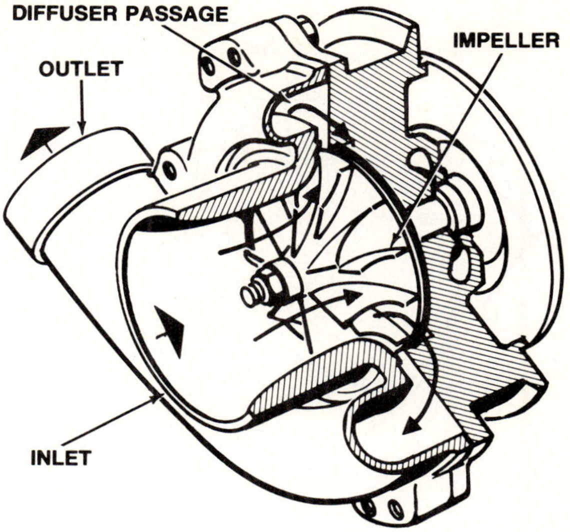

Figure 15-2. The centrifugal pump design accelerates and then slows the air in order to compress it.

TURBOCHARGERS

A **turbocharger** turbine, figure 15-3, looks much like the centrifugal pump. Hot exhaust gases flow from the combustion chamber to the turbine wheel. The gases are heated and expanded as they leave the engine. It is not the speed or force of the exhaust gases that forces the turbine wheel to turn, as is commonly thought, but the expansion of the hot gases against the turbine wheel's blades.

The turbine's main advantage over a mechanically driven supercharger is that the turbine does not drain power from the engine. In a normally aspirated engine, about half of the heat energy contained in the fuel goes out the

Supercharging: Use of an air pump to deliver an air-fuel mixture to the engine cylinders at a pressure greater than atmospheric pressure.

Boost: A measure of the amount of air pressurization, above atmospheric, that a supercharger can deliver.

Normally Aspirated: An engine that uses normal vacuum to draw in its air-fuel mixture. Not supercharged.

Turbocharger: A supercharging device that uses exhaust gases to turn a turbine that forces extra air into the cylinders.

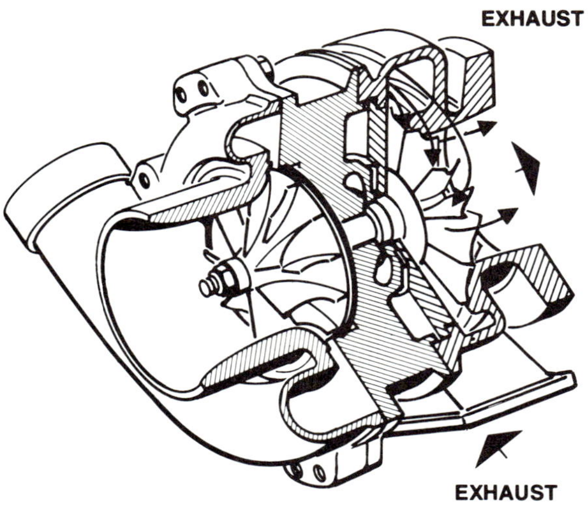

Figure 15-3. A turbine wheel is turned by the expansion of gases against its blades.

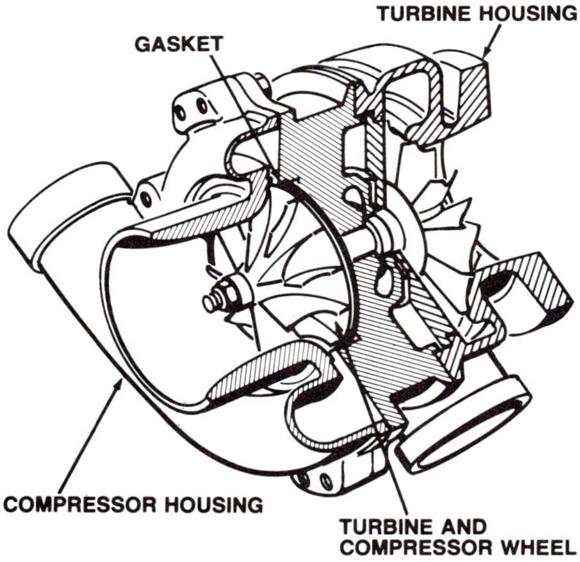

Figure 15-5. A cutaway view of a typical automotive turbocharger.

exhaust system, figure 15-4. Another 25 percent is lost through radiator cooling. Only about 25 percent is actually converted to mechanical power. A mechanically driven pump uses up some of this mechanical output, but a turbine gets its energy from the exhaust gases, converting more of the fuel's heat energy into mechanical energy.

Turbocharger Design

Most automobile turbochargers use a centrifugal pump because of cost, size, reliability, and

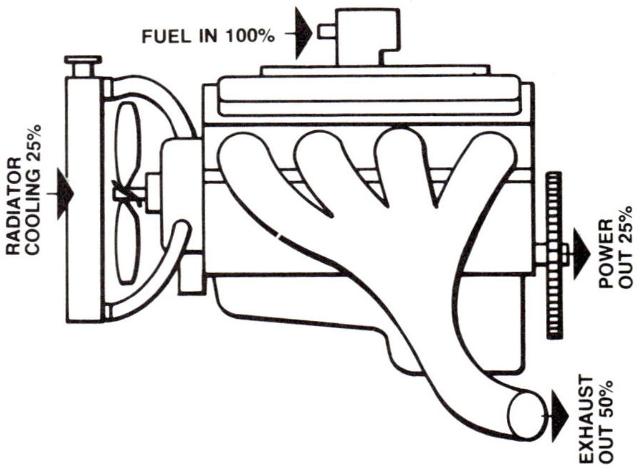

Figure 15-4. A turbine uses some of the heat energy that would normally be wasted.

efficiency. Figure 15-5 is a cutaway view of a typical automobile turbocharger. The turbine wheel and pump impeller are on the same shaft, so that they turn at the same speed. Because they also turn in the same direction, air flows as shown in figure 15-6.

Turbocharger Installation

A turbocharger can be installed in one of two ways:

- Between the air cleaner and the carburetor, figure 15-7.
- Between the carburetor and the engine, figure 15-8.

Each of these has its disadvantages. If compressed air is being forced through the carburetor, as in figure 15-7, the fuel pressure must be great enough so that fuel can enter the airflow. This also means that the carburetor and all fuel lines must be able to hold the increased pressure with no leaks.

If the air and fuel are mixed and then compressed, figure 15-8, the mixture can easily separate. Also, the carburetor must be mounted farther away from the engine, which can cause some problems in cold weather.

Diesel truck engines are often turbocharged. With a diesel fuel injection system, the installation problems just mentioned do not exist. The turbocharger delivers air to the induction system independently of the fuel injection system. The only modern gasoline engines using both a turbocharger and fuel injection are the German-built Porsche Turbo Carrera and the Swedish Saab turbo, which use the Bosch CIS injection mentioned in Chapter 14.

Engine Emissions

Turbochargers have just about the same effect as an increased compression ratio. Why, then,

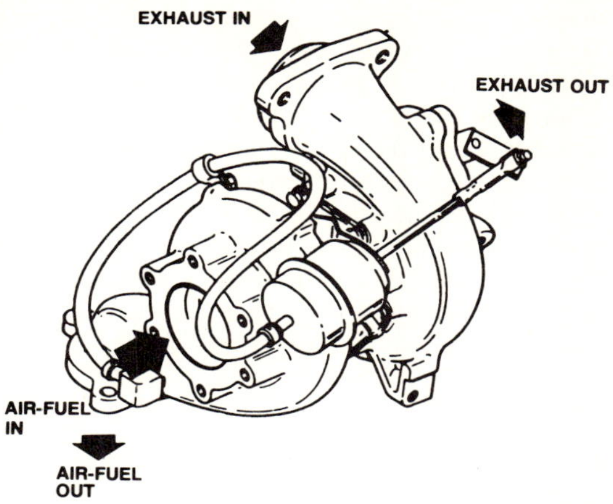

Figure 15-6. Airflow through a turbocharger. (Buick)

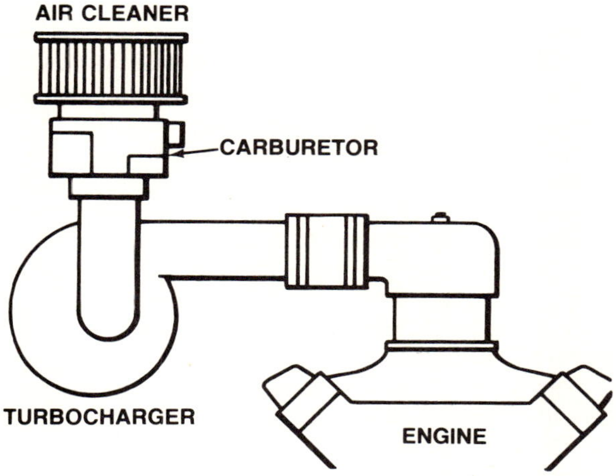

Figure 15-8. The turbocharger can force the air-fuel mixture directly into the intake manifold.

can turbochargers be used on emission-controlled engines when increased compression ratios cannot be used?

The advantage of a turbocharger is the way it controls compression. An engine built with a certain compression ratio will always have that ratio. As we said before, high compression engines tend to create high exhaust emissions during idle, deceleration, and choked operation. Turbocharger boost can be controlled so that the engine runs as a normally aspirated unit during these operating modes. Under acceleration or heavy load conditions, turbocharger boost is applied to the intake manifold.

The few turbocharged cars currently in production use essentially the same emission control devices as do nonturbocharged versions of the same engines.

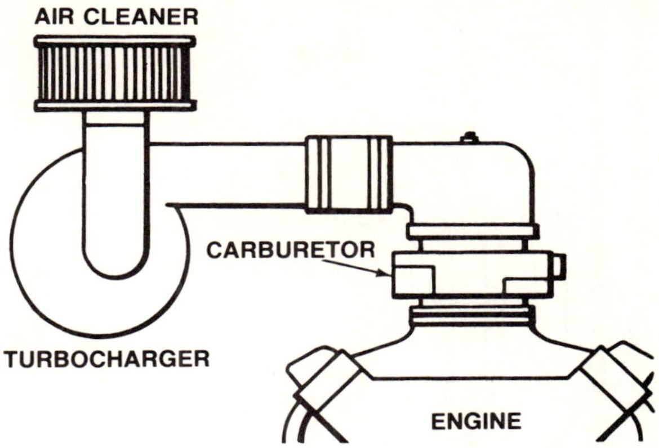

Figure 15-7. The turbocharger can force air into the carburetor.

Engine Detonation

Excessive heating of the air-fuel mixture can cause an explosion in the combustion chamber known as detonation. This wastes power and can damage the engine. Because a turbocharger compresses and therefore heats the air-fuel mixture, detonation is a common problem. Various ways have been devised to control or avoid detonation.

■ Turbocharging Vs. Emissions

Does turbocharging a conventional automotive engine reduce emission levels from that engine? Turbocharger manufacturers think so, and point to several closely controlled tests to prove their point. These tests show a 14 percent reduction of HC, 13 percent CO and 8 percent NO_x by simply bolting a turbocharger on a standard production engine. Four reasons explain this effect on exhaust emission:

1. A low-compression engine that runs satisfactorily on low-lead, low-octane gasoline is ideal for turbocharging.
2. The smaller displacement engine is more efficient, burning less fuel and giving off a smaller amount of combustion byproducts under normal use.
3. Turbocharging causes backpressure in the engine's exhaust system during periods of full speed and power. This increases the temperature and pressure of exhaust gases, leading to a more complete burning in the exhaust manifold.
4. Because the intake manifold and carburetor work under pressure instead of vacuum, the turbocharged engine maintains more precise air-fuel mixtures in each engine cylinder.

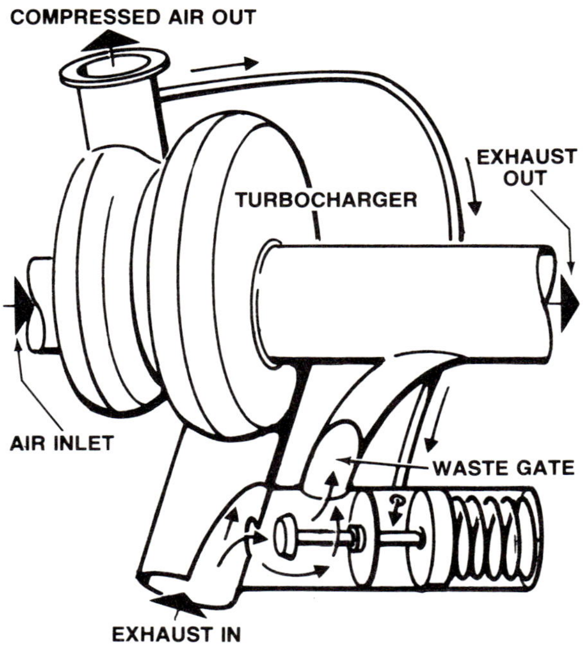

Figure 15-9. The waste gate reacts to intake manifold pressure and controls the amount of exhaust gas reaching the turbine wheel.

TURBOCHARGER CONTROLS

The action of a turbocharger must be controlled to avoid high exhaust emissions and detonation. The maximum boost pressure of a turbocharger must also be controlled, or the higher compression and power potential could damage the engine. There are three general ways of controlling a turbocharger system:

1. Changing the amount of boost
2. Cooling the compressed mixture
3. Altering the spark timing.

Each of these methods controls several aspects of a turbocharger's operation. The following paragraphs explain how the controls are achieved and what effects they have upon turbocharger operation.

■ Supercharging And The Indy 500

Many still think of supercharging or turbochargers as being specifically related to racing engines. Indeed, that's where supercharging got its first real start.

In 1924, in the 15th running of the famed Indianapolis 500 race, a Duesenberg with a centrifugal supercharger took first place. . .just barely. The driver was in the lead about a third of the way through the grueling race when a gear shattered in the supercharger. He pulled his ailing car into the pits and jumped into another supercharged "Dusey", roared back onto the track and finished first, with an average speed of 98.23 mph, an astounding record in those days.

The following year, the top 15 cars were supercharged, and in 1926 every car was using it. But in 1930, a rule change eliminated most of the superchargers. The concept did not make its way back to Gasoline Alley until the late 1960s, when turbochargers were introduced. By the 1970s, virtually every top Indy 500 racer was using turbocharging.

Boost Control

Boost control systems can either limit the amount of exhaust gas reaching the turbine, or vent off some of the compressed air mixture or air-fuel mixture before it reaches the combustion chamber. The mechanisms involved are waste gates and blowoff valves.

Waste gate

A waste gate, figure 15-9, controls the flow of exhaust gas to the turbine. It can allow all of the exhaust into the turbine, or it can route the exhaust directly to the muffler system.

Waste gates are normally controlled by a diaphragm linkage. The diaphragm is exposed to intake manifold pressure. When this pressure reaches a certain level, the diaphragm moves far enough so that the waste gate routes exhaust gases directly to the muffler and bypasses the turbine. Because waste gates are constantly exposed to corrosive exhaust gases, they must be made of high-quality alloys.

The waste gate action controls maximum boost pressure by limiting the speed of the exhaust turbine. It also controls detonation, to some extent, because the chances of detonation increase as the boost pressure increases.

Blowoff valve

A blowoff valve, or pressure control valve, affects the flow of compressed air between the turbocharger and the engine, figure 15-10. The valve can be operated in different ways depending upon what it is meant to control.

If the valve is to control the maximum boost pressure, then it is a simple spring-loaded unit. When boost pressure reaches a certain level, the spring tension is overcome, and some of the boost pressure is allowed to escape. If the turbocharger is compressing only air, as shown in figure 15-10, then the air can be vented to the atmosphere. If the turbocharger is compressing the combustible air-fuel mixture, then the mixture must be vented back into the turbocharger inlet.

If the blowoff valve is meant to control exhaust emissions, then it is operated by intake manifold vacuum. During idle, closed-throttle

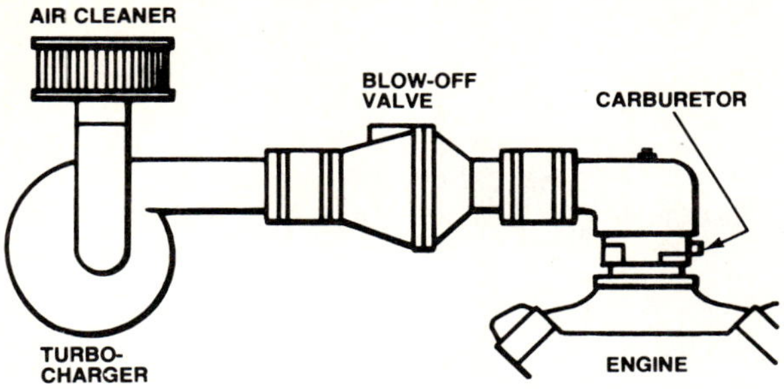

Figure 15-10. A blowoff valve can open to relieve intake manifold pressure.

Figure 15-11. The Buick turbocharged V-6. (Buick)

deceleration, and choked operation, vacuum in the intake manifold causes a diaphragm to move and open the valve. The engine operates as a normally-aspirated unit, avoiding the excessive emissions that the turbocharger compression would cause. Some blowoff valves are operated by vacuum routed to one side of the diaphragm and boost pressure routed to the other.

Mixture Cooling

Two methods are used to cool the compressed mixture before it reaches the combustion chamber: intercoolers and water injection systems. The main aim of these systems is to control detonation.

Intercoolers

Between the turbocharger and the engine, the compressed mixture flows through a radiator core which can contain water, air, or special chemicals. Although intercoolers require extra space and special plumbing connections, they are very efficient. Intercooling is often used on racing engines and on turbocharged diesel trucks.

Water injection

Water injection systems cool the compressed mixture by spraying a mist of water, or water and special chemicals, into the intake manifold. The injection can be powered by an electric pump or by boost pressure.

Spark Timing

Retarding the spark timing can control detonation by lowering the peak temperature of combustion. Spark control systems can be mechanically or electronically operated.

Mechanical operation

A special diaphragm at the ignition distributor is exposed to intake manifold pressure. As the boost increases, the diaphragm moves to retard the ignition timing.

Electronic operation

A sensor in the engine detects detonation and signals an electronic control module. The module's solid-state circuitry then adjusts ignition timing.

SPECIFIC TURBOCHARGERS

As we have said, few manufacturers have produced supercharged engines for everyday use. In the 1950's, some Ford, Packard, Studebaker, and Kaiser models were supercharged. In the early 1960's, some Oldsmobile F-85 Jetfires and Corvair Monzas and Corsas were turbocharged. The only turbocharged units in production today are the Porsche Turbo Carrera, the Saab

■ Switching The Emphasis On Turbos

Most drivers think of turbochargers as exotic trappings for souped-up street cars and racers. That's natural when you consider that automotive engineers normally bring a turbo into play high up on the engine's performance curve to deliver maximum power at top rpm. The result is additional peak power that has won many races, and further convinced the general driving public that turbine-operated superchargers are not for them.

The reappearance of turbocharging as a factory option on some 1978 V-6 engines approaches the subject from the opposite viewpoint. Engineers wanted 6-cylinder fuel economy with V-8 performance. The new turbo is used for low- and medium-speed passing, and acceleration from about 1,200 rpm up. As engine speed climbs, the waste gate begins to open, preventing an overload of the engine. This means that the turbocharger will be used for only about five percent of the time the engine is operating.

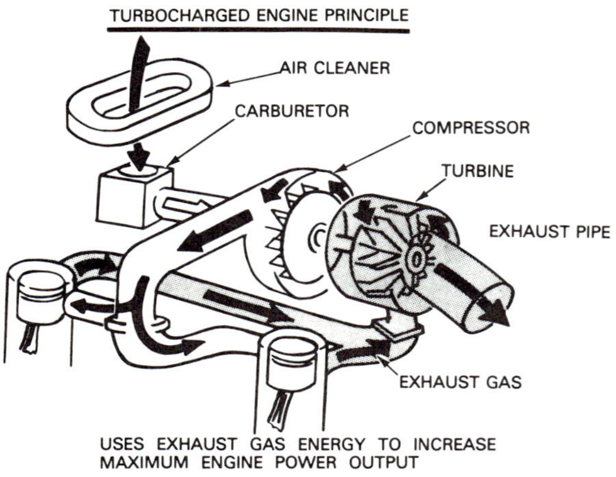

Figure 15-12. The Buick turbocharger is betweeen the carburetor and the engine. The white spaces indicate fresh air flow and the shaded portion is the exhaust flow driving the turbine. (Buick)

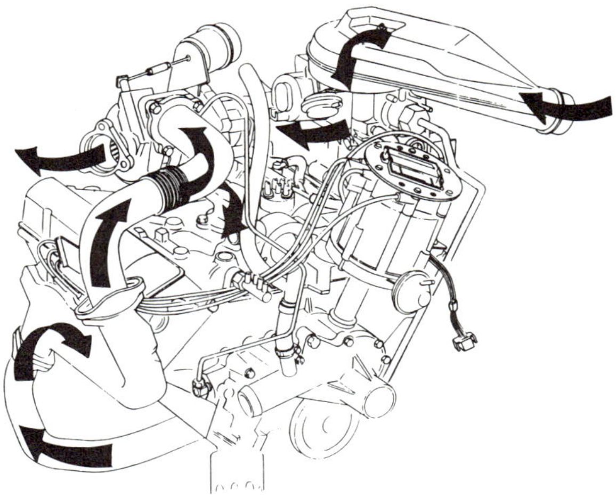

Figure 15-13. This schematic drawing shows the airflow through the Buick turbocharged V-6 engine. (Buick)

turbo, and some Buick V-6 engines. Because the Buick is based on a conventional domestic engine and is produced in greater numbers than the Porsche or Saab, we will concentrate on examining the Buick turbocharger system.

The turbocharged Buick V-6, introduced in 1978, is based on a normally aspirated production engine, figure 15-11. A few internal changes were made, such as strengthening the pistons and using a different piston ring design. The turbocharged and nonturbocharged versions share the same compression ratio and emission control devices.

The Buick turbocharger is mounted between the carburetor and the engine, figure 15-12. The carburetor air cleaner is remotely mounted in front of the engine to keep the engine as vertically short as possible. The turbocharger is used with either a 2-barrel or a 4-barrel carburetor. The 2-barrel version, with the M2ME-210 Dualjet carburetor, adds 45 horsepower (33 kW) to the 105 horsepower (78 kW) of the standard V-6. The 4-barrel version, with the M4ME Quadrajet, adds an additional 60 horsepower (45 kW).

The 2-barrel turbo operates with a maximum boost of 7 to 8 psi (48 to 55 kPa). The 4-barrel version provides a boost of 8 to 9 psi (55 to 62 kPa). Maximum boost is controlled by a manifold-pressure-operated wastegate. An actuator with a pressure-operated, spring-loaded diaphragm is tapped into the compressor outlet. Through mechanical linkage, the diaphragm opens and closes the exhaust wastegate in the turbine housing. When the wastegate is open, part of the exhaust bypasses the turbine to lower the turbocharger speed and limit boost pressure.

Exhaust gases are routed from the top front of the right-hand exhaust manifold to the turbine. The left-hand exhaust manifold is connected to the right-hand manifold by a crossover pipe, figure 15-13. Exhaust leaves the turbine through a single, large-diameter exhaust pipe that runs to the catalytic converter and muffler.

The turbocharger is mounted directly on top of a special intake manifold. Buick says this location reduces "turbo lag" (the delay between a rapid throttle opening and the delivery of increased boost). Lubrication is provided to the turbo by an external oil line that runs from the front of the engine to the center of the turbo housing.

The carburetor is mounted on a special plenum chamber casting that attaches to the compressor inlet, figure 15-14. By placing the carburetor on the inlet side of the compressor, the need for a pressurized carburetor and fuel delivery system was eliminated.

The air cleaner is forward of the left cylinder bank, near the left side of the car's grille. It is attached to the carburetor by flexible tubing and a metal shroud. Heated air is provided to the thermostatic air cleaner by a heat stove and tube on the left exhaust manifold. The air cleaner damper is controlled by a conventional vacuum motor.

Buick is using a modified high-energy ignition (HEI) system, which it calls electronic spark control (ESC). This controls engine detonation by retarding the ignition timing. The ESC system has four main parts:

- The intake manifold
- The detonation sensor

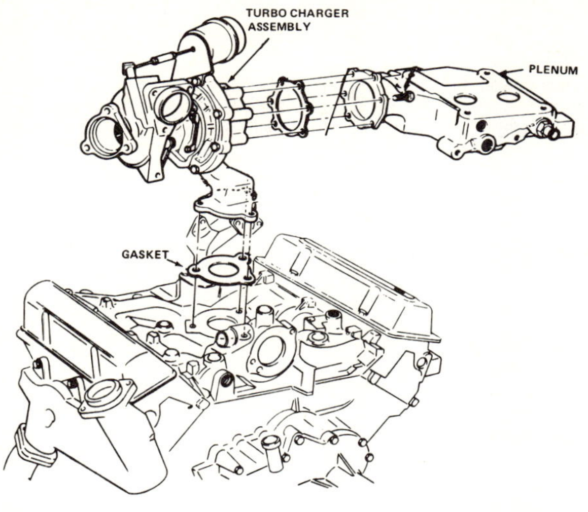

Figure 15-14. The carburetor for the turbocharged Buick V-6 engine is mounted on a special plenum chamber casting attached to the air compressor inlet. (Buick)

- The ESC control module
- The HEI distributor.

The control module is the heart of the system and is located on top of the engine fan shroud, figure 15-11.

The detonation sensor picks up detonation vibrations through the intake manifold and sends this information to the ESC control module. The controller then signals the HEI distributor to retard the timing. This is a constant monitoring process by the controller, which signals a retard in 4-degree jumps, up to a maximum of 22 degrees.

The detonation sensor detects the presence and intensity of the detonation explosions.

Battery voltage is applied to the controller through a wire from the ignition switch. Operation is controlled by a relay on the ground side of the controller. The relay is operated by voltage from the switch side of the starter solenoid.

The ESC controller is also connected, in parallel with the distributor pickup coil connections, to the ignition control module. The controller processes the signals from the detonation sensor and compares them to the timing signals from the pickup coil. It then sends a timing delay signal back to the ignition module when detonation occurs. Timing returns to its normal setting for speed and load conditions within 20 seconds after detonation stops.

SUMMARY

Like fuel injection, supercharging and turbochargers are not new. The concept of increasing mixture compression by providing the engine with an air-fuel mixture at greater than atmospheric pressure has been used on diesel and racing engines for years, and occasionally on passenger car engines.

Superchargers and turbochargers differ in the way they operate. The supercharger is an air pump mechanically driven by the engine. Turbochargers are driven by exhaust gas. They do not consume engine power for their operation.

As the automotive industry moves into the 1980's, the emphasis is on smaller cars with more fuel-efficient engines. Used as a boost device, turbocharging helps improve engine performance without affecting fuel economy or emission levels. This is possible because the engine will run normally under idle, deceleration, and choked conditions. Turbocharger boost is applied only during acceleration or heavy load conditions. The use of electronic spark timing control prevents detonation, a common problem with turbochargers.

■ Smoke Signals Mean A Lot

Even in these days of tough emission standards and a maze of electronics under the hood, the presence of visible smoke in the exhaust can still tell you a lot about the condition of the car's engine.

Experienced mechanics can still eyeball a car and give a pretty good analysis of what ails it, simply by looking at the exhaust. These are generally good indicators for gasoline-fueled engines:

- White smoke — moisture has entered the combustion chambers. Check for a blown head gasket or a cracked head.
- Black smoke — usually caused by an overly rich fuel mixture. Check for a closed choke, a flooded carburetor, or a float level that is too high. Also, if the carburetor power valve is a vacuum operated type, a stuck vacuum piston or a ruptured vacuum diaphragm can cause the power valve to stay open when it shouldn't.
- Blue or blue-gray smoke — oil has entered the combustion chambers or the exhaust system. If the blue smoke is seen during wide-open throttle operation, worn or stuck piston rings may be the cause. If the smoke is seen during deceleration, failed or worn valve stem seals or worn valve guides are most often the culprits.

Review Questions

Select the single most correct answer.
Compare your answers with the correct answers on page 280.

1. Boost is:
   a. The amount of pressurization above manifold pressure
   b. The amount of pressurization above atmospheric pressure
   c. The increased horsepower delivered by supercharging
   d. The increased torque delivered by supercharging

2. The Buick turbocharged V-6 was introduced in:
   a. 1950
   b. 1960
   c. 1977
   d. 1978

3. The main difference between a turbocharger and a supercharger is:
   a. The velocity of air leaving the turbocharger is greater
   b. The amount (volume) of air pumped by a turbocharger is greater
   c. A turbocharger uses exhaust gases to drive its impeller
   d. A turbocharger robs the engine of about half the power output

4. Detonation is a result of excessive heating of:
   a. The air-fuel mixture in the carburetor
   b. The air-fuel mixture in the intake manifold
   c. The air-fuel mixture in the combustion chamber
   d. The exhaust in the exhaust system

5. Supercharging delivers the air-fuel mixture to the cylinder at:
   a. Lower than atmospheric pressure
   b. Atmospheric pressure
   c. Higher than atmospheric pressure
   d. Three times the atmospheric pressure

6. The Buick V-6:
   a. Has its turbocharger mounted between the air cleaner and carburetor
   b. Has a waste gate control
   c. Uses mechanical spark timing control
   d. Uses a blowoff valve

7. Positive displacement pumps:
   a. Contain impellers
   b. Pump the same volume of air each revolution
   c. Increase air pressure by decelerating it
   d. Are highly efficient

8. A turbocharger turbine wheel turns because of:
   a. Centrifugal force
   b. Exhaust gas speed
   c. Manifold vacuum
   d. Expansion of hot exhaust gases

9. Of the heat energy contained in gasoline:
   a. 50 percent is converted to mechanical power
   b. 50 percent is lost to cooling
   c. 75 percent is converted to engine power
   d. 50 percent goes out the exhaust system.

10. Despite high compression pressures achieved, turbochargers can be used on emission-controlled engines because:
    a. They have fixed compression ratios
    b. Turbocharger boost can be varied to meet engine needs
    c. They are placed between the air cleaner and the carburetor
    d. They are placed between the carburetor and the engine

11. Which is *not* a method of controlling a turbocharger system:
    a. Changing the amount of boost
    b. Cooling the compressed mixture
    c. Readusting the carburetor idle
    d. Altering spark timing

12. The waste gate controls boost by:
    a. Controlling flow of exhaust gas
    b. Controlling compressed air
    c. Controlling the air-fuel mixture
    d. Controlling exhaust emissions

13. Retarding spark timing will control detonation by:
    a. Cooling the compressed mixture
    b. Cooling the exhaust manifold
    c. Reducing compression pressure
    d. Lowering peak temperature of combustion

PART FOUR

Emission Control Systems and Devices

Chapter

16

Positive Crankcase Ventilation

The problem of crankcase ventilation has existed since the beginning of the automobile. No piston ring, new or old, can provide a perfect seal between the piston and the cylinder wall. When an engine is running, the pressure of combustion forces the piston downward. This same pressure also forces gases and unburned fuel from the combustion chamber past the piston rings and into the crankcase. These gases are called crankcase vapors, or blowby, figure 16-1.

Blowby consists mainly of unburned hydrocarbons (HC), which must be removed from the crankcase. Up until 1961, the vapors were removed with a simple draft tube system, so that the blowby was vented directly to the atmosphere where it added to air pollution. After 1961, positive crankcase ventilation (PCV) systems were developed to keep from dumping these pollutants into the air.

The PCV valve, a major part of the PCV system, is often described as the most often overlooked and neglected part of a tune-up. A clogged or faulty valve can ruin any tune-up, and under severe conditions, can even cause an engine to fail.

PCV systems are not hard to service, but you must know what you're looking for, and how the various systems work. In this chapter, you will learn:

- The reasons for crankcase ventilation
- The differences between the open and the closed ventilation systems
- The parts of the modern PCV system
- How the PCV system ventilates the crankcase without polluting the air.

DRAFT TUBE VENTILATION

Blowby has three undesirable features:

1. It destroys the lubricating qualities of engine oil.
2. It causes sludge and varnish to form.
3. It helps cause formation of corrosive acids, which can damage engine parts.

After trying various ways of ventilating the engine crankcase, car makers at first settled on the road draft tube, figure 16-2. This is simply a tube connected to the engine crankcase that allows vapors to pass into the air. Fresh air to ventilate the crankcase enters through a vented oil filler cap. This air passes into the crankcase where it mixes with the vapors. Once the car is moving, a vacuum is created by the airflow past the road draft tube. This vacuum draws the crankcase vapors out into the atmosphere.

The road draft tube has three major shortcomings:

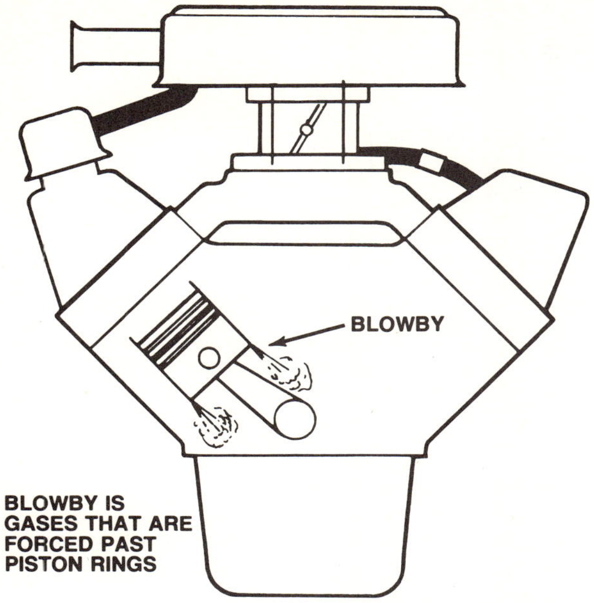

Figure 16-1. Piston rings do not provide a perfect seal. Combustion gases blow by the rings into the crankcase.

1. It works best only when there is a pressure difference between the oil filler cap and the draft tube. This pressure depends on car movement. When the car is moving slower than about 25 mph, there is not enough vacuum to remove the vapors from the crankcase.
2. It passes the crankcase vapors directly into the atmosphere where they add to air pollution.
3. At higher vehicle speeds, too much crankcase ventilation will increase engine oil consumption because oil droplets are drawn out through the road draft tube.

POSITIVE CRANKCASE VENTILATION (PCV) SYSTEMS

The drawbacks of the road draft tube were eliminated when the controlled crankcase ventilation system was introduced. Controlled or positive crankcase ventilation relies on intake manifold vacuum to draw the vapors from the crankcase and into the intake manifold. This results in a positive movement of air through the crankcase whenever the engine is running. The vapors are then returned to the combustion chambers, where they are burned. These systems may be classified as open or closed PCV systems, depending upon their design.

Open PCV Systems

Positive crankcase ventilation, as we know it today, received its first major use on 1961 California cars. Open systems were installed on many 1963 models nationwide. Open crankcase ventilation systems can be divided into three types.

Type 1 system operation
A hose connects the crankcase with the intake manifold, figure 16-3. When the engine is run-

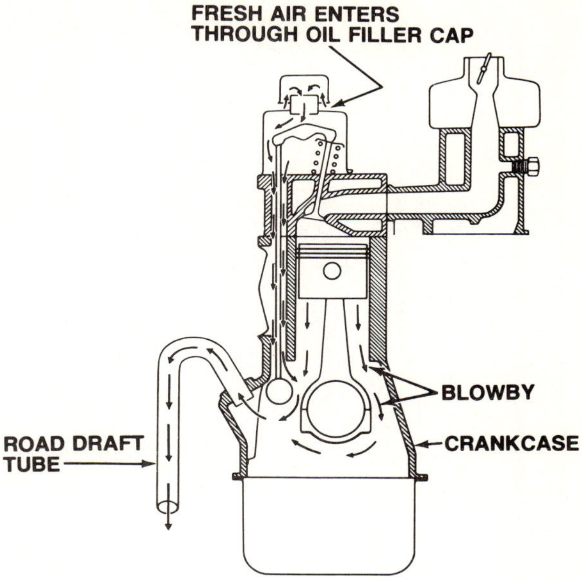

Figure 16-2. The road draft tube ventilates the crankcase to the atmosphere when the car is moving.

■ PCV System Service

When a PCV system becomes restricted or clogged, the cause is usually an engine problem or the lack of proper maintenance. For example, scored cylinder walls or badly worn rings and pistons will allow too much blowby to pass. Start-and-stop driving requires more frequent maintenance and causes PCV problems more quickly than highway driving, as will any condition allowing raw fuel to reach the crankcase. Using the wrong grade of oil or not changing the crankcase oil at periodic intervals will also cause the ventilation system to clog.

When a PCV system begins to clog, the engine tends to stall, idle roughly, or overheat. As ventilation becomes more restricted, burned plugs or valves, bearing failure, or scuffed pistons can result. Also look for an oil-soaked distributor or points, or leaking out around valve covers or other gaskets. Do not overlook the PCV system while troubleshooting. A partly or completely clogged PCV valve, or one of the incorrect capacity, may well be the cause of poor engine performance.

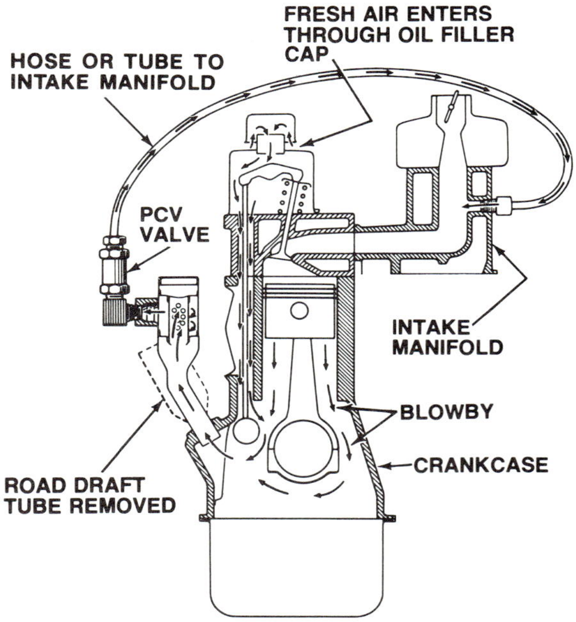

Figure 16-3. In a Type 1 open PCV system, fresh air enters through the oil filler cap. Crankcase vapors are returned to the intake manifold through a PCV valve and hose or tube.

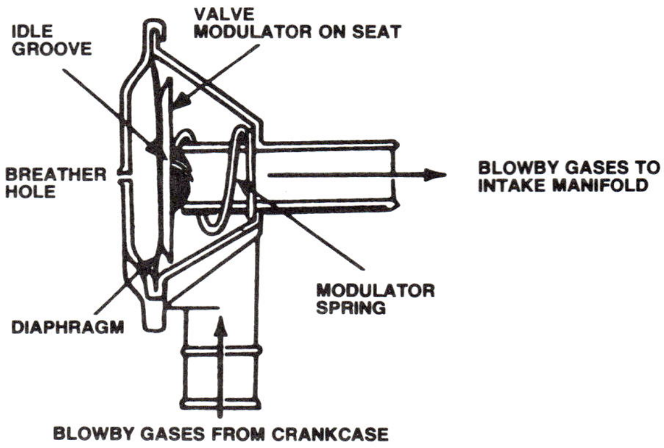

Figure 16-5. When the engine is at idle, the Type 2 PCV valve is closed by crankcase vacuum. Vapors flow through the idle groove at about 3 cubic feet per minute.

ning, fresh air is drawn into the crankcase through the vented oil filler cap. This air mixes with the crankcase vapors, travels to the intake manifold, and is drawn into the engine cylinders. Airflow from the crankcase is metered through a PCV valve, figure 16-4, that has a spring-operated plunger which controls the rate of airflow through the engine.

Type 1 open PCV systems were used as original equipment on many domestic passenger car engines through the mid-1960's.

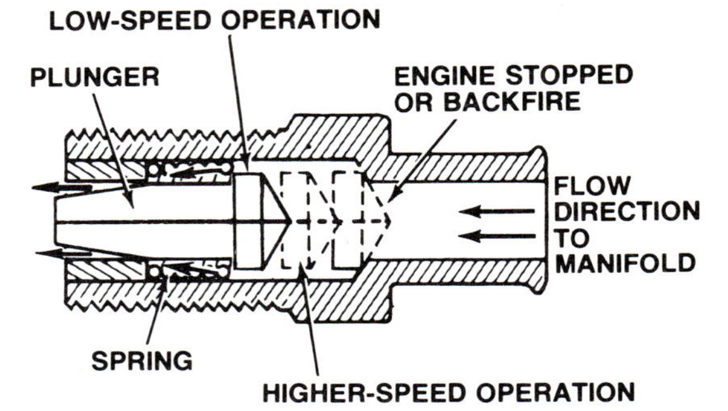

Figure 16-4. The PCV valve plunger responds to spring force and manifold vacuum to regulate the airflow rate.

Type 2 system operation

A Type 2 system is similar to a Type 1 system, but uses a special PCV valve and oil filler cap. The diaphragm-type PCV valve regulates the airflow according to crankcase vacuum. When a vacuum exists in the crankcase, the valve is closed, figure 16-5. When the crankcase is under pressure, figure 16-6, the valve opens. The oil filler cap has an orifice large enough to allow the right amount of air to enter but small enough to maintain crankcase vacuum. The orifice must remain open at all times for the system to work correctly.

Some imported cars, particularly British models in the mid-1960's, had Type 2 PCV systems as original equipment. Called a Smith's PCV valve, figure 16-7, it worked like the one we just described and could be opened for cleaning.

Type 3 system operation

In a Type 3 system, the crankcase is connected to the air cleaner by a hose, figure 16-8. Vapors are drawn from the crankcase by a slight suction in the air cleaner. No control valve is used. Fresh air enters through the oil filler cap, mixes with the crankcase vapors, and is drawn to the air cleaner. From the air cleaner, the vapors are then drawn through the carburetor and into the cylinders.

Because blowby reaches the carburetor *before* the air-fuel vaporization, this system tends to richen the air-fuel mixture. Carburetors used with a Type 3 system must be adjusted to make up for this richer mixture. Some imported cars use a "sealed" Type 3 system in which the oil filler cap is not vented.

Type 3 PCV systems were used as original equipment on some Ford and American Motors cars in the early 1960's. The systems on Ford

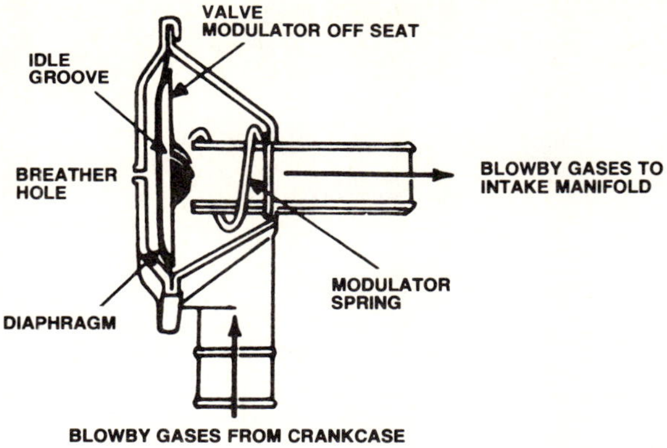

Figure 16-6. When the engine is at cruising speed, the Type 2 PCV valve is opened by crankcase pressure. The flow rate depends on the amount of blowby created by the engine.

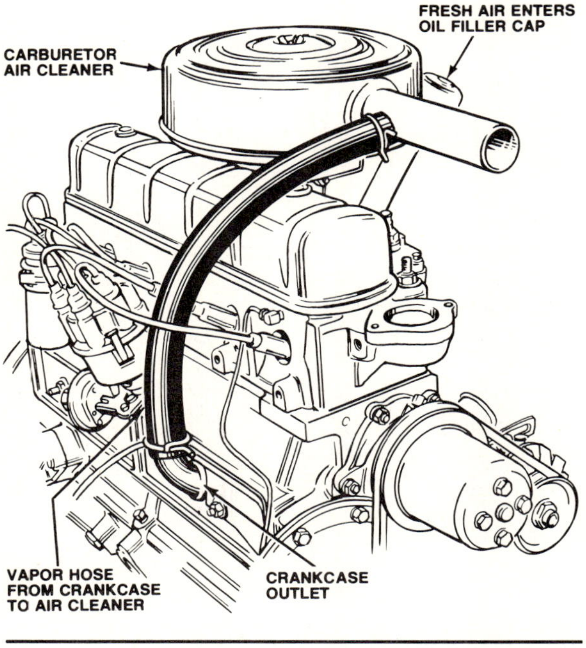

Figure 16-8. A Type 3 open PCV system has no PCV valve.

6-cylinder engines had oil separators at the crankcase outlets to minimize the amount of oil drawn through the PCV hose to the air cleaner, figure 16-9.

Limited efficiency of the open PCV system
Open PCV systems only partly control crankcase emissions. Since manifold vacuum decreases considerably under heavy load or acceleration, crankcase pressures build up. This forces some of the vapors into the atmosphere through the vented oil filler cap. Crankcase

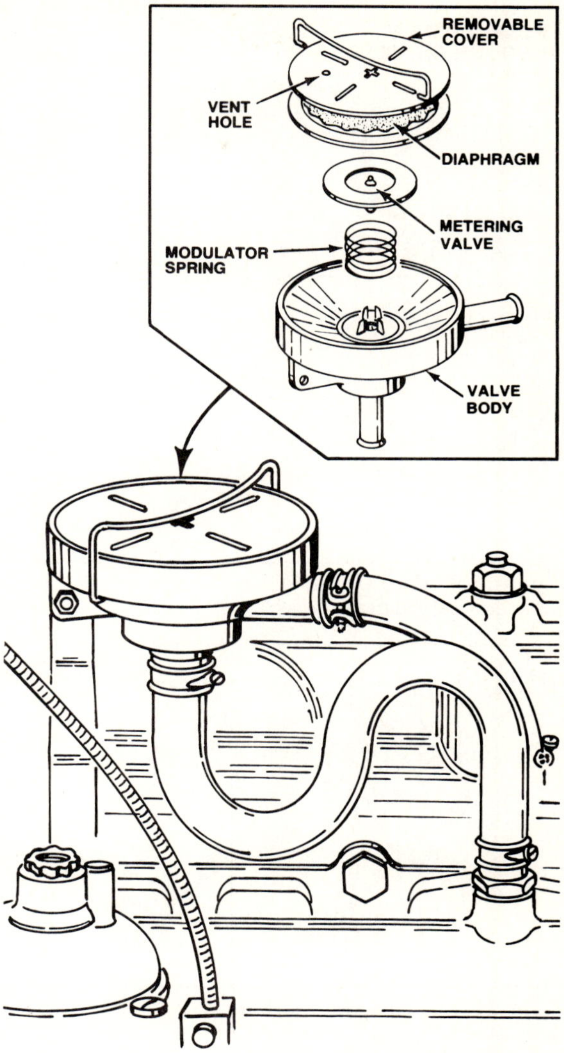

Figure 16-7. The Type 2 PCV valve used on some imported cars could be disassembled for cleaning.

vapors also pass through the vented cap into the air if the system becomes clogged.

Open PCV systems do provide some benefits:
1. They promote longer engine life by removing most harmful vapors from the crankcase.
2. They reduce the amount of crankcase vapors which pollute the air.

Closed PCV Systems

Closed, or Type 4, PCV systems were required on all new California passenger cars in 1964. They were standard nationwide by 1968, and are still used today on all new domestically built cars. In a closed PCV system, figure 16-10, the

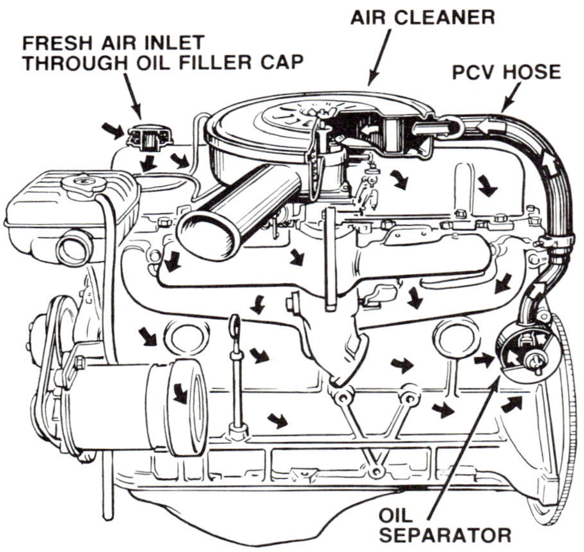

Figure 16-9. Ford Type 3 PCV systems had oil separators at the crankcase outlet.

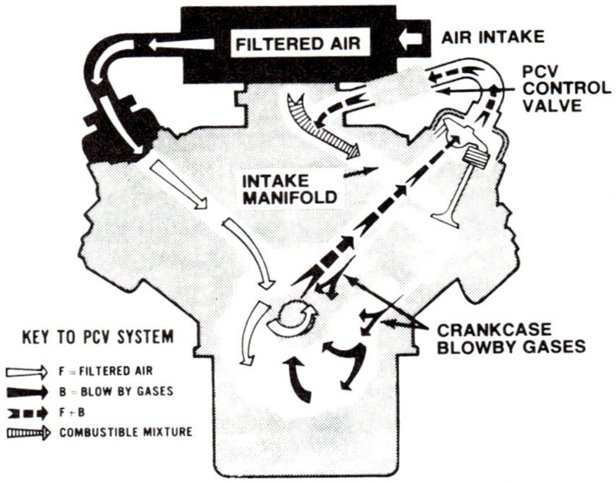

Figure 16-10. Type 4, sealed PCV system operation. (Chevrolet)

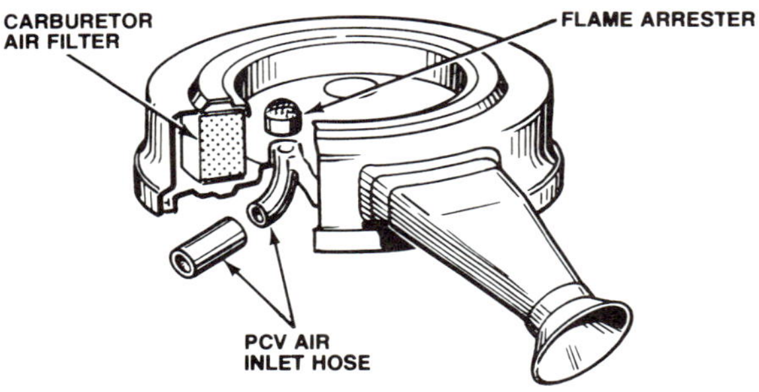

Figure 16-11. When PCV inlet air is drawn from the clean side of the air cleaner, a flame arrester is required in case of a backfire.

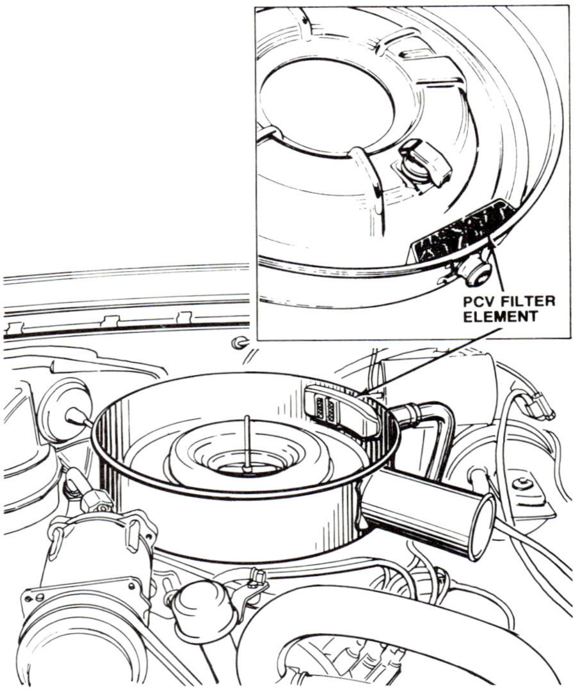

Figure 16-12. This PCV air filter is located in the air cleaner housing.

■ Check PCV Valve For Fuel Dilution

Winter driving, with its quick trips and cold starts, can cause fuel dilution of the engine oil. This will not only thin out the oil, but the gasoline vapors fed to the intake manifold through the PCV system will produce a rich idle mixture. If you are working on a car with a rich idle mixture and you suspect fuel dilution of the oil, you can check it quickly.

Remove the PCV valve from the valve cover and let it draw in fresh air with the engine running. If the idle mixture becomes normal or leaner than normal, the engine oil probably has too much fuel dilution. Drive the car at highway speeds for up to half an hour to purge fuel vapors from the crankcase. Then readjust the idle mixture. As an alternative, drain the crankcase and refill with fresh oil.

oil filler cap is not vented to the atmosphere. Air for the crankcase is drawn through a hose from the air cleaner to one of the valve covers or to a crankcase inlet below the intake manifold. The dipstick is also sealed to prevent air from leaking into the crankcase.

Crankcase ventilation air may come from either the clean side (inside) or the dirty side (outside) of the carburetor air filter. When air is

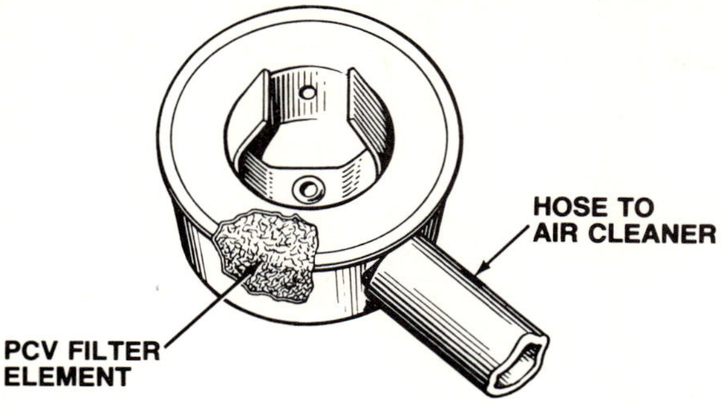

Figure 16-13. This PCV inlet air filter is located in the oil filler cap.

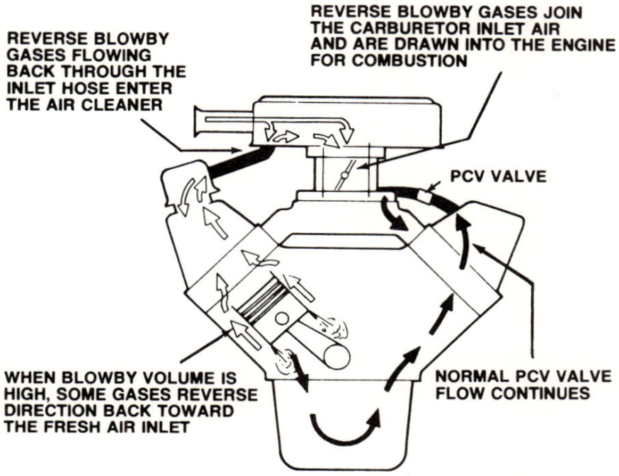

Figure 16-15. Closed PCV system operation under heavy load.

from the clean side using the air cleaner as a PCV filter, a flame arrester, figure 16-11, is used. This wire screen is in the PCV air intake line, either at the air cleaner or at the valve cover, to prevent a crankcase explosion if the engine backfires.

When air is drawn from the dirty side of the air cleaner, a separate PCV air filter is used. This can be located in the air cleaner, figure 16-12, in the oil filler cap, figure 16-13, or in the inlet air hose where it connects to the valve cover. A flame arrester is not required in this case, since the air cleaner filter does that.

Type 4 system operation

Under normal conditions, fresh air from the air cleaner passes through the inlet hose to the crankcase, figure 16-14. The fresh air mixes with the crankcase vapors, and passes through a PCV valve before being drawn into the intake

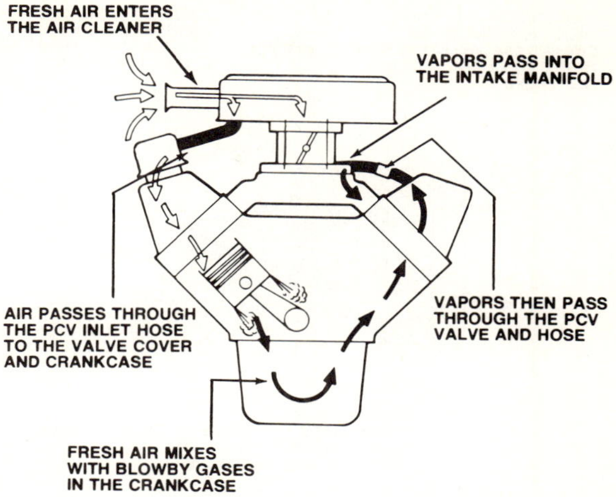

Figure 16-14. Closed PCV system operation under normal conditions.

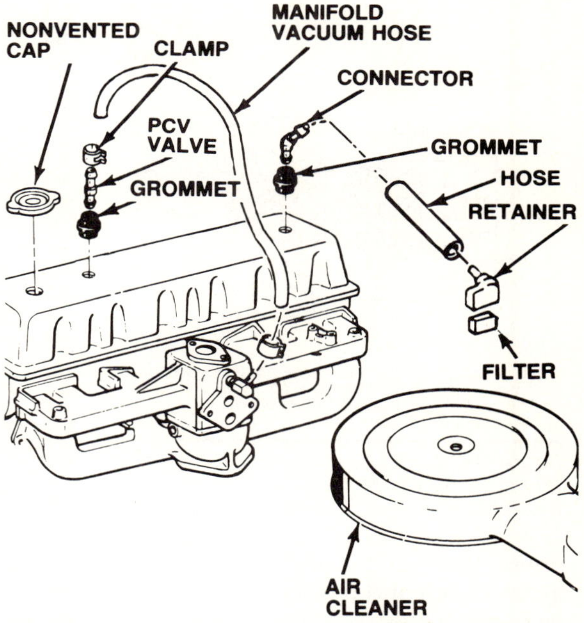

Figure 16-16. All late-model closed PCV systems share these common parts. (AMC)

manifold. Vapors that back up under certain conditions cannot escape from the closed system. If manifold vacuum drops, or if the system becomes clogged, extra crankcase vapors will reverse their direction. In the closed crankcase ventilation system, these excess vapors flow back to the air cleaner, figure 16-15, instead of passing out of the engine and into the atmos-

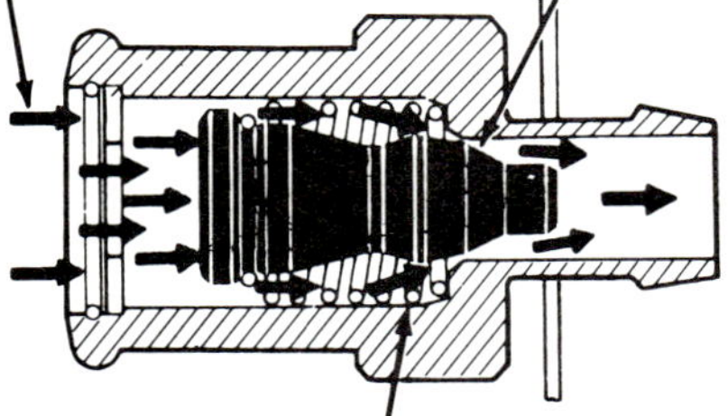

Figure 16-17. Spring force, crankcase pressure, and manifold vacuum work together to regulate PCV valve flow rate.

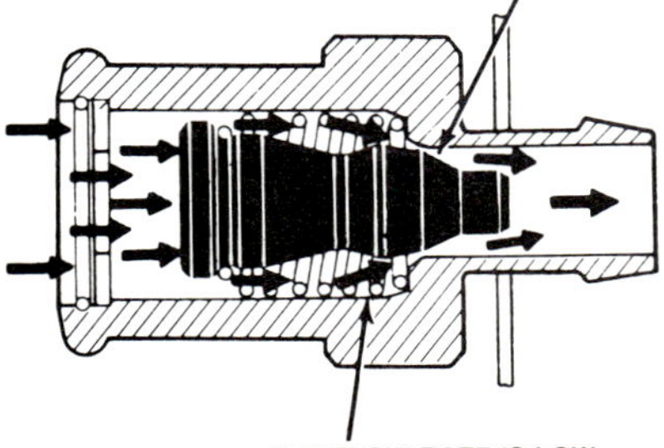

Figure 16-19. PCV valve airflow during cruising and light-load operation.

phere. Once in the air cleaner, they mix with incoming air and pass through the carburetor to be burned in the combustion chamber. This makes the closed system almost 100 percent effective in controlling crankcase emissions.

Closed PCV system efficiency
Closed crankcase ventilation provides three benefits:
1. It promotes longer engine life by removing harmful vapors from the crankcase.
2. It eliminates crankcase vapors that pollute the air.
3. It increases fuel economy by recirculating all unburned blowby back to the intake manifold.

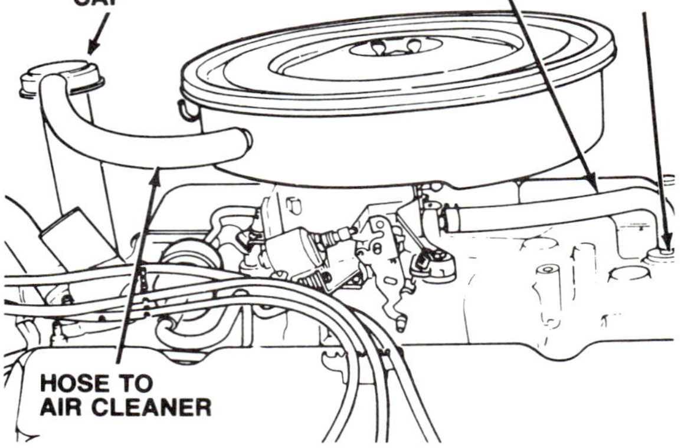

Figure 16-18. V-8 PCV system installation. (AMC)

ORIGINAL EQUIPMENT CLOSED PCV SYSTEMS

All new cars sold in the United States since 1968 have a Type 4, closed PCV system, figure 16-16. The design of closed PCV systems is essentially the same, regardless of the manufacturer. All use a PCV valve, an air inlet filter, and connecting hoses. The location of these parts may vary from one engine model to another, but all operate in the same manner.

Air Inlet Filter

PCV air inlet filters are usually installed in a retainer inside the air cleaner, figure 16-12, or in the oil filler cap, figure 16-13. They are made of wire gauze or polyurethane foam and are usually replaced at the same time as the air cleaner filter element.

Connecting Hoses

Two connecting hoses, figure 16-16, complete the closed PCV system. Fresh air travels from the air cleaner to the engine through the air inlet hose. Crankcase vapors travel from the engine to the intake manifold through the manifold vacuum hose. PCV hoses are made of special materials that resist oil vapors. Heater hose should *not* be used as a substitute.

PCV Valve

This one-way valve has a spring-operated plunger, figure 16-17, to control valve flow rate. Flow rate is set for each engine, and a valve for a

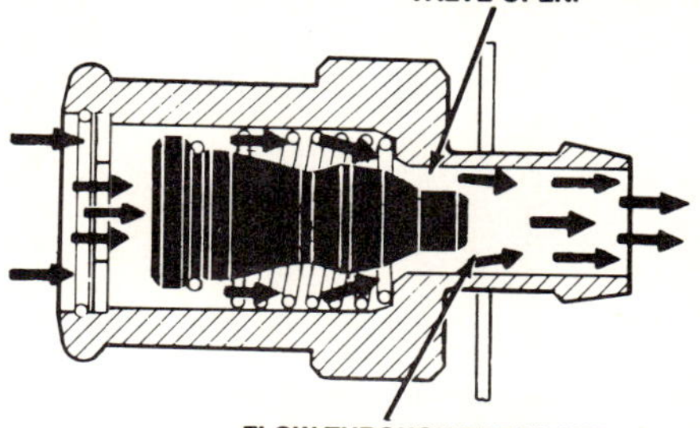

Figure 16-20. PCV valve airflow during acceleration and heavy-load operation.

different engine should not be substituted. This setting is determined by the size of the plunger and the holes inside the valve. PCV valves are usually either in the valve cover, figure 16-16, or in the intake manifold, figure 16-18.

PCV valve operation

When manifold vacuum is high, the PCV valve restricts the airflow, figure 16-19, to keep a balanced air-fuel ratio. It also prevents high intake manifold vacuum from pulling oil out of the crankcase and into the intake manifold.

The PCV valve regulates the airflow through the crankcase under all driving conditions and speeds. Under high speed or heavy loads, the valve opens and allows maximum airflow, figure 16-20. If the engine backfires, the valve will close instantly, figure 16-21, to prevent a crankcase explosion.

RETROFIT SYSTEMS

Used cars sold in California must have a PCV system approved by the state. Those 1955-67 cars originally sold outside California with open PCV systems must be converted to a closed-type system, when sold as used cars within the state.

SUMMARY

Pressure in the engine cylinders forces combustion gases past the pistons. These gases, called blowby, settle in the engine crankcase, where they contaminate the lubricating oil and create harmful acids. Ventilation is necessary to remove the vapors from the crankcase. The draft tube system was used until the 1960's but it doesn't work well at low speeds. It also allows the vapors to pollute the air.

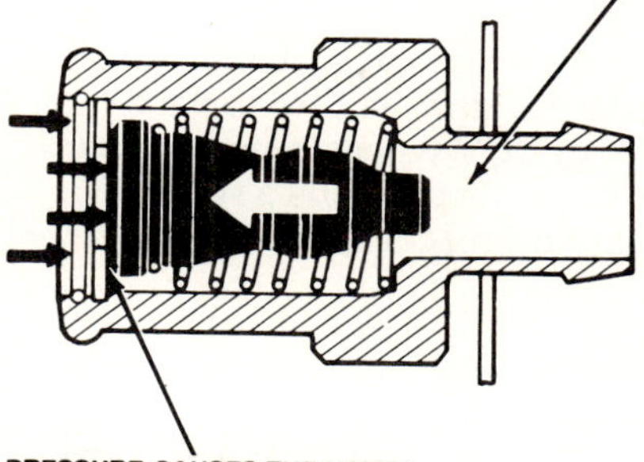

Figure 16-21. PCV valve operation in case of backfire.

The recirculation of crankcase vapors to the intake manifold is called positive crankcase ventilation. Four types of positive crankcase ventilation, or PCV systems, have been used. Types 1,

■ It Wasn't Always As Simple

The early PCV systems caused a good deal of grief and engine troubles for automakers. Many garages, even franchised dealers, ignored the PCV systems on 1963-64 cars. They required a lot of care and cleaning, and they clogged quickly when ignored. Contaminants remained in the crankcase, and sludge and moisture formed. This clogged oil lines and prevented adequate engine lubrication. The result was disaster for the engine, and major overhauls on engines still under warranty were often required.

The situation reached a crisis point for one major manufacturer, which stopped using PCV on its cars from the spring of 1964 until early in 1965. Auto engineers were frustrated by the problems PCV systems were creating. The systems had been designed to be simple and required only a minimum amount of service. But mechanics in the field completely ignored the emission control device, and engines began to fail.

These problems resulted in a crash project by the automakers. While engineers worked overtime developing a "better" PCV system, manufacturers started a program to educate dealers, servicemen, and car owners. The so-called "self-cleaning" PCV valve was developed and began appearing on mid-1965 models. This second-generation PCV system is practically the same one in use today.

2, and 3 are called open systems because vapors are forced into the atmosphere whenever manifold vacuum is low, or if the system is clogged. Type 4 is called a closed system because vapors cannot escape into the air under any conditions.

PCV systems use a fixed or variable metering device to regulate the airflow. The variable metering device, or PCV valve, is standard on modern PCV systems. It contains a spring-operated plunger and reacts to manifold vacuum.

It is important that the correct PCV valve be installed in an engine because of the lack of idle mixture adjustment on current carburetors. Differences in PCV valve idle airflow can greatly influence engine idle smoothness and vehicle driveability when incorrect valves are used.

Review Questions

Select the single most correct answer.
Compare your answers with the correct answers on page 280.

1. Which of the following is *not* true of blowby?
   a. It causes sludge and varnish to form
   b. It can damage engine parts
   c. It can lean out the incoming air-fuel mixture
   d. It can destroy the lubricating qualities of engine oil

2. The airflow through a road draft tube is caused by:
   a. Fresh air entering the carburetor
   b. Vacuum created by air rushing past the open tube
   c. Manifold vacuum
   d. Pressure from the pistons

3. Which of the following is a *closed* PCV system:
   a. Road draft tube
   b. Type 1 system
   c. Type 3 system
   d. Type 4 system

4. The illustration below shows a:

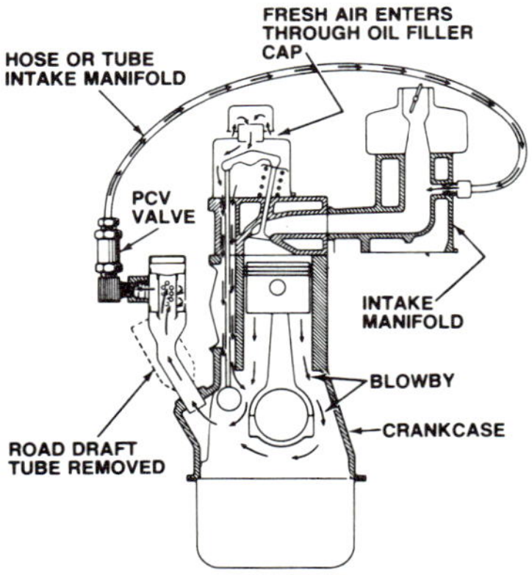

   a. Type 1 PCV system
   b. Type 2 PCV system
   c. Type 3 PCV system
   d. Type 4 PCV system

5. In a PCV system, crankcase vapors are recycled to the:
   a. Exhaust system
   b. Road draft tube
   c. Oil-filler breather cap
   d. Intake manifold

6. Controlled crankcase ventilation was introduced in California in:
   a. 1959
   b. 1960
   c. 1961
   d. 1962

7. Which of the following PCV systems has no PCV valve:
   a. Type 1
   b. Type 2
   c. Type 3
   d. Type 4

8. The kind of PCV valve that can be opened for cleaning is called a:
   a. Type 1 valve
   b. Smith's PCV valve
   c. Type 3 valve
   d. Type 4 valve

9. One of the major drawbacks to an open PCV system is that:
   a. Some vapors are vented out through the oil filler cap
   b. Engine life is significantly shortened because of crankcase wear
   c. It is expensive to install and maintain
   d. Vapors cannot escape under any conditions

10. In a Type 4 PCV system, the crankcase ventilation air may come from:
    a. The clean side of the carburetor air filter
    b. The dirty side of the carburetor air filter
    c. Neither A nor B
    d. Both A and B

11. Which of these is *not* part of a Type 4 PCV system:
    a. Nonvented oil filler cap
    b. Air inlet filter
    c. Oil separator
    d. PCV valve

12. Which of the following is true of a PCV valve:
    a. It regulates airflow through the crankcase under all conditions.
    b. It regulates airflow through the crankcase under low-speed conditions only
    c. It regulates airflow through the crankcase under high-speed conditions only
    d. It regulates airflow through the crankcase except in case of a backfire

13. The advantages of a closed PCV system are:
    a. It promotes longer engine life
    b. It eliminates crankcase vapors almost entirely
    c. It increases fuel economy
    d. All of the above

14. PCV system hoses are made from:
    a. Heater hose material
    b. Low-temperature-resistant nylon
    c. Oil resistant rubber material
    d. Fuel-resistant neoprene

Chapter

17

Air Injection

One of the first add-on devices to help combat high HC and CO exhaust emissions was **air injection**, figure 17-1, also known as the air pump system. By 1966, Chrysler was the only domestic automaker not using air injection on at least some engines. Chrysler, however, continued to develop engine modifications to meet the ever-more-stringent exhaust emission standards, and by the late 1960's, most other carmakers began to rely more on modifications and less on air injection. Then, in 1972, the balance tipped in the other direction again, and even Chrysler began installing air pumps on some engines.

The early air injection systems were bulky and suffered frequent breakdowns. They had masses of extra hoses, which meant the possibility of cracked or broken hoses was greatly increased.

The newer air pump systems are much less complicated than the older versions, but no less important to the smooth-running operation of a car. An air pump that is not working properly can cause the catalytic converter to clog up and wear out far faster than it normally would. Checking the air injection is an important part of diagnosing or tuning up a car. Although the newer-style air pumps are almost maintenance free, the entire system still is part of any routine troubleshooting process.

Air injection systems are not hard to understand or service. Even though they are made by different manufacturers, they all work about the same. In this chapter, you will learn:

- The principles of air injection
- The parts used in a typical air injection system
- The changes that have been made to create second-generation systems for use with catalytic converters.

Manufacturers use the following names for their air injection systems:

- American Motors — Air Guard
- Chrysler — Air Injection System
- Ford — Thermactor Air Injection System
- General Motors — Air Injector Reactor (AIR)

Regardless of the name, the systems are simple, and all systems are basically the same. A belt-driven air pump supplies fresh air to the injector nozzles in the exhaust manifold or cylinder head. The air mixes with the hot exhaust leaving the engine. This helps the oxidation, or burning, reaction necessary to reduce HC and CO emissions.

BASIC SYSTEM DESIGN AND OPERATION

The basic air injection system contains the following parts:

1. Air supply pump with filter

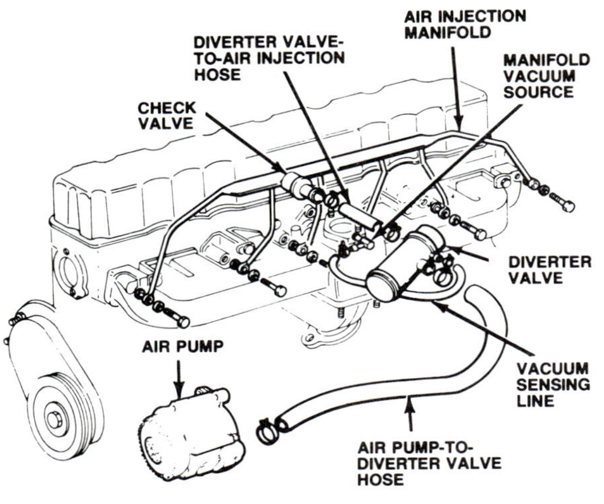

Figure 17-1. Typical late-model air injection system. (AMC)

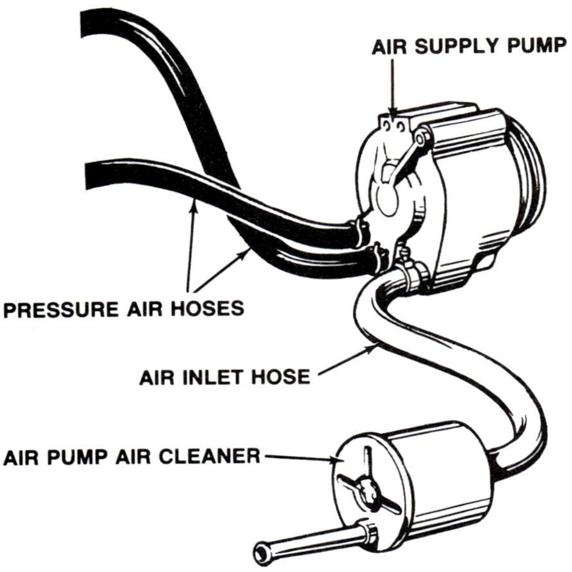

Figure 17-3. Older 3-vane pumps had separate inlet air cleaners.

2. Air manifolds and nozzles
3. Antibackfire valve
4. Check valve
5. Connecting hoses.

Air Supply Pump

The air pump, figure 17-2, is mounted at the front of the engine and driven by a belt from the crankshaft pulley. The pump pulls fresh air in through an external filter and pumps it under slight pressure to each exhaust port through

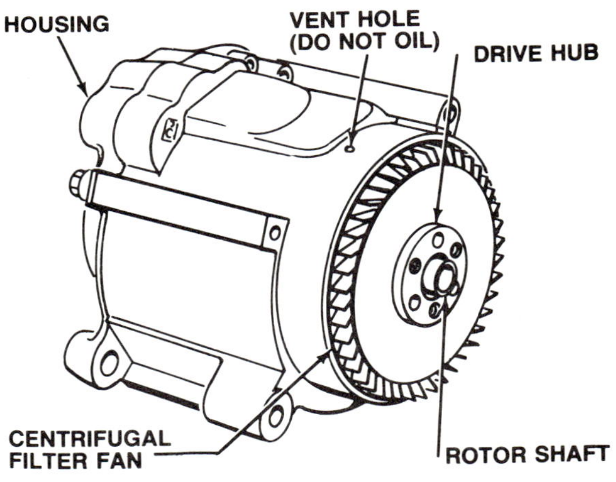

Figure 17-2. All air pumps used on domestic vehicles are made by the Saginaw Division of General Motors.

connecting hoses. Adding this extra air to the hot HC and CO emissions in the exhaust manifold causes oxidation to take place, which helps change the HC and CO into H_2O and CO_2.

The Saginaw Division of General Motors makes the air pumps used on all domestic cars. Early Saginaw pumps (1966-67) were a 3-vane design and could be rebuilt if necessary. A Saginaw 2-vane design replaced the 3-vane pump on 1968 and later models. The 2-vane pumps cannot be rebuilt but must be replaced if they fail. Also, most older 3-vane pumps still in use are usually replaced with new 2-vane pumps if they fail.

The main difference between the two pumps is the way they filter the intake air. The 3-vane pump drew its fresh air supply through a separate air filter, figure 17-3, or from the clean side of the air cleaner. The 2-vane pump uses an impeller-type, centrifugal air filter fan, figure 17-2, mounted on the air pump rotor shaft. This is not a true filter, but cleans the air entering the pump by centrifugal force, figure 17-4. The relatively heavy dust particles in the air are forced in the opposite direction to the inlet airflow. The lighter air is then drawn into the pump by the impeller-type fan.

To keep pump pressure from becoming too high, many pumps use a pressure relief valve, figure 17-5, which opens at high engine speed. Some late-model pumps use a replaceable plastic plug to control the pressure setting of the relief valve. This pressure setting can be changed by putting in a plug with a different pressure setting. Pumps without a pressure relief valve use a diverter valve that has a relief valve.

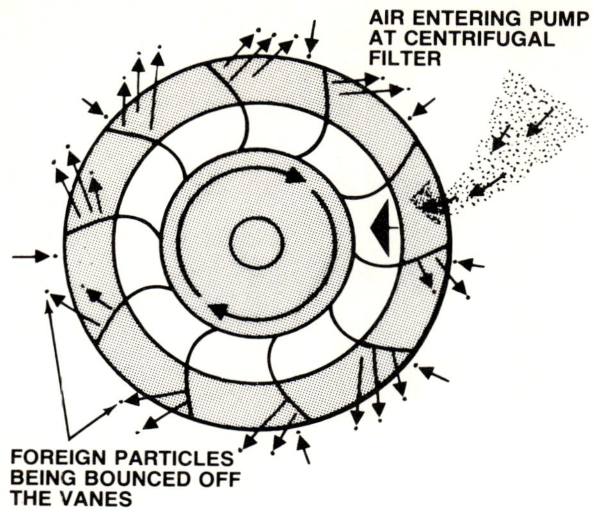

Figure 17-4. Dust and dirt are removed from the inlet air by centrifugal force. (Chevrolet)

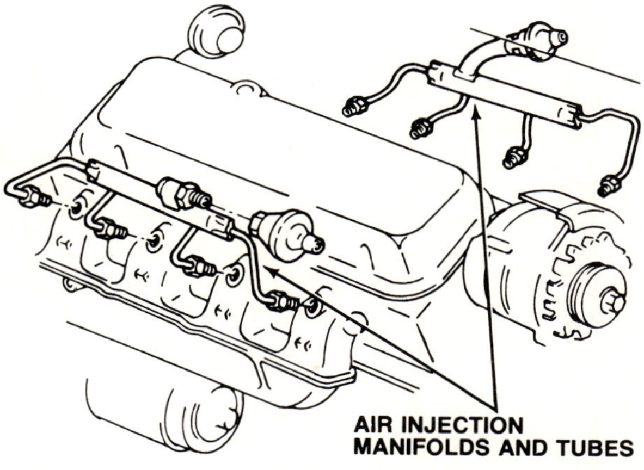

Figure 17-6. External air manifolds are used with many air injection systems. (Chevrolet)

Air Manifolds And Nozzles

In early air injection systems, air is delivered to the engine's exhaust system in one of two ways:
1. An external air manifold, figure 17-6, distributes the air through injection tubes to the exhaust port near each exhaust valve. This method is used mainly on smaller engines.
2. An internal air manifold, figure 17-7, distributes the air to the exhaust port near each exhaust valve through passages in the cylinder head or the exhaust manifold. This method is used mainly with larger engines.

The fresh air from the pump passes through the air injector tubes or the manifold to nozzles in the exhaust ports. These are made of stainless steel to resist the high exhaust temperatures.

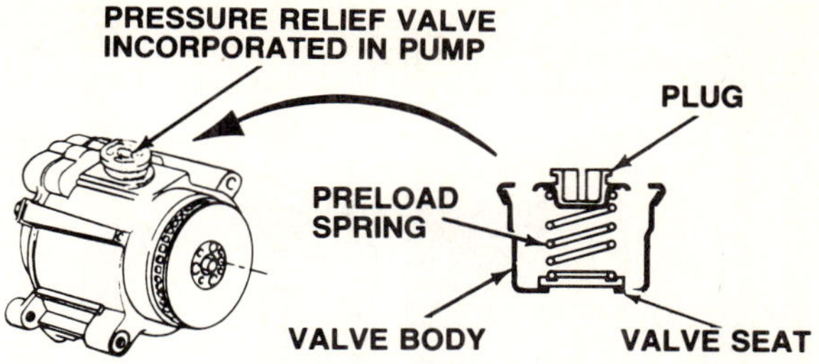

Figure 17-5. Some air pumps have built-in pressure relief valves. (AC-Delco)

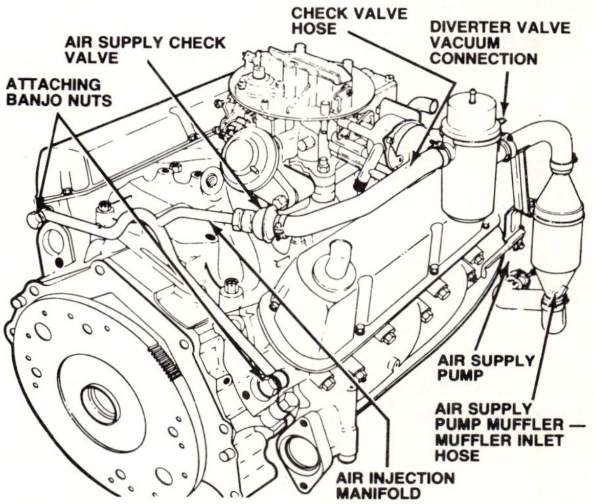

Figure 17-7. Some engines have air distribution passages built into the cylinder heads. (Ford)

Antibackfire Valve

During engine deceleration, high intake manifold vacuum richens the air-fuel mixture. If air is allowed to flow into the exhaust manifold during deceleration, it will combine with excess unburned fuel in the exhaust. The result is engine **backfire** — a rapid combustion of the

Air Injection: A way of reducing exhaust emissions by injecting air into each of the exhaust ports of an engine. It mixes with the hot exhaust and oxidizes the HC and CO to form H_2O and CO_2.

Backfire: The accidental combustion of gases in an engine's intake or exhaust system.

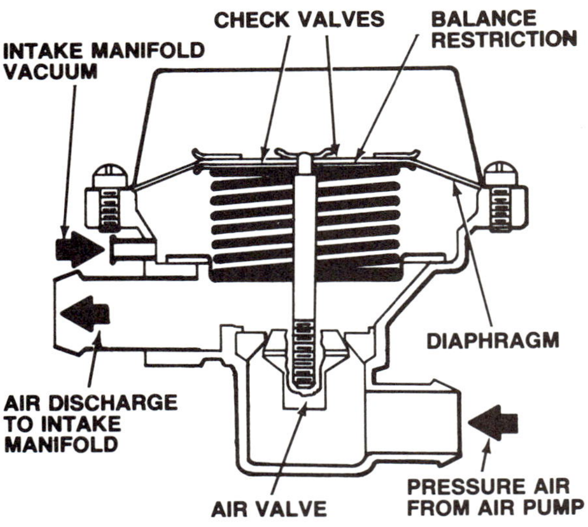

Figure 17-8. The gulp valve was used on early air injection systems.

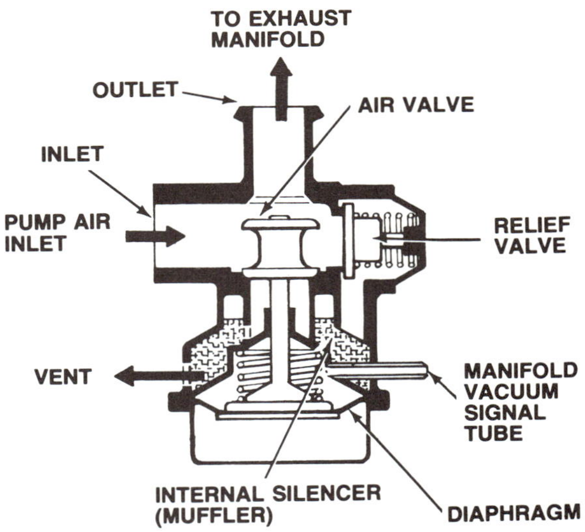

Figure 17-9. The diverter valve is used on most late-model air injection systems. This one also has a pressure relief valve.

unburned gases that can blow a muffler apart. To prevent engine backfire, the air pump flow must be shut off during deceleration. This is done by the antibackfire, or backfire suppressor, valve. Two kinds of valves have been used: the gulp valve and the diverter valve.

Gulp valve
Early air injection systems used a **gulp valve**, figure 17-8. When intake manifold vacuum is applied to the valve diaphragm, it causes the air valve to move. This redirects the pump air to the intake manifold to lean out the enriched air-fuel mixture during deceleration.

The gulp valve is connected to the intake manifold by two hoses. The large hose is the air discharge hose. The small hose is the sensing hose that sends manifold vacuum to the gulp valve to operate the diaphragm. There is a balance restriction, or bleed hole, inside the valve that equalizes pressure on both sides of the diaphragm after a few seconds. Even if manifold vacuum is high, the gulp valve only stays open for a few seconds until pressure equalizes.

Any sudden change in vacuum will operate the gulp valve. This is one of the undesirable features that led to its replacement. For example, the gulp valve will open when the engine starts. This can cause hard starting and a rough idle. Another problem with the gulp valve is that when the throttle is closed at high speeds when manifold vacuum is low, the valve may not open for a few seconds. The gulp valve has largely been replaced by the diverter valve on late-model air injection systems.

Diverter valve
The **diverter valve**, figure 17-9, is also called the dump valve or the bypass valve. Like the gulp valve, the diverter valve uses a diaphragm, operated by manifold vacuum, to redirect the airflow from the air pump. However, the pump air passes through the diverter valve continuously on its way to the air injection manifold. During deceleration, manifold vacuum operates the valve diaphragm to divert, or dump, the air directly to the atmosphere, not to the intake manifold. Some diverter valves vent the air to the engine air cleaner for muffling. Others vent it through a muffler and filter built into the valve. Because diverter valves do not affect the air-fuel mixture in the intake manifold, they are more troublefree than the gulp valves.

Some diverter valves also have a pressure relief valve to keep the pump from building up too much pressure in the system. This kind of diverter valve is used with an air pump that does not have a built-in relief valve.

Check Valve

A 1-way check valve, figure 17-10, protects the air pump from harmful reverse flow of exhaust gases from the engine. The check valve is between the air manifold and the diverter valve or the gulp valve. If exhaust pressure is higher than air injection pressure, or if the air pump fails, the check valve spring closes the valve to prevent the reverse flow of exhaust. A single check valve usually is installed on inline engines, and two valves (one per cylinder bank) usually are installed on V-type engines. However, some V-type engines have only one check valve, figure 17-7.

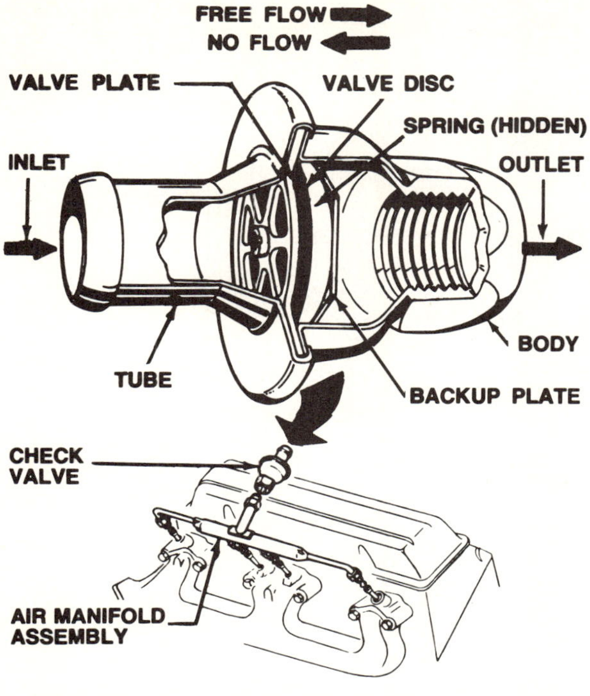

Figure 17-10. The check valve protects the system against reverse flow of exhaust gases. (AC-Delco)

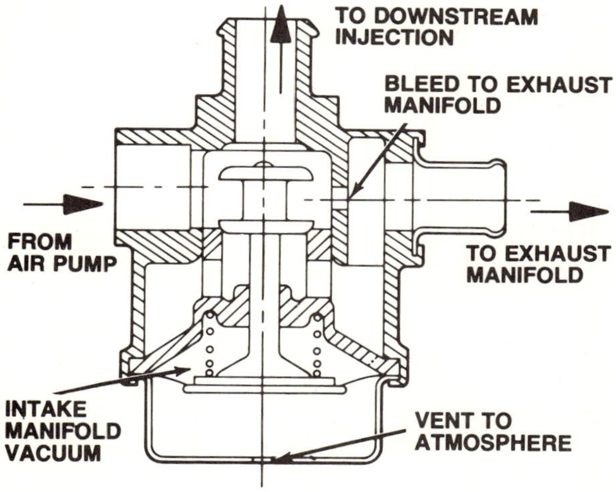

Figure 17-12. Cutaway of Chrysler's air switching valve. (Chrysler)

SECOND-GENERATION AIR INJECTION SYSTEMS

Air injection systems used with catalytic converters do the same thing as the basic air injection system just described. They help to oxidize the HC and CO in the exhaust gases by adding fresh air into the exhaust system. These newer, second-generation systems use many of

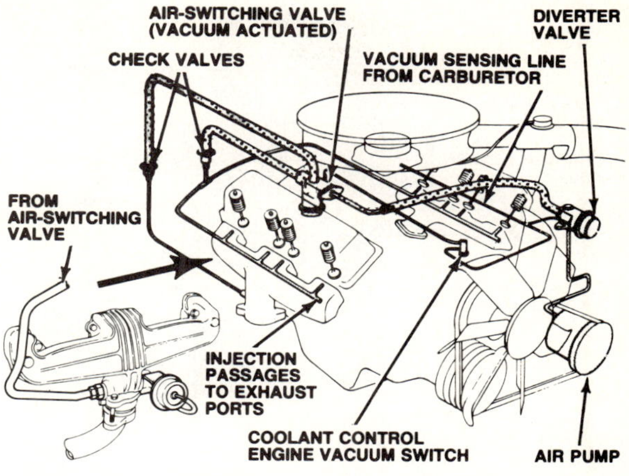

Figure 17-11. Chrysler introduced this air switching system in 1977. (Chrysler)

the same parts as the non-catalytic systems. However, more controls are needed for effective use with the catalytic converter. Specific automakers' systems are explained in the following paragraphs.

Chrysler Catalyst Air Injection System

Some 1977 California engines use an air switching valve, figure 17-11, to switch air pump output from the exhaust ports to the right exhaust manifold. The valve is operated by an engine coolant vacuum switch. When the coolant warms up enough, the vacuum switch routes vacuum to the air switching valve. This causes the valve to switch the fresh air from the exhaust ports "downstream" to the exhaust manifold. A bleed hole in the valve, figure 17-12, allows a small continuous airflow to the exhaust ports.

By switching the fresh airflow after engine warmup, the system helps oxidize HC and CO in the catalytic converter without interfering with the NO_x control by the EGR system.

Gulp Valve: A valve used in an air injection system to prevent backfire. During deceleration it redirects air from the air pump to the intake manifold where the air leans out the rich air-fuel mixture.

Diverter Valve: Also called a dump valve. A valve used in an air injection system to prevent backfire. During deceleration it "dumps" the air from the air pump into the atmosphere.

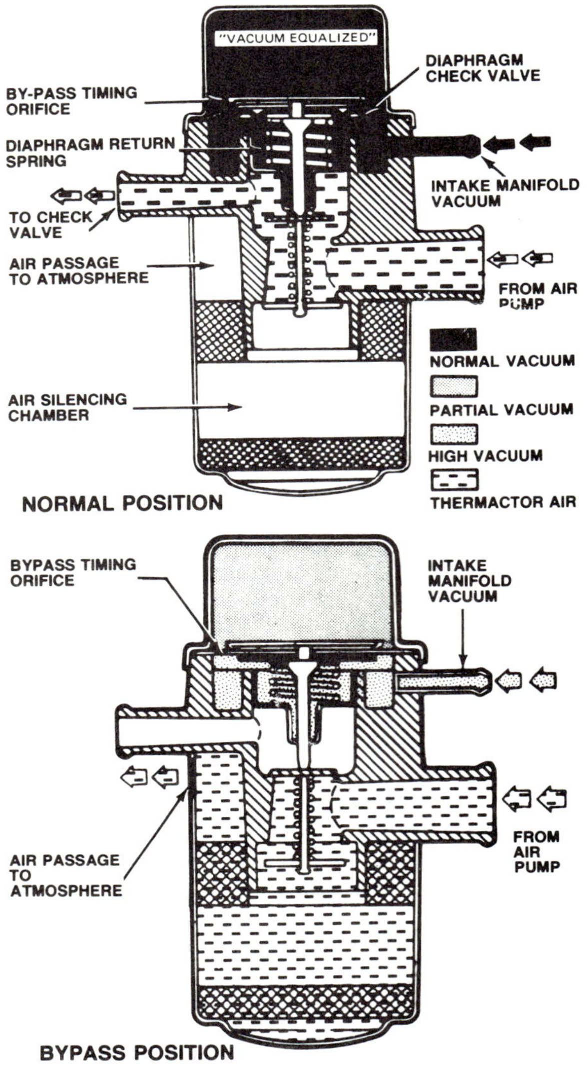

Figure 17-13. Ford diverter valve operation. (Ford)

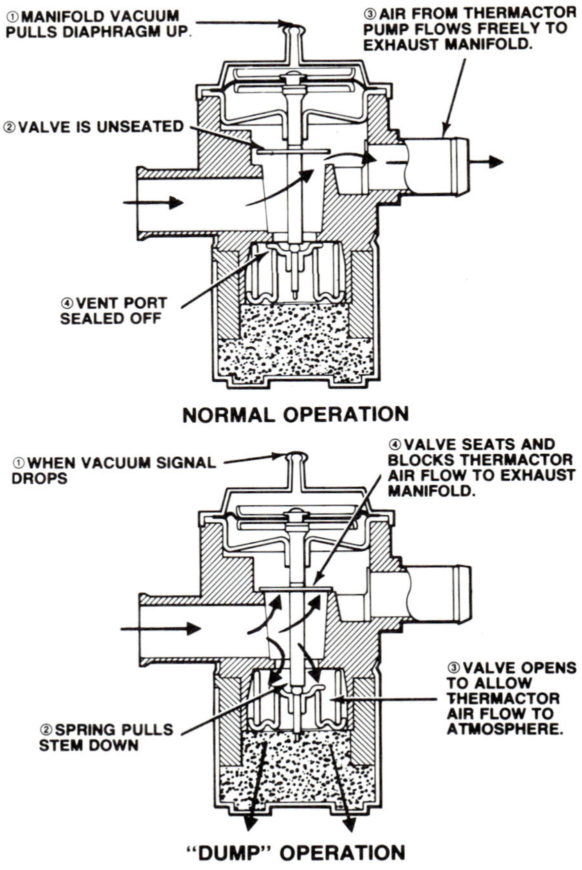

Figure 17-14. This Ford diverter valve, introduced in 1975, uses vacuum to hold the air passage *open* to the exhaust manifold. (Ford)

Ford Thermactor Air Injection System

Ford air injection systems used with catalytic converters have become quite complicated since 1975. We could not possibly cover all the variations here, but the major parts are described in the following paragraphs.

A large-capacity air pump without a relief valve is used on all 1975 and later Ford air injection systems. The system relief valve is built into the diverter valve. Ford uses three different kinds of diverter valves on late-model systems. (Ford calls its diverter valves "timed air bypass valves".)

One Ford valve, figure 17-13, works like the usual diverter valve we just described. Ford introduced another diverter valve in 1975 that uses manifold vacuum to hold the valve open and deliver air to the exhaust, figure 17-14. During cruising, manifold vacuum is applied to the valve diaphragm. During deceleration, long periods of idle, and cold engine operation, vacuum is cut off from the valve. This dumps the air pump output to the atmosphere. Various vacuum differential valves, vacuum delay valves, and bleed valves are used to control the vacuum to the diverter valve.

The vacuum differential valve, figure 17-15, is in the vacuum line to the diverter valve. Sudden changes in vacuum cause the differential valve to dump the vacuum that goes to the diverter valve. This causes the diverter valve to cut off the air injection and dump the air pump output to the atmosphere. This particular diverter valve, figure 17-14, can also be operated with a vacuum delay valve, an air temperature vacuum switch, and a ported vacuum signal taken above the throttle plates in the carburetor.

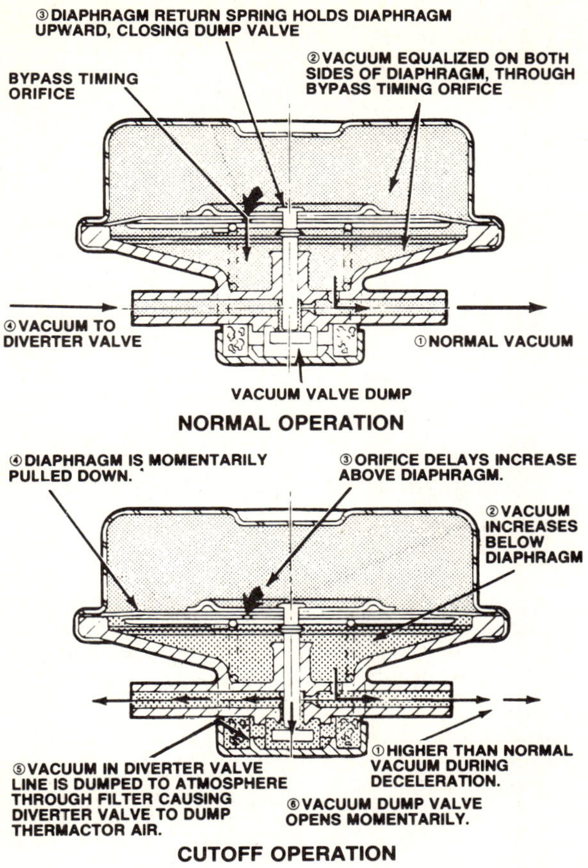

Figure 17-15. This vacuum differential valve dumps the vacuum from the diverter valve whenever vacuum increases suddenly. (Ford)

The third Ford diverter valve, figure 17-16, has a vacuum differential valve built into it. The top of the valve also has a vacuum vent. When the vent is closed, the valve acts as a normal diverter valve. When the vacuum vent is open, any manifold vacuum above 4 inches of mercury causes the diverter valve to dump the air pump output. An air temperature electric switch and an electric solenoid open the vent line at cold temperatures.

General Motors Air Injection Reactor System

General Motors air injection system has been modified on some converter-equipped engines. The newer system is called the converter air injection reactor (CAIR) or converter air injection (CAI) system. The AIR designation for the older system may also be called manifold air injection (MAI).

The air pump, the diverter valve, and the check valve are the same in both converter and

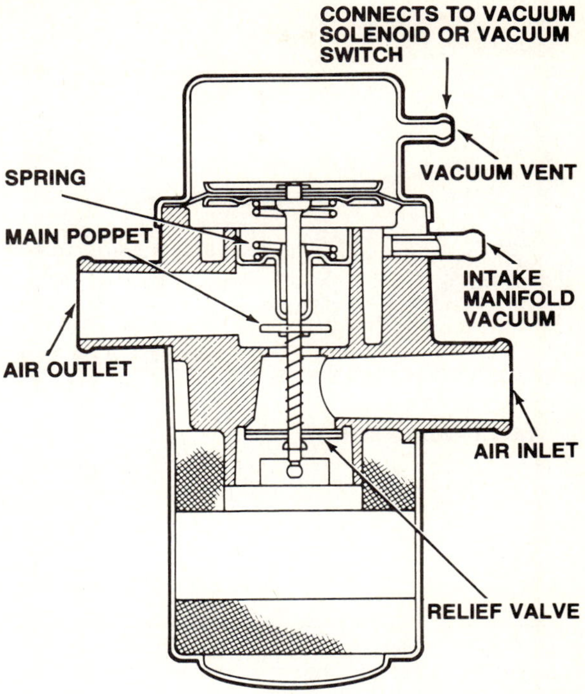

Figure 17-16. This Ford diverter valve has a vacuum vent that allows air to be dumped during cold engine operation. (Ford)

non-converter systems. In the CAI or CAIR system, air is injected into the exhaust pipe, downstream from the exhaust manifold, figure 17-17. This downstream injection delivers the air directly to the catalytic converter to aid HC and CO oxidation. Keeping the air out of the exhaust manifold also keeps the combustion chamber temperature lower, which reduces NO_x emissions.

On some late-model GM cars (mainly those without catalytic converters), the air is still injected into the exhaust manifold. On these engines, this is the hottest place in the exhaust system to oxidize HC and CO.

PULSE AIR INJECTION

Since 1975, the use of pulse air injection has increased on GM, Chrysler, and Ford products. Pulse air injection systems do not use an air pump. Instead, they use a pulse air valve, figure 17-18, which is similar to the check valve in an air pump system. It is a spring-loaded, diaphragm or reed valve that is connected to the exhaust system.

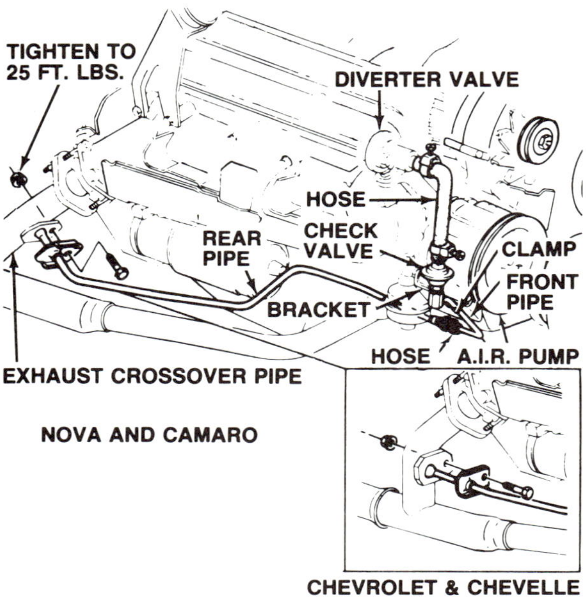

Figure 17-17. General Motors' converter air injection delivers the air to the exhaust pipe downstream from the exhaust manifold. (Chevrolet)

Figure 17-19. Pulse air injection was pioneered on Chevrolet's Cosworth Vega in 1975.

Each time an exhaust valve closes, there is a period when pressure in the manifold drops below atmospheric pressure. During these low pressure (vacuum) pulses, the air injection valve opens to admit fresh air to the exhaust. When exhaust pressure rises above atmospheric pressure, the valve acts as a check valve and closes.

The pulse air injection valve works best at low engine speed when extra air is needed most by the catalytic converter. The vacuum pulses occur too fast at high speed for the valve to follow, and the internal spring simply keeps the valve closed.

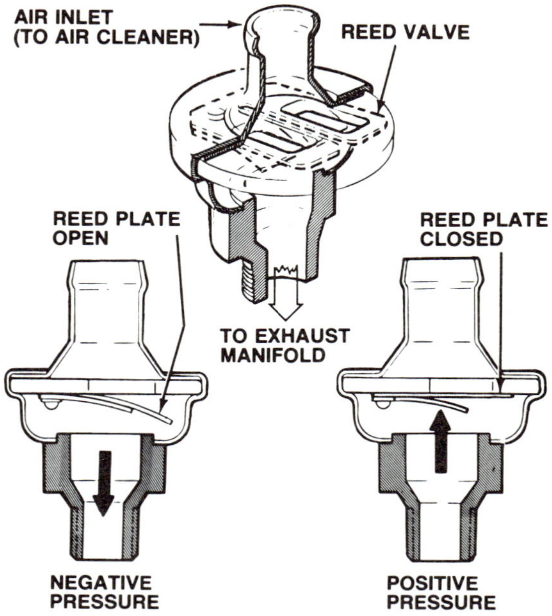

Figure 17-18. A pulse air injection valve opens when negative pressure occurs in the exhaust manifold. (Ford)

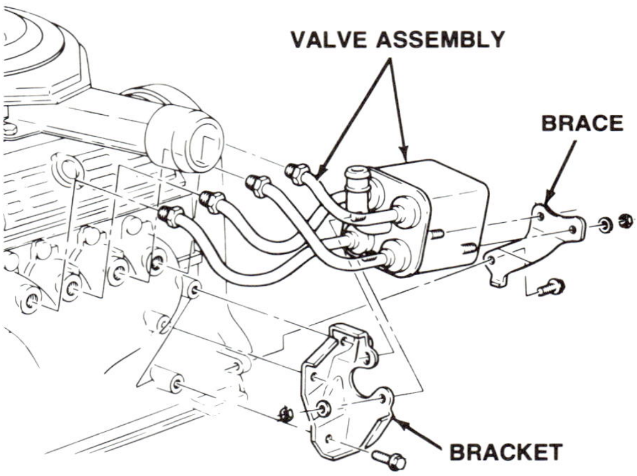

Figure 17-20. Pulse air injection was installed on the Chevette and other GM small cars in 1977. (Chevrolet)

Pulse air injection was first used on Chevrolet's Cosworth Vega in 1975 and 1976, figure 17-19. The system was added to other GM 4-cylinder engines in 1977, figure 17-20. Chrysler introduced a similar system in 1977, called an air aspirator system. The pulse injection (air aspirator) valve draws its fresh air from the clean side of the engine air filter, figure 17-21. In 1977, Ford installed a pulse air injection valve on the 300-cid 6-cylinder engines in some vans. This system was added to the 6-cylinder engines in Fairmonts and Zephyrs in 1978.

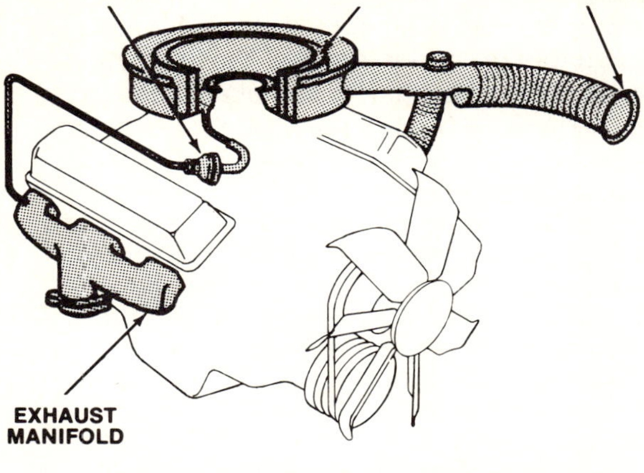

Figure 17-21. Chrysler's air aspirator system.

SUMMARY

Air injection is one of the oldest methods used to control HC and CO exhaust emissions. The injected air mixes with hot exhaust gas as it leaves the combustion chambers to further oxidize HC and CO emissions. All air injection systems used on domestic cars operate in essentially the same way, regardless of manufacturer. With slight modifications, the air injection system also is used on engines equipped with catalytic converters to promote converter action.

The pulse air injection system relies upon exhaust pulsations instead of an air pump to draw fresh air into the exhaust.

Review Questions

Select the single most correct answer.
Compare your answers with the correct answers on page 280.

1. Ford calls its air injection system:
   a. Thermactor Air Injection System
   b. Air Guard
   c. Air Injector Reactor (AIR)
   d. Air-Master

2. All air injection systems have the following parts *except*:
   a. Air supply pump
   b. Air manifolds and nozzles
   c. Manifold vacuum pump
   d. Antibackfire valve

3. Which of the following is *not* true of the air pump:
   a. Saginaw Division of GM makes all domestic air pumps
   b. 2-vane pumps use a separate air filter
   c. 3-vane pumps can be rebuilt if necessary
   d. The pump is driven by a belt from the crankshaft

4. Which of the following is true:
   a. External air manifolds are used mainly on larger engines
   b. Internal air manifolds are used mainly on larger engines
   c. Internal air manifolds are made of aluminum
   d. External air manifolds are made of tungsten alloy

5. Which of the following prevents engine backfire:
   a. A gulp valve
   b. A diverter valve
   c. A 1-way check valve
   d. Both A and B

6. The diverter valve is also called a:
   a. Dump valve
   b. Bypass valve
   c. Neither A nor B
   d. Both A and B

7. One of the disadvantages to a gulp valve is that:
   a. Changes in vacuum will not operate the valve
   b. Any sudden change in the air-fuel mixture will operate the valve
   c. Any sudden change in vacuum will operate the valve
   d. It allows air pump flow to continue during all operating conditions

8. The illustration below shows:

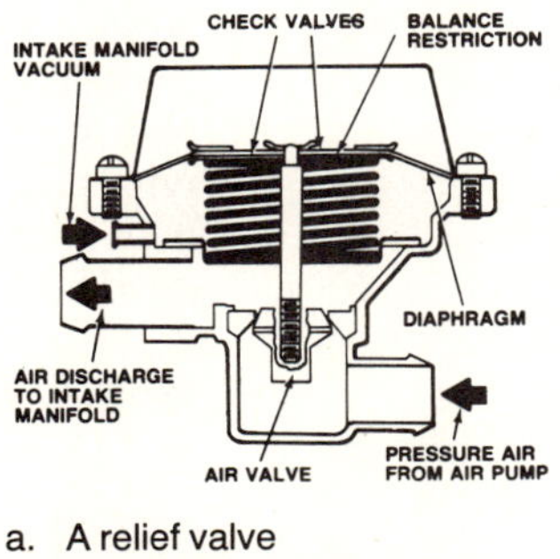

   a. A relief valve
   b. A solenoid operated vacuum valve
   c. A gulp valve
   d. A diverter valve

9. Chrysler's catalyst air injection system:
   a. Helps oxidize HC and CO
   b. Helps oxidize HC and NO_x
   c. Helps oxidize NO_x and HCO
   d. Helps oxidize CO and NO_x

10. On 1975 and later Ford air injection systems, relief capacity for the air pump is provided by:
    a. A relief valve built into the pump
    b. A relief valve built into the diverter valve
    c. A relief valve built into the gulp valve
    d. A relief valve built into the air manifold

11. New GM air injection systems are called:
    a. MAI
    b. CAIR
    c. Timed Bypass System
    d. Vacuum Differential System

12. Pulse air injection works best at:
    a. Low speeds
    b. High speeds
    c. Idle
    d. Deceleration

Chapter

18 Exhaust Gas Recirculation

Even though the elements of photochemical smog had been identified many years earlier, it was not until 1973 that the U.S. Federal Government told the nation's carmakers to drastically reduce the amount of oxides of nitrogen (NO_x) in car exhaust. In order to meet these new standards, the manufacturers developed **exhaust gas recirculation**, or **EGR**.

Under normal circumstances, nitrogen and oxygen do not combine to form NO_x unless temperatures are greater than 2,500° F (1,370° C). When it is hotter than this in the combustion chamber, large amounts of NO_x are produced. So, the job of the EGR system (as well as some other emission control systems, such as spark timing control) is to keep the combustion chamber temperatures below 2,500° F (1,370° C), but still hot enough to have nearly complete combustion.

Retarding the spark timing slightly will help to do the job, but EGR does it better. Exhaust gas recirculation works like this: the incoming air-fuel mixture is diluted with a small amount of **inert gas** to lower the combustion chamber temperature. Exhaust gases, because they are inert, or nearly so, are used to dilute the air-fuel mixture.

Since 1973, various valves and vacuum routings have been used by the carmakers. While these system parts may look different, they all work pretty much alike. Since the EGR system is important to good driveability, it is also an important part of a tune-up or troubleshooting procedure.

HOW EGR WORKS

Exhaust gas is recirculated by routing small amounts of gas (about 6 to 10 percent) from the engine's exhaust ports to the intake manifold, figure 18-1. This exhaust gas dilutes the incoming air-fuel mixture in the cylinder. Since the exhaust gas contains little or no oxygen, the resulting air-fuel mixture is not as powerful when ignited, and it does not create as much heat as the undiluted air-fuel mixture would produce.

How EGR Affects Combustion

Since it does not require much exhaust gas to cool down peak combustion temperatures, recirculation must be held to very low quantities. Even when the EGR valve used to reroute the exhaust gas is wide open, the orifice through which the gas passes is very small.

Because the amount of NO_x produced at low engine speeds is very small, exhaust recirculation is not needed or desirable at idle. It is also

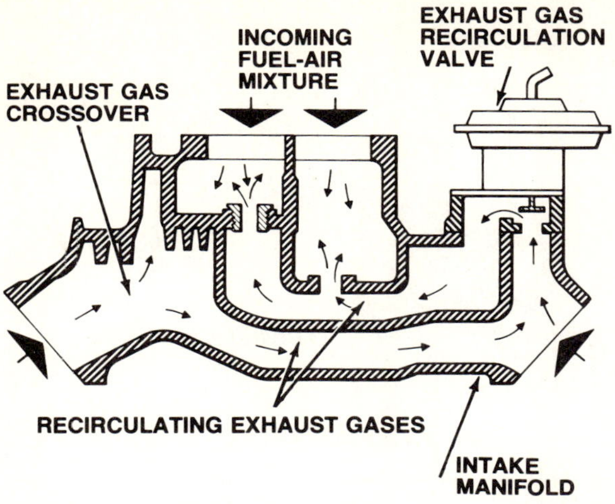

Figure 18-1. Basic exhaust gas recirculation methods.

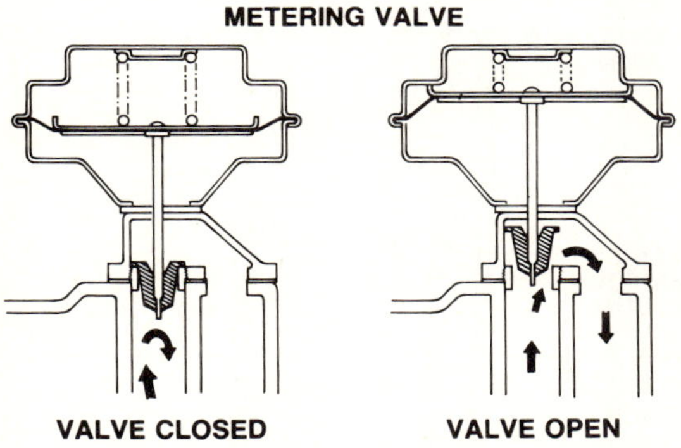

Figure 18-3. Typical EGR valve.

undesirable during high-speed driving at wide-open throttle, if efficient operation and good driveability are to be maintained. Maximum recirculation is necessary only during cruising and acceleration at speeds between 30 to 70 mph, when NO_x formation is greatest. Engine temperature is also important to recirculation. When engine temperature is low, NO_x formation is also low, and recirculation is eliminated to produce fast warmup and better driveability.

SYSTEM PARTS AND OPERATING PRINCIPLES

The devices used to recirculate exhaust gas are discussed below. Since system designs and controls differ from one manufacturer to another, we will discuss each of the carmakers later in this chapter.

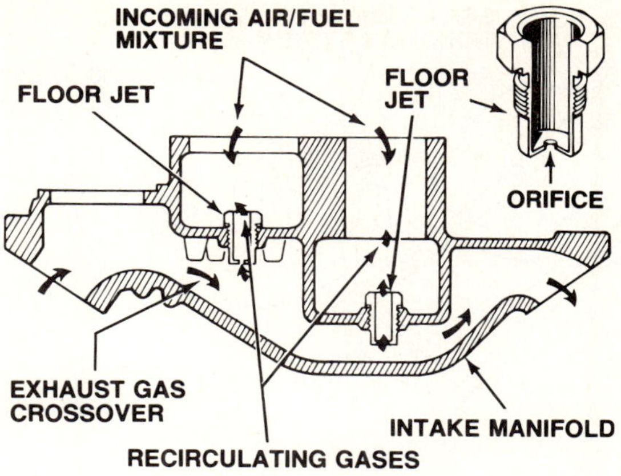

Figure 18-2. Chrysler's floor jet EGR system.

EGR Floor Jets

The floor jet system, figure 18-2, is used on some 1972-73 Chrysler-built engines. On V-8 engines, stainless steel jets are threaded into the floor of the intake manifold under the carburetor. With 6-cylinder engines, the jet is in the intake manifold hot spot beneath the carburetor. The jets provide an opening between the exhaust passage and the intake manifold. In this way, manifold vacuum controls how much exhaust is drawn into the intake system through the preset jet orifice. Floor jets are the simplest of all EGR system designs. However, they also are unsatisfactory because they allow exhaust gas to enter the intake manifold at all times. This causes rough engine operation at idle and during cold engine warmup.

EGR Valves

Introduced on 1972 Buicks, the EGR valve is a spring-loaded, vacuum-operated, poppet-type valve, figure 18-3. Modulating valves and tapered-stem valves are also used, but their operation is about the same as the poppet type. This valve meters the exhaust gas entering the intake system. The EGR valve is mounted on the intake manifold, figure 18-4, or on a plate under the carburetor, figure 18-5. The valve may be connected to the intake and exhaust

Exhaust Gas Recirculation (EGR): A way of reducing NO_x emissions by directing unburned exhaust back through an engine's intake.

Inert Gas: A gas that will not undergo chemical reaction.

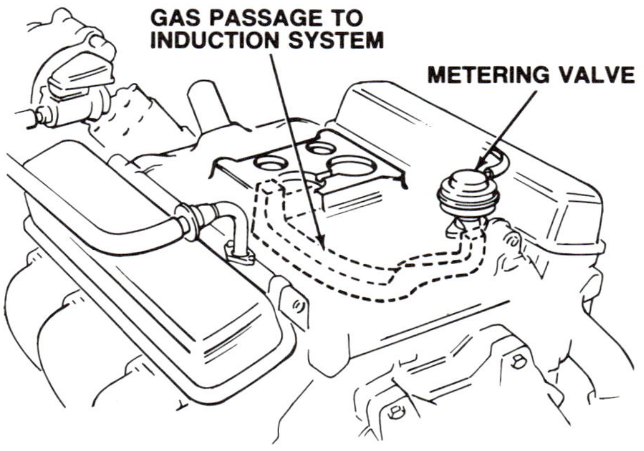

Figure 18-4. EGR metering valve mounted on intake manifold.

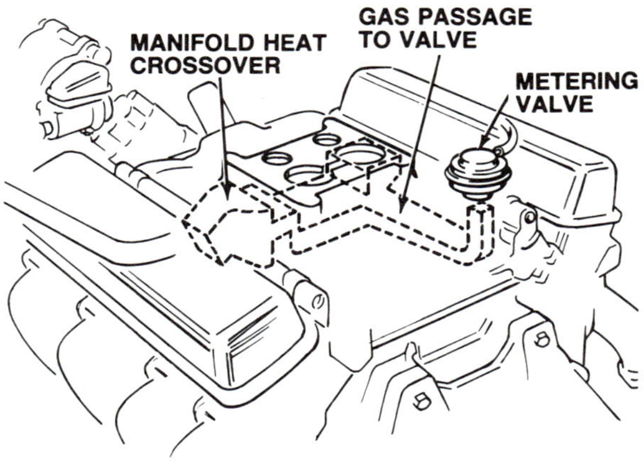

Figure 18-6. Exhaust gas passages to the EGR valve.

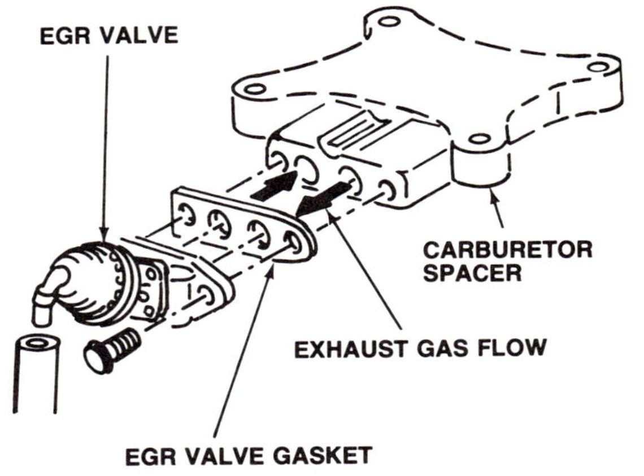

Figure 18-5. This EGR valve is mounted on a spacer that goes between the carburetor and the intake manifold.

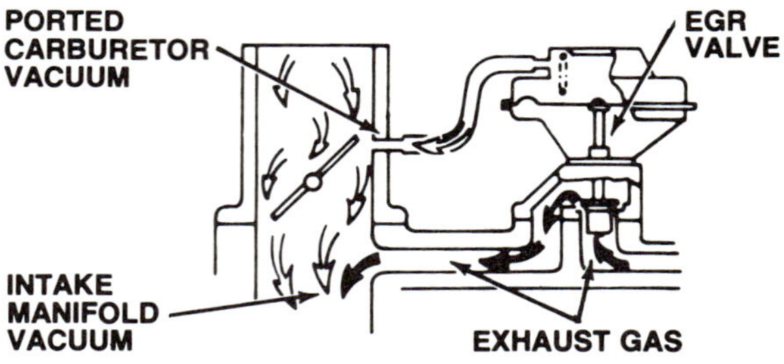

Figure 18-7. This EGR valve is operated by ported vacuum.

systems by internal passages in the intake manifold, figures 18-4 and 18-6, or in some cases by external steel tubing. The spring holds the EGR valve closed. The valve is opened by ported or venturi vacuum from the carburetor, depending upon the system design.

Ported Vacuum Systems

EGR systems controlled by ported vacuum use a slot-type port in the carburetor throttle body above the throttle plate, figure 18-7. This port is connected to the EGR valve by a vacuum line. With the throttle plate closed, no vacuum is transmitted. As the throttle plate opens, the port is exposed to increasing manifold vacuum. The amount of exhaust gas flow depends on manifold vacuum, throttle position, and exhaust gas backpressure. Since ported vacuum cannot be more than intake manifold vacuum, recirculation at wide-open throttle is prevented. The valve opens at a point greater than the relatively weak manifold vacuum produced during wide-open throttle operation.

Venturi Vacuum Systems

EGR systems controlled by venturi vacuum use a vacuum port at the throat of the carburetor venturi to provide a control vacuum, figure 18-8. Since this control vacuum is very weak, a vacuum amplifier boosts it enough to operate the EGR valve. The amount of exhaust gas flow depends mainly on engine intake airflow. It is also affected by intake vacuum and exhaust gas backpressure.

Recirculation at wide-open throttle is prevented by a relief valve, or dump diaphragm, that compares venturi and manifold vacuum. When the diaphragm senses that the throttle is wide open, the stored vacuum in the amplifier is vented to the atmosphere. This means that the vacuum output reaching the EGR valve is no more than that provided by manifold vacuum. Since the EGR valve opens at a point greater than the relatively weak manifold vac-

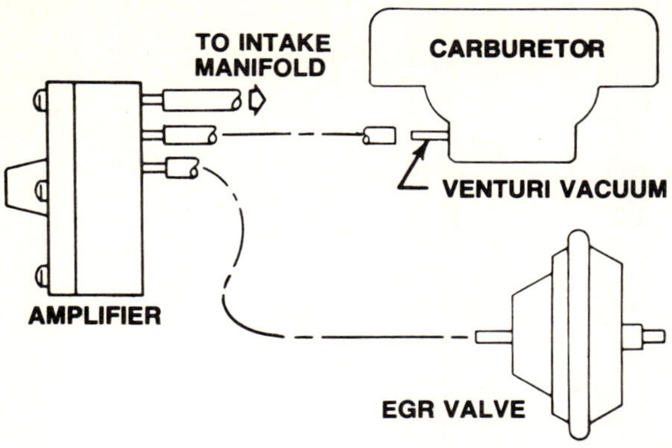

Figure 18-8. EGR systems controlled by venturi vacuum require a vacuum amplifier. (Ford)

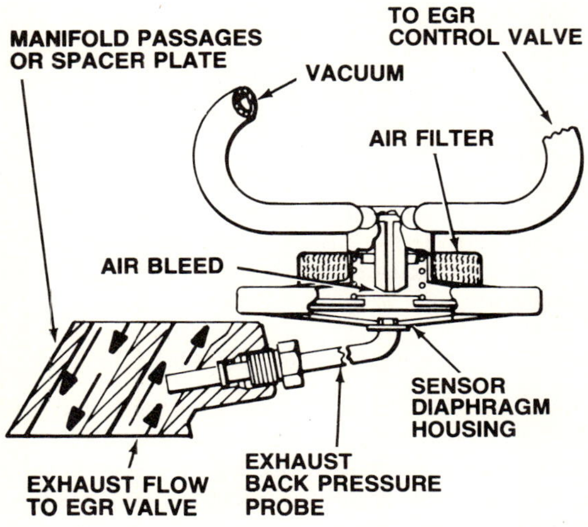

Figure 18-10. Typical EGR backpressure transducer.

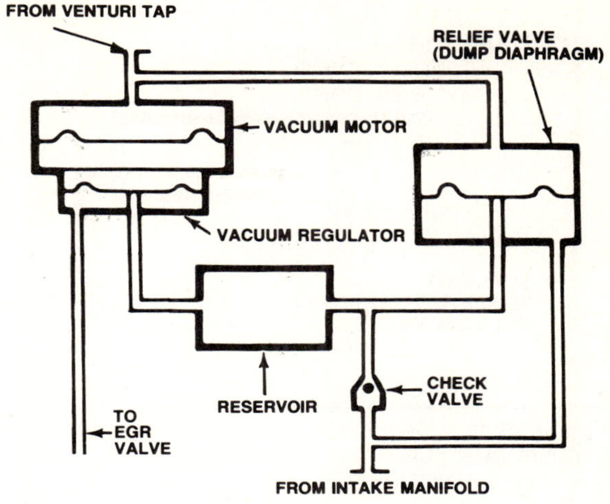

Figure 18-9. A vacuum amplifier is controlled by venturi vacuum and uses manifold vacuum to operate the EGR valve. (Ford)

Backpressure Transducer

This diaphragm-operated sensor unit, figure 18-10, has a tube that extends into an exhaust gas passage. When high exhaust backpressure is sensed through the tube, the diaphragm closes an air bleed hole in the EGR vacuum line. This provides maximum EGR during acceleration, when backpressure is high. As back-

■ Engine Modifications Can't Do The Whole Job

Since the automobile was discovered to be the biggest source of air pollution, many changes in engine design have been made to "clean it up." These engine modifications, such as EGR systems, have reduced exhaust emissions, but it is impossible to eliminate the major cause of HC emissions simply by changing the engine design. This is because the major cause of HC emissions is the effect of the "quench area" on combustion.

The quench area is the inner surface of the combustion chamber. When the ignition flame front passes through the combustion chamber, it burns the fuel charge as it goes until the quench area is reached. This is a thin layer between .002" and .010" thick at the edge of the combustion chamber. When the flame front reaches the quench area, it is snuffed out because the quench area is so close to the cylinder head water jacket that the temperature there is too low for combustion to continue. Consequently, hydrocarbons within the quench area do not burn. They are ejected from the cylinder on every exhaust stroke along with the exhaust gases formed by combustion and enter the atmosphere as pollutants.

uum, the valve remains closed at wide-open throttle.

Vacuum amplifiers

The vacuum amplifier, figure 18-9, converts the weak venturi control vacuum into one that is strong enough to operate the EGR valve. It does this by storing manifold vacuum in a reservoir inside the amplifier unit. This guarantees enough vacuum, regardless of variations in manifold vacuum. Whenever the venturi vacuum is equal to, or greater than, manifold vacuum, a relief valve (dump diaphragm) vents the reservoir, cancelling the output EGR signal.

If the vacuum amplifier works perfectly, it will produce an accurate, repeatable, and almost precise proportion between venturi airflow and EGR flow. But vacuum amplifiers are not as commonly used on late-model EGR systems because storage reservoirs tend to leak in actual use.

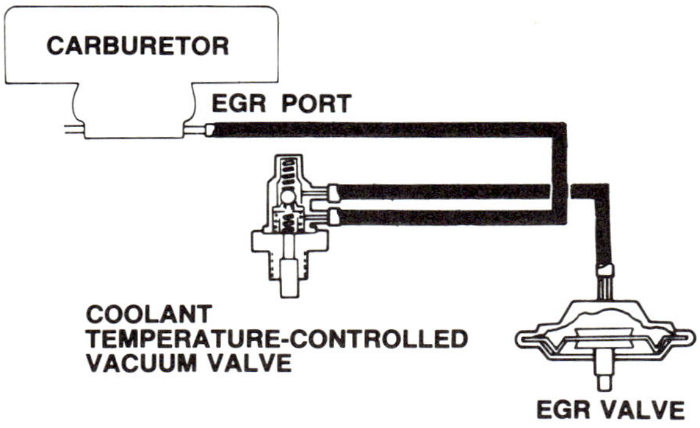

Figure 18-11. The coolant-temperature-controlled vacuum valve cuts off vacuum to the EGR valve when the engine is cold. (Ford)

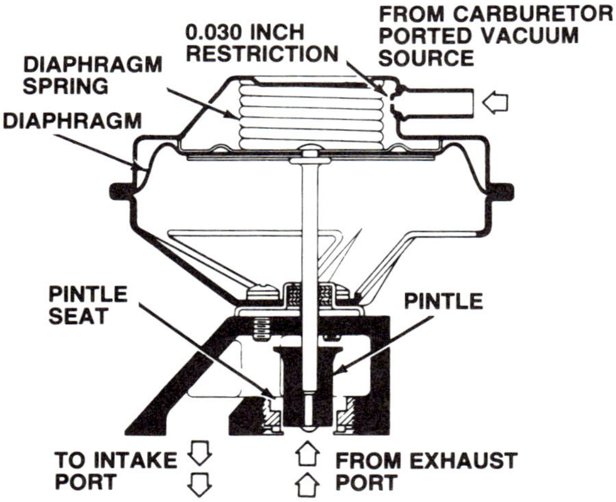

Figure 18-13. This American Motors EGR valve is typical of most single-diaphragm EGR valves. (AMC)

pressure drops, a spring moves the sensor diaphragm to reopen the vacuum line bleed. Since this decreases the vacuum at the EGR valve, the amount of exhaust gas recirculated is also reduced.

Modulating Devices

Engineers have developed various methods of modulating, or adjusting, EGR valve operation in relation to engine operating conditions. High and low ambient temperature vacuum modulators weaken the vacuum signal to the EGR valve. A coolant temperature override switch or valve, figure 18-11, may be used to eliminate EGR vacuum below certain engine operating temperatures. Some venturi-vacuum-controlled systems have a time delay solenoid to shut off vacuum to the EGR valve for about half a minute after the ignition is turned on.

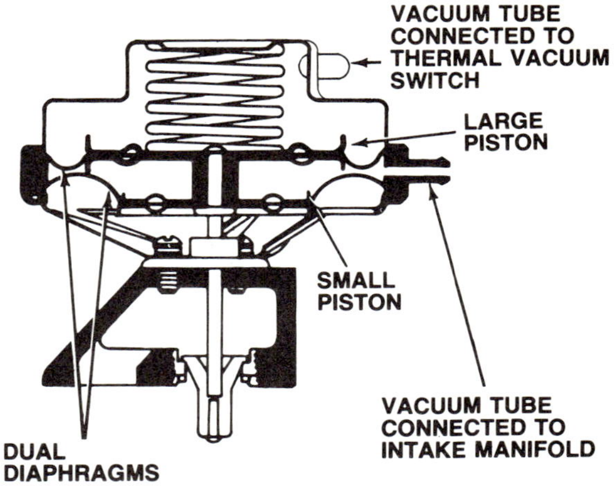

Figure 18-12. The dual-diaphragm EGR valve modulates the exhaust gas flow.

A vacuum-bias valve is used to bleed off part of the EGR vacuum signal under high-manifold-vacuum conditions to eliminate **high-speed surge**. A dual-diaphragm EGR valve, figure 18-12, is sometimes used. This uses manifold vacuum to help the valve spring offset the carburetor vacuum under certain cruising conditions.

AMERICAN MOTORS EGR SYSTEMS

The basic AMC EGR system, as introduced on 1973 models, uses a vacuum-operated EGR valve. When the carburetor throttle is opened beyond the idle position, ported vacuum is applied to the normally closed EGR valve diaphragm, figure 18-13. Moving upward against coil spring pressure, the diaphragm opens the **pintle valve**. This valve permits exhaust gas to be drawn into the engine intake from the manifold crossover exhaust passage on V-8 engines, figure 18-14, or from below the carburetor heat riser on 6-cylinder engines, figure 18-15. The EGR valve is closed during idle and deceleration to prevent a rough idle. California cars use an exhaust backpressure sensor, figure 18-10.

American Motors also uses the following override and modulator devices to control EGR system operation.

Coolant Temperature Override Switch

This EGR coolant temperature override (CTO) prevents normal EGR valve operation when engine coolant temperature is below 115° F (46° C). The 3-port switch originally used has been replaced by a 2-port switch, figure 18-16, on 1974 and later models. At coolant temperatures above 115° F (160° F on some models), vacuum is permitted to reach the EGR valve.

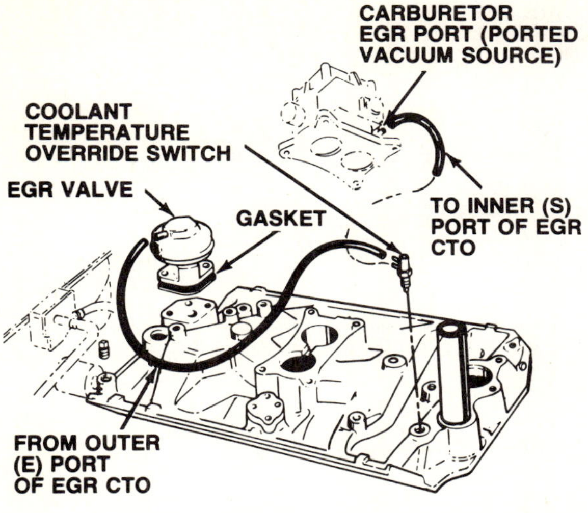

Figure 18-14. Basic AMC V-8 EGR installation. (AMC)

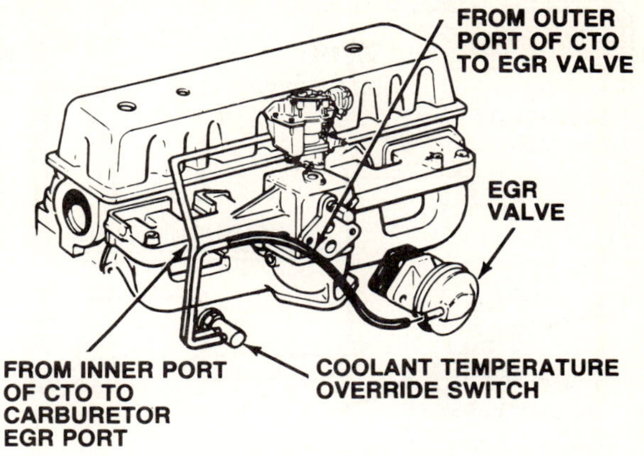

Figure 18-15. Basic AMC 6-cylinder EGR installation. (AMC)

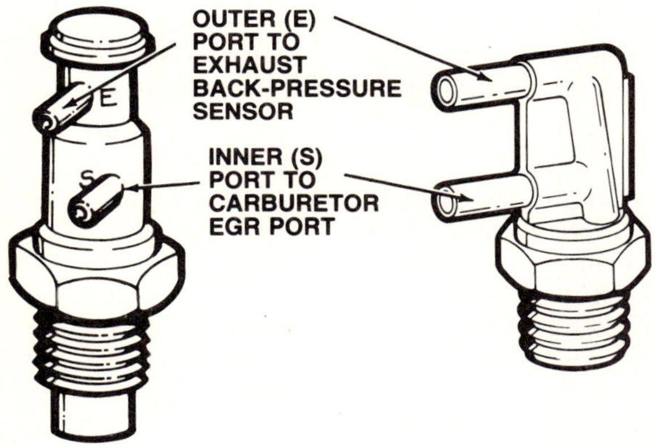

Figure 18-16. American Motors' 2-port EGR coolant temperature override (CTO) switch. (AMC)

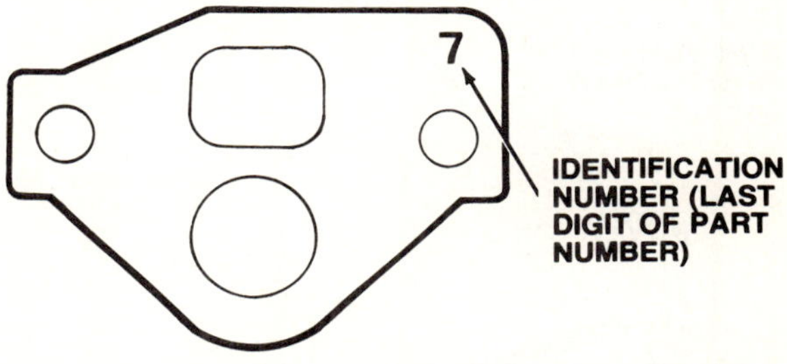

Figure 18-17. American Motors' EGR restrictor plate. (AMC)

Low- And High-Temperature Vacuum Signal Modulators

Used only on 1973 models, these modulators are connected to the EGR vacuum control line. When air temperature is under 60° F (16° C), the low-temperature modulator weakens the vacuum to the EGR valve. This decreases the amount of exhaust gas flow. At temperatures above 115° F (46° C), the high-temperature modulator restores vacuum to the EGR valve and increases exhaust gas flow.

Exhaust Backpressure Sensor

Fitted only on California cars during 1973-75, use of the backpressure sensor, figure 18-10, was extended to all 1976 and later models. The sensor permits EGR flow only when the engine is at normal operating temperature and exhaust backpressure is high. If backpressure is not high enough, and the EGR-CTO switch does not open, the EGR vacuum signal is vented to the atmosphere.

Restrictor Plate

A stainless steel restrictor plate, figure 18-17, is used on some 1974-75 California models, and on all 1976-77 cars. Located between the intake

High-Speed Surge: A sudden increase in engine speed caused by high manifold vacuum pulling in too much air-fuel mixture.

Pintle Valve: A valve shaped much like a hinge pin. In an EGR valve, the pintle is attached to a normally closed diaphragm. When ported vacuum is applied, the pintle rises from its seat and allows exhaust gas to be drawn into the engine's intake system.

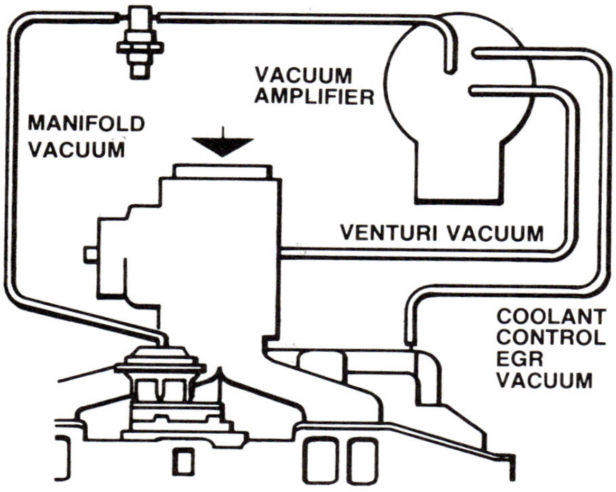

Figure 18-18. Chrysler's external vacuum reservoir.

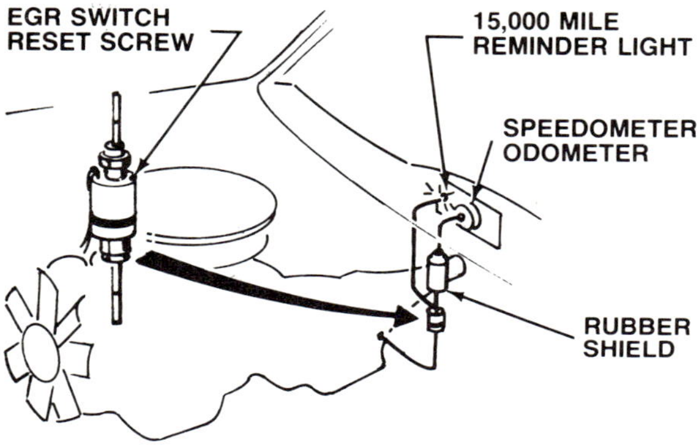

Figure 18-20. Chrysler's EGR maintenance reminder system.

manifold and spacer, the plate is adjusted for each engine and exhaust system. It limits EGR flow rate and improves driveability.

CHRYSLER EGR SYSTEMS

Chrysler used the floor jet system, figure 18-2, on some 1972-73 models. Stainless steel jets with a preset orifice are installed in the floor of the intake manifold beneath the carburetor. Engine vacuum draws a small amount of exhaust gas into the intake to dilute the air-fuel mixture. This system was supplemented in 1973 with both the ported and venturi vacuum systems. The floor jet system was discontinued after 1973.

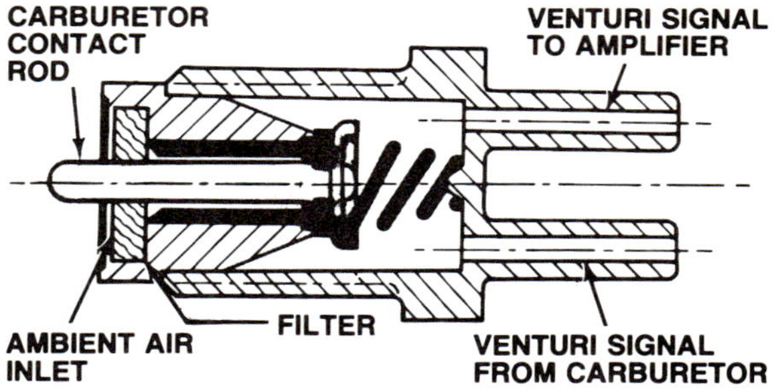

Figure 18-19. Wide-open-throttle EGR dump valve. (Chrysler)

Venturi Vacuum System

The venturi vacuum system was changed on some 1975 and later engines to include an external vacuum reservoir mounted on a bracket and attached to the vacuum amplifier, figure 18-18. After the internal vacuum reservoir has been vented, or dumped, the external reservoir supplies manifold vacuum for exhaust recirculation until the internal reservoir can be refilled.

Wide-Open-Throttle Dump Valve

A wide-open-throttle dump valve, figure 18-19, was added to some engines in March, 1976. These engines use a delay-dump amplifier, which can hold the EGR valve open too long at wide-open throttle. Mounted on the carburetor, the wide-open-throttle EGR dump valve overrides the delay dump amplifier at wide-open or near wide-open throttle by mechanically bleeding the vacuum to the amplifier. This closes the EGR valve immediately.

Coolant-Control EGR Valve

A coolant-control EGR valve (CCEGR) added to the 1974 system blocks vacuum flow to the EGR valve whenever coolant temperature is less than a certain preset level. Some 1974 and later engines use an EGR delay timer. This electric timer, mounted on the firewall, operates an engine-mounted vacuum solenoid to prevent exhaust gas recirculation for about half a minute after the ignition is turned on.

EGR Maintenance Reminder System

An EGR maintenance reminder system, figure 18-20, is on 1975 engines only. This mileage counting device signals the driver every 15,000 miles to have the EGR system inspected. An instrument panel reminder light comes on at 15,000-mile intervals, and remains lighted until the switch attached to the speedometer cable is reset manually.

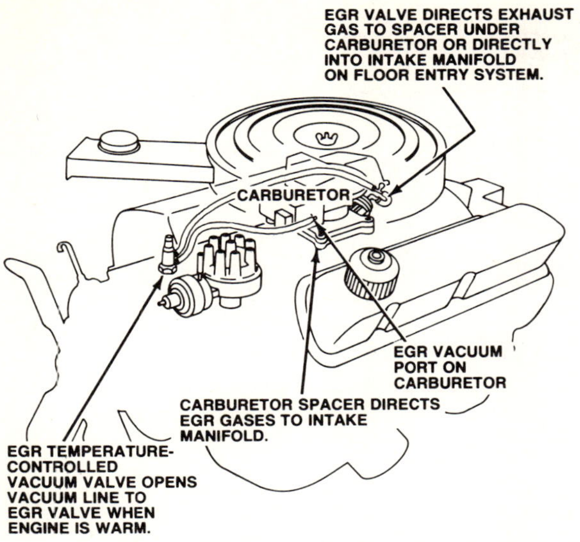

Figure 18-21. Basic Ford EGR system. (Ford)

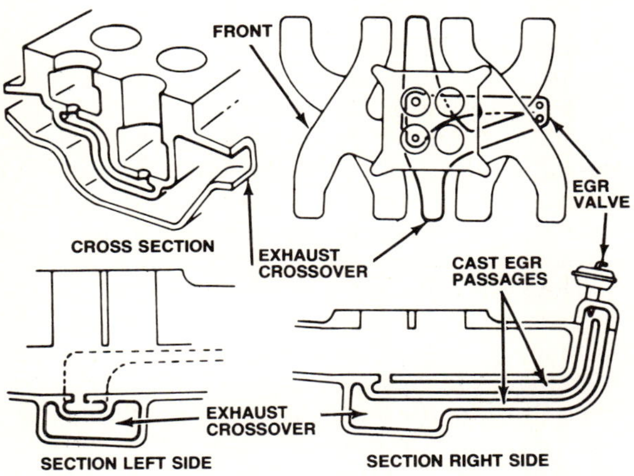

Figure 18-23. Ford 1974 floor entry EGR system. (Ford)

FORD EGR SYSTEM

The basic Ford EGR system, figure 18-21, was introduced on all Ford-built passenger cars in 1973, except those using the 1,600-cc and 2,000-cc engines. The basic system has three main parts: an EGR valve, a temperature-controlled vacuum valve, and a carburetor spacer.

EGR Valve

Three types of EGR valves are used: poppet, modulating, and tapered stem. The poppet valve contains a spring-loaded diaphragm, a valve and valve stem, and a flow restrictor.

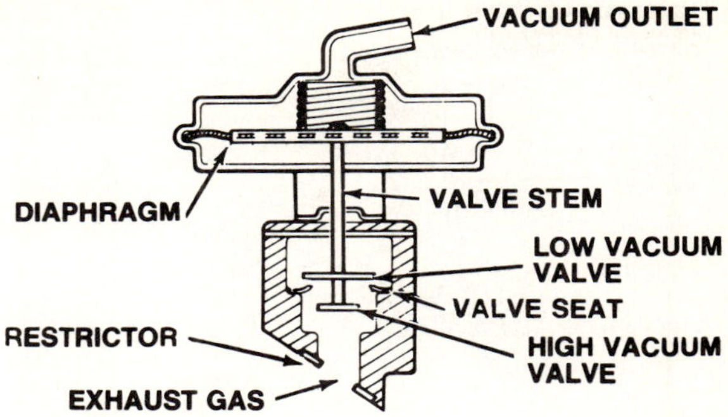

Figure 18-22. Ford modulating EGR valve. (Ford)

Ported carburetor vacuum opens the valve at a specific vacuum level and allows exhaust gas to enter the valve. Exhaust gas flow to the combustion chambers is controlled by the flow restrictor in the valve body inlet port.

The modulating valve, figure 18-22, has an extra disc valve on the stem below the main valve, and it operates much like a poppet valve. However, at a specific vacuum level, the lower disc valve restricts exhaust gas flow to the valve chamber. This modulating action improves the driveability of some engines.

The tapered-stem valve operates in the same way. Exhaust gas flow increases as the tapered stem gradually unseats.

The three valve types are not interchangeable between engine types or model years.

Ford uses various ways to deliver exhaust gas to the EGR valve. On many 4- and 6-cylinder engines, exhaust gas is drawn from the exhaust manifold through a stainless steel tube. Most V-8 engines use a passage in the exhaust crossover of the intake manifold. This routes the gas through the carburetor spacer to the EGR valve. The EGR valve then meters the gas back through a separate passage in the spacer to the carburetor primary venturis, where it is mixed with the air-fuel mixture.

Some 1974 engines use a floor-entry system, figure 18-23. The intake manifold has two passages cast in the floor under the intake runners. One connects the exhaust crossover to the EGR valve. The other connects the valve with two holes in the manifold floor directly under the carburetor primary venturis. This allows exhaust gas to mix with the air-fuel mixture before entering the combustion chambers. This is a valve-controlled system and is not the same as Chrysler's floor jet EGR system, which does not use a valve.

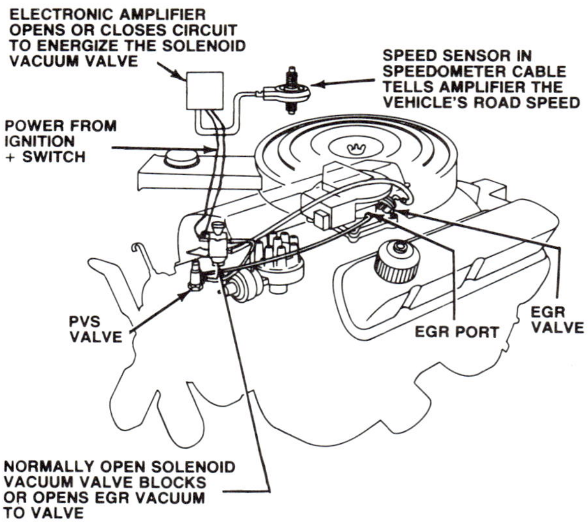

Figure 18-24. Ford high-speed EGR modulator subsystem. (Ford)

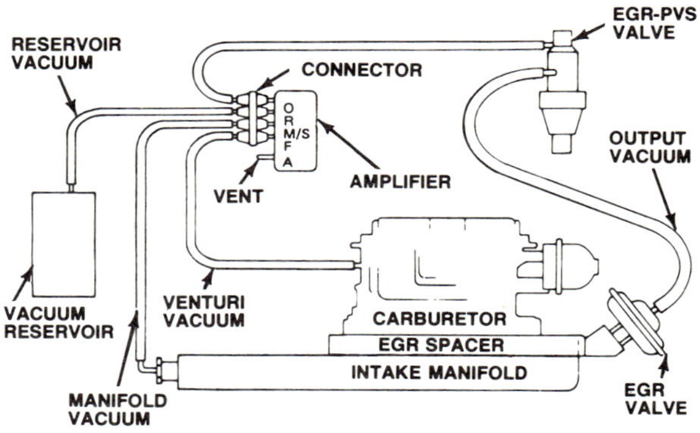

Figure 18-25. Ford venturi vacuum control subsystem with single-connector amplifier. (Ford)

Temperature-Controlled Vacuum Valve

A temperature-controlled vacuum valve in the engine cooling system cuts off vacuum to the EGR valve when the engine is cold, figure 18-21. Ford calls this valve a ported vacuum switch (PVS). It works the same as American Motors' coolant temperature override (CTO) switch and Chrysler's coolant-controlled EGR valve (CCEGR). The valve may be installed in a heater hose or in the intake manifold water jacket. Different temperature settings are used for different Ford engines. The valves are color coded for identification.

Carburetor Spacer

Ford's carburetor spacer plate contains separate inlet and outlet passages to the EGR valve. The valve is mounted at one end of the plate. The spacer is installed between the carburetor and the intake manifold with top and bottom gaskets to prevent leaks. Spacer plates are not interchangeable between engine types.

System Modifications

Basic EGR system operation is modified by the following subsystems.

High-speed EGR modulator subsystem

Some Ford V-8 engines have this subsystem, figure 18-24, to improve driveability at speeds above 64 mph. This subsystem shuts off carburetor vacuum to the EGR valve to prevent recirculation. A speed sensor driven by the speedometer cable signals an electronic module. When road speed exceeds 64 mph, the module closes the open solenoid valve. This, in turn, closes the EGR vacuum port and vents the vacuum outside the valve. As road speed drops below 64 mph, the module deenergizes the solenoid valve. This closes the vent and opens the EGR vacuum line.

Venturi vacuum control

Some 1974 6-cylinder engines use venturi vacuum rather than ported vacuum. A vacuum amplifier, figure 18-25, is connected between the EGR valve and intake manifold vacuum. Once airflow through the venturi is sufficient, venturi vacuum opens the amplifier. This allows manifold vacuum to open the EGR valve.

The EGR valve remains closed at idle, since there is no venturi vacuum produced at idle. Whenever venturi vacuum is equal to or greater than manifold vacuum, such as under wide-open throttle, the vacuum amplifier dumps the output vacuum to the EGR valve, causing the valve to close. Some of the amplifiers contain an output bias. This is a bleed designed for quicker EGR valve opening.

A single-connector amplifier with all ports on one side is used with 1974-75 systems, figure 18-25. A dual-connector amplifier with ports on both sides is used on 1975 and later systems, figure 18-26.

Backpressure transducer

Late 1975 and most later engines use a backpressure sensor, or transducer, figure 18-27. This is connected between the EGR valve and the intake manifold. It modulates exhaust gas flow by varying the vacuum to the EGR valve, according to exhaust backpressure.

On some 1977 models, the backpressure transducer is within the EGR valve, figure 18-28. This combination unit has an internal exhaust gas chamber and a transducer diaphragm to sense exhaust backpressure through a hollow valve stem. This combination EGR

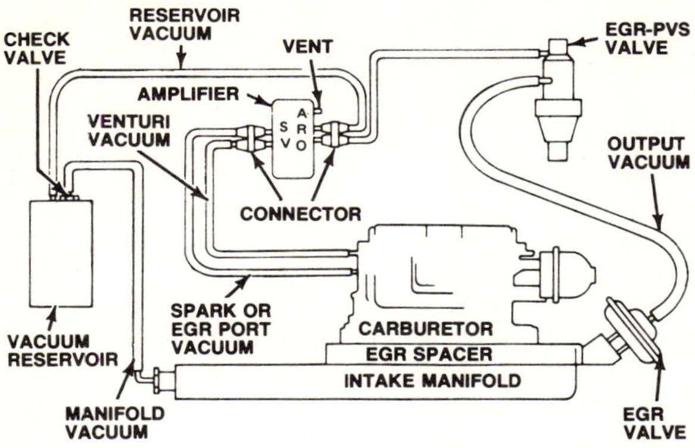

Figure 18-26. Ford venturi vacuum control subsystem with dual-connector amplifier. (Ford)

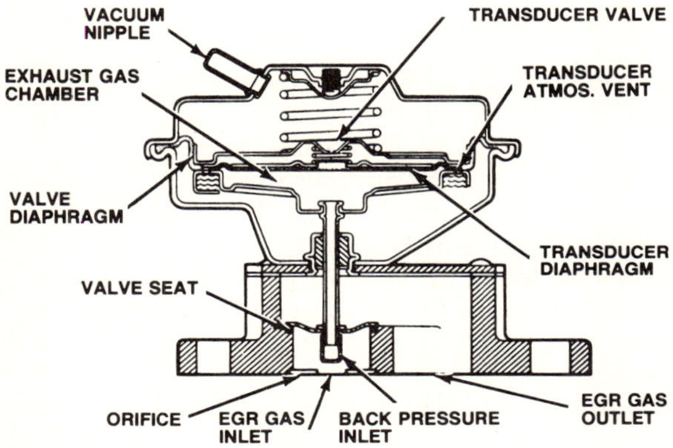

Figure 18-28. This Ford EGR valve has a built-in exhaust backpressure transducer. (Ford)

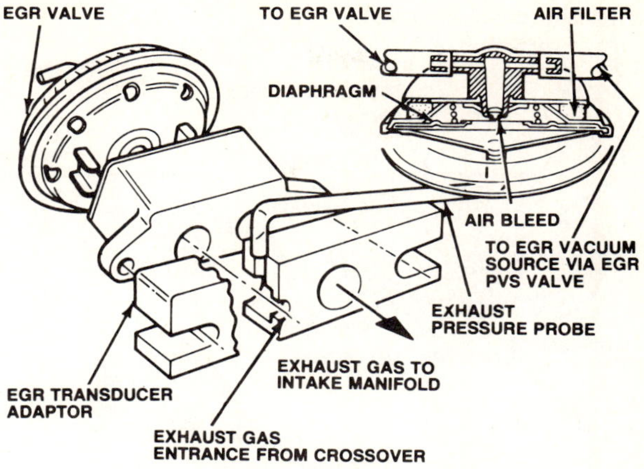

Figure 18-27. EGR backpressure transducer (sensor). (Ford)

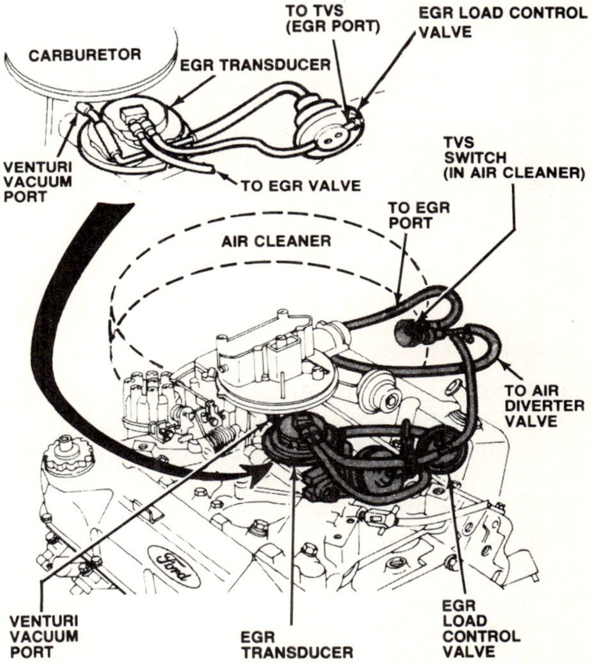

Figure 18-29. Ford air-temperature vacuum switch and EGR load control valve installation on a 400-cid V-8. (Ford)

valve-transducer constantly meters exhaust flow according to exhaust backpressure.

Temperature control

Some 1976 and later V-8 engines use an air-temperature vacuum switch (TVS) mounted in the air cleaner, figure 18-29, instead of the coolant PVS. The TVS also controls the vacuum flow to the diverter valve in the Thermactor air injection system. Since the switch is closed at temperatures below about 60° F (16° C), exhaust gas does not recirculate on a cold engine.

EGR load control

This valve is only on 1976 400-cid engines, figure 18-29. Normally, there is no vacuum at wide-open throttle, and the EGR valve would be closed. But the 400-cid V-8 is under-carbureted for fuel economy, using a 2-barrel instead of a 4-barrel carburetor. This means that vacuum still exists under wide-open throttle with this particular engine and carburetor combination. The load control valve is needed to keep the EGR valve closed for maximum power at wide-open throttle. Vacuum cannot reach the EGR valve because the venturi vacuum opens the load control valve to dump the vacuum signal. However, at less than wide-open throttle with this engine, venturi vacuum is not strong enough to open the load control valve. Vacuum is then routed to the EGR valve.

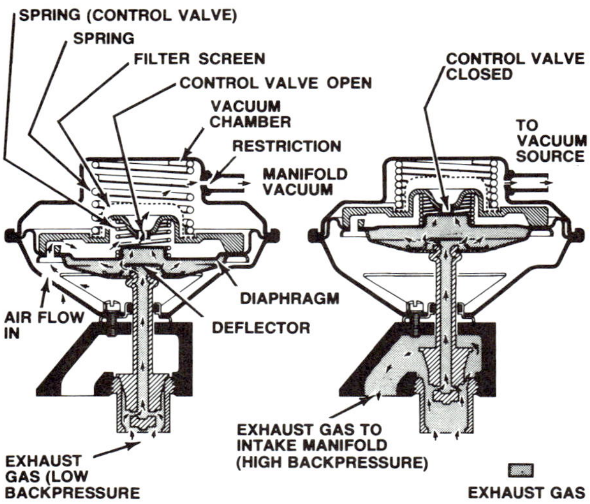

Figure 18-30. General Motors' EGR valve with built-in backpressure transducer. (Buick)

GENERAL MOTORS EGR SYSTEMS

Exhaust gas recirculation was introduced by Buick on 1972 cars with manual transmissions and on all 1972 cars sold in California. All 1973 and later GM engines use an EGR system to reduce NO_x. Although similar to the EGR systems used by other automakers, GM systems vary slightly among the company divisions.

Buick Division

The 1972 engines use ported vacuum to control the EGR valve. Exhaust gas is drawn from the manifold crossover exhaust channels and into the engine intake. Since ported vacuum is used as a control source, the valve is closed during idle and deceleration.

A temperature-control valve, in the carburetor-to-EGR valve vacuum line on 1973 models, prevents the valve from opening until underhood temperatures reach 60° F (16° C). The EPA ordered this removed, and a temperature vacuum switch (TVS) sensitive to engine coolant temperature replaced it on engines built after March 15, 1973. A dual-diaphragm EGR valve, figure 18-12, is used on some 1974 California engines.

An exhaust backpressure transducer is used on some 1976 models. The transducer is combined with the EGR valve, figure 18-30, for 1977 cars. EGR systems using the backpressure transducer, or the combined transducer and valve, regulate the timed vacuum to the EGR valve according to exhaust backpressure level.

Cadillac Division

The 1973 Cadillac EGR system is the same as that used by Buick. The inline temperature-sensitive switch was replaced after March 15, 1973, by a different temperature vacuum switch, which Cadillac calls a thermal vacuum valve (TVV). A backpressure transducer is on 1974 California cars, and on all 1975 models that have the air pump (AIR) system.

The backpressure transducer is used on all 1976 EFI-equipped cars and all California cars. For 1977, three valves are used:

1. The combined transducer-EGR valve, figure 18-30, on California cars
2. A standard EGR valve operated by ported vacuum on all non-California cars
3. A standard EGR valve and a separate backpressure transducer on all EFI cars.

Chevrolet Division

The basic 1973 EGR system was changed to add a TVS to the thermostat housing on 1974 models. The TVS improves cold engine starting by preventing recirculation until coolant temperature reaches a specific level. A dual-diaphragm EGR valve, figure 18-12, is on some engines. The combined transducer-EGR valve, figure 18-30, is on 1977 California and high-altitude cars.

Oldsmobile Division

The 1973-74 Oldsmobile EGR system is basically the same as that used by Cadillac, including the use of a backpressure transducer on California cars. A variety of temperature vacuum switches (EGR-TVS) and temperature control valves (EGR-TCV) are used on 1975 and later engines. Some are interconnected with the early fuel evaporation (EFE) system to switch vacuum from the EFE valve to the EGR valve at a specific temperature. An EGR check valve (EGR-CV) is used on some California engines with the backpressure transducer. The EGR-CV holds the highest ported vacuum reached. This keeps the EGR valve open during hard acceleration, when the backpressure transducer bleed is closed.

On some 1977 cars, the separate backpressure transducer has been replaced by the transducer-EGR valve, figure 18-30. A vacuum delay valve (EGR-VDV) is also used in the carburetor-to-EGR valve line on some V-8 engines to delay the vacuum from bleeding down too fast at the EGR valve. Metering the vacuum through a 0.005-inch orifice prevents its full loss for four seconds.

Pontiac Division

The basic 1973 Pontiac EGR system has three variations. Two of these are electrically interconnected with spark advance: the transmission-controlled spark (TCS) system on V-8 engines, and the combined emission control (CEC) system on 6-cylinder engines with manual transmissions. Those 6-cylinder engines with automatic transmissions use the normal ported vacuum EGR system. The TCS-EGR and CEC-EGR systems operate according to two basic rules:

1. When the TCS system is operating, there is no EGR.
2. When the EGR system is operating, there is no distributor vacuum advance.

TCS and EGR systems operate separately on 1973½ models, with a temperature vacuum valve (TVV) providing temperature control. A dual-diaphragm EGR valve, figure 18-12, is on some 1974 California engines, as well as a vacuum bias valve (VBV) to reduce surging. An exhaust backpressure transducer is used to modulate recirculation according to exhaust backpressure.

The system becomes far more complex on 1975 engines, with a variety of TVV, TCV, VDV, and backpressure transducers used. Virtually every engine type has different devices and operational temperatures. On 1976 engines, the TVV is replaced by a heat-sensitive snap disc valve. This is housed in a stamped steel shield and attached to the intake manifold on most engines. Since the snap disc valve senses engine radiant heat, it shuts off vacuum to the EGR valve on cold engine startup for improved driveability. For 1977, the primary change is the use of the combined transducer-EGR valve, figure 18-30, on some engines.

SUMMARY

Combustion chamber temperatures that exceed 2,500° F (1,370° C) cause nitrogen and oxygen to combine rapidly, forming large quantities of NO_x, a prime air pollutant. To reduce NO_x as much as possible, automakers use exhaust gas recirculation (EGR) systems to meter small quantities of exhaust gas into the incoming air-fuel mixture. This dilutes the fuel charge and results in lower combustion chamber temperatures.

Each automaker uses some variation of the basic system to achieve exhaust gas recirculation. Various temperature sensors, flow valves, and other control devices in the EGR system modulate the flow of exhaust gas to maintain driveability while reducing NO_x emissions.

Review Questions

Select the single most correct answer.
Compare your answers with the correct answers on page 280.

1. EGR systems came into general use in:
   a. 196l
   b. 1968
   c. 1973
   d. 1975

2. When exhaust gas is combined with the incoming air-fuel mixture, the result is:
   a. A less powerful air-fuel mixture
   b. A cooler air-fuel mixture
   c. Both A and B
   d. Neither A nor B

3. Which of the following is true:
   a. When engine temperature is low, NO_x formation is also low
   b. When engine temperature is high, NO_x formation is low
   c. When engine temperature is low, NO_x formation is high
   d. None of the above

4. The floor jet EGR system is used on:
   a. Some 1972-73 Ford-built engines
   b. Some 1972-73 GM-built engines
   c. Some 1972-73 Chrysler-built engines
   d. Some 1974-78 Chrysler-built engines

5. EGR valves were first used on:
   a. 1972 Fords
   b. 1973 Chevrolets
   c. 1969 Chryslers
   d. 1972 Buicks

6. Ported vacuum systems draw their vacuum from:
   a. Exhaust ports in the exhaust manifold
   b. Carburetor ports above the throttle plate
   c. Venturi ports at the throat of the carburetor
   d. EGR valve ports on the intake manifold

7. A vacuum amplifier is part of which EGR system:
   a. Floor jet system
   b. Backpressure transducer system
   c. Ported vacuum system
   d. Venturi vacuum system

8. Chrysler's EGR maintenance reminder system was used in:
   a. 1973
   b. 1975
   c. 1977
   d. 1972-77

9. The Ford ported vacuum switch works about the same as:
   a. AMC's coolant temperature override switch
   b. Chrysler's coolant-controlled EGR valve
   c. Both A and B
   d. Neither A nor B

10. The Ford temperature-controlled vacuum valve is called:
    a. CTO
    b. CCEGR
    c. PVS
    d. OSAC

11. The high-speed EGR modulator subsystem shuts off the EGR valve above:
    a. 24 mph
    b. 34 mph
    c. 64 mph
    d. 84 mph

Chapter

19

Catalytic Converters

In order to meet the U.S. Federal emission standards, carmakers in the mid-1970s turned to the **catalytic converter** to change the harmful hydrocarbons (HC) and carbon monoxide (CO) in the exhaust to harmless byproducts.

Catalytic converters are simple. They have no moving parts and no adjustments to them are necessary. Most converters have a guaranteed lifespan of 50,000 miles. And chances are you will never see the inside of a converter unless you cut one in half, since there is nothing inside that can be repaired. You can't even test a converter.

Still, the link between converters and a well-tuned engine is important. An engine that is misfiring or improperly tuned can destroy a catalytic converter, because the converter cannot accept exhaust temperatures above about 1,500° F (815° C), and it will not work right if the air-fuel mixture is too rich. Two spark plugs, misfiring in succession for a prolonged time, will raise the temperature in the converter to a point where it could shorten the life of the converter. The more you know about converters, the more you will understand the need for a well-tuned engine.

In this chapter, you will learn:

- How the catalytic converter is built
- How it works
- How to make sure it does its job properly.

REDUCING EMISSIONS

One way to reduce HC and CO is to raise the combustion temperature. But that also increases the formation of the third major element of exhaust emissions, oxides of nitrogen (NO_x). A better way to reduce all three pollutants is to use a catalytic converter, figure 19-1. By sending the exhaust gases through the converter in the presence of oxygen, the HC and CO compounds unite with the oxygen to form water vapor (H_2O) and carbon dioxide (CO_2).

The Oxidation Reaction

The **catalysts** in a converter are platinum and palladium. (It is the high cost of platinum that makes converter replacement so expensive.) These start the **oxidation**, or burning reaction. Oxidation is simply the addition of oxygen to the chemical reaction taking place. It adds the O to H_2O and CO_2.

Quite a lot of heat is generated by the oxidation process. The heat of the catalyst will range from 900° to 1,600° F (480° to 870° C), and the exhaust gas at the outlet end of the converter will be 50° to 200° F (10° to 90° C) higher than the

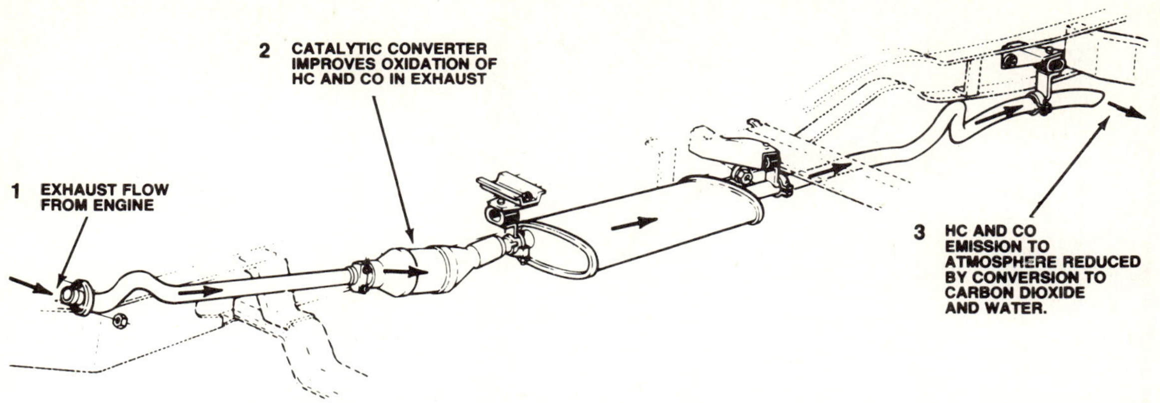

Figure 19-1. Typical catalytic converter installation. (Ford)

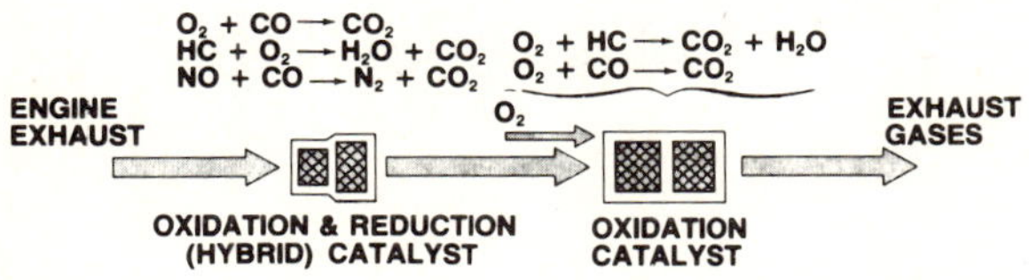

Figure 19-2. Catalytic oxidation and reduction reactions. (Chrysler)

inlet end. By 1978, automakers were able to reduce the outside, or "skin," temperature of converters by as much as 300° F (150° C). Using smaller engines and changing engine timing made the lower temperatures possible. In spite of the intense heat, however, oxidation does not generate the flame and radiant heat associated with a simple burning reaction.

The Reduction Reaction

The oxidation catalytic reaction we just mentioned has no effect on oxides of nitrogen. NO_x control requires a separate reaction, called reduction, rather than oxidation. **Reduction** is the opposite of oxidation. Reduction is the chemical removal of oxygen from a material. The reduction reaction changes NO_x to harmless nitrogen (N_2) and CO_2 by chemically promoting the switch of oxygen from the NO_x to the CO compound. The elements rhodium and platinum are used as reduction catalysts.

Because the oxidation and the reduction reactions oppose each other, it is hard for both to occur at the same time and in the same place. An oxidation and a reduction catalyst can be combined in the same converter. However, a second oxidation catalyst is also required for complete emission control. An oxidation-reduction catalyst is often called a 2-stage, a 3-way (because it works on all three major pollutants), or a hybrid, catalyst. A hybrid catalyst and a second oxidation catalyst can be installed in opposite ends of the same converter housing or in two separate converters, figure 19-2. The first hybrid converters were used by Volvo and Saab on a limited number of cars sold in California in 1977.

The hybrid converter works best to reduce NO_x when the CO level in the exhaust is between 0.8 and 1.5 percent. As the CO level increases or decreases from that percentage, the hybrid converter's efficiency decreases, figure 19-3. Exhaust gas first passes through the hybrid converter, where the NO_x is changed to N_2 and CO_2. It then goes through the oxidation converter, where the HC and CO are changed to H_2O and CO_2.

Oxidation Converter Design

Catalytic converters are made of two stamped metal pieces welded together to form a round or oval shell. Each contains the catalytic element, figure 19-4. The outer housing is made of alumi-

Catalytic Converter: A device installed in an exhaust system that converts pollutants to harmless byproducts through a catalytic chemical reaction.

Catalyst: A substance that causes a chemical reaction, without being changed by the reaction.

Oxidation: The combining of an element with oxygen in a chemical process that often produces extreme heat as a byproduct.

Reduction: A chemical process in which oxygen is taken away from a compound.

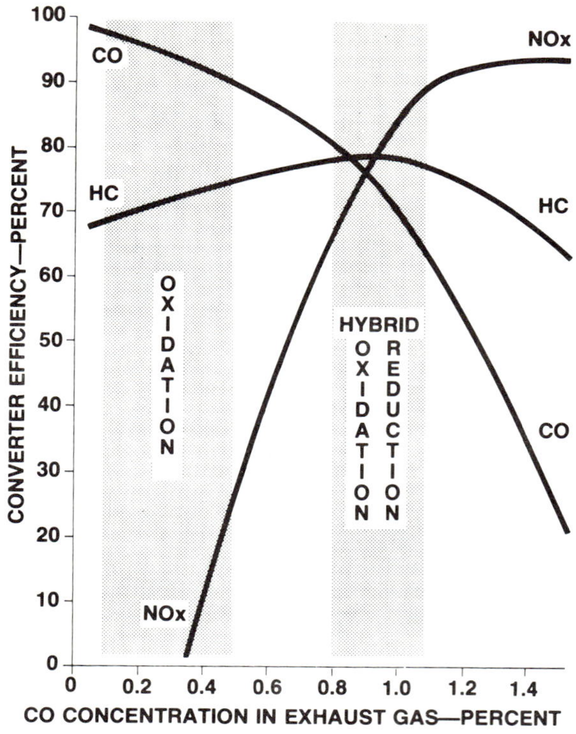

Figure 19-3. Catalyst operating characteristics. (Chrysler)

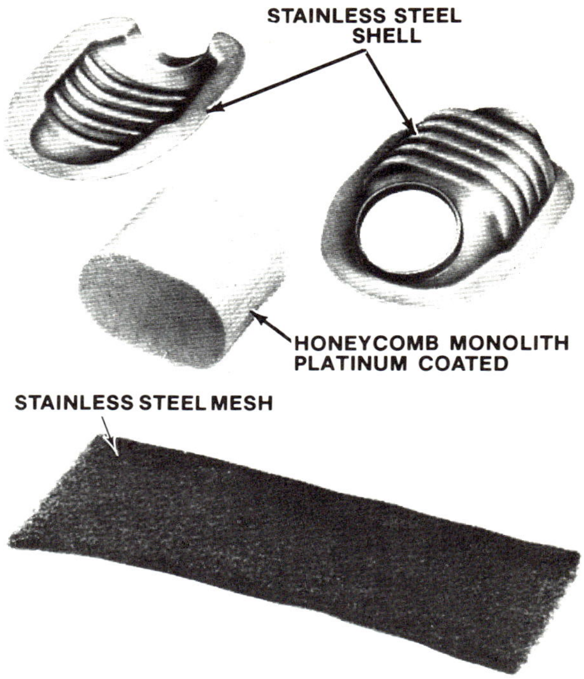

Figure 19-4. The parts of a typical catalytic converter. (Chrysler)

nized or stainless steel because it must be able to withstand the high temperatures of oxidation.

The catalytic element generally used is platinum, or a mixture of platinum and palladium. These two **noble metals** best meet the requirements of an effective catalyst: durability, operating temperature, and chemical activity. The catalyst is deposited on an aluminum oxide **substrate** through which the exhaust gas flows. This substrate must provide a catalyst support which can withstand high temperatures.

Two forms of substrate material are used:

1. Tiny pellets of aluminum oxide
2. A honeycomb **monolith** of aluminum oxide. It can be either a laminated (sandwiched) design or an **extruded** design.

The catalyst element is deposited on the surface of the substrate material. Both kinds of substrate material — pellets or monolith — provide several thousand square yards of catalyst surface area over which the exhaust gases flow.

In converters using a pellet substrate, the gases flow over the top and down through the substrate layers. With the monolithic substrate material, a diffuser inside the converter shell allows an even flow of exhaust gases over the entire area of the substrate. If a diffuser were not used, the gases would tend to flow only through a central portion of the substrate.

Monolithic substrate material can break easily when dropped or jolted. To prevent damage to the core, it is placed inside a stainless steel mesh, figure 19-4, which acts as a cushion. This also protects the core from thermal shock caused by temperature extremes, and keeps it properly positioned during final assembly of the converter shell.

Reduction Converter Design

A typical hybrid converter uses two substrates of different sizes coated with platinum and rhodium. These substrates are surrounded by the protective stainless steel mesh and are encased in a stainless steel shell. Other types of reduction converters are likely to be made available in the coming years. But, given the designs now available, the main differences between the oxidation and the reduction converters remains the type of catalyst used and the positioning of the reduction converter in the exhaust system ahead of the oxidation converter.

CONVERTER OPERATING PRECAUTIONS

There are several ways in which a catalytic converter can be damaged or destroyed. Using the wrong fuel, bad air-fuel mixtures, excessive combustion heat — all of these can damage a converter. We will look at these in detail.

Fuel Requirements

Cars with catalytic converters must use unleaded fuel. The lead and phosphorous additives used in leaded gasoline will greatly reduce the catalyst's efficiency. When leaded fuel is used, the lead covers the catalyst and forms a coating that prevents the exhaust gases from reaching the catalyst. Returning to unleaded gasoline will allow the converter to regain some, but not all, of its efficiency. How much it regains depends on how long leaded fuel is used. However, there is no known method at present of testing a converter's efficiency. Continued use of leaded fuel will eventually cause the catalyst to lose its ability to promote oxidation of HC and CO.

To prevent the use of leaded fuel in cars with catalytic converters, the unleaded gasoline pump nozzles in service stations and the fuel tank filler necks on cars are smaller in diameter than those using leaded fuel. In addition, cars which require unleaded fuel also have labels reading "Unleaded Gasoline Only" next to the gas tank filler and on the instrument panel in the driver's compartment. A United States Federal law makes it illegal to sell leaded fuel for use in converter-equipped cars.

Engine Condition

Excessively high temperatures can reduce converter life and destroy the core. Converters operate best at internal temperatures up to 1,500° F (815° C). At higher temperatures, the catalyst will start to break up or melt. Although temperature-protection systems are used on some vehicles, they work only under certain conditions. Proper engine maintenance is needed to prevent overheating the converter.

When an engine is in poor condition, or needs a tune-up, the exhaust gas fed to the converter will contain too much raw fuel. This will cause the converter to become a catalytic furnace. If more than two spark plugs misfire at the same time, raw fuel is pumped into the catalytic converter. This misfiring over a long period of time causes internal converter heat to rise rapidly.

Excessive temperatures may also be caused by improper use of an engine in good operating condition. Long idling periods are one of the worst running conditions, since more heat is developed when an engine runs at idle for long periods, than when driving at normal highway speeds. Idle periods of more than 10 minutes should be avoided. It is much better to shut off the engine and restart it when required.

Other Precautions

To avoid excessive catalyst temperatures and the possibility of fuel vapors reaching the converter, follow these rules:

1. Do not attempt to start the engine on compression by pushing the vehicle. Use jumper cables instead.
2. Do not crank an engine for more than 60 seconds when it is flooded or firing intermittently.
3. Do not turn off the ignition switch when the car is in motion.
4. Fix problems such as engine dieseling, heavy surging, repeated stalling or backfiring, and other such indications of below-normal performance immediately.
5. Avoid disconnecting spark plugs to test the ignition whenever possible. If an oscilloscope is not available, do not run the engine for more than 30 seconds with the plug wire off or shorted.

OXIDATION CONVERTER INSTALLATIONS

Federal regulations require that a catalytic converter last a minimum of 50,000 miles. No service should be required during this period. All oxidation converters are installed between the engine and the muffler. Reduction converters must be installed between the engine and the oxidation converter. Most converters are located under the passenger compartment. They are connected to the exhaust system in one of three ways:

1. A bolt-together flange, figure 19-5
2. A slip fit with a clamp, figure 19-5
3. Welded to the exhaust pipes.

Noble Metal: A metal (such as gold, silver, platinum, or palladium) or an alloy that will not easily corrode or oxidize.

Substrate: The layer, or honeycomb, of aluminum oxide upon which the catalyst (platimum, palladium, or rhodium) in a catalytic converter is deposited.

Monolith: A large block. In a catalytic converter, the monolith is made like a honeycomb to provide several thousand square yards of catalyst surface area.

Extruded: Shaped by forcing through a die.

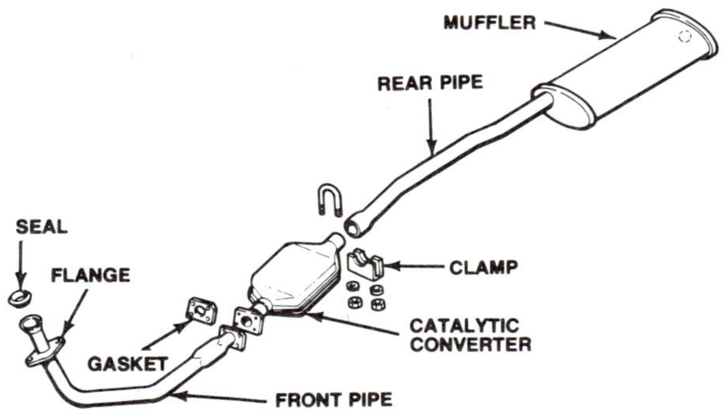

Figure 19-5. This catalytic converter is installed with a bolted flange at one end and a clamp at the other. (AMC)

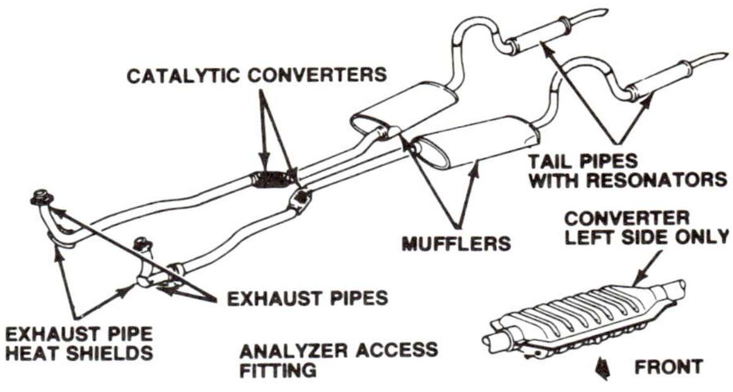

Figure 19-6. This dual exhaust system has two catalytic converters. (Chrysler)

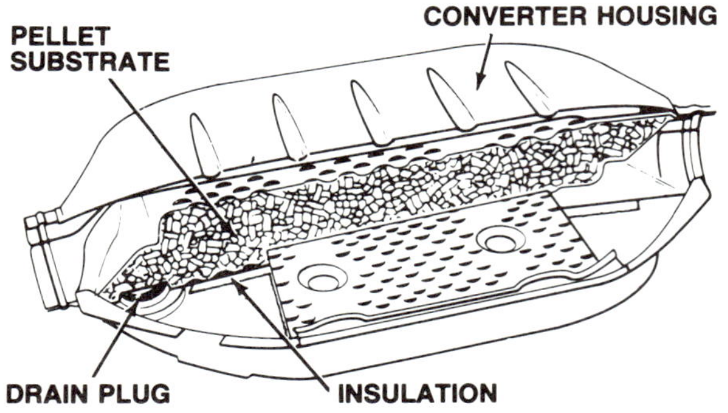

Figure 19-7. The AC pellet-type catalytic converter is used by American Motors and General Motors.

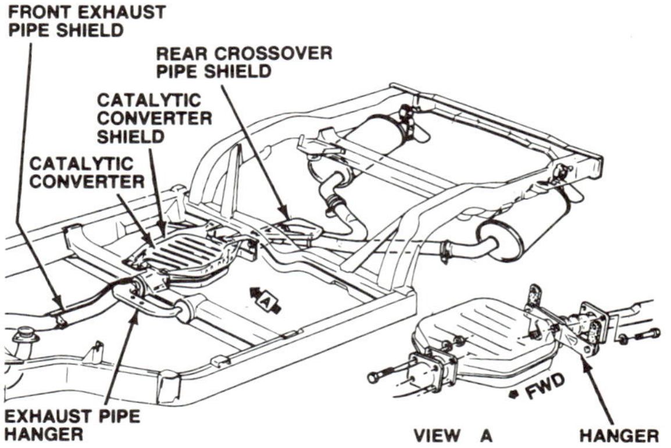

Figure 19-8. General Motors catalytic converter installation with a heat shield. (Chevrolet)

Most single exhaust systems have only one converter. Dual exhaust systems may have a converter on one side only or a separate converter on each side, figure 19-6. The catalytic converters used by domestic automakers are discussed below.

American Motors And General Motors Systems

AMC and GM cars use pellet converters made by the AC Division of General Motors, figure 19-7. Since the converter housing is insulated, it remains at about the same temperature as the outside of the muffler. Only a few cars, such as the Corvette, figure 19-8, require a heat shield. The pellet catalyst can be removed from the converter housing through a drain plug with special equipment. This permits replacement with a fresh catalyst charge after 50,000 miles or if the catalyst material is damaged.

Chrysler Systems

Chrysler products use the monolithic converter, figure 19-9. In case of damage or after 50,000 miles, the entire unit must be replaced. Since this type of converter transmits quite a lot of heat through the housing, heat shields, figure 19-10, are used to protect the passenger compartment, the automatic transmission and its cooler lines, the torsion bars, the rear shock absorbers, and other parts of the chassis.

Size, shape, and placement of these shields varies from car model to car model. The underbody floor pan above the converter is heavily shielded, and more heat insulation is placed under the floor mats. Those models with both a converter and an air pump system use even more shielding, since the air pump causes a converter to operate at a higher temperature than one without the pump. Air-pump-equipped models, as well as models sold in California, also use a protective grillwork under the converter. This guard prevents direct housing contact with burnable materials. It also helps cool the converter housing by radiating heat to the airflow beneath the car.

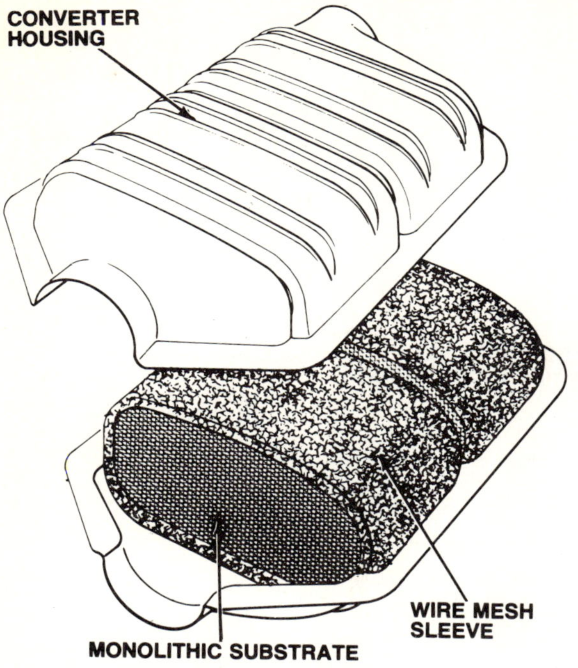

Figure 19-9. Chrysler catalytic converter with a monolithic substrate.

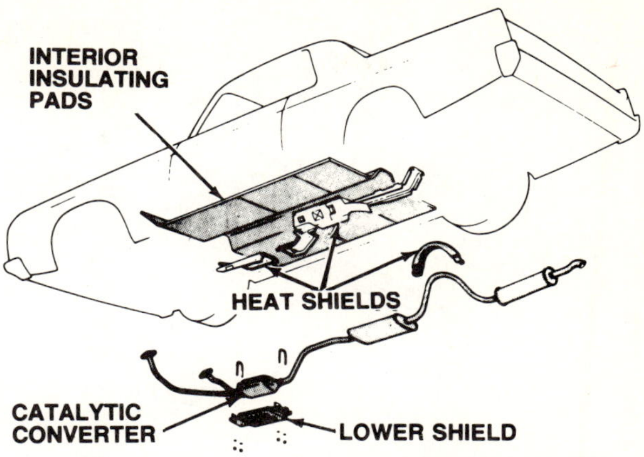

Figure 19-10. Chrysler catalytic converter heat shield installation. (Chrysler)

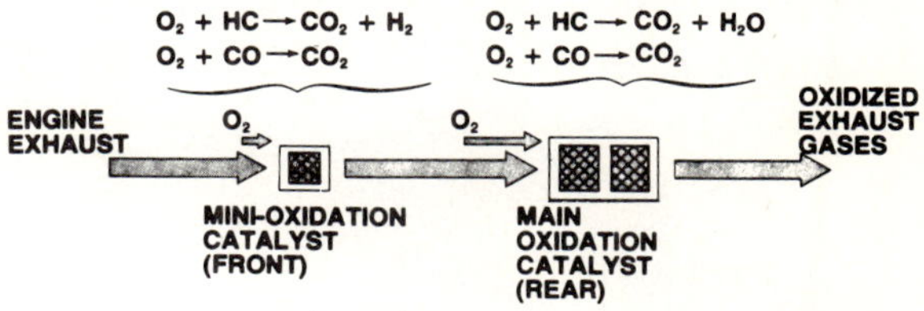

Figure 19-12. Two-stage oxidation reaction with Chrysler's mini-converter. (Chrysler)

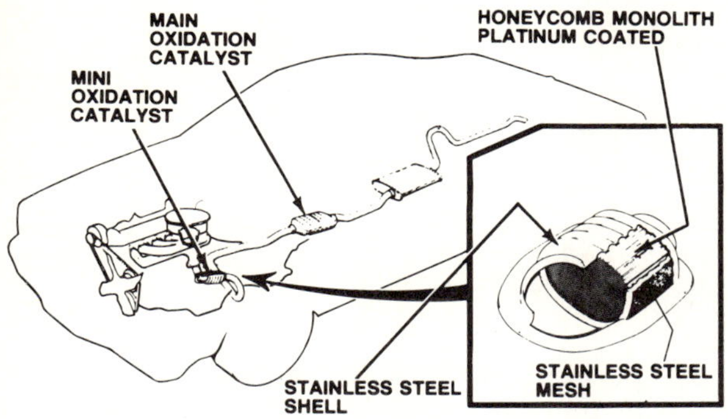

Figure 19-11. Chrysler mini-converter installation.

On some 1977 and later models, Chrysler uses a small oxidation-type converter welded into the engine exhaust pipe 6 to 12 inches from the exhaust manifold, figure 19-11. This mini-catalytic converter is called an underhood unit and is used along with the regular, or main underfloor, converter. The mini-converter is of the same design as the regular converter and begins the oxidation of HC and CO before they reach the main underfloor converter. This results in more complete oxidation of the exhaust gas, figure 19-12.

Some 1975 models use a catalyst protection system (CPS), figure 19-13. This prevents fuel from reaching the converter during closed-throttle deceleration after high-speed cruising. The CPS electronic speed switch determines

Catalytic Converter Odors

Although catalytic converters control HC and CO emissions, they also produce other undesirable emissions in small quantities. For example, gasoline has a little bit of sulfur in it. This reacts with the water vapor inside a converter to produce hydrogen sulfide. This toxic byproduct has the distinct odor of rotten eggs. The smell is usually noticed while the engine is warming up, or during deceleration.

When the odor is very strong at normal operating temperatures, it may mean that the engine is out-of-tune and is running too rich. In many such cases, the carburetor idle mixture screws are not set right, although the problem also may be caused by a high fuel level or some other carburetor problem. Since the odor does not necessarily mean an incorrect mixture adjustment, the carburetor should not be adjusted simply to get rid of the odor. Changing brands of gasoline may help control the odor in some cases, since the amount of sulfur present in gasoline varies from one brand to another.

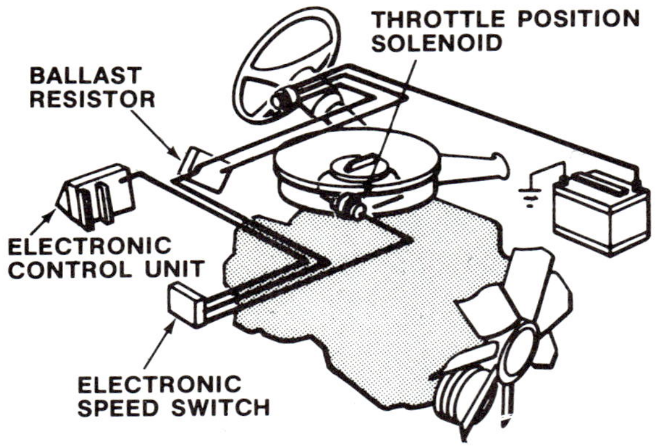

Figure 19-13. Chrysler's catalyst protection system.

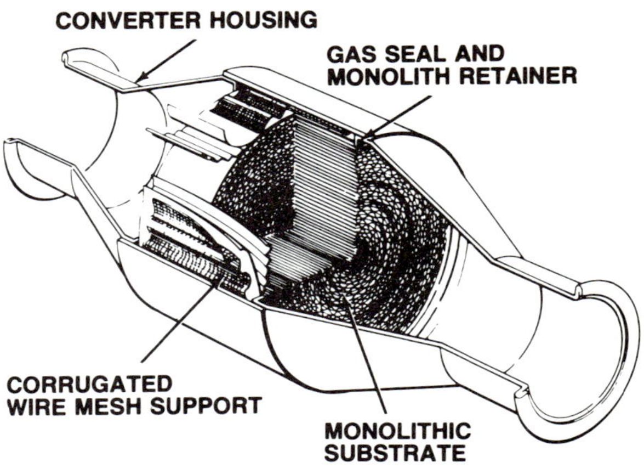

Figure 19-14. Ford's catalytic converter with a monolithic substrate.

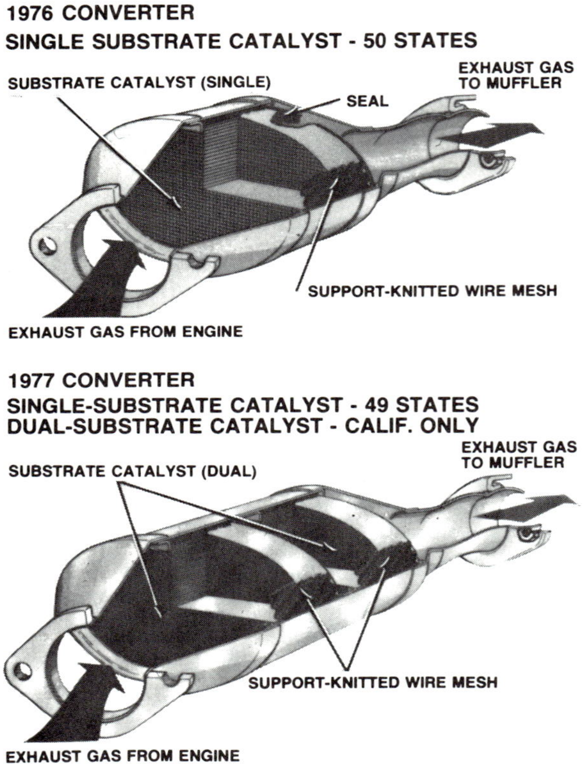

Figure 19-15. Ford's single- and dual-substrate catalytic converters. (Ford)

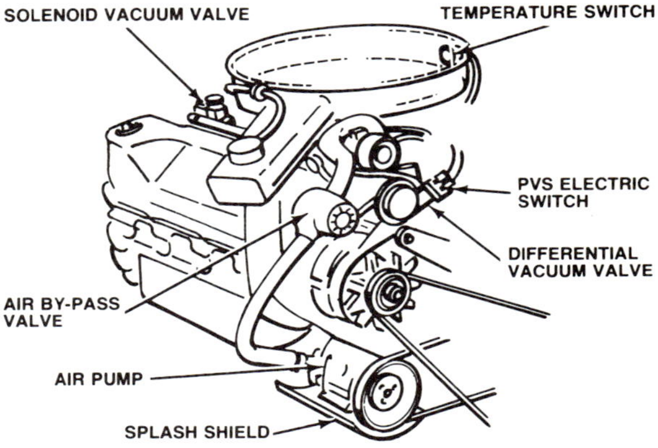

Figure 19-16. Ford's air injection system supplies extra air to the catalytic converter.

engine rpm. When engine speed reaches approximately 2,000 rpm, the switch energizes a throttle positioner solenoid to extend its plunger.

If the accelerator pedal is released when engine speed is above 2,000 rpm, the solenoid plunger holds the throttle valve open at the fast-idle position. This provides intake air to balance the air-fuel mixture and speed up combustion to protect the converter from overheating. Once engine speed drops below 2,000 rpm, the speed switch deenergizes the solenoid, and the throttle returns to a normal slow-idle position as the solenoid plunger retracts. The CPS system was discontinued as a running change on 1975 models.

Ford Systems

Ford products use a monolithic converter, figure 19-14. Some 1977 and later cars have converters with two monolithic substrates, figure 19-15. The entire unit must be replaced after 50,000 miles or in case of damage. Converters are located underneath the seating area floor pan on some models, and under the toeboard area on other models. Since this type of converter transmits a high level of heat through the

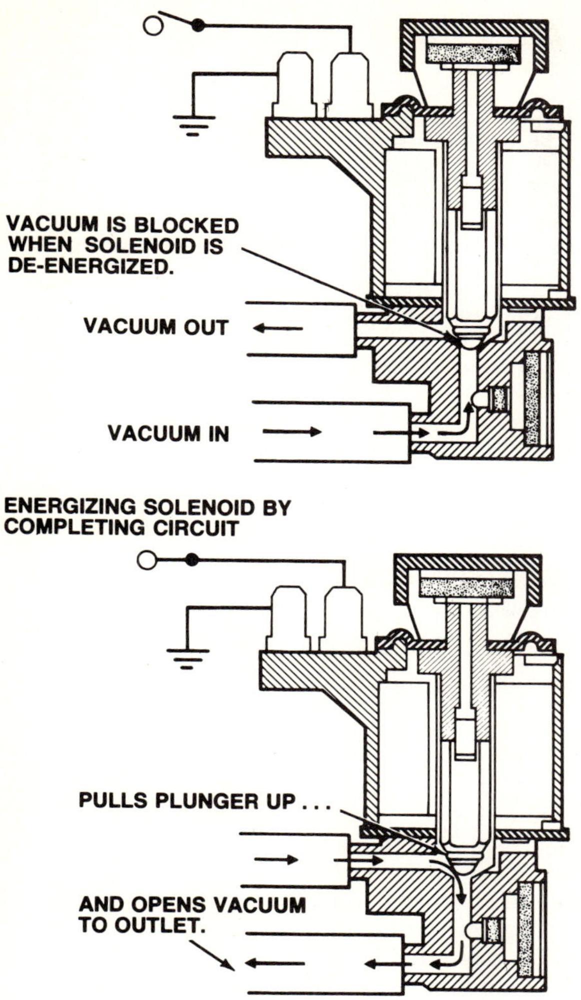

Figure 19-17. This solenoid valve controls vacuum to the air injection diverter valve on some Fords with catalytic converters.

housing, various formed and perforated exhaust shields are welded or clamped in place, according to the particular model.

The Ford catalytic converter system needs a secondary air source to keep enough oxygen in the exhaust for complete oxidation of the HC and CO. This is provided by the Thermactor air injection system, figure 19-16, which we discussed in Chapter 16. We are interested in the Thermactor system in this chapter because of the role it plays to protect the catalyst from overheating.

A 2-port solenoid vacuum valve, figure 19-17, which is normally closed, replaces a normally open valve of the same type used before 1975 with other Ford emission systems. This new valve does the same thing as the older one: it opens and closes the vacuum circuit according to an electrical signal. The solenoid vacuum valve controls the air-bypass operation of the diverter valve by controlling vacuum to a vacuum differential valve (VDV).

The solenoid vacuum valve may be energized in several different ways, depending on which car model it's on. Some vehicles use a temperature switch in the air cleaner. Those with catalyst protection use a floor-mounted switch, or either a normally open or a normally closed electric ported vacuum switch (PVS), figure 19-18. The switch works with the solenoid vacuum valve for bypass operation of the diverter valve if coolant temperature reaches or exceeds 235° F.

Whether a floor-mounted switch or a PVS is used, the result is the same. The switch or PVS activates the solenoid vacuum valve. This causes the vacuum differential valve to temporarily dump vacuum to the diverter valve during deceleration. It also prevents backfire when the exhaust mixture is too rich. In each case, the VDV prevents the converter from overheating and damaging the catalyst.

■ Volvo's Lambda-Sond Emission System

The 3-way catalyst can reduce emissions to a minimum without interfering with combustion inside the engine. To do this, it needs a precise mixture of exhaust gases to the converter. Volvo's Lambda-Sond system was the first use of 3-way converters. It has two main subsystems:

1. A 3-way catalytic converter that uses platinum to oxidize HC and CO, and rhodium to promote a reduction reaction for NO_x.
2. An exhaust gas oxygen sensor and an electronic control unit to monitor the amount of oxygen in the exhaust and send a feedback signal to the fuel injection system.

The sensor detects changes in the oxygen levels in the exhaust. It sends a continuous signal to the control unit, which relays the data to the fuel injectors. The injectors regulate the amount of fuel according to the sensor data, and even compensate for changes in altitude and barometric pressure.

Despite its success so far, the 3-way converter is probably not the final answer to solving our environmental and fuel supply problems. Designing the system to fit a 6- or 8-cylinder engine is far more difficult in terms of sensor location, the present known supply of rhodium will disappear quickly if used to fit some 15 million new vehicles per year, and the future of gasoline as a continued fuel for automobiles is in some doubt.

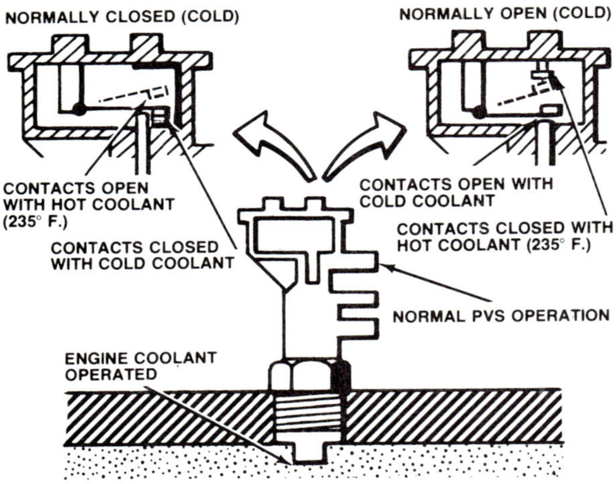

Figure 19-18. This electric ported vacuum switch is used in some Ford catalytic converter air injection systems.

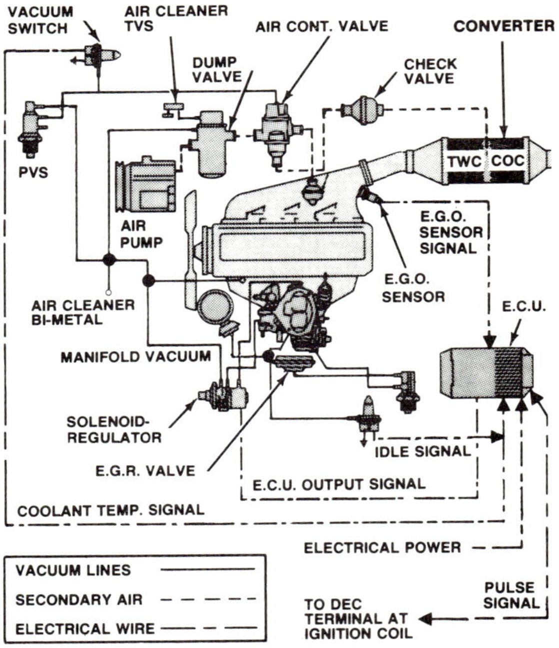

Figure 19-19. For 1978, Ford introduced this 3-way converter on some 4-cylinder engines. It works along with a closed-loop feedback system that uses an exhaust gas oxygen sensor and an electronically controlled carburetor. (Ford)

OXIDATION-REDUCTION CONVERTER INSTALLATIONS

The catalytic converters described in the previous section are all oxidation converters. They were first used on 1975 cars, and they are still being used. In 1977, Volvo and Saab introduced the first oxidation-reduction converters to control all three major exhaust emissions. In 1978, two domestic automakers — Ford and General Motors — began using hybrid, or 2-stage, converters to control HC, CO, and NO_x. These installations are described in the following paragraphs.

Ford Systems

The 2.3-litre, 4-cylinder engine used in California Pintos and Bobcats for 1978 has a closed-loop, feedback emission control system, figure 19-19. It has an oxidation-reduction catalytic converter, an electronically controlled carburetor, and an exhaust gas oxygen sensor.

The "closed-loop feedback" is provided by the oxygen sensor, which continuously monitors the unburned air-fuel ratio in the exhaust. This signal is sent to the electronic control module, which also receives signals from a throttle-angle switch, two vacuum switches, and the ignition control side of the ignition coil. The module sends a signal to the solenoid regulator that controls the vacuum supply to the carburetor power valve. This provides the precise control of the intake air-fuel ratio needed by the hybrid converter.

The front half of the converter has an oxidation-reduction catalyst that controls HC, CO, and NO_x. It uses platinum and rhodium as catalysts on a monolithic substrate. The rear half of the converter has only a platinum catalyst to further oxidize HC and CO.

During engine warmup, the air injection pump supplies extra air to the exhaust manifold so that it passes through the converter. When the engine reaches operating temperature, a vacuum signal from the temperature vacuum switch redirects the air injection to the middle of the converter. The extra air is then supplied *only* to the oxidation catalyst in the rear half of the converter.

In 1979, a variation of this system was introduced on some 351-cid V-8's by Ford.

General Motors Systems

Some of GM's 1978 California small cars also have a closed-loop feedback emission control system. It works about the same as the Ford system. It is installed on all 1978 California Sunbirds, Skyhawks, and Starfires with the 151-cid 4-cylinder engine, and is planned for some cars with the 231-cid V-6.

SUMMARY

Catalytic converters in the exhaust system promote a chemical reaction to change polluting byproducts of combustion into harmless water and carbon dioxide. Two types of converters are used: oxidation and reduction. The oxidation-type removes HC and CO emissions from the exhaust gases; reduction converters remove NO_x.

Converters use a monolithic or pellet-type substrate coated with platinum, palladium, or rhodium to provide a surface for the oxidation or reduction reaction to take place. These catalysts can be damaged by too much heat, the use of leaded fuel, or too much unburned fuel reaching their surface. There is no way to determine the efficiency of a given converter, nor do they require any service during the 50,000 mile lifespan required by U.S. Federal regulations.

Catalytic converters were first used in 1975. Some early converters have a temperature protection system to prevent converter damage from overheating. These were abandoned on most 1976 and later vehicles. All Chrysler and Ford converter systems use a series of heat shields to protect chassis components from heat damage. AMC and GM cars use the insulated pellet converter, whose housing does not transmit excessive heat, and does not require heat shields in most cases.

Hybrid oxidation-reduction catalytic converters were introduced on some 1977-78 cars to control all three major exhaust emissions: HC, CO, and NO_x.

Review Questions

Select the single most correct answer.
Compare your answers with the correct answers on page 280.

1. In a catalytic converter, the HC and CO combine with oxygen to form:
   a. Water vapor and carbon monoxide
   b. Water vapor and carbon dioxide
   c. Water vapor and lead oxide
   d. Water vapor and hydrogen oxide

2. Which of the following is true of the catalyst in a converter:
   a. It causes a chemical reaction
   b. It is not changed by a chemical reaction
   c. It is usually platinum, palladium, or rhodium
   d. All of the above

3. A hybrid catalytic converter:
   a. Combines an oxidation reaction with a reduction reaction
   b. Removes HC and CO but not NO_x from the exhaust
   c. Removes CO and NO_x but not HC from the exhaust
   d. Promotes a reduction reaction only

4. An oxidation-reduction catalytic converter is also known as:
   a. A 2-stage converter
   b. A 3-way converter
   c. A hybrid converter
   d. All of the above

5. Every catalytic converter has:
   a. A monolith
   b. A substrate
   c. Aluminum oxide pellets
   d. An extrusion

6. Cars with catalytic converters cannot use leaded gasoline because:
   a. Operating temperatures are too low for the catalytic reaction
   b. The engine will misfire
   c. The converter catalyst will become covered with lead
   d. Operating temperatures are too high for the catalytic reaction

7. Which of the following is true of Chrysler's monolithic converter system:
   a. The converter cannot be refilled after 50,000 miles
   b. Heat shields are needed to protect the chassis from the high heat
   c. Some cars use an air pump
   d. All of the above

8. The Thermactor air injection system is used by:
   a. Ford
   b. General Motors
   c. AMC
   d. Chrysler

9. Using an air pump system with a catalytic converter helps to:
   a. Reduce the amount of exhaust
   b. Add more oxygen to the exhaust
   c. Speed the flow of exhaust through the converter
   d. Lower the temperature of the exhaust inside the converter

10. Ford introduced the dual substrate catalytic converter:
    a. In California in 1976
    b. In California in 1977
    c. Nationally in 1976
    d. Nationally in 1977

11. Hybrid converters were first used on domestic cars in:
    a. 1975
    b. 1976
    c. 1977
    d. 1978

12. In a Ford closed-loop feedback system the air injector:
    a. Supplies extra air to the exhaust manifold during normal operation
    b. Supplies extra air only to the oxidation converter during normal operation
    c. Cuts air to the manifold during warmup
    d. All of the above

13. Cars with catalytic converters are required to use:
    a. Premium leaded fuel
    b. Unleaded fuel
    c. Low-lead fuel
    d. Any of the above

14. Which of the following will damage the converter:
    a. Push starting the engine
    b. Engine misfiring
    c. Long idling periods
    d. All of the above

15. The reduction converter must be located:
    a. Between the engine and the oxidation converter
    b. Between the oxidation converter and the muffler
    c. After the muffler
    d. Any of the above

NIASE Mechanic Certification Sample Test

This sample test is similar in format to the series of tests given by the National Institute for Automotive Service Excellence (NIASE). Each of these exams covers one of eight areas of automobile repair and service. The tests are given every fall and spring in about 250 cities throughout the United States.

For a mechanic to earn certification in a particular field, he or she must successfully complete one of these tests, and have at least two years of "hands-on" experience (or a combination of work experience and auto mechanics training). Successfully finishing all eight tests earns the mechanic the certification as a General Automobile Mechanic. More than 120,000 mechanics have taken these tests since the program was begun in 1972.

In the following sample test, some of the questions were provided by NIASE. *All* of the questions follow the form of the national exams. Learning to take this kind of test will help you if you plan to apply for certification later in your career.

For more information, write to:

National Institute for
Automotive Service Excellence
1825 K Street N.W.
Washington, D.C. 20006

1. Mechanic A says that the position of the carburetor idle mixture screws will affect HC emissions.

   Mechanic B says that the position of the carburetor idle mixture screws will affect CO emissions.

   Who is correct?

   a. A only
   b. B only
   c. Both A and B
   d. Neither A nor B

2. An engine's spark plugs are covered with a soft, fluffy black deposit. This is caused by:

   a. Plug heat range that is too hot
   b. Intake manifold leak
   c. Too much oil in the chamber
   d. Late ignition timing

3. An engine runs-on (diesels) when the ignition is turned off.

   Mechanic A says that this could be caused by the wrong idle speed.

   Mechanic B says that this could be caused by spark plugs with a too low (cold) heat range.

   Who is right?

   a. A only
   b. B only
   c. Both A and B
   d. Neither A nor B

4. Mechanic A says that with solid lifters, exhaust valves have a larger clearance than intake valves so that cylinder scavenging is complete.

   Mechanic B says that with solid lifters, exhaust valves have a larger clearance than intake valves to allow for heat expansion.

   Who is right?

   a. A only
   b. B only
   c. Both A and B
   d. Neither A nor B

5. If the vacuum diaphragm in a thermostatically controlled air cleaner is leaking, the car will probably have:

   a. Poor fuel economy
   b. Hard starting when cold
   c. Poor acceleration when cold
   d. Poor performance when warm

6. Mechanic A says that a rough idle can be caused by stepped (damaged) idle mixture screws.

   Mechanic B says that a rough idle can be caused by worn throttle plate bushings.

   Who is right?

   a. A only
   b. B only
   c. Both A and B
   d. Neither A nor B

7. A lean mixture at idle can be caused by:

   a. Varnish buildup near the idle ports
   b. Dirt in the idle passageways
   c. A blocked main jet
   d. All of the above

8. Mechanic A says that to use an infrared analyzer to adjust the idle on a converter-equipped car, you should put the probe in the tailpipe and disconnect the air pump.

   Mechanic B says that to get an accurate reading, you should put the probe in the exhaust source hose for the EGR valve.

   Who is right?

   a. A only
   b. B only
   c. Both A and B
   d. Neither A nor B

9. A car has a flat spot (hesitation) on acceleration.

   Mechanic A says it could be caused by a too-low float level.

   Mechanic B says it could be caused by a worn power valve.

   Who is right?

   a. A only
   b. B only
   c. Both A and B
   d. Neither A nor B

10. When an engine is run at 2,000 rpm, manifold vacuum reads 20 inches then falls to 14 inches. This could be caused by:
a. A restricted exhaust system
b. Worn valve guides
c. An intake manifold leak
d. A leaking carburetor power valve

11. A customer complains that the battery in his car is always low on water.

Mechanic A says that the starter current draw is one of the first things to check.

Mechanic B says the alternator output is one of the first things to check.

Who is right?
a. A only
b. B only
c. Both A and B
d. Neither A nor B

12. An alternator's current output meets specifications but the voltage is low.

Mechanic A says this could be caused by a bad regulator.

Mechanic B says this could be caused by a bad ammeter.

Who is right?
a. A only
b. B only
c. Both A and B
d. Neither A nor B

13. Mechanic A says that solid valve lifters have a lash adjustment to compensate for wear of the valve guides.

Mechanic B says that solid valve lifters have a lash adjustment to ensure correct valve seating.

Who is right?
a. A only
b. B only
c. Both A and B
d. Neither A nor B

14. All of the following are true EXCEPT:
a. When charging a battery, the electrolyte should not be allowed to get hotter than 125' F (52' C)
b. When charging a battery, you should not let the electrolyte boil over
c. You can charge a battery at a steady 3 amps until specific gravity meets specs
d. You can charge a battery at a steady 30 amps until it is fully charged

15. A mechanic uses the wrong polarity when hooking up a helper battery to start a car. Which will be hurt the most on the disabled car:
a. Battery
b. Starter
c. Ignition
d. Alternator

16. A customer complains of poor performance of his breaker-point ignition car. The distributor points are found to be burned.

Mechanic A says this could be caused by an open ballast resistor.

Mechanic B says this could be caused by a loose condenser connection.

Who is right?
a. A only
b. B only
c. Both A and B
d. Neither A nor B

17. At idle, an engine's dwell is set at 32 degrees. When the engine is accelerated, the dwell changes by 12 degrees, then returns to 32 when at idle. This is caused by:
a. A bad vacuum unit
b. Loose cam follower on points
c. Loose advance plate
d. Worn distributor shaft bushings

18. Mechanic A says that the centrifugal advance mechanism responds to changes in engine load.

Mechanic B says the centrifugal advance unit responds to changes in engine speed.

Who is right?
a. A only
b. B only
c. Both A and B
d. Neither A nor B

19. In an evaporative emission control system with a serviceable charcoal canister, Mechanic A says that the canister can be serviced by flushing it out and replacing the charcoal.

Mechanic B says it can be serviced by replacing the canister filter.

Who is right?
a. A only
b. B only
c. Both A and B
d. Neither A nor B

20. Mechanic A says a carburetor with 2 barrels can have secondary throttle linkages worked by vacuum.

Mechanic B says the secondary throttle linkages can be worked mechanically.

Who is right?
a. A only
b. B only
c. Both A and B
d. Neither A nor B

21. There are 12 volts present at the battery positive terminal and 0 volts present at the alternator battery terminal. This can be caused by:
a. An open diode
b. A shorted regulator
c. An open circuit
d. An open stator

22. The best way to keep an engine from starting during a cranking test is to:
a. Disconnect all spark plug cables
b. Ground the coil secondary cable
c. Disconnect the neutral start switch
d. Ground the ignition switch

23. A mechanic finds that there is a weak spark at all plugs. The most likely cause is:
a. Excessive primary resistance
b. Damaged ignition cables
c. An open coil secondary winding
d. None of the above

24. When reinstalling hydraulic valve lifters, some manufacturers recommend additional tightening after zero lash is achieved. What might be the reason for this:
a. Proper positioning of the plunger in the lifter
b. Quieter operation
c. Better high-speed performance
d. Better low-speed performance

25. When making a compression test, Mechanic A says the engine should be warm.

Mechanic B says the throttle should be held wide open.

Who is right?
a. A only
b. B only
c. Both A and B
d. Neither A nor B

26. During a coolant circulation test, bubbles appear in the radiator.

Mechanic A says the problem could be a blown head gasket.

Mechanic B says the problem could be a partially blocked bypass hose.

Who is right?
a. A only
b. B only
c. Both A and B
d. Neither A nor B

Glossary of Technical Terms

After Top Dead Center: The position of a piston after it has passed top dead center. Abbreviated: atdc. Usually expressed in degrees, such as 5° atdc.

Air-Fuel Ratio: The ratio of air to gasoline in the air-fuel mixture drawn into an engine.

Air Injection: A way of reducing exhaust emissions by injecting air into each of the exhaust ports of an engine. It mixes with the hot exhaust and oxidizes the HC and CO to form H_2O and CO_2.

Alternating Current: A flow of electricity through a conductor, first in one direction, then in the opposite direction.

Ambient Temperature: The temperature of the air surrounding a particular device or location.

Amperage: The amount of current flow through a conductor.

Ampere-Hour Rating: A battery rating based on the amperes of current that a battery can supply steadily for 20 hours, with no battery cell falling below 1.75 volts. Also called a 20-hour discharge rating.

Amperes: Units for measuring the rate of electrical current flow.

Analog: Describing a computer that uses similar (analogous) electrical signals to make its calculations.

Aneroid Bellows: An accordian-shaped bellows that responds to changes in coolant pressure or atmospheric pressure by expanding and contracting.

Antiknock Value: The characteristic of gasoline that helps prevent detonation or "knocking."

Antioxidants: Chemicals or compounds added to motor oil to reduce oil oxidation, which leaves carbon and varnish in the engine.

Antioxidant Inhibitors: A gasoline additive used to prevent the formation of gum.

API Service Classification: A system of letters signifying an oil's performance; assigned by the American Petroleum Institute.

Atmospheric Pressure: The pressure on the earth's surface caused by the weight of air in the atmosphere. At sea level, this pressure is 14.7 psi at 32° F (0° C).

Atomization: Breaking down into small particles or a fine mist.

Available Voltage: The peak voltage that a coil can produce.

Backfire: The accidental combustion of gases in an engine's intake or exhaust system.

Backpressure: A pressure that tends to slow the exit of exhaust gases from the combustion chamber; usually caused by restrictions in the exhaust system.

Baffle: A plate or obstruction that restricts the flow of air or liquids. The baffle in a fuel tank keeps the fuel from sloshing as the car moves.

Bakelite: A plastic material that is a good insulator. Distributor caps are often made of bakelite.

Ballast (Primary) Resistor: A resistor in the primary circuit that stabilizes ignition system voltage and current flow.

Base: The center layer of semiconductor material in a transistor.

Before Top Dead Center: The position of a piston as it nears top dead center. Abbreviated: btdc. Usually expressed in degrees, such as 5° btdc.

Bimetal Temperature Sensor: A sensor or switch that reacts to changes in temperature. Made of two strips of metal welded together that expand differently when heated or cooled, causing the strip to bend.

Blowby: Combustion gases that slip past the piston rings into the crankcase; composed of water vapor, acids, and unburned fuel.

Boost: A measure of the amount of air pressurization, above atmospheric, that a supercharger can deliver.

Bottom Dead Center: The exact bottom of a piston stroke. Abbreviated: bdc.

Breaker Points: The metal contact points that act as an electrical switch in a distributor. They open and close the ignition primary circuit.

Capacitor: A device that can store voltage without affecting the voltage in any way. Formed by bringing two conductive surfaces close together, separated only by an insulator. Also called condenser.

Carbon Monoxide: An odorless, colorless, tasteless poisonous gas. Abbreviated: CO. A major pollutant given off by an internal combustion engine.

Carburetor Icing: A condition that is the result of the rapid vaporization of fuel entering a carburetor; the temperature drops enough to freeze the water particles in the airflow.

Catalyst: A substance that causes a chemical reaction, without being changed by the reaction.

Catalytic Converter: A device installed in an exhaust system to convert pollutants into harmless byproducts through a catalytic chemical reaction.

Centrifugal (Mechanical) Advance: A method of advancing the ignition spark, using weights in the distributor that react to centrifugal force.

Centrifugal Force: A force applied to a rotating object, tending to move the object toward the outer edge of the circle of rotation.

Check Valve: A valve that permits flow in only one direction.

Circuit: A circle or unbroken path through which an electric current can flow.

Clearance Volume: The volume of a combustion chamber when the piston is at top dead center.

Cold-Cranking Rating: A battery rating based on the amperes of current that a battery can supply for 30 seconds at 0° F, with no battery cell falling below 1.2 volts.

Collector: The outside layer of semiconductor material, in a transistor, that conducts current away from the base.

Compression Ratio: The total volume of an engine cylinder divided by its clearance volume.

Concentric: Having the same center.

Condenser: A capacitor used in breaker-point ignition systems to prevent arcing across the points.

Conductors: Materials that allow easy electron flow because of their many free electrons.

Cross-Firing: Ignition voltage jumping from the distributor rotor to the wrong spark plug electrode inside the distributor cap. Also, ignition voltage jumping from one spark plug cable to another due to worn insulation.

Cycling: Battery electrochemical action. One complete cycle is the operation from fully charged to discharged and back to fully charged.

Detented: Positions in a switch that allow the switch to stay in that position. In an ignition switch, the On, Off, Lock, and Accessory positions are detented.

Detergent-Dispersant: A chemical added to motor oil to break down and disperse sludge and fuel particles picked up by the oil.

Detonation: Also called knocking, pinging. An unwanted explosion of an air-fuel mixture, caused by high heat and compression.

Diaphragm: A thin, flexible wall to separate two cavities, such as the diaphragm in a mechanical fuel pump or distributor vacuum advance unit.

Dielectric: The insulating material between the two plates of a condenser, or capacitor.

Dieseling: A condition in which extreme heat in an engine's combustion chamber continues to ignite excess fuel after the ignition has been turned off.

Digital: Describing a computer that makes calculations with quantities represented electronically as digits.

Diodes: Electronic devices made of semi-conductor material that allow current flow in one direction but block it in the other.

Direct Current: A flow of electricity in one direction through a conductor.

Displacement: A measurement of the volume of air displaced by a piston as it moves from bottom to top of its stroke. Engine displacement is the piston displacement multiplied by the number of pistons in an engine.

Diverter Valve: Also called a dump valve. A valve used in an air injection system to prevent backfire. During deceleration it "dumps" the air from the air pump into the atmosphere.

Dwell Angle: Also called cam angle, or dwell. The measurement in degrees of how far the distributor cam rotates while the breaker points are closed.

Dynamometer: A device used to measure the power of an engine, and to test the engine under simulated road conditions.

Electrolyte: The chemical solution in a battery that conducts electricity and reacts with the plate materials.

Electromagnet: A soft iron core wrapped in a coil of a current-carrying conductor. Current flow in the conductor induces a magnetic force around the core.

Emitter: The outside layer of semiconductor material, in a transistor, that conducts current to the base.

Evaporative Emission Control (EEC): A way of controlling HC emissions by collecting fuel vapors from the fuel tank and carburetor fuel bowl vents and directing them through an engine's intake system.

Exhaust Gas Recirculation (EGR): A way of reducing NO_x emissions by directing unburned exhaust back through an engine's intake.

Extended-Core Spark Plug: Also called power tip. The insulator core and the electrodes in this type of spark plug extend further into the combustion chamber than they do on other types.

Extruded: Shaped by forcing through a die.

Farad: The unit of measurement of a condenser's ability to store electrical energy.

Firing Order: The order in which combustion occurs in the cylinders of an engine.

Firing Voltage (Required Voltage): The voltage level that must be reached to ionize and create a spark in the air gap between the spark plug electrodes.

Flooding: A condition caused by heat expanding the fuel in a fuel line. The fuel pushes the carburetor inlet needle valve open and fills up the fuel bowl even when more fuel is not needed. Also, the presence of too much fuel in the intake manifold.

Four-Stroke Engine: The Otto cycle engine. An engine in which a piston must complete four strokes to make up one operating cycle. The strokes are: intake, compression, power, and exhaust.

Ground: The connection of an electrical circuit or unit to the engine or chassis to return electrical current to the battery.

Ground Cable: The battery cable that provides a ground connection from the vehicle chassis to the battery.

Gulp Valve: A valve used in an air injection system to prevent backfire. During deceleration it redirects air from the air pump to the intake.

Heat Range: The measure of a spark plug's ability to dissipate heat from its firing end.

High Speed Surge: A sudden increase in engine speed caused by high manifold vacuum pulling in too much air-fuel mixture.

Hot (Insulated) Cable: The battery cable that conducts battery current to the automotive electrical system.

Hydrocarbon: A chemical compound made up of hydrogen and carbon. Abbreviated: HC. A major pollutant given off by an internal combustion engine. Gasoline, itself, is a hydrocarbon compound.

Ignition Interval (Firing Interval): The number of degrees of crankshaft rotation between ignition sparks.

Induction: The production of an electrical current in a conductor or coil by moving the conductor or coil through a magnetic field, or by moving the magnetic field past the conductor or coil.

Inductive-Discharge Ignition: A method of igniting the air-fuel mixture in an engine cylinder. It is based on the induction of a high voltage in the secondary winding of the coil.

Inert Gas: A gas that will not undergo chemical reaction.

Insulators: Materials that will not conduct electron flow because of their many bound electrons.

Integrated Circuits: Electronic circuits in which the parts have been miniaturized and assembled together to reduce the number of parts in the circuit.

Ionize: To break up molecules into two or more oppositely charged ions. The air gap between the spark plug electrodes is ionized when the air-fuel mixture is changed from a non-conductor to a conductor.

Light-Emitting Diode: A diode that gives off visible light when current flows through it.

Liquid-Vapor Separator Valve: A valve in some EEC fuel systems that separates liquid fuel from fuel vapors.

Magnetic Saturation: The condition when a magnetic field reaches full strength and maximum flux density.

Manifold Vacuum: Low pressure in an engine's intake manifold, below the carburetor throttle.

Minus Rule: Minus metal on the minus side of the distributor breaker points means a minus-capacity condenser.

Misfire: When combustion does not occur during the power stroke.

Module: A self-contained, sealed unit that houses the solid-state circuits needed to control certain electrical or mechanical functions.

Monolith: A large block. In a catalytic converter, a monolith is made like a honeycomb to provide several thousand square yards of catalyst surface area.

Multigrade: An oil that has been tested at two different temperatures, and therefore has two SAE viscosity numbers.

Mutual Induction: The transfer of energy between two unconnected conductors, caused by the expanding or contracting magnetic flux lines of the current-carrying conductor.

Negative-Ground Electrical System: An automotive electrical system in which the battery negative terminal is connected to ground.

Negative Polarity: Also called ground polarity. A correct polarity of the ignition coil connections. Coil voltage is delivered to the spark plugs so that the center electrode of the plug is negatively charged and the grounded electrode is positively charged.

No-Load Oscillation: The rapid, back-and-forth, peak-to-peak oscillation of voltage in the ignition secondary circuit when the circuit is open.

Normally Aspirated: An engine that uses normal vacuum to draw in its air-fuel mixture. Not supercharged.

Octane Rating: The measurement of the anti-knock value of a gasoline.

Ohms: Units for measuring electrical resistance.

Oscillating: Moving back and forth with a steady rhythm.

Oxidation: The combining of an element with oxygen in a chemical process that often produces extreme heat as a byproduct.

Oxides of Nitrogen: Chemical compounds of nitrogen given off by an internal combustion engine. Abbreviated: NO_x. They combine with hydrocarbons to produce smog.

Particulates: Liquid or solid particles such as lead and carbon that are given off by an internal combustion engine as pollution.

Percolation: The bubbling and expansion of a liquid. Similar to boiling.

Photochemical Smog: A combination of pollutants, primarily HC and NO_x. When acted upon by sunlight, they form chemical compounds that are harmful to human, animal, and plant life.

Phototransistor: Also called a photocell. A type of solid-state device that will generate a voltage when exposed to light.

Pintle Valve: A valve shaped much like a hinge pin. In an EGR valve, the pintle is attached to a normally closed diaphragm. When ported vacuum is applied, the pintle rises from its seat and allows exhaust gas to be drawn into the engine's intake system.

Positive Crankcase Ventilation (PCV): Late-model crankcase ventilation systems that return blowby gases to the combustion chambers.

Positive-Ground Electrical System: An automotive electrical system in which the battery positive terminal is connected to ground.

Positive Polarity: Also called reverse polarity. An incorrect polarity of the ignition coil connections. Coil voltage is delivered to the spark plug so that the center electrode of the plug is positively charged and the grounded electrode is negatively charged.

Potentiometer: A variable resistor.

Ported Vacuum: Vacuum immediately above the throttle plate in a carburetor.

Preignition: Ignition of the air-fuel charge in the combustion chamber before the spark plug can ignite the mixture. A cause of "knocking."

Pressure Differential: A difference in pressure between two points.

Reach: The length of the spark plug shell from the seat to the bottom of the shell.

Reciprocating Engine: Also called a piston engine. An engine in which the pistons move up and down or back and forth, as a result of combustion in the top of the cylinder.

Rectified: Electrical current changed from alternating (a.c.) to direct (d.c.).

Reduction: A chemical process in which oxygen is taken away from a compound.

Relay: An electromagnetic switch. A relay uses a small amount of current flow to control the flow of a larger amount of current through a separate circuit.

Reluctance: An object's resistance to magnetic lines of force. Magnetic lines will concentrate in areas of low reluctance, and avoid areas of high reluctance.

Reserve-Capacity Rating: A battery rating based on the number of minutes a battery at 80° F can supply 25 amperes, with no battery cell falling below 1.75 volts.

Resistance: Opposition to electrical current flow.

Resistor-Type Spark Plug: A plug that has a resistor in the center electrode to reduce the inductive portion of the spark discharge.

Road Draft Tube: The earliest type of crankcase ventilation, which vented blowby gases to the atmosphere.

SAE Viscosity Rating: A system of numbers signifying an oil's viscosity at a certain temperature; assigned by the Society of Automotive Engineers.

Semiconductor: A material that is neither a good conductor nor a good insulator. Semiconductors are the raw material used to make solid-state devices such as diodes and transistors. Silicon and germanium are common semiconductors.

Series-Gap Spark Plug: Also called booster. A spark plug in which the center electrode has an extra gap to increase ionization voltage.

Silicon-Controlled Rectifier: Also called a thyristor. A solid-state device that normally blocks current flow in both directions, but which can be triggered to allow current flow in one direction while still blocking it in the opposite direction.

Single Grade (Straight Grade): An oil that has been tested at only one temperature, and so has only one SAE viscosity number.

Sintered: Welded together without using heat to form a porous material, such as the metal disc used in some vacuum delay valves.

Siphoning: The flowing of a liquid as a result of a pressure differential, without the aid of a mechanical pump.

Sludge: A thick, black deposit caused by the mixing of blowby and cold oil.

Solenoid: An iron core with a wire coil surrounding it. The core moves when electrical current is applied to the coil. Used to convert electrical energy to mechanical energy.

Solid-State: A method of controlling electrical current flow, in which the parts are primarily made of semiconductor materials.

Spark Voltage: The inductive portion of a spark that maintains the spark in the air gap between a spark plug's electrodes. Usually about one-quarter of the firing voltage level.

Specific Gravity: The weight of a volume of liquid divided by the weight of the same volume of water at a given temperature and pressure. Water has a specific gravity of 1.000.

Starting Bypass: A parallel circuit branch that bypasses the primary ballast resistor during cranking.

Stoichiometric Ratio: An ideal air-fuel mixture for combustion in which all oxygen and all fuel will be completely burned.

Stroke: One complete top-to-bottom or bottom-to-top movement of an engine piston.

Substrate: The layer, or honeycomb, or aluminum oxide upon which the catalyst (platinum, palladium, or rhodium) in a catalytic converter is deposited.

Sulfation: The crystalization of lead sulfate on the plates of a constantly discharged battery.

Sulfur Oxides: Chemical compounds given off by processing and burning gasoline and other fossil fuels. As they decompose, they combine with water to form sulfuric acid.

Supercharging: Use of an air pump to deliver an air-fuel mixture to the engine cylinders at a pressure greater than atmospheric pressure.

Synthetic Motor Oils: Lubricants formed by artificially combining molecules of petroleum and other materials.

Temperature Inversion: A weather pattern in which a layer or "lid" of warm air keeps the cooler air beneath it from rising.

Tetraethyl Lead: A gasoline additive used to help prevent detonation. Abbreviated: TEL.

Thermistor: An electrical conductor that changes its resistance as the surrounding temperature changes.

Tolerances: The clearances allowed between moving parts.

Top Dead Center: The exact top of a piston's stroke. Also a specification used when tuning an engine. Abbreviated: tdc.

Transducer: A device that changes one form of energy into another. In an ignition system, it may sense a mechanical movement and change it to an electrical signal.

Transistor: A 3-terminal semiconductor used for current switching, detection, and amplification. Low current flow between one pair of terminals can control high current flow between another pair of terminals, with one common terminal.

Turbocharger: A supercharging device that uses exhaust gases to turn a turbine that forces extra air into the cylinders.

Vacuum: A pressure less than atmospheric pressure.

Vacuum Advance: The use of engine vacuum to advance or retard ignition spark timing by moving the distributor breaker plate.

Vacuum Lock: A stoppage of fuel flow caused by insufficient air intake to the fuel tank.

Vaporization: Changing a liquid, such as gasoline, into a gaseous state.

Varnish: An undesirable deposit, usually on the engine pistons, formed by the oxidation of fuel and of motor oil.

Venturi: A restriction in an airflow, such as in a carburetor, that speeds the airflow and creates a vacuum.

Venturi Vacuum: Low pressure in the venturi of a carburetor, caused by fast airflow through the venturi.

Viscosity: The tendency of a liquid such as oil to resist flowing.

Volatility: The ability of a liquid to change from a liquid to a vapor.

Voltage: The electromotive force that causes current flow. The potential difference in electrical force between two points when one is negatively charged and the other is positively charged.

Voltage Decay: The rapid oscillation and dissipation of secondary voltage after the spark in a spark plug air gap has stopped.

Voltage Drop: The measurement of the loss of voltage caused by the resistance of a conductor or a circuit device.

Voltage Reserve: The amount of coil voltage available in excess of the voltage required to fire the spark plugs.

Volts: Units for measuring the amount of electrical force.

Volumetric Efficiency: The comparison of the *actual* volume of air-fuel mixture drawn into an engine to the *theoretical maximum* volume that could be drawn in. Written as a percentage.

Watts Rating: A battery rating based on the available cranking power of the battery. The rating can be found by multiplying the current available from the battery by the battery voltage at 0° F.

Index

Answers to Review and NIASE Questions

Chapter 1: Engine Operating Principles
1.(b) 2.(b) 3.(d) 4.(d) 5.(c) 6.(b) 7.(d) 8.(c) 9.(c) 10.(c)

Chapter 2: Engine Air-Fuel Requirements
1.(c) 2.(c) 3.(d) 4.(a) 5.(c) 6.(b) 7.(d) 8.(d) 9.(b)

Chapter 3: Engine Lubrication
1.(c) 2.(d) 3.(b) 4.(d) 5.(a) 6.(b) 7.(c) 8.(a) 9.(d) 10.(b) 11.(d) 12.(a) 13.(d) 14.(c) 15.(a) 16.(c)

Chapter 4: Cooling and Exhaust Systems
1.(c) 2.(a) 3.(c) 4.(b) 5.(d) 6.(c) 7.(d) 8.(a) 9.(b) 10.(c) 11.(d) 12.(b) 13.(a) 14.(d) 15.(c) 16.(d)

Chapter 5: Introduction to Emission Control
1.(c) 2.(c) 3.(d) 4.(a) 5.(c) 6.(d) 7.(b) 8.(d) 9.(c) 10.(b)

Chapter 6: What Is A Tune-Up
1.(b) 2.(c) 3.(a) 4.(d) 5.(c) 6.(d)

Chapter 7: The Battery, Starting, and Charging Systems
1.(c) 2.(b) 3.(b) 4.(d) 5.(b) 6.(d) 7.(c) 8.(d) 9.(c) 10.(b) 11.(a) 12.(b) 13.(b) 14.(d) 15.(d) 16.(d)

Chapter 8: The Ignition System
1.(d) 2.(c) 3.(a) 4.(c) 5.(c) 6.(d) 7.(b) 8.(c) 9.(b) 10.(d) 11.(b) 12.(b) 13.(d) 14.(c) 15.(c) 16.(d) 17.(d) 18.(c) 19.(d)

Chapter 9: Ignition Timing, Spark Advance, and Spark Timing Controls
1.(d) 2.(d) 3.(c) 4.(b) 5.(a) 6.(b) 7.(d) 8.(a) 9.(d) 10.(b) 11.(d) 12.(c)

Chapter 10: Ignition Coils, Condensers, and Breaker-Point Distributors
1.(d) 2.(d) 3.(c) 4.(a) 5.(d) 6.(c) 7.(c) 8.(c) 9.(d) 10.(b) 11.(d) 12.(a) 13.(d) 14.(a) 15.(c) 16.(b) 17.(b) 18.(a)

Chapter 11: Solid-State Ignition Systems
1.(d) 2.(b) 3.(c) 4.(c) 5.(d) 6.(a) 7.(c) 8.(d) 9.(c) 10.(a) 11.(c) 12.(b)

Chapter 12: Fuel System Operation
1.(a) 2.(c) 3.(d) 4.(b) 5.(a) 6.(c) 7.(a) 8.(c) 9.(d) 10.(b) 11.(c) 12.(b) 13.(a) 14.(c) 15.(b) 16.(a) 17.(d) 18.(c) 19.(b) 20.(c) 21.(b)

Chapter 13: Basic Carburetion and Manifolding
1.(c) 2.(b) 3.(a) 4.(d) 5.(c) 6.(a) 7.(b) 8.(d) 9.(c) 10.(c) 11.(b) 12.(a) 13.(c) 14.(d) 15.(a) 16.(b) 17.(c) 18.(c) 19.(d) 20.(b) 21.(b)

Chapter 14: Electronic Fuel Injection
1.(b) 2.(a) 3.(b) 4.(a) 5.(c) 6.(d) 7.(b) 8.(c) 9.(d) 10.(b)

Chapter 15: Supercharging and Turbochargers
1.(b) 2.(d) 3.(c) 4.(c) 5.(c) 6.(b) 7.(b) 8.(d) 9.(d) 10.(b) 11.(c) 12.(a) 13.(d)

Chapter 16: Positive Crankcase Ventilation
1.(c) 2.(b) 3.(d) 4.(a) 5.(d) 6.(c) 7.(c) 8.(b) 9.(a) 10.(d) 11.(c) 12.(a) 13.(d) 14.(c)

Chapter 17: Air Injection
1.(a) 2.(c) 3.(b) 4.(b) 5.(d) 6.(d) 7.(c) 8.(c) 9.(a) 10.(b) 11.(b) 12.(a)

Chapter 18: Exhaust Gas Recirculation
1.(c) 2.(c) 3.(a) 4.(c) 5.(d) 6.(b) 7.(d) 8.(b) 9.(c) 10.(c) 11.(c)

Chapter 19: Catalytic Converters
1.(b) 2.(d) 3.(a) 4.(d) 5.(b) 6.(c) 7.(d) 8.(a) 9.(b) 10.(b) 11.(d) 12.(b) 13.(b) 14.(d) 15.(a)

NIASE Mechanic Certification Sample Test
1.(c) 2.(d) 3.(a) 4.(b) 5.(c) 6.(c) 7.(d) 8.(d) 9.(d) 10.(a) 11.(b) 12.(a) 13.(b) 14.(d) 15.(d) 16.(b) 17.(c) 18.(b) 19.(b) 20.(c) 21.(c) 22.(b) 23.(a) 24.(a) 25.(c) 26.(a)